CELL AND MOLECULAR RESPONSES TO STRESS

Volume 3

Sensing, Signaling and Cell Adaptation

Cover illustration: Fig. 3.5 from Chapter 3 'A Profile of the Metabolic Responses to Anoxia in Marine Invertebrates', by Kevin Larade and Kenneth B. Storey.

SENSING, SIGNALING AND CELL ADAPTATION

Edited by

K.B. STOREY and J.M. STOREY

Institute of Biochemistry
Carleton University
Ottawa, Ontario
Canada

2002

ELSEVIER
Amsterdam – Boston – London – New York – Oxford – Paris
San Diego – San Francisco – Singapore – Sydney – Tokyo

ELSEVIER SCIENCE B.V.
Sara Burgerhartstraat 25
P.O. Box 211, 1000 AE Amsterdam, The Netherlands

First edition 2002

British Library Cataloguing in Publication Data

Sensing, signaling and cell adaptation. – (Cell and
   molecular responses to stress ; v. 3)
   1. Cell metabolism 2. Adaptation (Physiology)
   I. Storey, K. B. (Kenneth B.) II. Storey, J. (Janet)
   572.7

   ISBN 0444511474

Library of Congress Cataloging in Publication Data
A catalog record from the Library of Congress has been applied for.

ISBN: 0-444-51147-4

⊗ The paper used in this publication meets the requirements of ANSI/NISO Z39.48-1992 (Permanence of Paper).

Printed in The Netherlands.

Preface

"Science is the game we play with God to find out what His rules are."

This quote from Albert Einstein has perhaps never been more appropriate than it is today, early in the new millenium. Amazing recent advances in biochemistry and molecular biology have given rise to new fields including genomics, proteomics and bioinformatics. Discoveries in interlocking areas are synthesizing principles in the design and control of cellular functions including membrane transport, signal transduction, gene expression, macromolecular synthesis, and others. We are certainly coming closer to figuring out what the "rules" are. One clearly seems to be that protein kinases "rule"—the hundreds of protein kinases encoded on the genome and the involvement of reversible protein phosphorylation as a means of regulatory control in virtually every metabolic process show the universal importance of this mechanism.

This third and final volume of *Cell and Molecular Responses to Stress* has broad themes: sensing, signaling and cellular adaptation. Every chapter provides a unique view of how cells and organisms deal with stresses, great and small, on instantaneous versus evolutionary timelines, and provides examples of regulatory principles and their infinite variety that are used to preserve life on Earth. A group of chapters deals with the molecular mechanisms of sensing of external signals: pain, heat, cold, radiation, and nutrient levels. The many mechanisms of intracellular signal transduction and the regulation of gene expression are dealt with in others, with a major emphasis on the roles of protein kinases. Chapters on metabolic control analysis and the evolution of physiological adaptation consider important principles of metabolic regulation, whereas the vast potential of emerging analytical technologies such as cDNA array screening are analyzed in others. The adaptive significance of unique proteins such as uncoupling proteins and dehydrins are explored along with metabolic adaptations for dealing with ischemia, temperature change, freezing and infrared radiation.

We extend our thanks to all of the authors who worked so hard to write the excellent chapters in this volume. Their marvelous stories of biochemical adaptation make every chapter a pleasure to read.

Kenneth B. Storey
Janet M. Storey
Ottawa, May 2002

List of Contributors

Ian M. Adcock
National Heart and Lung Institute, Imperial College School of Medicine, Dovehouse Street, London SW3 6LY, UK
E-mail: ian.adcock@ic.ac.uk; Tel: +44 207 3528121 x 3061; Fax: +44 207 351 5675

Marie-Clotilde Alves-Guerra
Institut de Recherches Necker-Enfants Malades, rue de Vaugirard, 75015 Paris, France

Julien Averous
Unite de Nutrition Cellulaire et Moleculaire, INRA de Theix, 63122 Saint Genes Champanelle, France

Michael Bell
Center for Neuroscience Research, Children's Research Institute, Children's National Medical Center, Washington, DC 20010, USA
E-mail: mbell@cnmc.org; Tel: +1 202 8843597; Fax: +1 202 884 5724

Manuel Benito
Departamento de Bioquímica y Biología Molecular II, Universidad Complutensede Madrid, Ciudad Universitaria, 28040 Madrid, Spain

Tammie Bishop
MRC Dunn Human Nutrition Unit, Hills Road, Cambridge CB2 2XY, UK
E-mail: tammie@well.ox.ac.uk; Tel: +44 1223 252800; Fax: +44 1223 252806

Jennifer K. Bonnington
Department of Pharmacology, University of Cambridge, Tennis Court Rd, Cambridge, CB2 1PD, UK

Frédéric Bouillaud
Institut de Recherches Necker-Enfants Malades, rue de Vaugirard, 75015 Paris, France

Martin D. Brand
MRC Dunn Human Nutrition Unit, Hills Road, Cambridge CB2 2XY, UK
E-mail: martin.brand@mrc-dunn.cam.ac.uk; Tel: +44 1223 252800; Fax: +44 1223 252805

Alain Bruhat
Unite de Nutrition Cellulaire et Moleculaire, INRA de Theix, 63122 Saint Genes Champanelle, France

Neil R. Chapman
Division of Gene Regulation and Expression, School of Life Sciences, University of Dundee, Dundee, DD1 5EH, UK

Christine Clerici
Department of Physiology, Faculté de Médecine Xavier Bichat, Université Paris, 46 rue Henri Huchard, 75018 Paris, France
E-mail: christine.clerici@lmr.ap-hop-paris.fr; Tel: +33 1 44856270; Fax: +33 1 42281564

Timothy J. Close
Department of Botany & Plant Sciences, University of California, Riverside, CA 92521-0124, USA
E-mail: timothy.close@ucr.edu; Tel: +1 909 7873318; Fax: +1 909 7874437

Elodie Couplan
Institut de Recherches Necker-Enfants Malades, rue de Vaugirard, 75015 Paris, France

Douglas L. Crawford
Division of Molecular Biology & Biochemistry, University of Missouri-Kansas City, Kansas City, MO 64110-2499, USA
E-mail: crawforddo@umkc.edu; Tel: +1 816 2352565; Fax: +1 816 2355595

R. Keira Curtis
MRC Dunn Human Nutrition Unit, Hills Road, Cambridge CB2 2XY, UK
E-mail: keira.curtis@mrc-dunn.cam.ac.uk; Tel: +44 1223 252800; Fax: +44 1223 252806

Maria Del Mar Gonzalez Barroso
Institut de Recherches Necker-Enfants Malades, rue de Vaugirard, 75015 Paris, France

Kelly Drew
Department of Chemistry and Biochemsitry, Institute of Arctic Biology, Box 757000, University of Alaska Fairbanks, Fairbanks, AK 99775-7000, USA
E-mail: ffkld@uaf.edu; Tel: +1 907 4747190; Fax: +1 907 474 6050

Peter F. Dubbelhuis
Department of Biochemistry, University of Amsterdam, Meibergdreef 15, 1105 AZ Amsterdam, The Netherlands
E-mail: p.f.dubbelhuis@ams.uva.nl; Tel: +31 20 5665159; Fax: +31 20 6915519

Sean F. Eddy
Institute of Biochemistry and Department of Chemistry, Carleton University, 1125 Colonel By Drive, Ottawa, ON, Canada K1S 5B6
E-mail: seddy@ccs.carleton.ca

Pierre Fafournoux
Unite de Nutrition Cellulaire et Moleculaire, INRA de Theix, 63122 Saint Genes Champanelle, France
E-mail: fpierre@clermont.inra.fr; Tel: +33 4 73624562; Fax: +33 4 73624755

John Hallenbeck
Stroke Branch, NINDS, NIH, 36 Convent Drive MSC 4128, Bethesda, MD 20892-4128, USA
E-mail: hallenbj@ninds.nih.gov; Tel: +1 301 4966231; Fax: +1 301 4022769

Vidya Hebbar
Department of Pharmaceutics, Ernest Mario School of Pharmacy, Rutgers University, 160 Frelinghuysen Road, Piscataway, NJ 08854, USA

Norman P.A. Huner
Department of Plant Sciences, University of Western Ontario, London, Canada N6A 5B7
E-mail: nhuner@uwo.ca; Tel: +1 519 6612111 x86638; Fax: +1 519 6613935

Abdelbagi M. Ismail
International Rice Research Institute, DAPO 7777, Metro Manila, Philippines
E-mail: abdelbagi.ismail@cgiar.org; Tel: +63 2 762 0127; Fax: +63 2 761 2406

Alexander G. Ivanov
Department of Plant Sciences, University of Western Ontario, London, Canada N6A 5B7
E-mail: aivanov@uwo.ca; Tel: +1 519 6612111 x86488; Fax: +1 519 6613935

A.-N. Tony Kong
Department of Pharmaceutics, Ernest Mario School of Pharmacy, Rutgers University, 160 Frelinghuysen Road, Piscataway, NJ 08854, USA
E-mail: KongT@cop.rutgers.edu; Tel: +1 732 445 7335; Fax: +1 732 445 3134

Marianna Krol
Department of Plant Sciences, University of Western Ontario, London, Canada N6A 5B7
E-mail: mkrol@uwo.ca; Tel: +1 519 6612111 x86638; Fax: +1 519 6613935

Kevin Larade

Institute of Biochemistry and Department of Biology, Carleton University, 1125 Colonel By Drive, Ottawa, Canada K1S 5B6
E-mail: klarade@ccs.carleton.ca

Lee Laurent-Applegate

Laboratory of Oxidative Stress and Aging, University Hospital-Lausanne, CH-1011 Lausanne, Switzerland
E-mail: Lee.Laurent-Applegate@chuv.hospvd.ch; Tel: +41 21 314 3169; Fax: +41 21 314 3499

Hon Cheung Lee

Department of Pharmacology, University of Minnesota, 321 Church St. SE, 4-415 Jackson Hall, Minneapolis, MN 55455, USA
E-mail: leehc@tc.umn.edu; Tel: +1 612 6257120; Fax: +1 612 6250991

Joel M. Linden

Cardiovascular Research Center, University of Virginia School of Medicine, Charlottesville, VA 22908, USA
E-mail: jlinden@virginia.edu; Tel: +1 434 924 5600; Fax: +1 434 924 2828

Dmitry A. Los

Institute of Plant Physiology, 35 Botanicheskaya Street, 127276 Moscow, Russia
E-mail: losnet@ippras.ru: Tel/Fax: 7 095 977 9372

Bruno Maresca

University of Salerno, Faculty of Pharmacy, Via Ponte don Melillo, 80084 Fisciano, Salerno, Italy
E-mail: bmaresca@unisa.it; Tel/Fax: +39 081 7257302

Michael Matthay

Cardiovascular Research Institute and Departments of Medicine and Anesthesia, University of California, San Francisco,
CA 94143-0624, USA

Peter A. McNaughton

Department of Pharmacology, University of Cambridge, Tennis Court Rd, Cambridge, CB2 1PD, UK
E-mail: pam42@cam.ac.uk; Tel: +44 1223 334004; Fax: +44 1223 334040

Alfred J. Meijer

Department of Biochemistry, University of Amsterdam, Meibergdreef 15, 1105 AZ Amsterdam, The Netherlands
E-mail: a.j.meijer@amc.uva.nl; Tel: +31 20 5665159; Fax: +31 20 6915519

Bruno Miroux

Institut de Recherches Necker-Enfants Malades, rue de Vaugirard, 75015 Paris, France
E-mail: bmiroux@infobiogen.fr; Tel: +33 1 45075747; Fax: +33 1 45075890

Ewa Miskiewicz

Department of Plant Sciences, University of Western Ontario, London, Canada N6A 5B7
E-mail: emiskiew@uwo.ca; Tel: +1 519 6612111 x86475; Fax: +1 519 661 3935

Sylvie Mordier

Unite de Nutrition Cellulaire et Moleculaire, INRA de Theix, 63122 Saint Genes Champanelle, France

Norio Murata

National Institute for Basic Biology, 38 Nishigonaka, Myodaiji-cho, Okazaki 444 8585, Japan
E-mail: murata@nibb.ac.jp; Tel: +81 564 557600; Fax: +81 564 544866

Aaron Ngocky Nguyen

Section of Microbiology, University of California, Davis, CA 95616, USA

Gunnar Öquist

Department of Plant Physiology, Umeå University, S-907 81 Umeå, Sweden
Tel: +49 90 786 5416; Fax: +46 90 786 6676

E. Tapio Palva
Department of Biosciences, Division of Genetics, University of Helsinki, P.O. Box 56, 00014 Helsinki, Finland
E-mail: tapio.palva@helsinki.fi; Tel: +358 9 19159600; Fax: +358 9 191 59076

Claire Pecqueur
Institut de Recherches Necker-Enfants Malades, rue de Vaugirard, 75015 Paris, France

Neil D. Perkins
Division of Gene Regulation and Expression, School of Life Sciences, University of Dundee, Dundee, DD1 5EH, UK
E-mail: n.d.perkins@dundee.ac.uk; Tel: +44 1382 345606; Fax: +44 1382 348072

Carole Planès
Department of Physiology, Faculté de Médecine Xavier Bichat, Université Paris, 46 rue Henri Huchard, 75018 Paris, France

Almudena Porras
Departamento de Bioquímica y Biología Molecular II, Universidad Complutensede Madrid, Ciudad Universitaria, 28040 Madrid, Spain.
E-mail: maporras@farm.ucm.es; Tel: +34 91 3941627; Fax: +34 91 3941779

Daniel Ricquier
Institut de Recherches Necker-Enfants Malades, rue de Vaugirard, 75015 Paris, France

David R. Robinson
Department of Pharmacology, University of Cambridge, Tennis Court Rd, Cambridge, CB2 1PD, UK

Sonia Rocha
Division of Gene Regulation and Expression, School of Life Sciences, University of Dundee, Dundee, DD1 5EH, UK

Stéphanie Roques
Laboratory of Oxidative Stress and Aging, University Hospital-Lausanne, CH-1011 Lausanne, Switzerland

Alison Shaw
EMBL, Meyerhofstrasse 1, 69012 Heidelberg, Germany

Kazuhiro Shiozaki
Section of Microbiology, University of California, Davis, CA 95616, USA
E-mail: kshiozaki@ucdavis.edu; Tel: +1 530 7523628; Fax: +1 530 7529014

Mark Smith
Institute of Pathology, Case Western Reserve University, 2085 Adelbert Road, Cleveland, OH 44106, USA
E-mail: mas21@po.cwru.edu; Tel: +1 216 368 3670; Fax: 216 368 8964

Janet B. Storey
Institute of Biochemistry and Department of Biology, Carleton University, 1125 Colonel By Drive, Ottawa, Ontario, Canada K1S 5B6
E-mail: jstorey@ccs.carleton.ca; Tel: +1 613 520 3678; Fax: +1 613 520 2569

Kenneth B. Storey
Institute of Biochemistry and Department of Biology, Carleton University, 1125 Colonel By Drive, Ottawa, Ontario, Canada K1S 5B6
E-mail: kenneth_storey@carleton.ca; Tel: +1 613 520 3678; Fax: +1 613 520 2569

Gail W. Sullivan
Cardiovascular Research Center, University of Virginia School of Medicine, Charlottesville, VA 22908, USA
E-mail: gws3u@virginia.edu; Tel: +1 434 9249665; Fax: +1 434 9242828

Jan Svensson
Department of Biosciences, Division of Genetics, University of Helsinki, P.O. Box 56, 00014 Helsinki, Finland
E-mail: svensson@mappi.helsinki.fi; Tel: +358 9 19159600; Fax: +358 9 191 59076

Tokujiro Uchida
Department of Physiology, Faculté de Médecine Xavier Bichat, Université Paris, 46 rue Henri Huchard, 75018 Paris, France

Vittorio Vellani
Department of Pharmacology, University of Cambridge, Tennis Court Rd, Cambridge, CB2 1PD, UK

László Vígh
Institute of Biochemistry, Hungarian Academy of Sciences, Temesvári krt. 62, H-6726 Szeged, Hungary
E-mail: vigh@nucleus.szbk.u-szeged.hu; Tel/Fax: +36 62 432048

Kenneth E. Wilson
Department of Molecular Biology, University of Geneva, CH-1211 Geneva, Switzerland
E-mail: kenneth.wilson@molbio.unige.ch; Tel: 22 702 6187; Fax: 22 702 6868

Zhanguo Xin
Plant Stress and Germplasm Development Unit, USDA, Agricultural Research Service, 3810 4th Street, Lubbock, TX 79415, USA
E-mail: zxin@lbk.ars.usda.gov; Tel: +1 806 7495560; Fax: +1 806 7235272

Contents

Preface v

List of Contributors vii

Chapter 1. *Ischemic Tolerance in the Brain: Models and Mechanisms* – Michael Bell, Kelly Drew,
　　　　　　Mark Smith and John Hallenbeck. 1

1. Introduction 1
2. Myocardial preconditioning 1
　　2.1. Early preconditioning mechanisms 2
　　2.2. Late preconditioning mechanisms 3
3. Brain preconditioning 3
　　3.1. Ischemic preconditioning—clinical experience and animal models. 3
　　3.2. Alternative preconditioning stimuli 4
　　3.3. Proposed mechanisms of ischemic preconditioning in animal models 4
　　3.4. Cellular mechanisms—development of cell culture models. 5
4. Hibernation and hypoxia tolerance 6
5. Conclusions 9
References 9

Chapter 2. *Regulation of Gene Expression by Hypoxia in Lung Alveolar Epithelial Cells* – Christine Clerici,
　　　　　　Tokujiro Uchida, Carole Planès and Michael A. Matthay. 13

1. Introduction. 13
2. Effect of hypoxia on Na transport proteins. 14
3. Effect of hypoxia on proteins involved in glucose metabolism 16
　　3.1. Effect of hypoxia on glucose transporters 17
　　3.2. Effect of hypoxia on glycolytic enzymes. 18
4. Effect of hypoxia on vascular growth factor expression 18
5. Effect of hypoxia on adhesion molecules expression 20
6. Role of hypoxia-inducible factors in regulation of gene expression 20
7. Conclusion 22
References. 23

Chapter 3. *A Profile of the Metabolic Responses to Anoxia in Marine Invertebrates* – Kevin Larade and
　　　　　　Kenneth B. Storey 27

1. Introduction. 27
2. Metabolic response to anoxia. 27
　　2.1. Regulation of carbohydrate metabolism 27
　　2.2. Effects of pH on cellular metabolism. 29
　　2.3. Controlling glycolytic flux 29
3. Macromolecular synthesis 31
　　3.1. Economics of energy conservation 31
　　3.2. mRNA and protein synthesis 31
　　3.3. Mechanisms of translational control 33
　　3.4. Polysome analysis 34

3.5. Ribosomal proteins 34
3.6. Maintaining translatable mRNA pools 35
4. Gene expression 36
4.1. Anoxia-induced gene expression 36
4.2. Identification of differentially expressed genes 36
4.3. Pharmacology 39
5. Triggering the anoxic response 39
5.1. Oxygen sensing 39
5.2. Second messengers 40
5.3. The role of cGMP 40
6. Transcription factors 40
7. Perspectives 41
References 43

Chapter 4. *The Role of Adenosine in Tissue Protection During Ischemia-Reperfusion* – Gail W. Sullivan and
Joel Linden 47

1. Introduction 47
2. Endogenous adenosine is protective in ischemia-reperfusion injury 47
2.1. Adenosine metabolism 47
2.2. Adenosine localization and transport 49
3. Adenosine receptors 49
3.1. Adenosine receptor antagonists 49
3.2. Adenosine receptor agonists and allosteric enhancers 49
3.3. Adenosine receptor expression 50
3.4. Adenosine receptor down-regulation 51
4. Adenosine-stimulated cell pathways and effector activities 51
4.1. Blood flow 51
4.2. Preconditioning 52
4.3. Reperfusion 54
5. Conclusions and clinical implications 56
References 56

Chapter 5. *NF-κB Function in Inflammation, Cellular Stress and Disease* – Neil R. Chapman, Sonia Rocha,
Ian M. Adcock and Neil D. Perkins 61

1. Introduction 61
2. Activation of NF-κB 61
3. Attenuation of the NF-κB response 64
4. The inflammatory stress response 64
5. Rheumatoid arthritis 65
6. NF-κB and the airway 65
7. NF-κB and gastro-intestinal disease 66
8. NF-κB and reproductive function 67
8.1. The uterine cycle 67
8.2. Implantation and labor 67
8.3. Inflammatory diseases of the reproductive tract 67
9. Other NF-κB associated diseases 68
9.1. NF-κB, cancer and stress 68
9.2. Viral infection 69
9.3. Cardiovascular disease 69
9.4. Neurology 69
10. Conclusions 70
Acknowledgements 71
References 71

Chapter 6.   MAPping Stress Survival in Yeasts: From the Cell Surface to the Nucleus – Aaron Ngocky Nguyen and
Kazuhiro Shiozaki 75

1.   Introduction. 75
2.   The Hog1 MAPK cascade in budding yeast 76
    2.1.   Regulation of the Hog1 pathway by two independent osmosensing pathways 76
    2.2.   Nuclear translocation of Hog1 upon osmotic stress 77
    2.3.   Induction of adaptive responses to osmostress through Hog1 targets 79
3.   The Spc1 MAPK cascade in fission yeast 80
    3.1.   Multiple mechanisms of Spc1 activation 80
    3.2.   Regulation of Spc1 in response to heat shock 81
    3.3.   Oxidative stress signaling in fission yeast 82
    3.4.   Activation of Spc1 in response to high osmolarity stress 83
    3.5.   Recognition of Spc1 by its cognate MAPKK 84
    3.6.   Regulated nuclear translocation of Spc1 upon stress stimuli. 84
    3.7.   Crucial link between mitosis/meiosis and changes in the extracellular environment 85
4.   Concluding remarks 85
Acknowledgements 86
References. 86

Chapter 7.   Calcium Signaling Mediated by Cyclic ADP-Ribose and NAADP: Roles in Cellular Response to Stress –
Hon Cheung Lee 91

1.   Introduction. 91
2.   Calcium signaling mediated by cyclic ADP-ribose and NAADP 91
    2.1.   Structure of cyclic ADP-ribose. 91
    2.2.   Structure of NAADP 93
    2.3.   Calcium mobilization 93
    2.4.   Multiple Ca^{2+} stores in cells. 94
    2.5.   Enzymatic synthesis and degradation. 95
3.   cADPR and plant response to environmental stress 97
4.   cADPR and sponge response to heat stress. 98
5.   cADPR in bacterial infection and immune responses 99
6.   Conclusion. 100
References 101

Chapter 8.   The Cellular and Molecular Basis of the Detection of Pain – Jennifer K. Bonnington,
David R. Robinson, Vittorio Vellani and Peter A. McNaughton 105

1.   Introduction 105
2.   Ion channels involved in nociception 106
    2.1.   Ion channels gated by heat 106
    2.2.   ATP-gated ion channels 108
    2.3.   Proton-gated ion channels 108
    2.4.   Voltage-gated sodium channels 109
3.   Sensitisation of nociceptors 109
    3.1.   Sensitisation by PGE_2: the PKA pathway 110
    3.2.   Sensitisation by bradykinin: the PKC pathway 110
    3.3.   Sensitisation by nerve growth factor 112
4.   Pain detection in the viscera. 113
    4.1.   Visceral afferent neurons 113
    4.2.   Visceral hyperalgesia 114
    4.3.   Cardiac nociception 115
    4.4.   The peptide content of primary visceral neurons 115
Acknowledgements 116
References 116

Chapter 9. *Acquired Freezing Tolerance in Higher Plants: The Sensing and Molecular Responses to Low Nonfreezing Temperatures* – Zhanguo Xin 121

1. Introduction 121
2. Freezing injury in plants. 122
3. Biochemical changes associated with cold acclimation. 123
   3.1. Alteration in membrane structure and lipid composition 124
   3.2. Changes in soluble sugars. 124
   3.3. Changes in betaines 124
   3.4. Accumulation of proline. 125
4. Alteration of gene expression associated with cold acclimation 125
   4.1. Diversity of cold induced genes. 125
   4.2. COR genes. 126
5. CBF transcription factors define a major cold response pathway in flowering plants 126
6. Mutational analysis of freezing tolerance 127
7. Emerging signaling processes regulating cold acclimation 128
   7.1. Abscisic acid (ABA) as a plant hormone mediating cold acclimation 128
   7.2. Involvement of calcium in temperature sensing. 129
   7.3. Protein kinases and phosphatases 129
8. Molecular and genetic analysis of signaling transduction of stress responses 129
9. Multiple signal pathways mediate cold acclimation 130
10. Hunting for the temperature sensors 131
   10.1. Changes in membrane fluidity may serve as the basis for temperature sensing in higher plants 131
   10.2. Histidine kinases serve as one type of temperature sensors in Synechocystis 131
   10.3. Search for temperature sensors in higher plants. 133
11. Conclusions and perspectives. 133
Acknowledgements 134
Disclaimer 134
References 134

Chapter 10. *Sensing and Responses to Low Temperature in Cyanobacteria* – Dmitry A. Los and Norio Murata 139

1. Introduction 139
2. Cellular responses to low-temperature stress 140
3. Cold-inducible genes and their regulation 140
   3.1. Desaturases 142
   3.2. RNA-binding proteins. 145
   3.3. Cold-inducible RNA helicases 146
   3.4. Ribosomal proteins 146
   3.5. Caseinolytic proteases. 146
   3.6. Cytochrome c_M. 147
   3.7. Genes that are down-regulated by low temperature. 147
4. Membrane fluidity as a link between low temperatures and the induction of gene expression. 147
5. Sensors and transducers of low-temperature signals 149
6. Conclusions and perspectives. 149
References 150

Chapter 11. *Dehydrins* – Jan Svensson, Abdelbagi M. Ismail, E. Tapio Palva and Timothy J. Close 155

1. Introduction 155
2. Dehydrins 155
   2.1. Conserved sequence motifs 156
   2.2. Biochemical properties 158
   2.3. Post-translational modifications. 159
3. Dehydrin gene expression. 159
4. Association with abiotic stresses and anticipated functions. 160

4.1. Low temperature 160
4.2. Water deficit. 163
4.3. Salinity 163
5. Seed dehydrins 164
6. Localization of dehydrins 164
    6.1. Tissue distribution. 164
    6.2. Subcellular localization 166
7. Functional studies of dehydrins 166
8. Conclusions 167
References 167

Chapter 12. Dual Role of Membranes in Heat Stress: As Thermosensors They Modulate the Expression of Stress Genes and, by Interacting with Stress Proteins, Re-organize Their Own Lipid Order and Functionality – László Vígh and Bruno Maresca 173

1. Introduction 173
2. Membranes as "cellular thermometers". 174
3. Pre-existing membrane lipid composition and physical state determines the heat induced membrane damage and has a primary and major effect on cell viability 176
4. HSPs are actively involved in the repair of the damaged membranes in heat-stressed cells by association with the membranes 178
Acknowledgements 184
References 184

Chapter 13. Cellular Adaptation to Amino Acid Availability: Mechanisms Involved in the Regulation of Gene Expression and Protein Metabolism – Sylvie Mordier, Alain Bruhat, Julien Averous and Pierre Fafournoux 189

1. Introduction 189
2. Regulation of amino acid metabolism and homeostasis in the whole animal. 189
    2.1. Free amino acids pools 190
    2.2. Specific examples of the role of amino acids in the adaptation to protein deficiency 191
3. Amino acid control of gene expression 192
    3.1. Post-transcriptional regulation of genes expression by amino acid availability 192
    3.3. Transcriptional activation of mammalian genes by amino acid starvation 192
4. Amino acid control of protein metabolism 196
    4.1. Regulation of protein degradation by amino acids 196
    4.2. Regulation of protein synthesis by amino acids. 198
    4.3. Regulation of mTOR kinase activity by amino acids 200
5. Conclusion. 201
References 202

Chapter 14. Amino Acid-dependent Signal Transduction – Peter F. Dubbelhuis and Alfred J. Meijer. 207

1. Introduction: amino acid stimulation of S6 phosphorylation 207
2. Amino acids and p70S6 kinase activation 208
3. Amino acid stimulation of 4E-BP1 phosphorylation 209
4. Amino acid stimulation of eEF2kinase 209
5. Amino acid stimulation of eIF2α 209
6. Involvement of PI 3-kinase and protein kinase B in amino acid-dependent signaling? Amino acid/insulin synergy 210
7. Amino acids and mTOR activation 211
8. Amino acids and protein phosphatases 212
9. Negative feedback by amino acid signaling on insulin signaling. 212
10. mTOR as an ATP sensor. Involvement of AMP-dependent protein kinase? 213

11. Amino acid signaling in β-cells 214
12. Amino acid signaling *in vivo* 215
13. Amino acid signaling as a function of age 215
14. Mechanisms 215
15. Conclusions 217
Acknowledgement 217
References 217

Chapter 15. Signal Transduction Pathways Involved in the Regulation of Drug Metabolizing Enzymes –
          Vidya Hebbar and A.-N. Tony Kong. 221

1. Introduction 221
2. Drug Metabolizing Enzymes (DMEs). 222
    2.1. General considerations 222
    2.2. Phase I DMEs 222
    2.3. Phase II DMEs 224
3. The antioxidant response element (ARE) 224
4. Transacting factors regulating the expression of Phase II enzymes. 225
5. Mitogen-activated protein kinases (MAPKs) 227
6. MAPK activation by phenolic compounds and isothiocyanates: induction of Phase II enzymes . 228
7. MAPK activation and ARE-mediated gene expression via Nrf2-dependent mechanism. 229
8. Concluding remarks. 229
Acknowledgements 229
References 229

Chapter 16. Biological Actions of Infrared Radiation – Lee Laurent-Applegate and Stéphanie Roques. 233

1. Introduction 233
    1.1. Infrared radiation spectrum 233
    1.2. Penetration properties in tissue of infrared radiation 234
2. Electromagnetic radiation and human health 235
    2.1. Reactive oxygen species and their role in bio-regulation 235
    2.2. Free radical reactions and light production 235
3. Cellular interactions with IR radiation 235
    3.1. Cellular recognition and migration 235
    3.2. Sources of low dose infrared radiation 236
    3.3. Sources of high dose infrared radiation 236
4. Gene and protein induction 237
    4.1. Survival curves of human skin fibroblasts following IR-A radiation 237
    4.2. Induction of protective proteins 237
    4.3. Quantitative expression of ferritin *in vitro* following IR radiation 238
    4.4. Expression of ferritin in human skin *in vivo* following infrared-A radiation 239
    4.5. Expression of DNA damage, oxidative stress molecules and proteases of human skin *in vivo* following
          infrared-A radiation 240
5. Perspectives 240
Acknowledgements 240
References 240

Chapter 17. Energy Sensing and Photostasis in Photoautotrophs – Norman P.A. Huner, Alexander G. Ivanov,
          Kenneth E. Wilson, Ewa Miskiewicz, Marianna Krol and Gunnar Öquist 243

1. Introduction 243
2. Energy sensing and redox status 243
3. Photosynthetic electron transport 245
4. Excitation pressure and redox sensing 247

5. Photostasis 248
6. Excitation pressure, photoprotection and photostasis 248
    6.1. Photoacclimation 249
    6.2. Cold acclimation 250
    6.3. Nutrient limitations 252
    6.4. Drought 252
7. Summary 252
Acknowledgements 253
References 253

Chapter 18. The Uncoupling Proteins Family: From Thermogenesis to the Regulation of ROS – Marie-Clotilde
    Alves-Guerra, Claire Pecqueur, Alison Shaw, Elodie Couplan, Maria Del Mar Gonzalez Barroso,
    Daniel Ricquier, Frédéric Bouillaud and Bruno Miroux 257

1. Introduction 257
2. UCP1 belongs to the mitochondrial anion carriers family 257
    2.1. Historical background 257
    2.2. UCP1: the uncoupling protein of brown adipose tissue 258
    2.3. The mitochondrial carriers: a multifunctional protein family? 259
    2.4. Physiological relevance of a "mild uncoupling" 260
3. The discovery of the uncoupling proteins homologues 260
    3.1. Historical background 260
    3.2. *In vivo* distribution of the UCP homologues 261
    3.3. The uncoupling activity of the UCP1 homologues in heterologous expression systems 261
4. Lessons from the UCP knockout mice 263
    4.1. UCPs modify mitochondrial energy metabolism but not whole body energy expenditure 263
    4.2. Improvement of β cells function in Ucp2(–/–) mice 263
    4.3. Improvement of macrophage oxidative burst in Ucp2(–/–) mice 263
5. UCP family, from thermogenesis to the regulation of ROS 264
    5.1. UCP1, a misleading model 264
    5.2. UCP2 and UCP3: new players in the mitochondrial redox balance? 265
Acknowledgements 265
References 265

Chapter 19. Regulation of Proliferation, Differentiation and Apoptosis of Brown Adipocytes: Signal Transduction
    Pathways Involved – Almudena Porras and Manuel Benito 269

1. Introduction 269
    1.1. Function and characteristics of brown adipose tissue 269
    1.2. Uncoupling protein-1: function and activation 269
2. Development, differentiation and involution of brown adipose tissue 269
3. Transcriptional control of brown adipose tissue differentiation 270
    3.1. Regulation of adipogenesis 270
    3.2. Regulation of UCP-1 expression 271
4. Regulation of proliferation of brown adipocytes: signals and signal transduction pathways involved 272
5. Regulation of differentiation of brown adipocytes: signals and signal transduction pathways involved . . . 274
6. Apoptosis of brown adipocytes: signals and signal transduction pathways involved in its regulation 277
7. Conclusions 279
Acknowledgements 279
References 279

Chapter 20. Control Analysis of Metabolic Depression – R. Keira Curtis, Tammie Bishop and Martin Brand 283

1. Introduction 283
2. What is control analysis? 283

3.   Applications of control analysis. 283
4.   Theory 284
       4.1.   Definition of coefficients 284
       4.2.   Constructing the big picture 286
       4.3.   Calculation of coefficients 287
5.   Examples 288
       5.1.   Control analysis of mitochondrial respiration in aestivating snails 288
       5.2.   Control analysis of whole cell metabolism 289
       5.3.   Control analysis of cell signalling pathways 290
6.   Regulation analysis of cell signalling in pathways that initiate metabolic depression 291
       6.1.   Overview 291
       6.2.   Practicalities 292
       6.3.   How to interpret results from this type of experiment 294
7.   Summary 295
References 295

Chapter 21.   Evolution of Physiological Adaptation – Douglas L. Crawford 297

1.   Introduction 297
2.   Evolutionary physiology: appropriate comparisons. 297
3.   Variation in physiological and biochemical traits. 299
       3.1.   Enzyme expression 300
       3.2.   Fundulus 302
4.   Evolution of thermal acclimation 306
       4.1.   Molecular mechanism in Fundulus acclimation. 309
5.   Conclusions 310
Acknowledgement 310
References 310

*Chapter 22.   Dynamic Use of cDNA Arrays: Heterologous Probing for Gene Discovery and Exploration of
               Organismal Adaptation to Environmental Stress* – Sean F. Eddy and Kenneth B. Storey 315

1.   Introduction 315
2.   Differential gene expression: the early years 316
3.   DNA arrays: a brief history 317
4.   Evaluation of mammalian hibernation via cDNA array screening 318
5.   Reproducibility and reliability 323
6.   Outlook 323
References 324

Index 327

Sensing, Signaling and Cell Adaptation. Edited by K.B. Storey and J.M. Storey

CHAPTER 1

Ischemic Tolerance in the Brain: Models and Mechanisms

Michael Bell[1,4], Kelly Drew[2], Mark Smith[3] and John Hallenbeck[4]

[1]*Center for Neuroscience Research, Children's Research Institute, Children's National Medical Center, Washington, DC 20010, USA*
[2]*Alaskan Basic Neuroscience Program and Department of Chemistry and Biochemistry, Institute of Arctic Biology, University of Alaska Fairbanks, Fairbanks, AK 99775-7000, USA*
[3]*Institute of Pathology, Case Western Reserve University, Cleveland, OH 44106, USA*
[4]*Stroke Branch, NINDS, NIH, Bethesda, MD 20892-4128, USA*

1. Introduction

Ischemic brain cells produce a wide array of mediators that participate in progression of ischemic brain injury. As a noncomprehensive example, there is evidence for the participation of excitotoxins (Rothman and Olney, 1995), free radicals (Hall and Braughler, 1989), activated endothelium (Hallenbeck, 1996), leukocytes (del Zoppo et al., 1991), cytokines (Giulian and Robertson, 1990), platelets (Dougherty et al., 1977), platelet activating factor (Lindsberg et al., 1991), eicosanoids (Walker and Pickard, 1985), endothelin (Greenberg et al., 1992), altered gene expression (Nowak, 1990), edema (Klatzo, 1985), nitric oxide (Samdani et al., 1997), poly(ADP-ribose) polymerase (PARP) (Eliasson et al., 1997), growth factors (Finklestein et al., 1990), apoptosis (Choi, 1996), matrix metalloproteinases (Mun-Bryce and Rosenberg, 1998) and mitochondrial permeability pore transition (Duchen et al., 1993) in evolving brain injury. Disappointingly, clinical trials of agonists or antagonists of many of these mediators of brain injury have uniformly failed to confer significant cytoprotection in acute stroke. As an example, more than 80 clinical trials of over 30 different neuroprotective agents had been published by 1996 in the therapy of stroke and none has shown clear evidence of a benefit (Dorman and Sandercock, 1996). Study of endogenous neuroprotective mechanisms associated with tolerance to brain ischemia could guide the search for effective stroke therapy through this morass of interlaced mechanisms.

Preconditioning is the phenomenon whereby a stressful non-lethal stimulus sets in motion a cascade of biochemical events that renders cells, tissues or the whole organism tolerant to a future, more lethal stimulus. Ischemia has proven a potent primary stimulus for preconditioning both in the myocardium, where it was first described and in the brain. This chapter will begin with a review of myocardial preconditioning that emphasizes cellular mechanisms since these studies stimulated studies of preconditioning within neurological systems. Models of brain preconditioning in animals and cell cultures will then be reviewed with emphasis on the molecular mechanisms involved in the development of the neuroprotection. Finally a "natural model of tolerance", hibernation, will be discussed.

2. Myocardial preconditioning

Myocardial preconditioning in animal models has been studied since the mid-1980s and recent clinical studies suggest it may be operative in humans (Bahr et al., 2000; Noda et al., 1999; Ottani et al., 1995). Murry and colleagues were the first to develop a preconditioning paradigm and used infarct size as a measure of injury (Murry et al., 1986). Their paradigm involved four episodes of brief ischemia consisting of 5-min occlusion of circumflex coronary artery followed by 5-min of reperfusion. A longer ischemic insult was imposed immediately thereafter for varying amounts of time in separate groups of dogs (group 1 for 40 min and group 2 for 3 hours). Preconditioned dogs had a 75% reduction in infarct size in group 1 compared to shams while no protection was conferred to the preconditioned animals in group 2. This suggested that several brief episodes of ischemia created a brief time period of protection from more lethal ischemia. Other groups (Cohen et al., 1991; Shiki and Hearse, 1987) have used contractility, incidence of ischemia-induced arrhythmia or magnitude of ST-segment elevation as measures of injury and similarly found that brief preconditioning stimuli can confer immediate protection.

In 1993, two independent groups demonstrated that the protection achieved from brief episodes of ischemia is biphasic. A second "window" of protection from ischemia was found 24 h after the initial preconditioning stimuli (Kuzuya et al., 1993; Marber et al., 1993). Marber and colleagues found that a similar preconditioning stimulus of 4 episodes of 5 min coronary artery occlusion provided approximately 30% protection to a 30 min ischemic insult given 24 h after the preconditioned stimulus. This phenomenon was termed "late" preconditioning in contrast to "classic" or "early" preconditioning described above. Initial hypotheses were formulated suggesting that products of metabolism immediately released after ischemia may mediate "early" preconditioning while newly synthesized proteins may be required for "late" preconditioning cardioprotection.

2.1. Early preconditioning mechanisms

Early preconditioning was observed in several species including swine, rabbit, rat and dogs (Cohen et al., 1991; Murphy et al., 1993; Schott et al., 1990; Yao and Gross, 1993) and investigators strove to determine the biochemical mechanisms underlying this protective response. The rapidity with which the protection was afforded suggested that substances released from the ischemic myocardium were likely responsible for the cardioprotection observed rather than *de novo* synthesis of mediators. Adenosine, bradykinin and opioid peptides were identified as agents that were released from ischemic myocardium and could independently substitute for the brief periods of ischemia as an early preconditioning stimulus. Adenosine was found to exert its effects via binding to its A1 or A3 receptors within the myocardium in isolated cardiac myocytes (Eliasson et al., 1997), animal studies (Tracey et al., 1997) and human cardiac tissue (Carr et al., 1997). Binding of bradykinin to its B2 receptor protected cardiac myocytes (Parratt et al., 1995; Wall et al., 1994), although its marked hemodynamic effects make it difficult to exclude effects not related to preconditioning. Opioid peptides precondition by binding to their δ-receptor subtype and selective δ-opioid receptor agonists have been shown to induce cardioprotection in experimental models (Schultz et al., 1998).

With these mediators identified, efforts continued to define the intracellular events responsible for cardioprotection. Currently, a working theory of cardioprotection after preconditioning involves the ATP-sensitive potassium channel (K_{ATP}). This channel, thought to link myocardial metabolism to membrane electrical activity, is present in the sarcoplasmic reticulum membrane (sarc K_{ATP}) as well as the mitochondrial membrane (mt K_{ATP}). A role in preconditioning was postulated soon after its discovery (Noma, 1983). Noma noted that brief periods of ischemia caused an opening of sarc K_{ATP} channels and suggested a prominent role of sarc K_{ATP} channels in preconditioning. He theorized that opening of the sarc K_{ATP} channels would enhance the shortening of the cardiac action potential by accelerating phase 3 repolarization. This enhanced repolarization could inhibit calcium influx into the myocyte, prevent calcium overload within the cell and, therefore, enhance cellular viability. Evidence for a role of these channels in preconditioning was furnished by a series of studies. First, selective K_{ATP} antagonists (glibenclamide and sodium 5-hydroxydecanoate) blocked the protection while a K_{ATP}

opener (aprilkam) augmented the protection of preconditioning (Auchampach et al., 1992; Gross and Auchampach, 1992). Also, multiple mediators associated with ischemic preconditioning, including adenosine, regulate K_{ATP} channel function (Yao and Gross, 1994). Most convincingly, the development of specific openers of mt K_{ATP} channels, specifically diazoxide, demonstrated that opening of mt K_{ATP} channels can confer cardioprotection without involvement of sarc K_{ATP} channels (Garlid et al., 1997). While the regulation of mt K_{ATP} channel opening appears complex, preliminary studies suggest that multiple ligands can bind to a transmembrane G-protein leading to intracellular activation of protein kinase C. This ubiquitous intracellular messenger can lead to a conformational change in the mt K_{ATP} channel although further study is needed to fully define the mechanism underlying the cardioprotection (Marber, 2000). This work has culminated in clinical trials using nicorandil, a K_{ATP} channel opener with nitrate-like hemodynamic properties. Nicorandil improves left ventricular wall motion in ischemic heart disease and has improved EKG findings in patients undergoing angioplasty (Markham et al., 2000).

2.2. Late preconditioning mechanisms

Examination of the mechanisms involved in late preconditioning is in its infancy compared to the detailed delineation of key processes in early preconditioning. While some have cited adenosine release leading to protein kinase C activation as a potential mechanism of late preconditioning (Baxter et al., 1997), still others have suggested that the gradual onset and persistence of the protection suggests that *de novo* protein synthesis occurring subsequent to the initial stimulus represents a more probable mechanism. The heat shock response involves the production of novel proteins in response to a stress and was first described as a response of cells to thermal injury. Other stimuli including ischemia and oxidative stress have been shown to elicit this response and it has been conserved through evolution in species as diverse as fruit flies, yeast, plants and mammals. The synthesis of these proteins protects the organism from

subsequent events. In their initial study, Marber and colleagues noted that the cardioprotection of preconditioning was coincident with an upregulation of an inducible 70-kDa heat shock protein (HSP 70) (Marber et al., 1993) and this finding has been replicated by others. Moreover, overexpression of HSP 70 in transgenic mice also increases resistance to myocardial ischemia (Marber et al., 1995).

3. Brain preconditioning

Groups have attempted to replicate the preconditioned responses to ischemia observed in the myocardium in neurologic models. The similarities between the organs include the necessity of ion channels for function, the need for high-energy substrates to maintain homeostasis and a commonality of mediators known to alter the organ's function (adenosine, nitric oxide, for example). Important differences include the heterogeneity of cell types within the CNS (neurons, astrocytes, oligodendrocytes, microglia, in addition to blood vessels) compared to the relatively homogeneous population of myocardial myocytes within the heart. In work spanning more than a decade, groups of investigators have successfully demonstrated that the principles of preconditioning can be applied within the CNS and that ischemia need not be the sole initiating stimulus.

3.1. Ischemic preconditioning—clinical experience and animal models

Clinical evidence for preconditioning neuroprotection from transient ischemic attacks (TIA) exists (Moncayo et al., 2000) in that prior TIA has been associated with milder stroke symptoms after controlling for other cardiovascular risk factors. Yet most simply regard the occurrence of TIA as a risk factor for the development of stroke (Weih et al., 1999). The paucity of definitive clinical evidence of preconditioning notwithstanding, a number of animal models demonstrated that sublethal doses of ischemia could protect against a future, more lethal dose. As in the myocardium, multiple small preconditioning stimuli were found to be more effective than single stimuli and the interval between the preconditioned stimulus and the subsequent insult was found to be critical. However, brain models of preconditioning revealed key differences. Mainly, the interval between the preconditioning and lethal stimuli needed to be longer and the duration of the neuroprotection was increased relative to the observations in the myocardium. Furthermore, either global or focal preconditioning stimuli were found to induce protection against both global and focal lethal insults. These findings have led investigators to postulate novel mechanisms of protection within the brain.

Global preconditioning stimuli can protect against both global and focal injuries. Kitigawa and colleagues showed that a 5-min episode of bilateral carotid artery occlusion in gerbils caused selective damage to the CA1 region of the hippocampus, a region of brain selectively vulnerable to ischemia (Kitagawa et al., 1990). They demonstrated that two episodes of sublethal ischemia (2-min duration) performed at 24 h intervals for two days prior to the lethal ischemic injury provided almost complete protection of this vulnerable region. They also found that a single dose of the preconditioning stimulus was partially effective in providing protection but multiple smaller doses of preconditioning ischemia (1-min duration) were ineffective. Simon and colleagues found that 48 hours after a global ischemia, rats were protected against permanent middle cerebral artery occlusion, a focal injury, manifested as a decrease in the overall infarct size (Simon et al., 1993).

Similarly, focal preconditioning stimuli can protect against both focal and global injuries. Chen and colleagues used three 10-min episodes of MCA occlusion (separated by 45 min of reperfusion) as a preconditioning stimulus and showed decreased infarct size after a subsequent 100 min occlusion of the MCA in rats (Chen et al., 1996). Interestingly, the protection was only elicited 2–5 days after the preconditioning stimulus. At day 1 and day 7 after preconditioning, the protective effects of the preconditioning were not observed. Glazier and colleagues demonstrated that a focal preconditioning stimulus could protect against a global injury (Glazier et al., 1994). Using rats, the

middle cerebral artery (MCA) was occluded for 20-min followed by 24 hours of reperfusion. The rats were then subjected to bilateral common carotid artery occlusion with hemorrhagic hypotension and preconditioned animals showed decreased damage in ipsilateral and contralateral hippocampi. Others have observed essentially the same results using gerbils (Miyashita et al., 1994).

3.2. Alternative preconditioning stimuli

The study of preconditioning in animals has expanded to include models using alternative preconditioning stimuli. Hypoxia, defined as decreased oxygen delivery with no alteration in blood flow, is distinct from conditions in which blood flow is reduced. Using neonatal rats, 4 h of hypoxia (8% O_2 environment) administered 24 h prior to a hypoxic-ischemic insult (left carotid artery occlusion, 8% O_2 environment for 2 h) led to decreases in infarct volume and hippocampal cell loss (Ota et al., 1998). Inflammatory mediators including cytokines, particularly tumor necrosis factor-α (TNF-α), can also serve as a preconditioning stimulus in animal models. Systemic injection of lipopolysaccharide (LPS) induced protection from permanent MCA occlusion in spontaneously hypertensive rats and pharmacological blockade of the effects of TNF-α nullified this protection (Tasaki et al., 1997). In a follow-up study, preconditioning with intracisternal injection of TNF-α 24 h prior to ischemia decreased infarct size in mice after secondary permanent MCA occlusion in a dose- and time-dependent manner (Nawashiro et al., 1997a).

3.3. Proposed mechanisms of ischemic preconditioning in animal models

Several mechanisms have been postulated for preconditioning neuroprotection. Some evidence suggests that mediators released coincident with ischemia, such as excitatory amino acids and adenosine, may be involved in the ischemic preconditioning. Excitatory amino acids such as glutamate are released after ischemic injury and interact with the N-methyl-D-aspartate (NMDA) receptor.

An NMDA receptor antagonist (MK-801) administered prior to the ischemic preconditioning stimulus blunts the expected neuroprotective response in gerbils (Kato et al., 1992). However, others have failed to show a difference between interstitial brain concentrations of glutamate in preconditioned and sham animals using ischemic preconditioning models (Nakata et al., 1993). Some evidence suggests that adenosine also may play a role in ischemic preconditioning in brain. Adenosine A1 antagonists blunt the neuroprotection of ischemic preconditioning while A1 agonists have failed to show the degree of neuroprotection induced by preconditioning ischemia (Heurteaux et al., 1995).

By far, the most common early response noted in these animal models of preconditioning involves the heat shock response with models of both global and focal preconditioning stimuli increasing heat shock protein expression. HSP 72 was detected in some models (Kitagawa et al., 1991; Simon et al., 1993) while others have detected HSP 70 (Chen et al., 1996). A role in the neuroprotection has been postulated for these proteins because competitive inhibition of HSP 70 gene expression has blocked thermotolerance (Riabowol et al., 1988). Additional evidence for the heat shock response in preconditioning includes demonstration of a dose–response relationship between the dose of preconditioning stimuli and the expression of heat shock proteins. Kitigawa and colleagues (Kitagawa et al., 1991) found that 2- or 5-min doses of preconditioning stimuli increased HSP 72 protein and afforded protection while 1-min of stimulation was insufficient for expression of the protein and protection was not observed. Furthermore, the timing of HSP expression coincides with the time period of protection from subsequent ischemia. Expression of HSP 70 (Chen et al., 1996) in cortex was noted between day 1 to 5 after preconditioning at a time when protection was noted. The expression was undetectable by 7 days and protection was no longer afforded at this delayed time point.

Other proteins including the apoptosis regulating proteins bcl-2, bcl-x-long have also been implicated in preconditioning. Briefly, Bcl-2 is a naturally occurring neuroprotective protein that favors cellular survival over apoptotic cell death. Bcl-2 is strongly expressed in neurons that survive either focal or global ischemia (Shimazaki et al., 1994), is increased in a time course consistent with the induction of preconditioning and inhibition of bcl-2 protein exacerbates neuronal injury after ischemia (Chen et al., 2000). Although evidence from these studies suggests a number of potential cytoprotective mechanisms are activated in tolerance, the precise molecular regulation of the tolerant state remains under study.

3.4. Cellular mechanisms—development of cell culture models

Cell culture systems have been developed to further define the intracellular mechanisms of preconditioning since they can offer several advantages over whole animal models. Cell culture systems allow the study of isolated cell types within the CNS since differences might exist between neurons and glial cells with respect to responses to ischemia. The exposure of these cells to either preconditioning stimuli or therapeutic agents can be carefully controlled in these systems and gene transfer experiments can be conducted with specific constructs to enhance validity of the studies. Cell cultures also allow the completion of multiple experiments in a short time period, eliminate the need for blood brain barrier penetration of novel agents and eliminate the potential bias caused by unknown systemic side effects in pilot studies. Of course, results from cell culture studies need to be replicated in animal models before human trials can be considered.

Ischemia cannot be precisely replicated in cell culture models but hypoxia with deprivation of glucose has often been used as an effective surrogate. Bruer et al. (1997) developed a model of hypoxic preconditioning in neuronal cell culture which used a preconditioning stimulus of 1.5 h of oxygen-glucose deprivation (OGD) consisting of a humidified atmosphere with PO_2 of 2–4 mm Hg, 5% CO_2/95 %N_2, without glucose in medium. This stimulus was applied 48–72 h prior to a more lethal (3 h) dose of OGD. Controls (3 h OGD without preconditioning) showed an increase in LDH release, had 70–90% neuronal degeneration noted by phase contrast microscopy and showed a DNA fragmentation pattern consistent with apoptotic cell death. Preconditioned neurons showed decreased LDH release (attenuation between 30 and 60% compared to controls) and increased cell viability with many culture plates showing no signs of damage. In addition, the authors found that ouabain, a potent Na^+-K^+-ATPase inhibitor, given as a preconditioning stimulus offered similar degrees of neuroprotection. Recent studies have suggested a role for alterations in NMDA receptor properties (Aizenman et al., 2000). In addition, NMDA, nitric oxide, and p21 Ras/extracellular regulated kinase cascade signaling have been implicated in OGD preconditioning (Gonzalez-Zulueta et al., 2000). Up-regulation of adenosine A3 receptors is not associated with delayed tolerance (von Arnim et al., 2000) and hypothermia provides cytoprotection without stimulating expression of anti-apoptotic gene products and cell cycle regulatory components produced after hypoxic preconditioning (Bossenmeyer-Pourie et al., 2000).

The plethora of stresses that can initiate preconditioning raises the possibility that cytokine mediators of stress underlie the protective mechanisms. TNF-α can induce tolerance to brain ischemia in preclinical stroke models in rats and mice (Nawashiro et al., 1997b; Tasaki et al., 1997). In a multifaceted study (Liu et al., 2000), neurons were exposed to 20-min of hypoxia as a preconditioning stimulus. No morphological changes were noted with this treatment alone nor did it cause cell death as detected by ethidium homodimer fluorescence. Pretreated cells were protected against exposure to 2.5 h hypoxia or 2.5 h OGD at 24 h with cell loss inhibited by 50% in both sets of experiments, thereby demonstrating the classical preconditioning response. The authors then showed that neurons exposed to TNF-α for 24 h as a preconditioning stimulus had a similar degree of protection as those preconditioned with hypoxia and the addition of an antibody against TNF-α in the culture medium abolished the protective response. Immunostaining showed the number of TNF-α receptors was not increased in preconditioned

compared to naïve cells. These findings suggested that TNF-α released from neurons during preconditioning initiates a cascade of intracellular events that leads to neuroprotection.

Using this data, the authors suggested that ceramide, a sphingolipid known to mediate TNF-α effects in other cellular models, may be a mediator of TNF-α's preconditioning effects. The authors supported this theory by determining (1) that the presence of ceramide is protective, (2) that ceramide is increased at the appropriate time for preconditioning and (3) that blocking ceramide synthesis blocks preconditioning. Cell-permeable acetyl-ceramide (C2-ceramide) was added to cell cultures after which they were subjected to lethal hypoxic insult (2.5 h). The presence of C2-ceramide decreased cell death by over 50% in this paradigm. Secondly, after a 20-min hypoxic stimulus, ceramide levels within neurons were increased to 120-140% of control by 16 h and further increased to 180–200% by 24 h, the time course when protection against hypoxia is observed. This data was replicated by substituting TNF-α as the preconditioning stimulus. Lastly, cells that were preconditioned with 20-min of hypoxia in the presence of fumonisin B1, a ceramide synthesis inhibitor, lacked the protection from subsequent insults observed in cells treated without fumonisin B1. These results implicate TNF-α and *de novo* synthesized ceramide as potential mediators of the preconditioning response, but more study is needed to fully define the intracellular processes involved.

4. Hibernation and hypoxia tolerance

In addition to ischemic preconditioning, other models of tolerance such as neuroprotection in hibernation (Drew et al., 2001; Zhou et al., 2001) and hypoxia tolerance in fresh water turtles (Lutz and Nilsson, 1997) may point to novel, more efficacious management of ischemia. Hibernating animals tolerate dramatic decreases in cerebral blood flow during hibernation without neurological deficit and hippocampal slices from hibernating brain tolerate OGD better than rats or euthermic ground squirrels (Frerichs and Hallenbeck, 1998; Frerichs et al., 1994). Hibernating rodents also tolerate hypoxia better than rats (Bullard et al., 1960), in part due to hypoxia-induced metabolic suppression (Barros et al., 2001). Some fresh water turtles tolerate extensive periods of anoxia, submerged under ice during the winter season. Importantly, however, turtles do not resist ischemia because they rely on glycolysis during the early stages of hypoxia when plasma glucose concentrations and cerebral blood flow increase substantially (Lutz and Nilsson, 1997). Nonetheless, hypoxia tolerance is a common theme in models of tolerance. As discussed above, hypoxic preconditioning enhances tolerance to ischemia. Thus similar strategies and in some instances common mechanisms may exist between ischemic preconditioning, hibernation and hypoxia tolerance. Here we describe neuroprotective adaptations in hibernation and briefly compare adaptations associated with hibernation, hypoxia tolerance and ischemic preconditioning.

Hibernation is a unique physiological condition, known best for suppression of metabolism and body temperature and thought to promote survival during periods of food shortage (Lyman, 1948). Less well-recognized are the numerous other, potentially neuroprotective aspects of hibernation physiology such as leukocytopenia, immunosuppression, inhibition of protein synthesis, enhanced antioxidant defense and metabolic suppression (Snapp and Heller, 1981; ; McKenna, 1968; Frerichs, 1994;Frerichs, 1998; Gentile, 1996; Drew, 1999; Toien, 2001). Consistent with the hypothesis that hibernation is neuroprotective, Frerichs and Hallenbeck (Frerichs and Hallenbeck, 1998) reported increased tolerance to hypoxia/ aglycemia (OGD) in hippocampal slices in vitro from hibernating thirteen-lined ground squirrels. Moreover, the traumatic tissue response induced by insertion of microdialysis probes into hibernating and euthermic brains of Arctic ground squirrels was markedly suppressed in hibernating brain in vivo (Fig. 1.1) (Zhou et al., 2001). The significant degree of cellular resistance to injury observed during hibernation likely results from a combination of factors possibly stemming from a single regulatory mechanism.

Small decreases in intra-ischemic brain temperature are cytoprotective in ischemia and traumatic brain injury and attenuate ischemia-induced glutamate release, and free radical production (Busto et al., 1987; Globus et al., 1995; Huh et al., 2000). Hypothermia by itself, however, provides only limited neuronal protection and recently failed to improve outcome after acute brain injury in a large clinical trial (Clifton et al., 2001). Also, extreme hypothermia is not well tolerated by non-hibernating species and may exacerbate tissue injury (Wass and Lanier, 1996). More importantly, cytoprotection in hippocampal slices from hibernating ground squirrels is preserved at 36°C indicating involvement of additional neuroprotective mechanisms (Frerichs and Hallenbeck, 1998).

Profound leukocytopenia as reported in ground squirrels is another potentially neuroprotective hallmark of hibernation (Frerichs et al., 1994; Toien et al., 2001). Neutrophil adhesion and macrophage infiltration at sites of injury promote cytotoxic reactions (Bowes et al., 1993; Hallenbeck et al., 1986; Weiss, 1989; Whalen et al., 2000). It is uncertain whether the observed decrease in antibody formation (McKenna and Musacchia, 1968) also protects hibernating brain tissue.

Plasma and cerebrospinal fluid concentrations of ascorbate increase significantly during hibernation (Drew et al., 1999; Toien et al., 2001) as do other antioxidant defense systems (Buzadzic et al., 1990). Increased antioxidant defense systems, therefore, may further protect hibernating brain tissue. Following penetrating brain injury, expression of heme oxygenase-1 (HO-1), a heme metabolizing protein induced in the presence of oxidative stress, increased in euthermic but not in hibernating brain supporting the hypothesis that hibernation attenuates injury-induced oxidative stress (Zhou et al., 2001).

Inhibition of protein synthesis shown to be neuroprotective in a rodent model of focal ischemia-reperfusion (Du et al., 1996) also occurs during hibernation (Frerichs et al., 1998; Knight et al., 2000). Thus, a number of potentially neuroprotective adaptations, in addition to profound hypothermia, leukocytopenia and metabolic

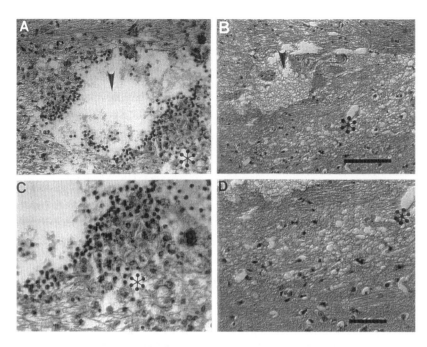

Fig. 1.1. A greater tissue reaction around a microdialysis probe is seen in hematoxylin and eosin stained sections in euthermic animals (A, C) as compared with hibernating animals (B, D). C and D are pictures with higher magnification of the area marked with (*) in A and B. Arrowheads point to the probe track. Microdialysis probes were inserted through indwelling guide cannulae while animals were euthermic or hibernating and left in place for three days. Sections from euthermic animals (A and C) demonstrate the probe cavity, accompanied by mononuclear inflammatory infiltrate, accumulation of lipid-laden macrophages, axonal swellings and fibrillary gliosis. In contrast, brain from hibernating animals (B and D) showed the probe cavity with no discernible inflammatory or reparative reaction. Specifically, no macrophages were seen, and there were no swollen axons or histological evidence of astrocytosis. There was simply a cavity surrounded by slight pallor of the surrounding neural parenchyma. Scale bar: A, B = 100 μm, C, D = 50 μm. (Reprinted from Zhou et al., 2001 with permission from American Society for Investigative Pathology, copyright holder of The American Journal of Pathology).

suppression characterize hibernation and likely produce additive or synergistic neuroprotective effects.

During the 6–8 month hibernation season, ground squirrels enter prolonged states of torpor (1–3 weeks) from which they periodically re-warm for brief periods of euthermia. Speculation still exists as to what drives periodic arousal or why it is necessary. However, it is tempting to hypothesize that a combination of adaptations have evolved in parallel with hibernation to protect vulnerable tissues during long bouts of profoundly reduced blood flow and capacity to deliver oxygen interspersed with frequent periodic re-warming and accompanying reperfusion of metabolically active tissues.

As discussed above, ischemic brain cells produce a variety of mediators that participate in progression of ischemic brain injury. Many of these signaling events are associated with ischemic preconditioning, hibernation or anoxia tolerance suggesting the possibility of common or related mechanisms. Suppression of brain energy consumption, in part via channel arrest, is one adaptation common to hypoxia tolerance and hibernation. NMDA receptors, necessary for ischemic preconditioning (Kato et al., 1992) are "silenced" in response to hypoxia in turtles (Bickler et al., 2000). Preliminary studies suggest NMDA receptors may likewise be down-regulated in hibernation (Richardson and Drew, unpublished data). Furthermore, adenosine, an inhibitory neuromodulator released in response to energy failure, increases during the early period of anoxia in turtles. Evidence suggests adenosine mediates increased cerebral blood flow and suppresses neuronal activity until other mechanisms of metabolic arrest supercede. Importantly, adenosine is an effective preconditioning stimulus suggesting that preconditioning may facilitate adenosine release during subsequent ischemia. Evidence further suggests that K_{ATP} channel activation contributes to anoxia tolerance in turtles (Pek-Scott and Lutz, 1998). A role for K_{ATP} channels during hibernation remains an area for further investigation. Regarding signaling molecules, an as yet unidentified, 88 kD protein isolated from plasma of hibernating ground squirrels has δ-opiate agonist activity (Horton et al., 1998). In contrast with evidence for a role of heat shock proteins in ischemic preconditioning, changes in HSP 70 mRNA or protein expression in brain of hibernating animals has not been found to change throughout the hibernation cycle despite dramatic changes in body temperature and metabolic rate (Bitting et al., 1999; Storey, 1997).

5. Conclusions

Preconditioning and tolerance are endogenous neuroprotective phenomena. Discovering the mechanisms responsible for this neuroprotection may be fruitful for developing clinical strategies to treat cerebral ischemia and may offer an alternative to the paradigm of agonism/antagonism of a single mediator that has been the basis for experimental studies in the past. Adenosine, excitatory amino acids and newly synthesized proteins are produced within ischemic brain and appear to be associated with preconditioning within the brain. Also, intracellular events related to the development of the heat shock response, the synthesis of genes from the bcl-2 family, nitric oxide, p21 Ras/ERK and the TNF-α/ceramide axis are interesting potential participants in this neuroprotection. Unraveling the intracellular mechanisms of this important protective response may offer hope in clinical situations in the future. The challenge is to discover and characterize the master regulatory switches and molecular mechanisms that simultaneously counteract multiple mediators of injury and confer resistance to ischemic damage. Comparison of models of tolerance such as ischemic preconditioning, hibernation and hypoxia tolerance may point to similar and potentially synergistic adaptations that can be developed for therapeutic benefit.

References

Aizenman, E., Sinor, J.D., Brimecombe, J.C. and Herin, G.A. (2000). Alterations of N-methyl-D-aspartate receptor properties after chemical ischemia. J. Pharmacol. Exp. Ther. 295, 572–577.

Auchampach, J.A., Grover, G.J. and Gross, G.J. (1992). Blockade of ischaemic preconditioning in dogs by the novel ATP dependent potassium channel antagonist sodium 5-hydroxydecanoate. Cardiovasc. Res. 26, 1054–1062.

Bahr, R.D., Leino, E.V. and Christenson, R.H. (2000). Prodromal unstable angina in acute myocardial infarction: prognostic value of short-and long-term outcome and predictor of infarct size. Am. Heart J. 140, 126–133.

Barros, R.C., Zimmer, M.E., Branco, L.G. and Milsom, W.K. (2001). Hypoxic metabolic response of the golden-mantled ground squirrel. J. Appl. Physiol. 91, 603–612.

Baxter, G.F., Goma, F.M. and Yellon, D.M. (1997). Characterisation of the infarct-limiting effect of delayed preconditioning: timecourse and dose-dependency studies in rabbit myocardium. Basic Res. Cardiol. 92, 159–167.

Bickler, P.E., Donohoe, P.H. and Buck, L.T. (2000). Hypoxia-induced silencing of NMDA receptors in turtle neurons. J. Neurosci. 20, 3522–3528.

Bitting, L., Watson, F.L., O'Hara, B.F., Kilduff, T.S. and Heller, H.C. (1999). HSP70 expression is increased during the day in a diurnal animal, the golden-mantled ground squirrel *Spermophilus lateralis*. Mol. Cell. Biochem. 199, 25–34.

Bossenmeyer-Pourie, C., Koziel, V. and Daval, J.L. (2000). Effects of hypothermia on hypoxia-induced apoptosis in cultured neurons from developing rat forebrain: comparison with preconditioning. Pediatr. Res. 47, 385–391.

Bowes, M.P., Zivin, J.A. and Rothlein, R. (1993). Monoclonal antibody to the ICAM-1 adhesion site reduces neurological damage in a rabbit cerebral embolism stroke model. Exp. Neurol. 119, 215–219.

Bruer, U., Weih, M.K., Isaev, N.K., Meisel, A., Ruscher, K., Bergk, A., Trendelenburg, G., Wiegand, F., Victorov, I.V. and Dirnagl, U. (1997). Induction of tolerance in rat cortical neurons: hypoxic preconditioning. FEBS Lett. 414, 117–121.

Bullard, R., David, G. and Nichols, C. (1960). The mechanisms of hypoxic tolerance in hibernating and non-hibernating mammals. In: Mammalian Hibernation (Dawe, A., Ed.), pp. 322–335. Cambridge, MA.

Busto, R., Dietrich, W.D., Globus, M.Y., Valdes, I., Scheinberg, P. and Ginsberg, M.D. (1987). Small differences in intraischemic brain temperature critically determine the extent of ischemic neuronal injury. J. Cereb. Blood Flow Metab. 7, 729–738.

Buzadzic, B., Spasic, M., Saicic, Z.S., Radojicic, R., Petrovic, V.M. and Halliwell, B. (1990). Antioxidant defenses in the ground squirrel *Citellus citellus*. 2. The effect of hibernation. Free Radic. Biol. Med. 9, 407–413.

Carr, C.S., Hill, R.J., Masamune, H., Kennedy, S.P., Knight, D.R., Tracey, W.R. and Yellon, D.M. (1997). Evidence for a role for both the adenosine A1 and A3 receptors in protection of isolated human atrial muscle against simulated ischaemia. Cardiovasc. Res. 36, 52–59.

Chen, J., Graham, S.H., Zhu, R.L. and Simon, R.P. (1996). Stress proteins and tolerance to focal cerebral ischemia. J. Cereb. Blood Flow Metab. 16, 566–577.

Chen, J., Simon, R.P., Nagayama, T., Zhu, R., Loeffert, J.E., Watkins, S.C. and Graham, S.H. (2000). Suppression of endogenous bcl-2 expression by antisense treatment exacerbates ischemic neuronal death. J. Cereb. Blood Flow Metab. 20, 1033–1039.

Choi, D.W. (1996). Ischemia-induced neuronal apoptosis. Curr. Opin. Neurobiol. 6, 667–672.

Clifton, G.L., Miller, E.R., Choi, S.C., Levin, H.S., McCauley, S., Smith, K.R., Muizelaar, J.P., Wagner, F.C., Marion, D.W., Luerssen, T.G., et al. (2001). Lack of effect of induction of hypothermia after acute brain injury. N. Engl. J. Med. 344, 556–563.

Cohen, M.V., Liu, G.S. and Downey, J.M. (1991). Preconditioning causes improved wall motion as well as smaller infarcts after transient coronary occlusion in rabbits. Circulation 84, 341–349.

del Zoppo, G.J., Schmid-Schonbein, G.W., Mori, E., Copeland, B.R. and Chang, C.M. (1991). Polymorphonuclear leukocytes occlude capillaries following middle cerebral artery occlusion and reperfusion in baboons. Stroke 22, 1276–1283.

Dorman, P.J. and Sandercock, P.A. (1996). Considerations in the design of clinical trials of neuroprotective therapy in acute stroke. Stroke 27, 1507–1515.

Dougherty, J.H., Levy, D.E. and Weksler, B.B. (1977). Platelet activation in acute cerebral ischaemia. Serial measurements of platelet function in cerebrovascular disease. Lancet 1, 821–824.

Drew, K.L., Osborne, P.G., Frerichs, K.U., Hu, Y., Koren, R.E., Hallenbeck, J.M. and Rice, M.E. (1999). Ascorbate and glutathione regulation in hibernating ground squirrels. Brain Res. 851, 1–8.

Drew, K.L., Rice, M.E., Kuhn, T.B. and Smith, M.A. (2001). Neuroprotective adaptations in hibernation: therapeutic implications for ischemia-reperfusion, traumatic brain injury and neurodegenerative diseases. Free Radic. Biol. Med. 31, 563–573.

Du, C., Hu, R., Csernansky, C.A., Liu, X.Z., Hsu, C.Y. and Choi, D.W. (1996). Additive neuroprotective effects of dextrorphan and cycloheximide in rats subjected to transient focal cerebral ischemia. Brain Res. 718, 233–236.

Duchen, M.R., McGuinness, O., Brown, L.A. and Crompton, M. (1993). On the involvement of a cyclosporin A sensitive mitochondrial pore in myocardial

reperfusion injury. Cardiovasc. Res. 27, 1790–1794.

Eliasson, M.J., Sampei, K., Mandir, A.S., Hurn, P.D., Traystman, R.J., Bao, J., Pieper, A., Wang, Z.Q., Dawson, T.M., Snyder, S.H. and Dawson, V.L. (1997). Poly (ADP-ribose) polymerase gene disruption renders mice resistant to cerebral ischemia. Nat. Med. 3, 1089–1095.

Finklestein, S.P., Caday, C.G., Kano, M., Berlove, D.J., Hsu, C.Y., Moskowitz, M. and Klagsbrun, M. (1990). Growth factor expression after stroke. Stroke 21, III 122–124.

Frerichs, K.U. and Hallenbeck, J.M. (1998). Hibernation in ground squirrels induces state and species-specific tolerance to hypoxia and aglycemia: an in vitro study in hippocampal slices. J. Cereb. Blood Flow Metab. 18, 168–175.

Frerichs, K.U., Kennedy, C., Sokoloff, L. and Hallenbeck, J.M. (1994). Local cerebral blood flow during hibernation, a model of natural tolerance to "cerebral ischemia". J. Cereb. Blood Flow Metab. 14, 193–205.

Frerichs, K.U., Smith, C.B., Brenner, M., DeGracia, D.J., Krause, G.S., Marrone, L., Dever, T.E. and Hallenbeck, J.M. (1998). Suppression of protein synthesis in brain during hibernation involves inhibition of protein initiation and elongation. Proc. Natl. Acad. Sci. USA 95, 14511–14516.

Garlid, K.D., Paucek, P., Yarov-Yarovoy, V., Murray, H.N., Darbenzio, R.B., D'Alonzo, A.J., Lodge, N.J., Smith, M.A. and Grover, G.J. (1997). Cardioprotective effect of diazoxide and its interaction with mitochondrial ATP-sensitive K+ channels. Possible mechanism of cardioprotection. Circ. Res. 81, 1072–1082.

Giulian, D. and Robertson, C. (1990). Inhibition of mononuclear phagocytes reduces ischemic injury in the spinal cord. Ann. Neurol. 27, 33–42.

Glazier, S.S., O'Rourke, D.M., Graham, D.I. and Welsh, F.A. (1994). Induction of ischemic tolerance following brief focal ischemia in rat brain. J. Cereb. Blood Flow Metab. 14, 545–553.

Globus, M.Y., Alonso, O., Dietrich, W.D., Busto, R. and Ginsberg, M.D. (1995). Glutamate release and free radical production following brain injury: effects of posttraumatic hypothermia. J. Neurochem. 65, 1704–1711.

Gonzalez-Zulueta, M., Feldman, A.B., Klesse, L.J., Kalb, R.G., Dillman, J.F., Parada, L.F., Dawson, T.M. and Dawson, V.L. (2000). Requirement for nitric oxide activation of p21(ras)/extracellular regulated kinase in neuronal ischemic preconditioning. Proc. Natl. Acad. Sci. USA 97, 436–441.

Greenberg, D.A., Chan, J. and Sampson, H.A. (1992). Endothelins and the nervous system. Neurology 42, 25–31.

Gross, G.J. and Auchampach, J.A. (1992). Blockade of ATP-sensitive potassium channels prevents myocardial preconditioning in dogs. Circ. Res. 70, 223–233.

Hall, E.D. and Braughler, J.M. (1989). Central nervous system trauma and stroke. II. Physiological and pharmacological evidence for involvement of oxygen radicals and lipid peroxidation. Free Radic. Biol. Med. 6, 303–313.

Hallenbeck, J.M. (1996). Inflammatory reactions at the blood-endothelial interface in acute stroke. Adv. Neurol. 71, 281–297; discussion 297–300.

Hallenbeck, J.M., Dutka, A.J., Tanishima, T., Kochanek, P.M., Kumaroo, K.K., Thompson, C.B., Obrenovitch, T.P. and Contreras, T. J. (1986). Polymorphonuclear leukocyte accumulation in brain regions with low blood flow during the early postischemic period. Stroke 17, 246–253.

Heurteaux, C., Lauritzen, I., Widmann, C. and Lazdunski, M. (1995). Essential role of adenosine, adenosine A1 receptors, and ATP-sensitive K+ channels in cerebral ischemic preconditioning. Proc. Natl. Acad. Sci. USA 92, 4666–4670.

Horton, N.D., Kaftani, D.J., Bruce, D.S., Bailey, E.C., Krober, A.S., Jones, J.R., Turker, M., Khattar, N., Su, T.P., Bolling, S.F. and Oeltgen, P.R. (1998). Isolation and partial characterization of an opioid-like 88 kDa hibernation-related protein. Comp. Biochem. Physiol. B 119, 787–805.

Huh, P.W., Belayev, L., Zhao, W., Koch, S., Busto, R. and Ginsberg, M.D. (2000). Comparative neuroprotective efficacy of prolonged moderate intraischemic and postischemic hypothermia in focal cerebral ischemia. J. Neurosurg. 92, 91–99.

Kato, H., Liu, Y., Araki, T. and Kogure, K. (1992). MK-801, but not anisomycin, inhibits the induction of tolerance to ischemia in the gerbil hippocampus. Neurosci. Lett. 139, 118–121.

Kitagawa, K., Matsumoto, M., Kuwabara, K., Tagaya, M., Ohtsuki, T., Hata, R., Ueda, H., Handa, N., Kimura, K. and Kamada, T. (1991). 'Ischemic tolerance' phenomenon detected in various brain regions. Brain Res. 561, 203–211.

Kitagawa, K., Matsumoto, M., Tagaya, M., Hata, R., Ueda, H., Niinobe, M., Handa, N., Fukunaga, R., Kimura, K., Mikoshiba, K. et al. (1990). 'Ischemic tolerance' phenomenon found in the brain. Brain Res. 528, 21–24.

Klatzo, I. (1985). Concepts of ischemic injury associated with brain edema. In: Brain Edema (Spatz, M., Ed.), pp. 1–5. Springer Verlag, New York.

Knight, J.E., Narus, E.N., Martin, S.L., Jacobson, A., Barnes, B.M. and Boyer, B.B. (2000). mRNA stability and polysome loss in hibernating Arctic ground squirrels (*Spermophilus parryii*). Mol. Cell. Biol. 20, 6374–6379.

Kuzuya, T., Hoshida, S., Yamashita, N., Fuji, H., Oe, H., Hori, M., Kamada, T. and Tada, M. (1993). Delayed effects of sublethal ischemia on the acquisition of tolerance to ischemia. Circ. Res. 72, 1293–1299.

Lindsberg, P.J., Hallenbeck, J.M. and Feuerstein, G. (1991). Platelet-activating factor in stroke and brain injury. Ann. Neurol. 30, 117–129.

Liu, J., Ginis, I., Spatz, M. and Hallenbeck, J.M. (2000).

Hypoxic preconditioning protects cultured neurons against hypoxic stress via TNF-alpha and ceramide. Am. J. Physiol. 278, C144–153.

Lutz, P.L. and Nilsson, G.E. (1997). Contrasting strategies for anoxic brain survival—glycolysis up or down. J. Exp. Biol. 200, 411–419.

Lyman, C. (1948). The oxygen consumption and temperature regulation of hibernating hamsters. J. Exp. Zool. 109, 55–78.

Marber, M.S. (2000). Ischemic preconditioning in isolated cells. Circ. Res. 86, 926–931.

Marber, M.S., Latchman, D.S., Walker, J.M. and Yellon, D.M. (1993). Cardiac stress protein elevation 24 hours after brief ischemia or heat stress is associated with resistance to myocardial infarction. Circulation 88, 1264–1272.

Marber, M.S., Mestril, R., Chi, S.H., Sayen, M.R., Yellon, D.M. and Dillmann, W.H. (1995). Overexpression of the rat inducible 70-kD heat stress protein in a transgenic mouse increases the resistance of the heart to ischemic injury. J. Clin. Invest. 95, 1446–1456.

Markham, A., Plosker, G.L. and Goa, K.L. (2000). Nicorandil. An updated review of its use in ischaemic heart disease with emphasis on its cardioprotective effects. Drugs 60, 955–974.

McKenna, J.M. and Musacchia, X.J. (1968). Antibody formation in hibernating ground squirrels (*Citellus tridecemlineatus*). Proc. Soc. Exp. Biol. Med. 129, 720–724.

Miyashita, K., Abe, H., Nakajima, T., Ishikawa, A., Nishiura, M., Sawada, T. and Naritomi, H. (1994). Induction of ischaemic tolerance in gerbil hippocampus by pretreatment with focal ischaemia. Neuroreport 6, 46–48.

Moncayo, J., de Freitas, G.R., Bogousslavsky, J., Altieri, M. and van Melle, G. (2000). Do transient ischemic attacks have a neuroprotective effect? Neurology 54, 2089–2094.

Mun-Bryce, S. and Rosenberg, G.A. (1998). Matrix metalloproteinases in cerebrovascular disease. J. Cereb. Blood Flow Metab. 18, 1163–1172.

Murphy, E., Fralix, T.A., London, R.E. and Steenbergen, C. (1993). Effects of adenosine antagonists on hexose uptake and preconditioning in perfused rat heart. Am. J. Physiol. 265, C1146–1155.

Murry, C.E., Jennings, R.B. and Reimer, K.A. (1986). Preconditioning with ischemia: a delay of lethal cell injury in ischemic myocardium. Circulation 74, 1124–1136.

Nakata, N., Kato, H. and Kogure, K. (1993). Effects of repeated cerebral ischemia on extracellular amino acid concentrations measured with intracerebral microdialysis in the gerbil hippocampus. Stroke 24, 458–463; discussion 463–454.

Nawashiro, H., Martin, D. and Hallenbeck, J.M. (1997a). Neuroprotective effects of TNF binding protein in focal cerebral ischemia. Brain Res. 778, 265–271.

Nawashiro, H., Tasaki, K., Ruetzler, C.A. and Hallenbeck,

J.M. (1997b). TNF-alpha pretreatment induces protective effects against focal cerebral ischemia in mice. J. Cereb. Blood Flow Metab. 17, 483–490.

Noda, T., Minatoguchi, S., Fujii, K., Hori, M., Ito, T., Kanmatsuse, K., Matsuzaki, M., Miura, T., Nonogi, H., Tada, M. et al. (1999). Evidence for the delayed effect in human ischemic preconditioning: prospective multicenter study for preconditioning in acute myocardial infarction. J. Am. Coll. Cardiol. 34, 1966–1974.

Noma, A. (1983). ATP-regulated K$^+$ channels in cardiac muscle. Nature 305, 147–148.

Nowak, T. S. (1990). Protein synthesis and the heart shock/ stress response after ischemia. Cerebrovasc. Brain Metab. Rev. 2, 345–366.

Ota, A., Ikeda, T., Abe, K., Sameshima, H., Xia, X.Y., Xia, Y.X. and Ikenoue, T. (1998). Hypoxic-ischemic tolerance phenomenon observed in neonatal rat brain. Am. J. Obstet. Gynecol. 179, 1075–1078.

Ottani, F., Galvani, M., Ferrini, D., Sorbello, F., Limonetti, P., Pantoli, D. and Rusticali, F. (1995). Prodromal angina limits infarct size. A role for ischemic preconditioning. Circulation 91, 291–297.

Parratt, J.R., Vegh, A. and Papp, J.G. (1995). Bradykinin as an endogenous myocardial protective substance with particular reference to ischemic preconditioning—a brief review of the evidence. Can. J. Physiol. Pharmacol. 73, 837–842.

Pek-Scott, M. and Lutz, P.L. (1998). ATP-sensitive K+ channel activation provides transient protection to the anoxic turtle brain. Am. J. Physiol. 275, R2023–2027.

Riabowol, K.T., Mizzen, L.A. and Welch, W.J. (1988). Heat shock is lethal to fibroblasts microinjected with antibodies against hsp70. Science 242, 433–436.

Rothman, S.M. and Olney, J.W. (1995). Excitotoxicity and the NMDA receptor—still lethal after eight years. Trends Neurosci. 18, 57–58.

Samdani, A.F., Dawson, T.M. and Dawson, V.L. (1997). Nitric oxide synthase in models of focal ischemia. Stroke 28, 1283–1288.

Schott, R.J., Rohmann, S., Braun, E.R. and Schaper, W. (1990). Ischemic preconditioning reduces infarct size in swine myocardium. Circ. Res. 66, 1133–1142.

Schultz, J.E.-J., Hsu, A.K., Nagase, H. and Gross, G.J. (1998). TAN-67, a delta 1-opioid receptor agonist, reduces infarct size via activation of Gi/o proteins and KATP channels. Am. J. Physiol. 274, H909–H914.

Shiki, K. and Hearse, D.J. (1987). Preconditioning of ischemic myocardium: reperfusion-induced arrhythmias. Am. J. Physiol. 253, H1470–H1476.

Shimazaki, K., Ishida, A. and Kawai, N. (1994). Increase in bcl-2 oncoprotein and the tolerance to ischemia-induced neuronal death in the gerbil hippocampus. Neurosci. Res. 20, 95–99.

Simon, R.P., Niiro, M. and Gwinn, R. (1993). Prior ischemic stress protects against experimental stroke.

Neurosci. Lett. 163, 135–137.

Storey, K.B. (1997). Metabolic regulation in mammalian hibernation: enzyme and protein adaptations. Comp. Biochem. Physiol. A 118, 1115–1124.

Tasaki, K., Ruetzler, C.A., Ohtsuki, T., Martin, D., Nawashiro, H. and Hallenbeck, J.M. (1997). Lipopolysaccharide pre-treatment induces resistance against subsequent focal cerebral ischemic damage in spontaneously hypertensive rats. Brain Res. 748, 267–270.

Toien, O., Drew, K.L., Chao, M.L. and Rice, M.E. (2001). Ascorbate dynamics and oxygen consumption during arousal from hibernation in Arctic ground squirrels. Am. J. Physiol. 281, R572–R583.

Tracey, W.R., Magee, W., Masamune, H., Kennedy, S.P., Knight, D.R., Buchholz, R.A. and Hill, R.J. (1997). Selective adenosine A3 receptor stimulation reduces ischemic myocardial injury in the rabbit heart. Cardiovasc. Res. 33, 410–415.

von Arnim, C.A., Timmler, M., Ludolph, A.C. and Riepe, M.W. (2000). Adenosine receptor up-regulation: initiated upon preconditioning but not upheld. Neuroreport 11, 1223–1226.

Walker, V. and Pickard, J.D. (1985). Prostaglandins, thromboxane, leukotrienes and the cerebral circulation in health and disease. Adv. Tech. Stand. Neurosurg. 12, 3–90.

Wall, T.M., Sheehy, R. and Hartman, J.C. (1994). Role of bradykinin in myocardial preconditioning. J. Pharmacol. Exp. Ther. 270, 681–689.

Wass, C.T. and Lanier, W.L. (1996). Hypothermia-associated protection from ischemic brain injury: implications for patient management. Int. Anesthesiol. Clin. 34, 95–111.

Weih, M., Kallenberg, K., Bergk, A., Dirnagl, U., Harms, L., Wernecke, K.D. and Einhaupl, K.M. (1999). Attenuated stroke severity after prodromal TIA: a role for ischemic tolerance in the brain? Stroke 30, 1851–1854.

Weiss, S.J. (1989). Tissue destruction by neutrophils. N. Engl. J. Med. 320, 365–376.

Whalen, M.J., Carlos, T.M., Dixon, C.E., Robichaud, P., Clark, R.S., Marion, D.W. and Kochanek, P.M. (2000). Reduced brain edema after traumatic brain injury in mice deficient in P-selectin and intercellular adhesion molecule-1. J. Leukoc. Biol. 67, 160–168.

Yao, Z. and Gross, G.J. (1993). Glibenclamide antagonizes adenosine A1 receptor-mediated cardioprotection in stunned canine myocardium. Circulation 88, 235–244.

Yao, Z. and Gross, G.J. (1994). A comparison of adenosine-induced cardioprotection and ischemic preconditioning in dogs. Efficacy, time course, and role of KATP channels. Circulation 89, 1229–1236.

Zhou, F., Zhu, X., Castellani, R.J., Stimmelmayr, R., Perry, G., Smith, M.A. and Drew, K.L. (2001). Hibernation, a model of neuroprotection. Am. J. Pathol. 158, 2145–2151.

Sensing, Signaling and Cell Adaptation. Edited by K.B. Storey and J.M. Storey
© 2002 Elsevier Science B.V. All rights reserved.

CHAPTER 2

Regulation of Gene Expression by Hypoxia in Lung Alveolar Epithelial Cells

Christine Clerici[1], Tokujiro Uchida[1], Carole Planès[1] and Michael A. Matthay[2]
[1]*Department of Physiology, INSERM U 426 Faculté de Médecine Xavier Bichat, Université Paris 7, France*
[2]*Cardiovascular Research Institute and Departments of Medicine and Anesthesia, University of California, San Francisco, CA 94143-0624, USA*

1. Introduction

For humans, hypoxia is a life-threatening stress that must be dealt with at both cellular and systemic levels. Whereas hypoxia is the natural consequence of some environments (e.g., high altitude, diving), it is also a common feature of many clinical diseases or syndromes resulting from inadequate respiration, e.g., sleep apnea syndrome, chronic obstructive pulmonary disease or pulmonary edema from heart failure or acute lung injury. Because of a relatively high incidence of obstructive disease in the United States and Europe, many humans live with hypoxia as an everyday occurrence, and express a multitude of adaptive responses at the cellular, molecular and systemic levels as strategies to minimize the deleterious effects of hypoxia. Each organ responds to hypoxia by regulating expression of specific genes that control organ specific functions. For instance, the endothelial cells of pulmonary circulation respond to hypoxia stress by regulating the expression of vasoactive substances and matrix proteins involved in modulating vascular tone and remodeling the vasculature, whereas the kidney responds by production of erythropoietin. The responses to hypoxia of lung alveolar epithelium are primordial because alveolar epithelial cells are directly exposed to acute and chronic changes in alveolar O_2 tension. Until the last decade, there has been little

information regarding the specific responses of alveolar epithelial cells to low O_2 tension. However, *in vivo* and *in vitro* studies reported that lung cells are relatively tolerant to severe and prolonged hypoxia with no change in the ultrastructural characteristics or cell viability suggesting that adaptive mechanisms take place in order to cope with oxygen deprivation (Ouiddir et al., 1999). Recently, considerable work has been done to investigate molecular and cellular responses of the alveolar epithelium to acute and chronic O_2 deprivation. The first part of this review focuses on the responses of alveolar epithelial cells to hypoxia at the cellular and molecular levels and the second part discusses the role of hypoxia-inducible factors known to be responsible for orchestrating a large number of hypoxia-sensitive genes.

The airways and alveoli constitute the interface between the lung parenchyma and external environment and are lined with a continuous epithelium. The alveolar epithelium covers more than 95% of the air space surface area in the lung and contains thin, squamous type I cells and cuboidal type II cells. The close apposition between the alveolar epithelium and the vascular endothelium facilitates efficient exchanges of gases, but also forms a tight barrier to movement of liquid and proteins from the interstitial and vascular spaces, thus assisting in maintaining relatively dry alveoli. Alveolar type II (ATII) cells are more numerous and have a number of critical and specific functions (Matthay et al., 2002). First, they serve as progenitors of ATI cells during development as well as in the reparative phase of the alveolar epithelium following injury. In addition, they produce and store surfactant, a complex mixture of phospholipids and specific apoproteins. Also, ATII cells participate in active reabsorption of alveolar fluid because of their ability to transport Na$^+$ from the apical to the basolateral surface with water following passively, probably across both ATI and ATII cells. ATII cells express a variety of proteins required for these specific functions: (i) transport proteins such as Na$^+$ channels and Na,K-ATPase that are involved in the vectorial transport of sodium (Na$^+$) from the alveolar to interstitial spaces and therefore drive the resorption of alveolar fluid; and (ii) surfactant apoproteins which, in association with phospholipids, decrease alveolar surface tension and participate in host antibacterial defences. Besides these specific functions, ATII cells also express proteins such adhesion molecules or growth factors that play a role in a variety of immune mechanisms and in vascular remodeling.

2. Effect of hypoxia on Na transport proteins

The question of whether hypoxia can regulate Na transport proteins involved in the vectorial transport of Na has only been addressed recently in both *in vitro* and *in vivo* studies. Amiloride-sensitive Na channels are the major pathway for apical Na entry in alveolar cells and may be the rate-limiting step in alveolar transepithelial Na transport (Matalon and O'Brodovich, 1999). A number of studies using patch clamp techniques, measurements of ^{22}Na uptake and of binding of amiloride analogs have demonstrated the presence of amiloride-sensitive sodium channels in alveolar type II cells. Based on the biophysical and pharmacological characterization, at least three types of amiloride sensitive sodium channels have been described in the apical membrane of cultured adult alveolar type II cells. The most frequently observed in both inside-out and cell-attached configurations is a

non-selective cation (NSC) channel with a conductance of ~21 pS that was equally permeable to sodium and potassium (PNa/PK ~1), voltage-independent calcium-activated and completely blocked by 1 μM amiloride. Also Yue et al. (1995) recorded a channel with a conductance of 27 pS with a highest relative permeability of sodium to potassium (7:1), completely blocked by 1 μM amiloride or its analog EIPA. Finally, the presence of highly selective cation (HSC) channels has been identified in adult alveolar type II cells cultured on permeable support, in air–liquid interface with aldosterone. These channels recorded on cell-attached configuration have a low conductance (5–7 pS) and were inhibited by low concentration of amiloride (<0.1 μM) (Jain et al., 2001). A variety of other channels has been recorded in alveolar type II cells including ones regulated by G proteins (Matalon and O'Brodovich, 1999).

Molecular identification of the proteins involved in amiloride-sensitive sodium influx has been achieved in recent years. Three homologous subunits, entitled α, β, and γENaC (Epithelial Na Channel), correspond to the pore forming subunits. When co-expressed, the three subunits reproduce the expected properties of the highly selective cation (HSC) channel (Canessa et al., 1994). Recent molecular studies demonstrated that rat adult ATII cells express mRNA transcripts encoding the α, β, γ subunits of the rat epithelial Na channel (rENaC) (Planes et al., 1997; O'Brodovich et al., 1993). The critical role of ENaC in the absorption of salt and fluid by lung epithelia has been confirmed by the generation of knockout mice with inactivated subunits of ENaC. After inactivation of murine αENaC, deficient neonates develop respiratory distress syndrome and die within 40 hours of birth, from failure to clear their lungs of fluid (Hummler et al., 1996). By contrast, β or γENaC knockout mice were able to clear fluid from lungs at birth, although at a slower rate that in wild type controls (Barker et al., 1998; McDonald et al., 1999). In conjunction with apical Na channels, basolateral Na,K-ATPase represents the major protein involved in transepithelial Na transport by alveolar cells. In the alveolar epithelium, the Na,K-ATPase protein is detected primarily in type II cells

Table 2.1. Hypoxic gene regulation in lung alveolar epithelial cells

Hypoxic-regulated genes	mRNA and protein level	References
Na transport proteins *in vitro*		Planes et al. (1997); Wodopia et al. (2000)
Epithelial Na channels	Decreased α-, β-, γ mRNA subunits	
Na,K-ATPase	Decreased α_1-, β_1- mRNA subunits	
Na transport proteins *in vivo*		Vivona et al. (2001)
Epithelial Na channels	Increased α subunit mRNA and protein	
Na,K-ATPase	Increased β_1 subunit mRNA	
VEGF	Increased mRNA and protein levels	Pham et al. (2002); Boussat et al. (2000)
Glucose transporter 1	Increased mRNA and protein levels	Ouiddir et al. (1999)
GAPDH enzyme	Increased mRNA and protein levels	Escoubet (1999); Wodopia et al. (2000)
Hexokinase II enzyme	Increased mRNA and protein levels	Riddle et al. (2000)
Adhesion molecule ICAM-1	Increased mRNA and protein levels	Beck-Schimmer et al. (2001)
Adhesion molecule VCAM-1	Increased mRNA and protein levels	Beck-Schimmer et al. (2001)

VEGF: Vascular endothelial growth factor; GAPDH: glyceraldehyde-3-phosphate dehydrogenase.

although there is expression in type I cells. The Na,K-ATPase is a heterodimeric transmembrane protein composed of one α- and one β-subunit in a 1:1 ratio. Although there are multiple isoforms of both subunits usually expressed in a tissue-specific pattern, the heterodimeric form comprised of the α_1 and β_1 subunits is the predominant Na,K-ATPase isoform in alveolar epithelial cells. The functional activity of Na,K-ATPase is usually tightly coupled with apical Na entry in order to ensure efficient vectorial Na transport.

Several studies have indicated that hypoxia alters Na transport in ATII cells (Table 2.1). Both *in vitro* and *in vivo* studies clearly show that decreased O_2 tension reduces the capacity of alveolar epithelial cells to actively transport sodium across alveolar epithelium. In alveolar type II cells, hypoxia (0% and 3% O_2) inhibits dome formation, decreases both amiloride-sensitive ^{22}Na influx and Na,K-ATPase activity (Planes et al., 1996; Planes et al., 1997; Mairbaurl et al., 1997) and reduces amiloride-sensitive short circuit current (Mairbaurl et al., 2002; Planes et al., 2002), suggesting that the transepithelial sodium transport is impaired. The mechanisms whereby hypoxia induces decreased activity of sodium transport proteins depends on the severity and the length of hypoxia exposure. For long exposure times (>12 h), the decrease in amiloride-sensitive sodium channel and Na,K-

ATPase activities was associated with a parallel decline in mRNA levels of the three subunits, α, β, and γ of rENaC and two subunits α_1 and β_1 of Na,K-ATPase and the rate of α-rENaC protein synthesis, indicating a transcriptional or post-transcriptional regulation. For short exposure time (3 h), the decrease in ^{22}Na influx and Na,K-ATPase activity preceded any detectable change in mRNA levels, a finding that suggests that other mechanisms may be involved in regulation, including decreased efficiency in the translation of rENaC mRNA or in apical membrane trafficking of rENaC subunits, abnormal degradation or internalization of the channel protein or hypoxia-induced modification of intracellular signals (Planes et al., 1997).

The decrease in gene expression of sodium transport proteins likely represents the direct effect of hypoxia and raises the question of whether gene expression for sodium channels and Na,K-ATPase is regulated by ambient PO_2 in alveolar type II cells. In support of this hypothesis, an increase in O_2 tension upregulated the level of ENaC mRNA transcripts in alveolar type II cells. In fetal distal lung epithelial (FDLE) cells in culture, the transfer of the cells from 3% O_2 to 21% O_2 concentrations upregulated α-, β-, and γ-rENaC mRNA subunits as well as sodium channel activity (Pitkanen et al., 1996; Baines et al., 2001). Exposure of rats to

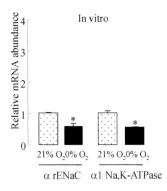

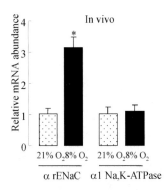

Fig. 2.1. Effect of hypoxia on αrENaC and $α_1$ Na,K-ATPase mRNAs in rat ATII cells. *In vitro*, ATII cells were exposed to hypoxia (0% O_2 for 18 h) and *in vivo*, freshly ATII cells were isolated from the lungs of rats exposed to 21% or 8% O_2 for 24 hours. αrENaC and $α_1$ Na,K-ATPase mRNA expression was normalized to the corresponding β-actin mRNA. Statistical difference of values from normoxia group is indicated by * $p < 0.01$. (From Planes et al. (1997) and Vivona et al. (2001)).

sublethal hyperoxia (85% O_2) increased alveolar type II cell α-rENaC mRNA levels (Yue et al., 1995). In addition, it has been reported that $α_1$- and $β_1$ Na,K-ATPase mRNA transcript levels both increased in alveolar type II cells from rats exposed to hyperoxia (Harris et al., 1996). The mechanism whereby O_2 tension regulates sodium channels and Na,K-ATPase gene expression is thought to be transcriptional. Activation of ENaC by increased O_2 concentration is associated with NF-κB activation consistent with the identification of NF-κB transcription binding sites in the α-ENaC promoter region (Rafii et al., 1998). In addition to the potential role for an O_2 responsive element in the ENaC promoter, it is also possible that there are other genes that are O_2 responsive and alter ENaC

expression through their expressed protein or metabolic products.

Whatever the mechanism, the hypoxia-induced down-regulation of Na transport proteins *in vitro* may have important pathophysiological implications *in vivo*. Indeed, in some humans, a decrease in ambient O_2 tension at high altitude can be associated with development of pulmonary edema. Although the initial cause of the edema may be related to altered hemodynamics or an increase in microvascular permeability (Maggiorini et al., 2001), alteration of vectorial Na transport by the alveolar epithelium may hamper the resorption of alveolar fluid and the resolution of hypoxia-induced pulmonary edema.

The effect of hypoxia under *in vivo* conditions has been studied primarily in rats. In anesthetized rats, as well as in isolated perfused lungs, hypoxia decreased alveolar liquid clearance by inhibition of the amiloride sensitive component (Suzuki et al., 1999a; Vivona et al., 2001). In contrast to the *in vitro* studies, hypoxia increased α-rENaC and $β_1$-Na,K-ATPase mRNA transcripts with little increase or no change in protein amounts, suggesting a post-translational mechanism such as a direct change of sodium transporter protein activity or internalization (Vivona et al., 2001). This latter hypothesis was supported by the normalization of alveolar fluid clearance following intratracheal administration of the cAMP agonist (terbutaline), which is known to increase the trafficking of sodium transporter proteins from the cytoplasm to the membrane (Snyder, 2000). However, the hypoxia-induced decrease in alveolar fluid clearance *in vivo* seems related to the degree and the duration of hypoxia since Sakuma and colleagues reported that 10% O_2 for up to 120 h increased alveolar fluid clearance (Sakuma et al., 2001).

3. Effect of hypoxia on proteins involved in glucose metabolism

Under basal conditions, ATII cells display a high metabolic rate and use glucose for cellular oxidative metabolism, synthesis of surfactant, growth and differentiation (Simon et al., 1977). *In vivo* and *in vitro* studies have indicated that lung tissue and alveolar epithelial cells tolerate hypoxia remarkably well surviving more than 24 hours without cellular damage. In diverse tissues and cell types, tolerance to hypoxia is the result of maintenance of adequate energy supply under hypoxia despite the reduction of oxidative metabolism. In lung hypoxic cells, ATP content was maintained close to the level in normoxia both in lung tissue of rats exposed to 10% O_2 (Suzuki et al., 1999b), and in ATII cells exposed to 0% O_2 for 24 h (Ouiddir et al., 1999). The maintenance of ATP level under hypoxia is related to the ability of ATII cells to increase anaerobic glycolysis by upregulating the activity and expression of glycolytic enzymes (Simon et al., 1978) and by increasing glucose transport at the membrane level (Ouiddir et al., 1999).

3.1. Effect of hypoxia on glucose transporters

Glucose transport in alveolar epithelial cells takes place through different pathways. Based on *in vivo* and *in vitro* studies, glucose enters alveolar epithelial cells mostly by a Na^+-independent carrier-mediated process (Clerici et al., 1991). The ubiquitous glucose transporter GLUT1, which is responsible for basal glucose uptake in most tissues, is predominant in ATII cells (Saumon and Makhloufi, 1997). However, the presence of an apical Na^+-dependent D-glucose transporter in alveolar epithelial cells was reported in *in vivo* studies but the results are less consistent *in vitro* (O'Brodovich et al., 1991; Clerici et al., 1991).

In alveolar epithelial cells, hypoxia increases anaerobic glycolysis by upregulating both glucose transport at the membrane level (Ouiddir et al., 1999) and glycolytic enzymes. In cultured ATII cells, hypoxia induced an increase in glucose influx, measured by the uptake of deoxy-D-glucose (DG), a glucose analog that is transported by the facilitative glucose transporter. This stimulation of glucose influx was associated with an increase in mRNA and protein levels of GLUT 1. Several lines of evidence suggest that the hypoxia-induced increase of DG uptake was related to upregulation of GLUT 1: (i) the time course for the hypoxia-

induced increase in DG influx, with no substantial change before 6 h of hypoxia; (ii) the prevention of this effect by cycloheximide, an inhibitor of translation; and (iii) the parallel recovery during reoxygenation of DG uptake and the level of GLUT 1 mRNA. Comparison of ATII cells with other hypoxic tolerant cells showed that they resemble lung endothelial cells (Loike et al., 1992), but differ from heart and skeletal muscle in which upregulation of glucose transport occurs in less than one hour with no change of GLUT 1 mRNA level (Cartee et al., 1991). In ATII cells, Na-dependent glucose transport is weakly expressed in normoxia, and Ouiddir and colleagues showed that it was not stimulated by hypoxia (Ouiddir et al., 1999) (Fig. 2.2).

The O_2 sensing mechanisms whereby hypoxia regulates the level GLUT1 mRNA in ATII cells are not univocal and result from both a reduced O_2 concentration *per se* and inhibition of oxidative phosphorylation (Ouiddir et al., 1999). This distinction has been made through studies with specific chemical agents that mimic the actions of the different components of the hypoxic response

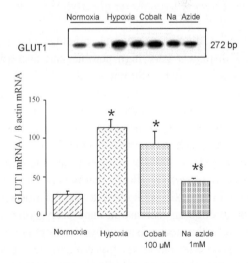

Fig. 2.2. Effect of hypoxia, cobalt chloride and sodium azide on GLUT1 mRNA in cultured alveolar epithelial cells exposed to hypoxia (0% O_2) or normoxia (21% O_2) in the presence of cobalt chloride or sodium azide for 24 h. GLUT1 mRNA expression was normalized to the corresponding β-actin mRNA. Statistical difference of values from normoxia group is indicated by * $p < 0.01$ and from hypoxia group § $p < 0.01$. (From Ouiddir et al., 1999).

(Behrooz and Ismail, 1997). Most hypoxia-regulated genes involve a ferroprotein sensor and an important characteristic of this system is that the inducible response to hypoxia is mimicked by exposure to particular transition metals. For instance, in the presence of O_2, cobalt chloride simulates the effect of lowered O_2 concentration since it substitutes for O_2 in the heme protein. In normoxic ATII cells, GLUT1 mRNA levels and glucose uptake are increased by cobalt chloride, strongly supporting the hypothesis that upregulation of GLUT1 gene is dependent on O_2 deprivation itself. However, inhibition of oxidative phosphorylation by sodium azide in normoxic ATII cells is as effective as hypoxia for stimulating glucose transport and increasing GLUT1 mRNA levels. That hypoxia-induced GLUT 1 gene upregulation results from two different mechanisms was also previously reported in a cell clone derived from a hepatoma (Behrooz and Ismail, 1997). However, the temporal and spatial contribution of these two mechanisms in hypoxia-induced upregulation of GLUT1 mRNA level in ATII cells remains to be determined.

3.2. Effect of hypoxia on glycolytic enzymes

Under normoxic conditions, alveolar epithelial cells display a very high aerobic and anaerobic glycolytic activity (Simon et al., 1978). Many years ago, Simon and colleagues reported that exposure of alveolar epithelial cells to hypoxia upregulates the activity of several glycolytic enzymes, such as pyruvate kinase and lactate dehydrogenase. This hypoxia-induced increase in glycolytic enzymes allows maintenance of ATP levels in hypoxic cells close to that of normoxic cells through an increase in anaerobic glycolysis. Recently, the mechanisms of hypoxic regulation of two glycolytic enzymes, glyceraldehyde-3-phosphate dehydrogenase (GAPDH) and hexokinase (HK), have been evaluated.

The induction of the glycolytic enzyme GAPDH by hypoxia is probably a distinctive feature of hypoxia-tolerant mammalian cells because it has been reported only in skeletal muscle cells, myoblasts and endothelial cells that are able to

survive to prolonged hypoxia, but not in the hypoxia-sensitive renal cells, lung fibroblasts or smooth muscle cells (Graven et al., 1994). For the first time in native epithelial cells, Escoubet et al. (1999) reported a hypoxic induction of GAPDH mRNA expression in ATII cells associated with a protein synthesis and enzymatic activity increases. They demonstrated that hypoxia-induced GAPDH upregulation is due to increased GAPDH gene transcription with no change in mRNA stability. The effects of hypoxia were mimicked by cobalt chloride suggesting that GAPDH regulation was directly related to O_2 deprivation. Although the promoter region of the human GAPDH gene contains several hypoxia responsive elements (HREs), it has not been firmly established that hypoxic induction of GAPDH occurred through HIF-1 activation. However, this hypothesis is strongly supported by a recent study showing that in mutant Chinese hamster ovary cells defective in HIF-1α subunit, hypoxic induction of GAPDH was completely abrogated (Wood et al., 1996).

Hexokinase (HK) is the rate limiting enzyme for glucose utilization in the lung. Three isoforms are expressed in lung: HK-I, HK-II and HK-III. HK-I is the predominant isoform in human A549 alveolar epithelial cells under normoxic conditions but interestingly this isoform is not regulated by hypoxia (Riddle et al., 2000). In contrast, HK-II is less expressed in normoxia but its mRNA expression as well as protein expression and enzymatic activity are dramatically induced by hypoxia. The mechanism of hypoxic induction of HK-II is quite similar to that described for GAPDH: it is a transcriptionally mediated mechanism, directly related to O_2 deprivation. Although potential HREs have been identified on the HK promoter gene, further study is needed to fully characterise the potential HRE sites in the human HK-II gene.

4. Effect of hypoxia on vascular growth factor expression

Vascular endothelial growth factor (VEGF), initially identified as vascular permeability factor, is an important growth and permeability factor for endothelial cells. VEGF is expressed in many tissues of healthy adult rats and mice, with the most abundant expression in the lung (Marti and Risau, 1998; Monacci et al., 1993). In normal lung, *in situ* hybridization studies have located VEGF transcripts at the luminal surface of alveolar walls mostly in alveolar epithelial type II cells (ATII) (Tuder et al., 1995). In accordance with the *in vivo* studies, VEGF transcripts were also detected *in vitro* in a human alveolar epithelial cell line and in isolated and cultured rat alveolar epithelial cells (Boussat et al., 2000; Pham et al., 2002). The transcripts correspond to the three major VEGF isoforms previously described: $VEGF_{121}$ and $VEGF_{165}$ which are secreted in a soluble form and $VEGF_{189}$ which remains cell surface associated or is primarily deposited in the extracellular matrix (Boussat et al., 2000).

In many tissues, hypoxia is the best characterised potent inducer of VEGF mRNA expression. Also, in the lung, hypoxia induced an increase in VEGF expression in both *in vivo* and *in vitro* studies. Exposure of alveolar epithelial cells to hypoxia induces an upregulation of VEGF mRNA due to transcriptional activation of the VEGF gene without change in mRNA stability. Interestingly, this upregulation of VEGF mRNA was associated with an increase in VEGF proteins levels and protein secretion in the extracellular space. *In vitro*, the secretion in normoxia occurred both at the apical and basolateral sides but was significantly less at the apical side. The presence of VEGF in the bronchoalveolar fluid suggested that, *in vivo*, VEGF is secreted at least in part at the apical surface. The functional role of VEGF secretion by alveolar epithelial cells remains unclear. One hypothesis is that VEGF secreted at the basolateral side targets the capillary endothelium and acts through VEGF receptors located on endothelial cells (Ferrara et al., 1992; Thomas, 1996). The second possibility is that VEGF exerts an autocrine or paracrine regulatory role since the functional VEGF receptors neuropilin-1, that binds only $VEGF_{165}$, and KDR are present on distal airway epithelial cells of human fetal lung explants (Brown et al., 2001). Addition of exogenous VEGF in these explants induced epithelial cell proliferation and increased the

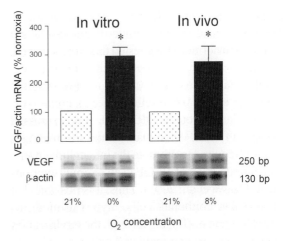

Fig. 2.3. Induction of VEGF mRNA expression by hypoxia in rat ATII cells *in vitro* (primary cultures at day 3) and *in vivo* (freshly isolated ATII cells). ATII cells in primary culture were exposed to 21% or 0% O_2 for 18 h. Freshly ATII cells were isolated from the lungs of rats exposed to 21% or 8% O_2 for 24 h. VEGF mRNA expression was normalized to the corresponding β-actin mRNA. Data are the means ± s.d. for $n = 4$ for cultured ATII cells and $n = 4$ for freshly isolated ATII cells. *$p < 0.001$ vs 21% O_2 (personal data).

expression of surfactant proteins while inhibition of VEGF receptors in the rat causes lung cell apoptosis and enlargement of the air spaces (Kasahara et al., 2000). Finally, because VEGF receptor-1 (Flt-1) has been reported to be present on monocytes/macrophages, the activation of which led to monocyte activation and chemotaxis (Clauss et al., 1996), it could be hypothesized that apical secretion of VEGF may play a role in the recruitment of immune cells in the alveolar space.

Exposure to hypoxia increased apical secretion of VEGF both *in vitro* and *in vivo*. The increase of VEGF in the apical compartment was not related to a passage of VEGF from the basolateral to the apical side because permeability of alveolar epithelial cells grown on filters, *in vitro*, as well as permeability of alveolar epithelium, *in vivo*, were unchanged after exposure to hypoxia. The observation that VEGF secretion only increased on the apical side during hypoxia is unexpected. One explanation could be that, in our culture conditions, secreted VEGF remained partly bound to the extracellular matrix and that the value measured in the culture medium did not reflect all of the basolateral secretion. However, VEGF$_{165}$ is a soluble isoform and the magnitude of the basolateral secretion in normoxia suggested that, if it occurred, binding of VEGF$_{165}$ to the extracellular matrix was small. Finally, a recent study has reported that the inducible endoplasmic reticulum chaperone oxygen-regulated protein, ORP 150, was expressed along with VEGF in human wound macrophages and was required to promote the intracellular transport and secretion of VEGF in an hypoxic environment (Ozawa et al., 2001). Thus, it is possible that in alveolar epithelial cells, hypoxia modifies VEGF protein trafficking through the modification of chaperone oxygen-regulated proteins.

5. Effect of hypoxia on adhesion molecules expression

Alveolar epithelial cells constitutively express an intercellular adhesion molecule-1 (ICAM-1). This transmembrane adhesion protein is expressed on the apical surface membrane of type I cells *in vivo* and on type II cells *in vitro* as they spread in culture assuming type I cell characteristics (Beck-Schimmer et al., 2001). In alveolar epithelial cells, ICAM-1 is associated with the cytoskeleton and may provide a fixed intermediary between mobile inflammatory cells and the alveolar surface (Barton et al., 1996). ICAM-1 interacts with two β_2 integrin ligands on leukocytes: LFA-1 and Mac-1 and augments the adhesiveness of neutrophils to alveolar epithelial cells. Another protein adhesion molecule, the vascular cell adhesion molecule-1 (VCAM-1), is also expressed by alveolar epithelial cells (Rosseau et al., 2000). The primary role of this molecule is the promotion of lymphocyte and macrophage adhesion. The expression of ICAM-1 or VCAM-1 is stimulated by tumor necrosis fac-tor-α, lipopolysaccharide (LPS), interferon γ, rhinovirus infection as well as mechanical stress.

In hypoxic lung injury, both *in vivo* and *in vitro* studies have demonstrated that enhanced adherence of neutrophils to endothelial cells play a major role in tissue damage that accompanies hypoxia/reoxygenation. Beck-Schimmer et al. (2001) reported that, in alveolar epithelial cells,

expression of ICAM-1 and VCAM-1 were upregulated by moderate and acute hypoxia. Exposure of alveolar type II cells to 10% and 5% O$_2$ for 1 h induced a two-fold increase in mRNA followed by a modest increase in protein expression between 2 and 4 h that was mediated through an increase in transcription. Hypoxia-induced upregulation was considerably less intense compared with LPS-induced induction of ICAM-1 and VCAM-1. Interestingly, a strong TNF-α dependency was still demonstrable. The biological function of the hypoxia-induced upregulation of the adhesion molecules was supported by the demonstration that neutrophil adherence and macrophage adherence were increased in hypoxic alveolar epithelial cells and were blocked by anti-ICAM-1 and anti-VCAM-1 antibodies, respectively. These results support a role of adhesion molecules during hypoxia promoting adherence of neutrophils and macrophages to alveolar epithelial cells (Table 2.1).

6. Role of hypoxia-inducible factors in regulation of gene expression

Hypoxia-inducible factors (HIF) are central to the regulation of genes involved in hypoxic responses. HIF-1 is a heterodimeric DNA binding complex, consisting of 2 subunits, HIF-1α which belongs to basic-helix–loop-helix (bHLH)-PAS proteins and HIF-1β, which is also known as aryl hydrocarbon receptor nuclear translocator (Arnt) (Wang et al., 1995). The regulation of HIF itself occurs through modifications of its α subunit, whereas the β subunit is a constitutive nuclear protein. The HIF-1α subunit is rapidly degraded in normoxia by a ubiquitin–proteasome pathway, involving the von Hippel–Lindau tumor suppressor protein (Maxwell et al., 1999; Tanimoto et al., 2000; Ohh et al., 2000). In hypoxia, this pathway is inhibited and HIF-1α is stabilized. Endothelial PAS domain protein-1 (EPAS 1), also designated as HIF-1α-like factor (HLF), HIF-related factor (HRF) or HIF-2α, is a recently identified bHLH-PAS protein. EPAS-1 shares 48% sequence identity with HIF-1α (Tian et al., 1997; Ema et al., 1997).

HIF-1α expression is ubiquitous whereas EPAS 1 is mainly restricted to endothelial cells (Wiesener et al., 1998). Except for the major difference in abundance and distribution between HIF-1α and EPAS 1, the regulatory characteristics of these proteins are not much different. After activation by hypoxia, both factors are translocated to the nucleus, can heterodimerize with Arnt and can bind to the same hypoxia responsive element (HRE) located in the 3'-flanking region of hypoxia-inducible genes.

The presence of HIF-1 has been reported in the hypoxic lung of several species, such as mouse, rat and ferret (Wiener et al., 1996; Jain et al., 1998; Yu et al., 1998). Immunochemistry of lung sections revealed no detectable HIF-1α expression in normoxic lungs whereas HIF-1α protein was detected throughout the hypoxic lung, with the highest expression occurring in the bronchial and alveolar epithelium, bronchial smooth muscle and vascular endothelium. In contrast, HIF-1β was detected in normoxic and hypoxic lung particularly in bronchial epithelium and vascular endothelium with a modest expression in alveolar epithelium (Yu et al., 1998). Protein expression was induced maximally when lungs were ventilated with 0 or 1% O_2 for 4 h and reoxygenation induced a rapid degradation with a half life of less than 1 min. *In vitro*, cultured alveolar epithelial cell lines exhibited minimal or no expression of HIF-1α under normoxic conditions but hypoxia induced a temporal and spatial increase in HIF-1α expression in nuclear extracts (Yu et al., 1998). In contrast to HIF-1α, HIF-1β is not hypoxia-sensitive and was expressed at 20% O_2. Only a modest increase was observed under hypoxic conditions. Interestingly, the temporal characteristics and responses of HIF-1α expression *in vivo* to varying O_2 concentrations were remarkably similar to those of tissue culture cells.

EPAS 1 expression, in contrast to the ubiquitous mRNA expression of HIF-1α, was reported to be largely restricted to endothelial cells (Tian et al., 1997). However, analysis of lung tissues and *in situ* hybridization showed that EPAS 1 was also expressed in alveolar epithelium (Ema et al., 1997). In agreement with this observation, we and others reported that EPAS 1 mRNA was expressed in A549 cells, a cell line derived from pulmonary adenocarcinoma which has characteristics of alveolar type II cells, in normoxic conditions and was increased under hypoxia (Sato et al., 2002; Uchida et al., 2002).

The regulation of hypoxia inducible factors could occur at multiple levels, including mRNA expression (Wiener et al., 1996; Yu et al., 1998), protein expression (Huang et al., 1996; Salceda and Caro, 1997; Jiang et al., 1996), nuclear localization and transactivation (Huang et al., 1996; Jiang et al., 1997). In ferret lung, the hypoxia-induced increase in HIF-1α protein was associated with an increase of HIF-1α mRNA which was noticeable at 4 h, peaked at 6 h after hypoxic stimulation, and then declined after 8 h of continuous hypoxia. In A549 cells, hypoxia induced an increase in HIF-1α protein levels up to 6 hours with no change in mRNA levels. Concerning EPAS 1, Uchida et al. (2002) reported that hypoxia induced upregulation only at the post-transcriptional level. In contrast, Sato et al. (2002) showed that regulation of EPAS1 occurred at the transcriptional level and was mediated by Src family kinases. Of interest is the observation that increasing EPAS by adenoviral transfection increased endogenous EPAS1 mRNA levels suggesting a positive auto-regulation of EPAS gene transcription by EPAS 1 itself (Sato et al., 2002).

The means by which cells sense reduced O_2 tensions and subsequently induce changes in HIF activity has remained elusive until recently. In normoxia, the HIFα subunit is rapidly degraded so that almost no HIF protein accumulates. Under hypoxic conditions, degradation of the α-subunit is blocked, allowing HIF-1α to accumulate within the nucleus where, upon binding to HIF-1β it recognizes HIF-responsive elements within the promoter of hypoxia-responsive genes (Huang et al., 1996; Salceda and Caro, 1997). Until recently, circumstantial evidence implicated the participation of a heme protein in the hypoxic response pathway since besides low O_2 tensions, HIF can also be activated by transition metals cations (Co^{2+}, Ni^{2+} and Mn^{2+}) and also by reagents that chelate iron (Goldberg et al., 1988; Semenza, 1999). However, several recent studies have provided evidence in support of the hypothesis that degradation of HIF-1α in normoxic conditions is triggered by post-translational hydroxylation of proline residues within a polypeptide segment of HIF-1α called the oxygen dependent domain (Jaakkola et al., 2001; Bruick and McKnight, 2001). This critical regulatory event is carried out by an iron dependent prolyl hydroxylase enzyme that used O_2 as its substrate to catalyse hydroxylation of the target proline residues. The hydroxylated proline residues in this sequence are recognized by the product of the tumor suppressor von Hippel Lindau gene, a component of a ubiquitin protein E3 ligase complex that tags the α subunit for degradation by the proteasome pathway. When cells are hypoxic or are treated with iron chelator, the proline is not hydroxylated and so HIF-1α escapes degradation and forms a heterodimer with HIF-1β. Other consequences of changes in O_2 concentration, such as alteration of intracellular redox potentials or amounts of reactive oxygen species (ROS), may influence HIF induction. However, there are conflicting data about these pathways of oxygen sensing. Acker and colleagues have postulated that a low output NADPH oxidase, similar to that one present in neutrophils, could act as an oxygen sensor (Acker, 1994). In this model, hydrogen peroxide would be continuously generated by the oxidase in an oxygen dependent manner and would have a continuous negative effect on HIF-1α survival. However, evidence against this model is the finding that diphenylene iodonium (DPI), an NADPH inhibitor, decreases rather than stimulates the hypoxia response (Gleadle et al., 1995). Chandel and Schumacker (2000) have proposed another redox model of oxygen sensing based on the production of ROS by the mitochondria. In this model, hypoxia increased superoxide production at the level of complex III of the respiratory electron transfer chain. The increase in mitochondrial ROS would be proportional to the degree of hypoxia and would be the starting point of the hypoxic signal. Experimental support for this model is the unresponsiveness to hypoxia of cells lacking mitochondrial DNA. However, Srinivas et al. (2001) reported that other type of mitochondrial DNA-

less cells have a normal response to hypoxia. These results suggest that the response is dependent on cell types or on procedures used to establish the mitochondrial DNA-less cells.

In alveolar epithelial cells, HIFα and EPAS proteins as well as hypoxia regulated genes, glycolytic enzymes, GLUT1 and VEGF are upregulated by O_2 deprivation as well as Co^{2+} and the iron chelator, desferrioxamine, suggesting that the oxygen sensing pathway involves an oxygen and iron dependent enzyme which could be the prolyl hydroxylase enzyme (Uchida et al., 2002; Escoubet et al., 1999; Ouiddir et al., 1999; Pham et al., 2002). The role of mitochondrial ROS in alveolar epithelial cells has been addressed directly in an alveolar epithelial cell line, A549 cells, by using mutant cells lacking mitochondrial DNA (ρ0 cells). Vaux et al. (2001) showed that HIF induction by hypoxia was essentially normal in wild type cells as well as ρ0 cells and that hydrogen peroxide production was reduced in ρ0 cells versus wild type cells and decreased by hypoxia in both ρ0 and wild type cells. They also reported that the hypoxic induction of GLUT1 mRNA was similar in the ρ0 and wild type cells. Taken together, these data indicate that a functional respiratory chain is not necessary for a regulation of HIF by oxygen in alveolar epithelial cells. In accordance with this observation, we found that, in rat alveolar type II cells, neither DPI nor *N*-acetyl cysteine, a non-mitochondrial ROS inhibitor, prevent hypoxic induction of VEGF expression.

7. Conclusion

Genes regulated by hypoxia in the lungs can serve as physiological signals, but how this occurs is incompletely understood. Therefore, basic mechanisms by which low O_2 tension directly activates or down-regulates proteins in the alveolar epithelium needs to be explored. At present, only a few genes have been reported to be regulated by hypoxia in alveolar epithelial cells and much work still needs to be done to determine the regulation of other genes such as surfactant proteins which are involved in specific functions. Several questions are unsolved and should be addressed in future studies. For example, what is the functional significance of hypoxia-induced downregulation of Na transport proteins or upregulation of VEGF and adhesion molecules? What is the role of the transcriptional factors such as HIF and/or EPAS in the response to hypoxia?

References

Acker, H. (1994). Cellular oxygen sensors. Ann. N.Y. Acad Sci. 718, 3–10.

Baines, D., Ramminger, S., Collett, A., Haddad, J., Best, O., Land, S., Olver, R. and Wilson, S. (2001). Oxygen-evoked Na transport in rat fetal distal lung epithelial cells. J. Physiol. (London) 532, 105–113.

Barker, P.M., Nguyen, M.S., Gatzy, J.T., Grubb, B., Norman, H., Hummler, E., Rossier, B., Boucher, R.C. and Koller, B. (1998). Role of gammaENaC subunit in lung liquid clearance and electrolyte balance in newborn mice. Insights into perinatal adaptation and pseudohypoaldosteronism. J. Clin. Invest. 102, 1634–1640.

Barton, W.W., Wilcoxen, S.E., Christensen, P.J. and Paine, R. (1996). Association of ICAM-1 with the cytoskeleton in rat alveolar epithelial cells in primary culture. Am. J. Physiol. 271, L707–L718.

Beck-Schimmer, B., Schimmer, R.C., Madjdpour, C., Bonvini, J.M., Pasch, T., Ward, P.A., Oertli, B. and Ziegler, U. (2001). Hypoxia mediates increased neutrophil and macrophage adhesiveness to alveolar epithelial cells. Am. J. Respir. Cell Mol. Biol. 25, 780–787.

Behrooz, A. and Ismail, B.F. (1997). Dual control of glut1 glucose transporter gene expression by hypoxia and by inhibition of oxidative phosphorylation. J. Biol. Chem. 272, 5555–5562.

Boussat, S., Eddahibi, S., Coste, A., Fataccioli, V., Gouge, M., Housset, B., Adnot, S. and Maitre, B. (2000). Expression and regulation of vascular endothelial growth factor in human pulmonary epithelial cells. Am. J. Physiol. 279, L371–L378.

Brown, K.R., England, K.M., Goss, K.L., Snyder, J.M. and Acarregui, M.J. (2001). VEGF induces airway epithelial cell proliferation in human fetal lung in vitro. Am. J. Physiol. 281, L1001–L1010.

Bruick, R.K. and McKnight, S.L. (2001). A conserved family of prolyl-4-hydroxylases that modify HIF. Science 294, 1337–1340.

Canessa, C.M., Schild, L., Buell, G., Thorens, B., Gautschi, I., Horisberger, J.D. and Rossier, B.C. (1994). Amiloride-sensitive epithelial Na+ channel is made of three homologous subunits. Nature 367, 463–467.

Cartee, G.D., Douen, A.G., Ramlal, T., Klip, A. and Holloszy, J.O. (1991). Stimulation of glucose transport in skeletal muscle by hypoxia. J. Appl. Physiol. 70, 1593–1600.

Chandel, N.S. and Schumacker, P.T. (2000). Cellular oxygen sensing by mitochondria: old questions, new insight. J. Appl. Physiol. 88, 1880–1889.

Clauss, M., Weich, H., Breier, G., Knies, U., Rolkl, W., Waltenberger,J. and Rissau, W. (1996). The vascular endothelial growth factor receptor Flt-1 mediates biological activities. Implication for a functional role of placenta growth factor in monocyte activation and chemotaxis. J. Biol. Chem. 271, 17629–17634.

Clerici, C., Soler, P. and Saumon, G. (1991). Sodium-dependent phosphate and alanine transports but sodium-independent hexose transport in type II alveolar epithelial cells in culture. Biochim. Biophys. Acta. 1063, 27–35.

Ema, M., Taya, S., Yokotani, N., Sogawa, K., Matsuda, Y. and Fujii-Kuriyama, Y. (1997). A novel bHLH-PAS factor with close sequence similarity to hypoxia-inducible factor 1α regulates the VEGF expression and is potentially involved in lung and vascular development. Proc. Natl. Acad. Sci. USA 94, 4273–4278.

Escoubet, B., Planes, C. and Clerici, C. (1999). Hypoxia increases glyceraldehyde-3-phosphate dehydrogenase transcription in rat alveolar epithelial cells. Biochem Biophys. Res. Commun. 266, 156–161.

Ferrara, N., Jakeman, K. and Leung, D. (1992). Molecular and biological properties of the vascular endothelial growth factor family of proteins. Endocr. Rev. 13, 18–32.

Gleadle, J.M., Ebert, B.L. and Ratcliffe, P.J. (1995). Diphenylene iodonium inhibits the induction of erythropoietin and other mammalian genes by hypoxia. Implications for the mechanism of oxygen sensing. Eur. J. Biochem. 234, 92–99.

Goldberg, M., Dunning, S. and Bunn, H. (1988). Regulation for erythropoietin gene: evidence for the oxygen sensor is a heme protein. Science 242, 1412–1415.

Graven, K.K., Troxler, R.F., Kornfeld, H., Panchenko, M.V. and Farber, H.W. (1994). Regulation of endothelial cell glyceraldehyde-3-phosphate dehydrogenase expression by hypoxia. J. Biol. Chem. 269, 24446–24453.

Harris, Z., Ridge, K., Gonzalez-Flecha, B., Gottlieb, L., Zucker, A. and Sznajder, J. (1996). Hyperbaric oxygenation upregulates rat lung Na,K-ATPase. Eur. Respir. J. 9, 472–477.

Huang, L.E., Arany, Z., Livingston, D.M. and Bunn, H.F. (1996). Activation of hypoxia-inducible transcription factor depends primarily upon redox-sensitive stabilization of its alpha subunit. J. Biol. Chem. 271, 32253–32259.

Hummler, E., Barker, P., Gatzy, J., Beermann, F., Verdumo, C., Schmidt, A., Boucher, R. and Rossier, B.C. (1996). Early death due to defective neonatal lung liquid clearance in α-ENaC-deficient mice. Nature Genet. 12, 325–328.

Jaakkola, P., Mole, D.R., Tian, Y.M., Wilson, M.I., Gielbert, J., Gaskell, S.J., Kriegsheim, A., Hebestreit, H.F., Mukherji, M., Schofield, C.J., Maxwell, P.H., Pugh, C.W. and Ratcliffe, P.J. (2001). Targeting of HIF-α to the von Hippel-Lindau ubiquitylation complex by O_2-regulated prolyl hydroxylation. Science 292, 468–472.

Jain, L., Chen, X.J., Ramosevac, S., Brown, L.A. and Eaton, D.C. (2001). Expression of highly selective sodium channels in alveolar type II cells is determined by culture conditions. Am. J. Physiol. 280, L646–L658.

Jain, S., Maltepe, E., Lu, M.M., Simon, C. and Bradfield, C. A. (1998). Expression of ARNT, ARNT2, HIF1α, HIF2α and Ah receptor mRNAs in the developing mouse. Mech. Dev. 73, 117–123.

Jiang, B.H., Semenza, G.L., Bauer, C. and Marti, H.H. (1996). Hypoxia-inducible factor 1 levels vary exponentially over a physiologically relevant range of O_2 tension. J. Biol. Chem. 271, 32253–32259.

Jiang, B.H., Zheng, J.Z., Leung, S.W., Roe, R. and Semenza, G.L. (1997). Transactivation and inhibitory domains of hypoxia-inducible factor 1α. Modulation of transcriptional activity by oxygen tension. J. Biol. Chem. 272, 19253–19260.

Kasahara, Y., Tuder, R., Taraseviene-Stewart, L., Le Cras, T., Abman, S., Hirth, P., Waltenberger, J. and Voelkel, N. (2000). Inhibition of VEGF receptors causes lung cell apoptosis and emphysema. J. Clin. Invest. 106, 1311–1319.

Loike, J.D., Cao, L., Brett, J., Ogawa, O., Silverstein, S.C. and Stern, D. (1992). Hypoxia induces glucose transporter expression in endothelial cells. Am. J. Physiol. 263, C326–C333.

Maggiorini, M., Melot, C., Pierre, S., Pfeiffer, F., Greve, I., Sratori, C., Lepori, M., Hauser, M., Scherrer, U. and Naeije, R. (2001). High-altitude pulmonary edema is initially caused by an increase in capillary pressure. Circulation 103, 2078–2083.

Mairbaurl, H., Mayer, K., Kim, K., Borok, Z., Bartsch, P. and Crandall, E. (2002). Hypoxia decreases active Na^+ transport across primary rat alveolar epithelial cell monolayers. Am. J. Physiol. 282, L659–665.

Mairbaurl, H., Wodopia, R., Eckes, S., Schulz, S. and Bartsch, P. (1997). Impairment of cation transport in A549 cells and rat alveolar epithelial cells by hypoxia. Am. J. Physiol. 273, L797–L806.

Marti, H.H. and Risau, W. (1998). Systemic hypoxia changes the organ-specific distribution of vascular endothelial growth factor and its receptors. Proc. Natl. Acad. Sci. USA 95, 15809–15814.

Matalon, S. and O'Brodovich, H. (1999). Sodium channels in alveolar epithelial cells: molecular characterization, biophysical properties, and physiological significance. Annu. Rev. Physiol. 61, 627–661.

Matthay, M., Folkesson, H. and Clerici, C. (2002). Epithelial fluid transport and the resolution of pulmonary edema. Physiol. Rev. (in press).

Maxwell, P.H., Wiesener, M.S., Chang, G.W., Clifford, S.C., Vaux, E.C., Cockman, M.E., Wykoff, C.C., Pugh, C.W., Maher, E.R. and Ratcliffe, P.J. (1999). The tumour suppressor protein VHL targets hypoxia-inducible factors for oxygen-dependent proteolysis. Nature 399, 271–275.

McDonald, F.J., Yang, B., Hrstka, R.F., Drummond, H.A., Tarr, D.E., McCray, P.B., Stokes, J.B., Welsh, M.J. and Williamson, R.A. (1999). Disruption of the beta subunit of the epithelial Na^+ channel in mice: hyperkalemia and neonatal death associated with a pseudohypoaldosteronism phenotype. Proc. Natl. Acad. Sci. USA 96, 1727–1731.

Monacci, W., Merrill, M. and Oldfield, E. (1993). Expression of vascular permeability factor/vascular endothelial growth factor in normal rat tissues. Am. J. Physiol. 264, C995–C1002.

O'Brodovich, H., Canessa, C., Ueda, J., Rafii, B., Rossier, B. and Edelson, J. (1993). Expression of the epithelial Na^+ channel subunits are differentially regulated during development and by steroids. Am. J. Physiol. 269, C805–C812.

O'Brodovich, H., Hannam, V. and Rafii, B. (1991). Sodium channel but neither Na(+)-H+ nor Na-glucose symport inhibitors slow neonatal lung water clearance. Am. J. Respir. Cell Mol. Biol. 5, 377–384.

Ohh, M., Park, C.W., Ivan, M., Hoffman, M.A., Kim, T.Y., Huang, L.E., Pavletich, N., Chau, V. and Kaelin, W.G. (2000). Ubiquitination of hypoxia-inducible factor requires direct binding to the β-domain of the von Hippel-Lindau protein. Nature Cell Biol. 2, 423–427.

Ouiddir, A., Planes, C., Fernandes, I., VanHesse, A. and Clerici, C. (1999). Hypoxia upregulates activity and expression of the glucose transporter GLUT1 in alveolar epithelial cells. Am. J. Respir. Cell Mol. Biol. 21, 710–718.

Ozawa, K., Kondo, T., Hori, O., Kitao, Y., Stern, D., Eisenmenger, W., Ogawa, S. and Ohshima, T. (2001). Expression of the oxygen-regulated protein ORP 150 accelerates wound healing by modulating intracellular VEGF transport. J. Clin. Invest. 108, 41–50.

Pham, I., Uchida, T., Planes, C., Matthay, M. and Clerici, C. (2002). Hypoxia upregulates VEGF expression in alveolar epithelial cells in vitro and in vivo. Am. J. Physiol. (in revision).

Pitkanen, O., Transwell, A.K., Downey, G. and O'Brodovich, H. (1996). Increased Po_2 alters the bioelectric properties of fetal distal lung epithelium. Am. J. Physiol. 270, L1060–L1066.

Planes, C., Blot-Chabaud, M., Couette, S., Matthay, M. and Clerici, C. (2002). The hypoxia-induced decrease in sodium transport in alveolar type II cells is mediated by decreased delivery of ENaC to the apical membrane. FASEB J: (abstract).

Planes, C., Escoubet, B., Blot-Chabaud, M., Friedlander, G., Farman, N. and Clerici, C. (1997). Hypoxia down-regulates expression and activity of epithelial sodium channels in rat alveolar epithelial cells. Am. J. Respir. Cell Mol. Biol. 17, 508–518.

Planes, C., Friedlander, G., Loiseau, A., Amiel, C. and Clerici, C. (1996). Inhibition of Na-K-ATPase activity after prolonged hypoxia in an alveolar epithelial cell line. Am. J. Physiol. 271, L70–L78.

Rafii, B., Tanswell, A.K., Otulakowski, G., Pitkanen, O., Belcastro, T.R. and O'Brodovich, H. (1998). O$_2$-induced ENaC expression is associated with NF-κB activation and blocked by superoxide scavenger. Am. J. Physiol. 275, L764–L770.

Riddle, S.R., Ahmad, A., Ahmad, S., Deeb, S.S., Malkki, M., Schneider, B.K., Allen, C.B. and White, C.W. (2000). Hypoxia induces hexokinase II gene expression in human lung cell line A549. Am. J. Physiol. 278, L407–L416.

Rosseau, S., Selhorst, J., Wiechmann, K., Leissner, K., Maus, U., Mayer, K., Grimminger, F., Seeger, W. and Lohmeyer, J. (2000). Monocyte migration through the alveolar epithelial barrier: adhesion molecule mechanisms and impact of chemokines. J. Immunol. 164, 427–435.

Sakuma, T., Hida, M., Nambu, Y., Osanai, K., Takahashi, K., Ohya, N., Inoue, M. and Watanabe, Y. (2001). Effects of hypoxia on alveolar fluid transport capacity in rat lungs. J. Appl. Physiol. 91, 1766–1774.

Salceda, S. and Caro, J. (1997). Hypoxia-inducible factor 1α (HIF-1α) protein is rapidly degraded by the ubiquitin-proteasome system under normoxic conditions. Its stabilization by hypoxia depends on redox-induced changes. J. Biol. Chem. 272, 22642–22647.

Sato, M., Tanaka, T., Maeno, T., Sando, Y., Suga, T., Maeno, Y., Sato, H., Nagai, R. and Kurabayashi, M. (2002). Inducible expression of endothelial PAS domain protein-1 by hypoxia in human lung adenocarcinoma A 549 cells. Am. J. Respir. Cell Mol. Biol. 26, 127–134.

Saumon, G. and Makhloufi, Y. (1997). Glucose transporter gene expression in rat lung and alveolar type II cells in primary culture. Am. J. Respir. Crit. Care Med. 155, A831.

Semenza, G. (1999). Perspectives on oxygen sensing. Cell 98, 281–284.

Simon, L.M., Robin, E.D., Phillips, J.R., Acevedo, J., Axline, S.G. and Theodore, J. (1977). Enzymatic basis for bioenergetic differences of alveolar versus peritoneal macrophages and enzyme regulation by molecular O$_2$. J. Clin. Invest. 59, 443–448.

Simon, L.M., Robin, E.D., Taffin, T., Theodore, J. and Douglas, W.H.J. (1978). Bioenergetic pattern of isolated type II pneumocytes in air and during hypoxia. J. Clin. Invest. 61, 1232–1239.

Snyder, P. (2000). Liddle's syndrome mutations disrupt cAMP-mediated translocation of the epithelial Na channel to the cell surface. J. Clin. Invest. 105, 45–53.

Srinivas, V., Leshchinsky, I., Sang, N., King, M.P., Minchenko, A. and Caro, J. (2001). Oxygen sensing and HIF-1 activation does not require an active mitochondrial respiratory chain electron-transfer pathway. J. Biol. Chem. 276, 21995–21998.

Suzuki, S., Noda, M., Sugita, M., Ono, S., Koike, K. and Fujimura, S. (1999a). Impairment of transalveolar fluid transport and lung Na(+)-K(+)-ATPase function by hypoxia in rats. J. Appl. Physiol. 87, 962–968.

Suzuki, S., Sugita, M., Noda, M., Tsubochi, H. and Fujimura, S. (1999b). Effects of intraalveolar oxygen concentration on alveolar fluid absorption and metabolism in isolated rat lungs. Respir. Physiol. 115, 325–332.

Tanimoto, K., Makino, Y., Pereira, T. and Poellinger, L. (2000). Mechanism of regulation of the hypoxia-inducible factor-1α by the von Hippel-Lindau tumor suppressor protein. Embo J. 19, 4298–4309.

Thomas, K. (1996). Vascular endothelial growth factor, a potent and selective angiogenic agent. J. Biol. Chem. 271, 603–606.

Tian, H., McKnight, S.L. and Russell, D.W. (1997). Endothelial PAS domain protein 1 (EPAS1), a transcription factor selectively expressed in endothelial cells. Genes Dev. 11, 72.

Tuder, R.M., Flook, B.E. and Voelkel, N.F. (1995). Increased gene expression for VEGF and the VEGF receptors KDR/Flk and Flt in lungs exposed to acute or to chronic hypoxia. Modulation of gene expression by nitric oxide. J. Clin. Invest. 95, 1798–1807.

Uchida, T., Matthay, M. and Clerici, C. (2002). Exposure to hypoxia activates the transcriptional factor EPAS1 expressed by lung alveolar epithelial cells. FASEB J. (abstract).

Vaux, E.C., Metzen, E., Yeates, K.M. and Ratcliffe, P.J. (2001). Regulation of hypoxia-inducible factor is preserved in the absence of a functioning mitochondrial respiratory chain. Blood 98, 296–302.

Vivona, M., Matthay, M., Blot Chabaud, M., Friedlander, G. and Clerici, C. (2001). Hypoxia reduces alveolar epithelial sodium and fluid transport in rats: reversal by β adrenergic agonist. Am. Rev. Respir. Cell Mol. Biol. 25, 1–9.

Wang, G.L., Jiang, B.H., Rue, E.A. and Semenza, G.L. (1995). Hypoxia-inducible factor 1 is a basic-helix–loop-helix-PAS heterodimer regulated by cellular O$_2$ tension. Proc. Natl. Acad. Sci. USA 92, 5510–5514.

Wiener, C.M., Booth, G. and Semenza, G.L. (1996). In vivo expression of mRNAs encoding hypoxia-inducible factor 1. Biochem. Biophys. Res. Commun. 225, 485–488.

Wiesener, M.S., Turley, H., Allen, W.E., Willam, C., Eckardt, K.U., Talks, K.L., Wood, S.M., Gatter, K.C., Harris, A.L., Pugh, C.W., Ratcliffe, P.J. and Maxwell, P.H. (1998). Induction of endothelial PAS domain protein-1 by hypoxia: characterization and comparison with hypoxia-inducible factor-1α. Blood 92, 2260–2268.

Wodopia, R., Ko, H., Billian, J., Wiesner, R., Bartsch, P. and Mairbaurl, H. (2000). Hypoxia decreases proteins involved in epithelial electrolyte transport in A549 cells and rat lung. Am. J. Physiol. 279, L1110–L1119.

Wood, S.M., Gleadle, J.M., Pugh, C.W., Hankinson, O. and Ratcliffe, P.J. (1996). The role of the aryl hydrocarbon receptor nuclear translocator (ARNT) in hypoxic induction of gene expression. Studies in ARNT-deficient cells. J. Biol. Chem. 271, 15117–15123.

Yu, A.Y., Frid, M.G., Shimoda, L.A., Wiener, C.M., Stenmark, K. and Semenza, G.L. (1998). Temporal, spatial, and oxygen-regulated expression of hypoxia-inducible factor-1 in the lung. Am. J. Physiol. 275, L818–L826.

Yue, G., Russell, W., Benos, D., Jackson, R., Olman, M. and Matalon, S. (1995). Increased expression and activity of sodium channels in alveolar type II cells of hyperoxic rats. Proc. Natl. Acad. Sci. USA 92, 8418–8422.

Sensing, Signaling and Cell Adaptation. Edited by K.B. Storey and J.M. Storey
© *2002 Elsevier Science B.V. All rights reserved.*

CHAPTER 3

A Profile of the Metabolic Responses to Anoxia in Marine Invertebrates

Kevin Larade and Kenneth B. Storey*
Institute of Biochemistry and Department of Biology, Carleton University, Ottawa, Canada K1S 5B6

1. Introduction

A capacity for long-term survival without oxygen is well developed among many invertebrate species as well as in selected ectothermic vertebrates. Anoxia tolerance has been particularly well studied in various species of marine molluscs, including both bivalves (e.g., mussels, clams, oysters) and gastropods (e.g., littorine snails, whelks). These can encounter low environmental oxygen as a result of multiple factors: (1) gill-breathing intertidal species are deprived of oxygen when the waters retreat with every low tide; (2) burrowing and benthic species can encounter anoxic bottom sediments; (3) high silt or toxin levels in the water as well as predator harassment can force shell valve closure, leading to substantial periods of "self-imposed" anoxia; (4) animals in small tidepools can be oxygen-limited when animal and plant respiration depletes oxygen supplies in the water; and (5) freeze-tolerant intertidal species face oxygen deprivation whenever their body fluids freeze (Truchot and Duhamel-Jouve, 1980; de Zwaan and Putzer, 1985; Grieshaber et al., 1994; Loomis, 1995). Life in the intertidal zone is particularly challenging since, in addition to the cyclic availability of oxygenated water (each tide cycle lasts 12.4 h), organisms can also be challenged with multiple other stresses including desiccation, changes in salinity, and changes in temperature,

sometimes including freezing; all can potentially change rapidly over the course of a single tidal cycle of immersion and emmersion (Bridges, 1994; Loomis, 1995). For this reason, various residents of the intertidal zone have been used extensively as model systems of stress tolerance, the most widely studied species being the sessile bivalve, the blue mussel *Mytilus edulis*. Littoral snails that graze on rocks in the high intertidal zone are also an excellent model system for studies of both anoxia tolerance and freeze tolerance. This chapter reviews recent advances in our understanding of the biochemistry and molecular biology of anoxia tolerance in marine molluscs. Our emphasis is on the molecular biology of the phenomenon, particularly our extensive recent work with the periwinkle snail, *Littorina littorea*, to analyze the role of gene expression in anoxia tolerance and the mechanisms that regulate anoxia-induced changes in transcription and translation.

2. Metabolic response to anoxia

2.1. Regulation of carbohydrate metabolism

Under aerobic conditions, organisms can make use of lipid, carbohydrate or amino acid fuels for respiration with considerable variation between species and between organs in the relative importance of different fuel types. Under anoxic conditions, however, carbohydrates become the primary substrate because the oxidation of hexose phosphates

---

*Corresponding author

(derived from glucose or glycogen) to pyruvate via glycolysis produces ATP in substrate-level phosphorylation events. Although the yield of ATP is low (2 or 3 moles ATP/mole glucose or glucosyl unit from glycogen, respectively) compared with that available from the complete oxidation to CO_2 and H_2O by the tricarboxylic acid cycle (36 or 38 moles ATP/mole, respectively), anoxia tolerant species have capitalized on this pathway with adaptations that maximize the length of time that fermentative metabolism can sustain survival. Among anoxia-tolerant molluscs, these adaptive strategies include: (1) large tissue stores of fermentable fuels (mainly glycogen but also selected amino acids); (2) coupling of glycolysis to additional substrate-level phosphorylation reactions to increase the ATP output per hexose phosphate; (3) production of alternative end products to lactic acid that are either volatile or less acidic so that cellular homeostasis is minimally perturbed by acid build-up during long-term anoxia; and (4) strong metabolic rate depression that greatly lowers the rate of ATP utilization by tissues to a rate that can be sustained over the long term by the ATP output of fermentation reactions alone. Thus, the Pasteur effect —a large increase in glycolytic rate when oxygen is limiting—is not seen in anoxia-tolerant species.

Anoxia-tolerant molluscs show a two-phase response to declining oxygen tension. As tissue oxygen is depleted (such as during shell valve closure or during emersion at low tide), organisms first enter a period of hypoxia. During this period, a graded increase in carbohydrate catabolism can occur that allows a compensatory increase in fermentative ATP output in order to maintain normal rates of ATP turnover. However, as hypoxia deepens, a critical low oxygen tension is exceeded and further attempts at compensation are abandoned in favour of the initiation of conservation strategies. In this phase of severe hypoxia or anoxia, the rates of ATP production and ATP utilization are strongly suppressed and net metabolic rate drops to below 10% of the corresponding aerobic metabolic rate at the same temperature (Storey and Storey, 1990). The critical pO_2 values that stimulate these transitions differ between species and contribute to the differential success of various species in hypoxic or polluted environments (de Zwaan et al., 1992). Metabolic rate depression greatly extends the time that a fixed reserve of internal fuels can support survival and many marine molluscs can survive days or weeks of anoxia exposure.

Metabolic rate depression is quantitatively the most important factor in anoxia survival. However, the use of modified pathways of fermentative metabolism substantially enhances the ATP output in anoxia and leads to the formation of non-acidic and/or volatile end products that are compatible with the maintenance of long-term homeostasis in the anoxic state. The initial response to anoxia is typically the coupled fermentation of glycogen and aspartate substrates to produce the end products alanine and succinate, respectively. Glycogen is catabolized to pyruvate via glycolysis and then, instead of reduction to lactate, the pyruvate undergoes a transamination reaction to form alanine using an amino group transferred from aspartate. The product of aspartate deamination, oxaloacetate, is reduced to malate (using the NADH that would otherwise have been used by the lactate dehydrogenase reaction). This malate is converted to fumarate and then succinate in mitochondrial reactions that constitute a partial reversal of the tricarboxylic acid cycle. Fumarate conversion to succinate generates ATP and the further conversion of succinate to propionate in some species is linked with additional ATP synthesis. As aspartate pools become depleted, a metabolic shift takes place that directs the output of glycolysis, phosphoenolpyruvate (PEP), into the synthesis of oxaloacetate. This is accomplished via inhibition of the enzyme pyruvate kinase (PK) so that PEP is carboxylated instead via the PEP carboxykinase (PEPCK) reaction to produce the 4-carbon intermediate, oxaloacetate, that then feeds into the reactions of succinate synthesis. Fermentation of glucose to succinate produces 4 ATP/mol of glucose whereas the glucose to propionate conversion produces 6 ATP/mol glucose, compared with only 2 ATP/mol when lactate is the product.

The tricarboxylic acid (TCA) cycle is coupled to electron transport via the electron-transferring enzyme complex succinate dehydrogenase (complex II of the electron transport chain; ETC), which reduces malate to succinate. An extensive review of the ETC of anaerobically functioning eukaryotes has been compiled by Tielens and Van Hellemond (1998). They state that, depending on whether the system is aerobic or anaerobic, the reducing equivalents of complex II are transferred to ubiquinone or rhodoquinone, respectively. Both compounds have been detected in marine intertidal molluscs, as would be expected since molluscs are subject to aerial exposure at regular intervals (Van Hellemond et al., 1995). During immersion, molluscs rely on aerobic energy metabolism via the Krebs cycle, with electrons being transferred from succinate to ubiquinone through complex II in the ETC. During emersion, molluscs function anaerobically, with electron transfer likely occurring through rhodoquinone to fumarate producing succinate (Tielens and Van Hellemond, 1998). Fumarate acts as the electron acceptor when oxygen is not available, producing succinate. Increased succinate levels during anaerobiosis allow mitochondria to remain metabolically active and intact until oxygen again becomes available, initiating respiration (Bacchiocchi and Principato, 2000).

2.2. *Effects of pH on cellular metabolism*

The precise relationship between pH and metabolic rate depression in molluscs has yet to be established and has been reviewed by a number of researchers (Storey, 1993; Guppy et al., 1994; Hand and Hardewig, 1996). Both intracellular and extracellular pH decrease during anaerobiosis in marine molluscs (Ellington, 1983; Walsh et al., 1984), but it has yet to be determined whether changes in pH play a role in initiating metabolic suppression. The change in pH during anoxia is typically a steady decline over a long time, sometimes extending to days, whereas the transition into the hypometabolic state during anaerobiosis occurs shortly after anoxia begins (Storey and Storey, 1990 and references therein). Since pH change takes place gradually, it is unlikely to play a role in signaling metabolic suppression during anaerobiosis. Furthermore, anoxia-tolerant molluscs use strategies to minimize the acid load during

anaerobiosis (accumulating neutral or volatile products, using shell bicarbonate for buffering) (de Zwaan, 1977; Storey and Storey, 1990), which argues against a signaling role for low pH. However, a moderate decrease in pH during anoxia probably helps to create a metabolic context that favors metabolic depression (Busa and Nuccitelli, 1984). For instance, a lower pH environment favours the catabolism of phosphoenolpyruvate (PEP) via the PEP carboxykinase (PEPCK) reaction rather than the pyruvate kinase (PK) reaction and, therefore, contributes to the diversion of glycolytic carbon into the reactions of succinate synthesis (Hochachka and Somero, 1984). A lower pH environment during anoxia may also facilitate other actions such as enzyme binding with subcellular structural elements, enzyme reaction rates, the relative activities of protein kinases versus protein phosphatases, and changing protein stability (Storey, 1988; Hand and Hardewig, 1996; Schmidt and Kamp, 1996; Sokolova et al., 2000).

2.3. *Controlling glycolytic flux*

Several mechanisms contribute to the anoxia-induced regulation of glycolytic rate in marine molluscs. Allosteric controls by metabolites affect specific enzyme loci during anoxia; for example, reduced levels of fructose-2,6-bisphosphate, a potent activator of 6-phosphofructo-1-kinase (PFK-1), and increased levels of the L-alanine, a strong inhibitor of pyruvate kinase (PK) contribute to anoxia-induced inhibition of these two enzymes. Reduced cellular pH in anoxia also favours PEP processing via PEPCK versus PK. Seasonality also affects the response to anoxia. The effects of anoxia exposure on the maximal activities of selected glycolytic enzymes were generally stronger in winter versus summer in oyster tissues (Greenway and Storey, 1999) and for PK this appeared to be due to seasonal changes in PK isoform present, the winter isoform showing much more pronounced changes in kinetic properties in response to anoxia than the summer form (Greenway and Storey, 2000). For example, winter PK in oyster gill showed a 4-fold increase in the K_m for PEP and a 7-fold increase in the K_a for the activator, fructose-1,6-bisphosphate after anoxia exposure (both changes that would reduce PK activity under anoxia) whereas these parameters were unaffected by anoxia exposure of summer animals (Greenway and Storey, 2000). Anoxia exposure had similar strong effects on the kinetic properties of PK from hepatopancreas of winter *L. littorea* (also including a strong increase in enzyme sensitivity to L-alanine inhibition) but again had little effect on the summer enzyme (Greenway and Storey, 2001). However, *L. littorea* muscle PK showed similar sensitivity to anoxia in both seasons (Greenway and Storey, 2001). Seasonal changes in isoform and in anoxia effects on enzymes may aid prolonged periods of cold-induced inactivity with closed valves or opercula and self-imposed hypoxia/anoxia that triggers strong metabolic rate depression.

Although the above mechanisms contribute to glycolytic control during anoxia, the primary

mechanism of overriding importance appears to be the reversible phosphorylation of enzymes catalyzed by the actions of protein kinases and protein phosphatases (Storey, 1992). Anoxia-induced covalent modification of enzymes strongly reduces maximal activities, alters kinetic properties and strongly suppresses overall glycolytic rate. The major kinases involved in the reversible control of enzymes of intermediary metabolism are cAMP-dependent protein kinase (PKA), cGMP-dependent protein kinase (PKG), and Ca^{2+} and phospholipid-dependent protein kinase (PKC), their actions mediated by their respective second messengers, adenosine 3′,5′-cyclic monophosphate (cAMP), guanosine 3′,5′-cyclic phosphate (cGMP), and Ca^{2+} and phospholipids (Fig. 3.1). Several glycolytic enzymes in marine molluscs are subject to anoxia-induced phosphorylation that suppresses their activity. These include glycogen phosphorylase (GP), PFK-1, 6-phosphofructo-2-

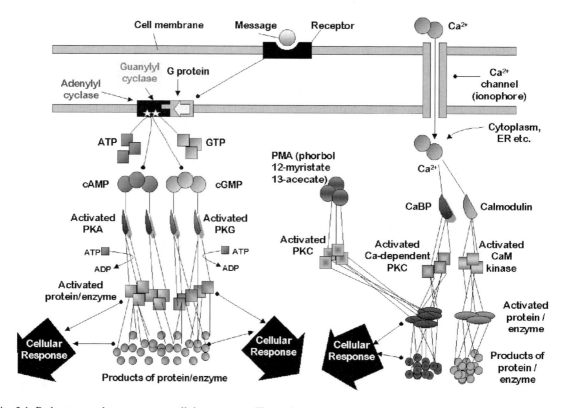

Fig. 3.1. Pathways used to generate a cellular response. The major mechanism for regulating protein phosphorylation is via protein kinases, which are mediated by their respective second messengers. The general pathway involves a "message" that is sensed by a receptor and this signal is transmitted to the cytoplasm in the form of a second messenger. The second messenger activates the protein kinase, which triggers a cellular response.

kinase (PFK-2), and PK (Storey, 1993). Control over GP regulates the availability of the primary fuel, glycogen, which is present in high quantities in all tissues of anoxia-tolerant marine molluscs. PFK-1 activity is a major determinant of glycolytic flux and inhibitory control of PFK-2 (that synthesizes fructose-2,6-P_2, a potent activator of PFK-1) relieves PFK-1 from responding to anabolic demands during anoxia by suppressing levels of the activator (F2,6P_2) that normally regulates the anabolic use of carbohydrate reserves. As mentioned above, regulation of PK controls the catabolism of PEP, directing glycolytic flux away from aerobic routes (via PK and into mitochondrial oxidation reactions) and into anaerobic routes (via PEPCK and into the synthesis of succinate or propionate) (de Zwaan, 1977). In molluscs, anoxia-induced PK phosphorylation appears to be stimulated by cGMP, implying a key role for PKG in the aerobic/anoxic transition (Brooks and Storey, 1990).

Anoxia-induced phosphorylation produces major changes in PK kinetic properties that converts it to a much less active enzyme. For example, anoxia-induced phosphorylation of PK from radular retractor muscle of *Busycon canaliculatum* results in a 12-fold increase in the K_m for PEP, a 24-fold increase in the K_a for fructose-1,6-bis-phosphate, and a 490-fold decrease in the K_i for alanine (Plaxton and Storey, 1984). Similarly, phosphorylation of PFK-1 alters enzyme kinetic properties to convert the enzyme to a less active form and in the anterior byssus retractor muscle (ABRM) of *M. edulis* this again appears to be mediated via cGMP (Michaelidis and Storey, 1990). Interestingly, an increase in cGMP occurs in the ABRM in response to acetylcholine, which is the neurotransmitter that stimulates the contraction of this catch muscle to close the shell (Kohler and Lindl, 1980). For bivalve molluscs, shell valve closure is the proximal event that soon leads to internal oxygen depletion and necessitates the induction of anaerobic adaptations. Further assessment of PKG control of PK found that it is indirect with PKG apparently stimulating a specific PK-kinase that, in turn, phosphorylates and inactivates PK (Brooks and Storey, 1991). The potential

for wide-ranging effects by cGMP and PKG in the control of other metabolic responses to anoxia is high but remains an unexplored area.

3. Macromolecular synthesis

3.1. Economics of energy conservation

Until recently, the majority of research on molluscan anaerobiosis has focused on the regulation of intermediary energy metabolism and very little was known about the molecular biology of anoxia tolerance. Of the variety of biochemical mechanisms involved in molluscan anaerobiosis (Grieshaber et al., 1994), the capacity for metabolic rate suppression stands out as crucial (Storey, 1993; Guppy and Withers, 1999). Metabolic rate is reduced to a level that can be sustained over the long term by fermentative ATP production alone and in species with well-developed anoxia tolerance, anoxic metabolic rate is only 1–10% of the normoxic value at the same temperature (de Zwaan and Putzer, 1985). Metabolic rate is a measure of net ATP turnover and it is obvious, then, that to maintain long-term homeostasis under anoxic conditions that the rates of all ATP utilizing reactions in cells must be strongly suppressed to a level that matches the anoxic rate of ATP production. A hierarchy of energy-consuming processes exists in cells, with pathways of macromolecule biosynthesis (protein/RNA/DNA synthesis) being the most sensitive to ATP supply (Buttgereit and Brand, 1995). It can be proposed, therefore, that these biosynthetic pathways are obvious targets for strong suppression during molluscan anaerobiosis and, indeed, recent research (described below) is confirming this.

3.2. mRNA and protein synthesis

Stress-induced suppression of macromolecular synthesis has been examined recently in several invertebrate systems (Hand, 1998; Marsh et al., 2001; Robertson et al., 2001), including one anoxia-tolerant marine mollusc, the periwinkle *L. littorea* (Larade and Storey, 2002a). In *L. littorea*

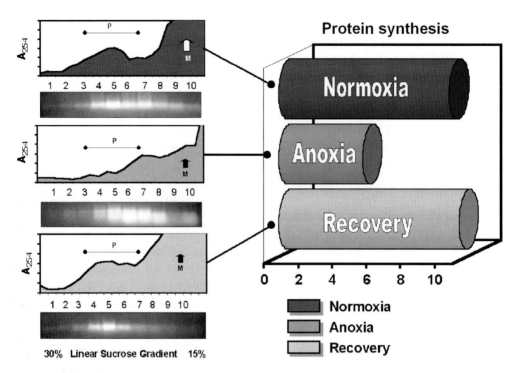

Fig. 3.2. Examination of translation and protein synthesis in the marine mollusc *L. littorea*. Polysome profiles of hepatopancreas extracts from normoxic, anoxic (72 h) and aerobic recovered (72 h anoxia followed by 3 h recovery) snails are shown on the left side of the figure. Post-mitochondrial supernatants were centrifuged on 15–30% continuous sucrose density gradients and fractions were collected with RNA detection by two methods: absorbance monitored at 254 nm (upper panels) and (b) agarose gel electrophoresis of samples from each fraction followed by ethidium bromide staining (lower panels). Polysome region [P]; Monosome peak [M]. The right side of the figure shows the relative rates of protein synthesis in *L. Littorea* hepatopancreas measured as [^{3}H] leucine incorporation into trichloroacetic acid-precipitable material in cell-free-lysates sampled from normoxic, anoxic (0.5 h), and aerobic recovered (12 h) animals. Modified from Larade and Storey (2002b).

hepatopancreas, a ~50% reduction in the rate of protein synthesis (measured as ^{3}H-leucine incorporation into protein) was evident within 30 minutes when snails were placed under an N$_2$ gas atmosphere and was sustained over 48 h of anoxia exposure (Larade and Storey, 2002a). Following a return to oxygenated conditions, protein synthesis was restored to control values within hours (Fig. 3.2). Complementing these findings, *L. littorea* hepatopancreas also showed a dramatic reduction in nuclear transcription rates during anoxia. The rate of mRNA elongation (measured as ^{32}P-UTP incorporation into nascent mRNAs by isolated nuclei) dropped to less than one-third of the normoxic rate (Larade and Storey, unpublished data). Hence, the mRNA substrate available for translation may be reduced under anoxic conditions.

Gene expression and protein synthesis are costly processes that require many resources, not only large amounts of ATP for energy but supplies of nucleotide and amino acid substrates and sustained amounts of the transcriptional and translational machinery (enzymes, ribosomes, tRNAs). ATP limitation during anoxia is probably the first and strongest reason for the suppression of nuclear transcription rates but other components of the transcriptional and translational machinery may also become limiting, particularly if anoxia exposure is prolonged. It is understandable, therefore, that cells and organisms would conserve their resources during anaerobiosis by strongly suppressing the rates of transcription and translation of most genes. Against a background of generally reduced gene transcription, those genes whose transcription is specifically

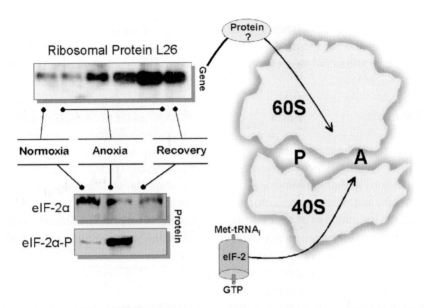

Fig. 3.3. Analysis of ribosomal components during anoxia exposure in *L. littorea*. The top left panel shows an increase in the expression of ribosomal protein L26 mRNA in hepatopancreas over the course of anoxia exposure (4 lanes representing 12, 24, 96 and 120 h anoxia, respectively) and aerobic recovery (1 h recovery after 120 h anoxia). The bottom left panel shows changes in the phosphorylation of eIF-2a after 24 h anoxia or 1 h recovery after anoxia. Both L26 and eIF-2 play roles in efficient translation; the right panel is a schematic diagram that displays the location of both components on the intact ribosome. L26 is located at the subunit interface and functions during the transfer from the A (aminoacyl) site to the P (peptidyl) site, while eIF-2, which performs a role in initiation, is associated with the A site. Modified from Larade et al. (2001) and Larade and Storey (2002).

up-regulated during anoxia stand out as genes whose protein products are likely to play very important roles in anoxia. We explore this idea more in Sections 4.5 and 5 with new research on anoxia-induced gene expression.

3.3. Mechanisms of translational control

Protein synthesis is controlled by the efficiency of the translational apparatus, which is determined by the factors that influence translation initiation (Kaufman, 1994). Initiation of translation involves consecutive recruitment of the small and large subunits of ribosomes to specific mRNAs, with the formation of an active ribosome at the initiation site. The predominant mechanism for control of protein synthesis appears to be reversible phosphorylation, under the control of selected protein kinases and protein phosphatases. The targets of covalent modification in this case are translational components, specifically the initiation and elongation factors (Hershey, 1991). The eukaryotic

Initiation Factor 2 alpha (eIF-2α), which promotes the binding of initiator tRNA to the 40S ribosomal subunit, is an example of a factor that is regulated in this manner. Phosphorylation of eIF-2α is correlated with inhibition of protein synthesis in a range of eukaryotes (Rhoads, 1993) and our recent studies with *L. littorea* concur (Larade and Storey, 2002a). In *L. littorea* hepatopancreas, the total content of eIF-2α was constant in three groups: aerobic control snails, snails exposed to 24 h under a N$_2$ gas atmosphere, and snails given 1 h aerated recovery after 24 h anoxia exposure. However, in response to anoxia exposure, the content of phosphorylated eIF-2α rose ~15-fold compared with aerobic controls (Fig. 3.3) (Larade and Storey, 2002a). This was reversed rapidly during aerobic recovery with phosphoprotein content reduced again to control levels within 1 h post-anoxia. These data support the concept that anoxia exposure in *L. littorea*, and likely also in other anoxia-tolerant molluscs, stimulates a substantial suppression of protein synthesis, a proposal that is further supported by the direct measurements of the protein biosynthesis rates (discussed earlier) and by the changes in the distribution of ribosomes between polysomes versus monosomes (discussed below).

3.4. Polysome analysis

Protein translation requires the sequential accumulation of specific aminoacyl-tRNAs by active ribosomes. This recruits corresponding amino acids to the peptidyl transferase site where protein elongation occurs through the formation of peptide bonds. The ribosome travels down the message, liberating previously occupied codons in the process and allowing additional ribosomes to initiate translation on the 5′ end of the message. Depending on the length of a particular mRNA, transcripts have the capacity to retain several ribosomes, creating a structure known as a polyribosome or polysome. When not translationally active, these polysome aggregates dissociate again into monosomes. In general, the activity state of the protein-synthesizing machinery in a cell/tissue can be inferred from the state of ribosomal assembly. Hence, an effective way of gauging the effects of a stress on cellular protein synthesis is to assess the relative proportions of polysomes versus monosomes in control versus stressed states, these two ribosomal states being readily separable on a sucrose gradient (Surks and Berkowitz, 1971). Indeed, several recent studies have documented a strong reduction in polysome content and increase in monosomes in other situations of facultative metabolic rate depression (e.g., in hibernating mammals) (Frerichs et al., 1998; Martin et al., 2000; Hittel and Storey, 2002).

In an aerobic environment, mRNAs are generally associated with polysomes indicating active protein synthesis. When ribosome distribution patterns were examined in extracts of *L. littorea* hepatopancreas, a high proportion of ribosomes were found associated with polysomes in extracts from normoxic control snails (Fig. 3.2). This is consistent with active translation in the normoxic state. After anoxia exposure, however, there was little evidence of polysomes remaining in hepatopancreas and most of the ribosomal RNA occurred

in the monosome peak. This indicates a significant suppression of the activity of the protein synthesizing machinery during anoxia, consistent with the other lines of evidence discussed above. Such a decrease in polysome size (ie. the number of ribosomes attached to a mRNA transcript) could result from either a decrease in initiation or stimulation of elongation and termination (discussed in Mathews et al., 1996). To determine which is the trigger for polysome breakdown, the overall rate of protein synthesis must also be examined. Suppression of protein synthesis, combined with a decrease in polysome size, is indicative of blocked initiation. The observed decrease in the polysome population and rate of protein synthesis in *L. littorea* during anoxia indicates regulation of translation at the level of initiation. This is consistent with observations by Guppy et al. (1994) who suggested that regulation of protein synthesis is at the level of initiation for systems where the rate of protein synthesis is down-and up-regulated in a global manner. When 24 h anoxic snails were returned to aerobic conditions, a strong shift back towards polysomes was observed, with the polysome peak appearing similar to that in normoxic control profiles (Fig. 3.2). This corresponds well with the results from the *in vitro* protein synthesis experiments, described earlier, which showed that the rate of protein synthesis returned to normoxic control values during aerobic recovery.

3.5. Ribosomal proteins

Ribosomes are composed of four types of ribosomal RNAs and over 80 ribosomal proteins. The expression of genes encoding ribosomal proteins and the synthesis of the proteins is tightly regulated and coordinated and has been linked with various physiological conditions suggesting an active role for ribosomal proteins during adaptation and cellular responses to stress (Teem et al., 1984; Mager, 1988; Li et al., 1999; Meyuhas, 2000; Larade et al., 2001; Mitsumoto et al., 2002). Many ribosomal proteins play a major role in the activity and stability of ribosomes, but few have been examined in marine invertebrates. Those that have been studied were isolated from marine invertebrates in various developmental stages (Rhodes and Van Beneden, 1997; Watanabe, 1998; Snyder, 1999) or stresses (Larade et al., 2001). Of particular interest are those ribosomal proteins located at the interface of the large and small subunit, since this site represents the active center of the ribosome. Proteins found here often play a role in stabilizing the mature ribosome through interactions with other ribosomal proteins on the adjacent ribosomal subunit, thereby allowing translation to occur. Ribosomal proteins located at the peptidyl transferase center, specifically ribosomal protein L26 (Villarreal and Lee, 1998; Marzouki et al., 1990), can regulate initiation of translation via interaction with elongation factor-2 (Nygard et al., 1987). Lee and Horowitz (1992) suggested that L26 is likely to be involved in subunit interactions, based on the observation that it undergoes structural re-arrangement as the ribosomal subunits associate. Other authors have implicated L26 as a protein involved in forming the region that binds EF-2 to the 60S ribosomal subunit preceding translocation of peptidyl-tRNA from the A to the P site during peptide bond formation (Yeh et al., 1986; Nygard et al., 1987). In either role, L26 is central to the formation and function of the intact ribosome. Analysis of L26 gene expression during anoxia in *L. littorea* hepatopancreas revealed a steady increase in transcript levels over the course of anoxia exposure, reaching about five-fold higher than in controls and remaining high during the early hours of aerobic recovery (Fig. 3.3). Similarly, L26 transcripts in foot muscle rose by about 3-fold over the course of anoxia (Larade et al., 2001). These increases in L26 transcript levels suggest that a similar increase in the synthesis of L26 protein may occur during anoxia and/or recovery.

It has been demonstrated that during stress, cells are able to suppress the biosynthesis of their translational apparatus, regulating gene expression at the translational level (for review, see Meyuhas, 2000). Therefore, ribosomal protein transcripts may increase during anoxic exposure and be sequestered or stored in some manner, in preparation for normoxic recovery, which for *L. littorea* in a natural environment would occur during re-immersion of the snails in water at the end of a low

tide cycle. High levels of transcript would then be in position to be translated once oxygen was re-introduced into the system. Translation of L26 when aerobic conditions are re-established, likely in concert with various other ribosomal proteins because these are coordinately expressed, will provide a quick increase in the capacity of the translational apparatus to cope with a demand for increased protein synthesis upon the re-introduction of oxygen.

3.6. *Maintaining translatable mRNA pools*

In addition to its role as a translatable message, mRNA also appears to function as a regulator of protein translation. It accomplishes this via specific nucleotide sequences (and in some cases secondary structures) that are recognized by either general or specific factors. To be translated, mRNA must be exported from the nucleus, a process generally mediated by specific mRNA binding factors known as heterogeneous nuclear ribonucleoproteins (hnRNPs). Specific hnRNPs bind to mRNA and help to export it from the nucleus and, in some cases, transport mRNA to its final destination. The majority of mRNAs in a cell are usually present in mRNP (messenger ribonucleoprotein) complexes, often localized as a stress granule, representing a stable intermediary for untranslated messages consisting of a core of both mRNA and RNA-binding proteins such as hnRNPs. Specific RNA binding domains, including those of hnRNPs, are described in detail by Derrigo et al. (2000). In most cases, hnRNPs bind to pre-mRNAs as they are synthesized in the nucleus, although some associate later as a consequence of mRNA processing reactions (Dreyfuss et al., 2002). It has been reported that hnRNPs regulate mRNA localization, translation, and turnover (reviewed by van der Houven van Oordt et al., 2000). Kedersha et al. (1999) report that additional RNA-binding proteins, TIA-1 and TIAR, associate with stress granules in the cytoplasm, an assembly triggered by the phosphorylation of eIF-2a. As discussed in Section 3.3, this leads to inhibition of translation at the initiation stage; active transcripts continue to be processed until all ribosomes "fall off", resulting in a corresponding decrease in polysome number. Studies in plant cells have demonstrated the existence of stress granules that store untranslated mRNAs during induced stress (Nover et al., 1983; Scharf et al., 1998), while supporting the hypothesis that stress-induced mRNAs are kept liberated (Collier et al., 1988; Nover et al., 1989).

Recent studies have supported the idea that untranslated mRNA transcripts (sometimes described as latent mRNA) are maintained in cells when metabolic rate is depressed. These mRNA transcripts are sequestered into mRNP complexes, where they are hidden from the translational apparatus (Spirin, 1996; Ruan et al., 1997) and remain untranslated until the stress is removed and cells return to normal metabolic function. The mRNA transcripts involved may include those produced prior to metabolic depression as well as some transcript types that are up-regulated by the stress. Indeed, Laine et al. (1994) suggest that mRNA transcription under stress conditions can be an anticipatory response that is important for the ultimate recovery after stress is removed. Thus, transcripts can continue to accumulate during the period of metabolic arrest but their translation is prevented by sequestering them into mRNPs, which lowers the number of active polysomes. When stress conditions are reversed (e.g., a return to aerobic life), these stored mRNAs are immediately available for translation so synthesis of their protein products can be very rapidly reinstated. This model was verified for ribosomal protein S25 by Adilakshmi and Laine (2002).

4. Gene expression

4.1. Anoxia-induced gene expression

Most energy-consuming processes are strongly suppressed in organisms that show long-term endurance of oxygen deprivation and this includes general protein synthesis, as described above. However, selected specific RNA transcripts are up-regulated by anoxia exposure, presumably providing for the synthesis of specific protein products that enhance survival. Recent work from our laboratory has described anoxia-induced up-regulation

of several genes in *L. littorea* (Larade et al., 2001; Larade and Storey, 2002b; Larade and Storey, unpublished data) and also in organs of anoxia-tolerant turtles (Cai and Storey, 1996; Willmore et al., 2001).

Gene expression responses to hypoxia (low oxygen) have been extensively explored for the last ten years or more and are typically mediated via the hypoxia-inducible transcription factor, HIF-1. Hypoxia-responsive genes are typically those whose gene products mediate compensatory responses to overcome low oxygen stress. For example, HIF-1 mediates the increased synthesis of glycolytic enzymes to enhance the capacity for fermentative ATP production while at the same time stimulating red blood cell production (via induction of erythropoietin) and the proliferation of capillaries (via induction of vascular endothelial growth factor) (Semenza, 2000).

Gene expression responses to anoxia by anoxia-tolerant animals are different. They are not compensatory in nature but rather are involved in altering metabolism to ensure long-term survival in the absence of oxygen. Furthermore, the transcripts of some anoxia-induced genes may be accumulated during anoxia but not immediately translated. Rather they may be stored to allow translation to be rapidly re-initiated when oxygenated conditions return.

4.2. Identification of differentially expressed genes

To identify genes that are up-regulated during anoxia exposure, a number of options are available. Numerous methods, including differential display PCR, cDNA libraries screening (normalized or subtracted), Serial Analysis of Gene Expression (SAGE), and more recently cDNA array technology, have been used to screen for differentially expressed genes, with each method offering advantages and disadvantages. Due to the large number of methods and even larger range of applications, this review will not discuss all of the available methods, since usage often depends on the project of interest or the preferences of the particular researcher. However, we will highlight our recent experiences with cDNA array technology, in particular the value of heterologous screening as a tool for the initial screening and putative identification of stress-specific genes up-regulated in unusual animal model systems.

Array technology provides the opportunity to simultaneously screen the expression profiles of hundreds of genes, representing multiple protein families and many kinds of cellular functions. The data provides broad insight into the coordinated genetic responses by cells to specific stresses, identifying responses that may or may not have been anticipated. At present, commercially-available arrays have been produced for several of the major animal model systems (e.g., human, rat, *Drosophila, Caenorhabditis elegans*) and many labs are using these for homologous screening to identify cell and tissue specific gene expression profiles associated with multiple metabolic events (e.g., growth, development, stress, response to hormones, carcinogenesis). Our lab has become interested in the potential for using arrays in heterologous or "cross-species" screening and we have found excellent cross-reactivity that has allowed us to use human and rat cDNA arrays to effectively assess

hibernation-specific gene expression in ground squirrels and bats (Hittel and Storey, 2001; Eddy and Storey, 2002). Success with cross-species screening between different mammal species is perhaps not unexpected as there is upwards of 80% sequence identity between homologous genes in placental mammals. Use of mammalian cDNA arrays for screening a molluscan system has a much greater chance of failure due to the probability that sequence identity between mammalian and molluscan gene homologues could be very low. Nonetheless, for those genes that do show significant cross-reaction, the potential benefit is enormous, allowing a one-step survey of the responses to anoxia stress by hundreds of genes, and the potential identification of many new genes that respond to anoxia, including many whose role in anoxia survival may never before have been suspected.

Human 19K cDNA arrays were screened with cDNA probe made from mRNA isolated from hepatopancreas of control versus anoxic (12 h) *L. littorea*. In general, cross-reactivity was low; *L. littorea* cDNA showed significant reactivity with only 18.35% of the human cDNA sequences (Fig. 3.4). This can be compared with the ~85% cross-

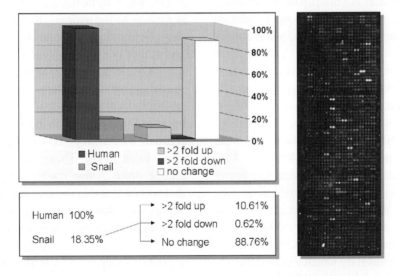

Fig. 3.4. Cross-reactivity of genes on human 19k cDNA arrays screened with mRNA isolated from normoxic and anoxic *L. littorea* hepatopancreas. The left side of the graph shows the percentage of *L. littorea* cDNA clones that showed significant reactivity with the human cDNA sequences on the array. The right side of the graph evaluates changes in mRNA transcript levels in response to anoxia exposure by those snail genes that showed cross-reactivity with the human array. Levels of mRNA transcripts were compared in aerobic control versus 12 h anoxic snails to identify genes that showed greater than two-fold up-regulation during anoxia, greater than two-fold down-regulation in anoxia, or no change.

reactivity seen when the same arrays were screened with cDNA made from mRNA of hibernating bats (Eddy and Storey, 2002). This points out the obvious limitations of heterologous probing and is perhaps not surprising given the phylogenetic distance between gastropods and humans. Nonetheless, of those genes that did show significant cross-reaction due to high sequence identity, a number of very interesting results were observed. Despite the metabolic rate depression of the anoxic state, nearly 11% of the cross-reacting genes were apparently up-regulated by two-fold or more during anoxia. These included selected protein phosphatases and kinases, MAP kinase-interacting factors, translation factors, antioxidant enzymes, and nuclear receptors (Larade and Storey, unpublished data). Virtually all of these represented proteins that have never before been implicated in anoxia adaptation and, hence, we now have a much broader view of the potential gene expression responses that may play significant roles in anoxia survival. Hence, this use of heterologous probing of cDNA arrays has provided multiple new avenues of research to pursue. Although all of these candidates must be confirmed as anoxia up-regulated in mollusc tissues via RT-PCR or Northern blotting, and some may not hold up to this scrutiny, this heterologous screening technique is sure to lead to novel discoveries about molluscan anoxia tolerance.

An interesting outcome of the array analysis was the distribution of responses by snail genes to anoxia exposure. Whereas 10.6% of those that cross-reacted were up-regulated by anoxia, only 0.6% of genes showed significant down-regulation (by more than two-fold) in anoxia and mRNA levels of all the rest were unchanged under anoxia exposure. Recalling that the rate of protein synthesis is suppressed by about 50% during anoxia in *L. littorea* hepatopancreas (Larade and Storey, 2002a), it is obvious that a suppression of mRNA transcript levels is not the mechanism by which this is achieved. More likely, as discussed earlier, mRNA transcripts are retained throughout the anoxic episode but sequestered in translationally inactive pools due to the break-up of active polysomes.

Current commercially-available cDNA arrays are made from the genomes of anoxia-intolerant organisms (e.g., humans) and, therefore, one of the key limitations in their use with anoxia-tolerant organisms is that they may not be able to detect novel genes that are specific for anoxia survival. This problem might be remedied by using a *C. elegans* array or one produced from another invertebrate species, but until that time, differential screening of a cDNA library provides the best way to scan for genes that are differentially expressed in response to anoxia or any other stress. Indeed, using a cDNA library constructed from hepatopancreas of anoxic *L. littorea*, we have identified a number of genes that are up-regulated during anoxia in the snails, including both identifiable genes and novel genes.

Genes that are up-regulated during anoxia exposure in *L. littorina* include ribosomal protein L26 (Larade et al., 2001), ferritin heavy chain, cytochrome c oxidase subunit II (COII), granulin/epithelin, a novel gene that we have named *kvn* (Larade and Storey, 2002b), and various unidentified genes (Larade and Storey, unpublished data). It is not difficult to draw a parallel for some of these genes between their known roles in other organisms and their possible roles in anoxia tolerance. For example, ferritin sequesters iron and is actively regulated during hypoxia in many organisms. COII plays an integral role in the electron transport chain and may play a role in oxygen signaling. However, the rationale for the up-regulation of granulin (a growth factor) is a mystery at present. Other clones encoded novel proteins that could not be identified by sequence searches in BLASTX or tBLASTX (Larade and Storey, 2002b; Larade and Storey, unpublished data). These may encode a suite of unique proteins that confer anoxia tolerance and that are not found in the genomes of anoxia intolerant species. Novel stress-specific genes have been identified in a range of stress-tolerant invertebrates (Bogdanov et al., 1994; Ma et al., 1999; Balaban et al., 2001; Goto, 2001; Larade and Storey, 2002b) and often prove to perform unique functions. In this case, it is likely that genes transcribed during anoxia exposure help the organism to survive without oxygen, or prepare it for the inevitable return of oxygen into the system.

4.3. Pharmacology

L. littorea has proven to be a suitable organism for performing organ culture experiments, allowing tissues to be maintained *in vitro* for short periods of time in order to analyze influences on anoxia-induced gene expression. For example, this analysis was applied to evaluate the influences of various protein kinases on the expression of the ribosomal protein, L26, that is anoxia-induced in *L. littorea* hepatopancreas and foot muscle (Larade et al., 2001). Hepatopancreas explants were incubated *in vitro* under aerobic conditions with various second messengers including dibutyryl cAMP, dibutyryl cGMP, calcium ionophore (A23187), and phorbol 12-myristate 13-acecate (PMA), second messengers that should stimulate protein kinases A, G, B and C, respectively (Fig. 3.1). The expression of several genes (each up-regulated by anoxia exposure *in vivo* in *L. littorea*) was assessed after tissues were exposed to each of the second messengers *in vitro* for 2 h incubations. For comparison, the effects of *in vitro* anoxia (incubation in N_2 bubbled medium for 12 h) and *in vitro* freezing (at $-7°C$ for 12 h; *L. littorea* is a freeze-tolerant species) were tested. Transcript levels of all genes up-regulated by anoxia exposure *in vivo* in *L. littorea* also increased under anoxia exposure *in vitro,* and most also responded to freezing, an ischemic stress that imposes an oxygen limitation and a strong osmotic stress on tissues. L26 transcript levels were strongly induced by anoxia exposure *in vitro* and also by incubation with db-cGMP but did not respond to incubation with cAMP, PMA or Ca^{2+} ionophore/Ca^{2+} (Larade et al., 2001). A novel gene of unknown function, *kvn*, was also up-regulated by anoxia and freezing exposure *in vitro* and this was mimicked by incubation with db-cGMP (Larade and Storey, 2002b). Indeed, all of the *L. littorea* anoxia-responsive genes tested were up-regulated by cGMP but showed little or no response to the other second messenger treatments (Larade and Storey, unpublished results). This implicates a cGMP-mediated signaling cascade in the gene expression response to anoxia in this marine mollusc. Little can be concluded at this time about other elements of the

signal transduction pathway involved in the up-regulation of these genes but additional evidence for the role of a cGMP-mediated process is discussed in Section 5.3 below.

5. Triggering the anoxic response

5.1. Oxygen sensing

A topic of interest to many labs is the specific signal that initiates and/or modulates the metabolic responses to hypoxic or anoxic conditions. One obvious signal is the decrease in pO_2. The level of available oxygen is gauged by an as yet unknown mechanism, with organisms, cells, and pathways responding accordingly. The hunt for a cellular oxygen sensor encompasses a wide range of approaches and model systems that is beyond the scope of this review. Bunn and Poyton (1996) pieced together an exhaustive summary of the research in this area. The most recent research suggests that the oxygen sensor is a prolyl hydroxylase enzyme, which oxidatively modifies (hydroxylates) a highly conserved proline residue found in the oxygen-dependent degradation domain of HIF-1, the central transcription factor involved in the regulation of gene transcription by oxygen (Ivan et al., 2001; Jaakkola et al., 2001). Whether or not this prolyl hydroxylase sensor is universal remains to be determined, although it is likely to be conserved across most phyla (Zhu and Bunn, 2001). By contrast, the secondary target(s) of the oxygen sensor, HIF-1 or others, are most likely different between hypoxia-sensitive cells/tissues and anoxia-tolerant ones. HIF-1 is well known in hypoxia-sensitive mammalian systems for its actions in triggering various compensatory responses that increase oxygen delivery to tissues and improve their glycolytic capacity. Anoxia-tolerate animals do none of these things but instead they initiate a series of actions that suppress metabolic rate and reversibly shut down multiple energy-expensive metabolic processes for the duration of the anoxic excursion. It is likely, therefore, that a transcription factor different to HIF-1 is involved in mediating any necessary gene expression responses to oxygen deprivation in anoxia-tolerant species. Indeed, using antibodies to mammalian HIF-1, we have been unable to detect HIF-1 in either anoxia-tolerant vertebrates (freshwater turtles) or in invertebrates (*L. littorea*) (Storey, unpublished data).

5.2. *Second messengers*

So how exactly is the low oxygen signal mediated in molluscs? The parallel responses of tissues to anoxia exposure *in vivo* vs. *in vitro* (both metabolic and gene responses) argues against hormonal or nervous mediation of signals (Storey, 1993; Larade et al., 2001). It is probable, therefore, that each cell detects the low oxygen signal by itself and from this triggers one or more intracellular signal transduction cascades. These, in turn, initiate adaptive responses that suppress metabolic rate, redirect carbohydrate flux into fermentative pathways, slow overall protein biosynthesis, and initiate the expression of selected genes. A candidate for a primary role in anoxia-responsive signal transduction is cyclic guanosine 3′,5′ monophosphate (cGMP) and its protein kinase (cGMP-dependent protein kinase; PKG). PKG is now known to mediate both the anoxia-induced phosphorylation of selected enzymes (e.g., pyruvate kinase) involved in carbohydrate fermentation in many molluscan species (Brooks and Storey, 1990, 1997) and the expression of anoxia-responsive genes (Larade et al., 2001; Larade and Storey, 2002b). The PKG I pathway has also been implicated in the control of gene expression of various promoter response elements (Gudi et al., 2000).

5.3. *The role of cGMP*

The accumulation of cGMP, via stimulation of guanylyl cyclases, regulates complex signaling cascades through downstream effectors (Fig. 3.1). Specific guanylyl cyclases, which are activated through the actions of coupled receptors and/or associated cofactors, convert GTP to cGMP, which is presumed to propagate the anoxic signal. It has been demonstrated that cGMP activates various protein kinases, directly gates specific ion channels, and alters intracellular concentrations of cyclic nucleotides through regulation of phosphodiesterases (for a recent review, see Lucas et al., 2000). A well-known activator of guanylyl cyclases, the diffusible signal molecule nitric oxide (NO) exerts its effects through direct binding, association or interaction with target proteins (Stamler et al., 1992). Little is known about the exact mode of action for NO and NO-related metabolites. In molluscs, intracellular cGMP levels have been shown to increase in the nervous system of various species due to activation of soluble guanylyl cyclase by NO and nitric oxide synthase (NOS) activity has been confirmed in a number of gastropod species, with putative NOS activity localized in ~30 molluscan genera (Moroz, 2000). It is interesting to note that the reaction that produces NO, that is catalyzed by the NOS family of enzymes, requires molecular oxygen (Griffin and Stuehr, 1995). Recent studies have shown that NO is involved in low oxygen signaling in *Drosophila* (Wingrove and O'Farrell, 1999) and this, coupled with demonstrations of cGMP involvement in anoxia-induced protein phosphorylation and gene responses by anoxia-tolerant molluscs (Brooks and Storey, 1997; Larade et al., 2001; Larade and Storey, 2002b) suggests that the nitric oxide/cGMP signaling pathway may be centrally involved in the response to oxygen deprivation in anoxia-tolerant species. Preliminary results are promising, but the mechanism by which cGMP up-regulates genes in molluscs remains unknown at present. A number of questions remain unanswered: How does cGMP operate? Is NO involved? Are cGMP protein kinases involved? Are transcription factors involved? If so, which ones? To date, however, there is little information on the intracellular levels of cGMP in marine molluscs and how these respond to anoxia (Higgins and Greenberg, 1974; Kohler and Lindl, 1980; Holwerda et al., 1981), so much remains to be investigated.

6. **Transcription factors**

Different model systems are now being explored using the wide variety of commercial antibodies available, in the hopes of determining which protein kinases mediate adaptive responses to environmental stress. The initial step in a signaling cascade generally involves the activation of a target protein that has either kinase activity or activates protein kinases in the cytoplasm. This signal then travels to the nucleus to activate transcription factors that regulate gene expression. It has been previously reported that environmental stresses activate MAP kinases (Raingeaud et al., 1995; Karin and Hunter, 1995), which function primarily to regulate gene expression. This suggests that links may exist between the actions of these kinases and the stress-induced expression of selected genes, including those that are responsive to anoxia signals.

One class of MAPKs is the stress-activated protein kinases (SAPKs), which includes the p38 MAPK. These MAPKs are regulated by tyrosine and threonine phosphorylation mediated by MKKs (MAP Kinase Kinases), which are, in turn, phosphorylated and activated by specific MKKKs (MKK Kinases). This cascade has been outlined in numerous review articles (Cohen, 1997a; Martin-Blanco, 2000). Target substrates of p38 MAPK include various transcription factors (Cohen, 1997b) and protein kinases, such as the serine/threonine specific kinase MAPKAP2, which phosphorylates the small heat shock protein HSP27 (Rouse et al., 1994), and MNK (Fukunaga and Hunter, 1997), which phosphorylates the translation initiation factor eIF-4E (Waskiewicz et al., 1997; Pyronnet et al., 1999). Canesi et al. (2000) have suggested a relationship between cellular redox balance and tyrosine kinase-mediated cell signaling in molluscs, specifically involving MAPK activation. Such a relationship has been demonstrated in mammalian cells, involving the mitogen and stress-activated protein kinase, MSK-1. This kinase is activated by oxidative stress and its effects are mediated via the p38 and ERK pathways (Deak et al., 1998). Although it has not yet been proven, it appears that NO and cGMP may also play a role in transcriptional activation in these systems. Browning et al. (2000) found that nitric oxide activated p38 MAPK via PKG, whereas Gudi et al. (2000) demonstrated that the cAMP-response element binding protein (CREB) is phosphorylated by PKG both *in vitro* and *in vivo*, when PKG is activated by cGMP or by NO. This data must be interpreted cautiously since the response of any given pathway can be highly tissue- or cell-specific.

p38 MAPK signaling pathways, known to play a role in regulating transcription and translation, are currently being investigated in *L. littorea* (Fig. 3.5) (Larade and Storey, unpublished results). These studies use two types of antibodies (a polyclonal to provide an estimate of total protein and a specific antibody raised against the phosphopeptide to estimate the amount of phosphoprotein) to evaluate the effect of stress on both the total amount of a given protein and the relative level of protein phosphorylation. Preliminary results on the effects of 12 h anoxia exposure on hepatopancreas indicate an increase in the amount of phosphorylated p38, CREB and HSP27 with little or no change in the total content of these proteins (Fig. 3.5). These data, although preliminary, suggest that the p38 MAPK pathway may play a role in the response to anoxia by marine molluscs. A great deal of research remains to be completed to confirm any of the pathways involved.

7. Perspectives

The material covered in this chapter demonstrates the value of basic science and attempts to link genomic and cellular responses with physiological adaptation. Molluscan model systems promote both *in vivo* and *in vitro* research at the level of cell, organ, and whole organism, allowing experimental results to be confirmed at various levels. The use of molluscan models with new molecular techniques (e.g., cDNA arrays), standard molecular techniques (e.g., cDNA library screening), and the development and adaptation of techniques (e.g., organ culture with pharmacological agents; polysome profiling) have provided new information on the effects of hypoxia and anoxia on metabolic response and regulation of gene expression.

Metabolic depression can be examined from a variety of viewpoints, with each new development

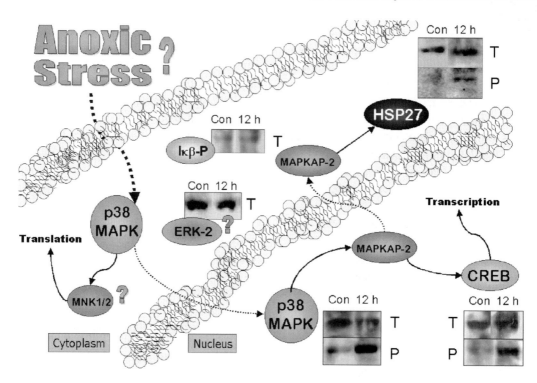

Fig. 3.5. Stress-induced signaling in *L. littorea*. MAPK signaling pathways, previously reported to play a role in regulating transcription and translation, have been investigated in the marine snail. The relative amount and phosphorylation state of various target substrates, including transcription factors and protein kinases, was assessed in *L. littorea* hepatopancreas and preliminary results suggest that the p38 MAPK pathway may play a role in the response to anoxia in molluscs. Con = control; 12 h = 12 h anoxic exposure; T = total; P = phosphorylated.

extending in many different directions (Fig. 3.6). Identifying the role of anoxia-induced genes and proteins (genomics and proteomics) is in vogue at the present time. In particular, there is high interest in genes that are activated during anoxia, producing proteins responsible for generating (or regulating) many of the observed adaptations. Studies of gene-protein expression during anoxia are particularly interesting because of the high energetic cost associated with biosynthetic activities. Genes that are actively transcribed and translated when energy supply is restricted should, predictably, be only those whose protein products perform important roles in anaerobiosis, or that facilitate metabolic recovery when oxygen is again available.

Expressed genes that represent known homologues can often fit into an existing scheme, or be justified based on the established function of the gene. Novel genes, however, pose a considerable challenge, since little is known about them other than the nucleotide sequence. This "small" amount

of information does, however, allow the protein sequence to be decoded which permits study of the protein itself via homology analysis, domain searching and evaluation with specific antibodies. Future research should involve not only the actual production of proteins, but also the modification to

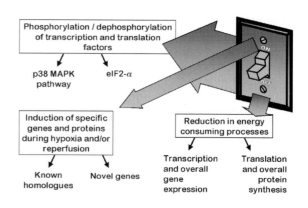

Fig. 3.6. Summary of the different cellular mechanisms that are differentially regulated as a response to anoxia exposure in molluscs.

proteins, both of which are proving to be exciting and dynamic fields of research. Examination of promoter regions of novel genes may uncover conserved promoter elements in the suite of anoxia-induced genes, which will help determine the transcription factors involved in adaptive responses to anoxia. The exact function of cGMP during anoxia is not yet known, although this classical second messenger appears to play a significant role in triggering metabolic depression at the level of transcription, translation, and glycolysis. Future research on marine molluscs will hopefully elucidate functional relationships that exist between anoxia-induced genes and proteins, providing insight into their mechanism of regulation and involvement in metabolic depression.

References

Adilakshmi, T. and Laine, R. (2002). Ribosomal protein S25 mRNA partners with MTF-1 and La to provide a p53-mediated mechanism for survival or death. J. Biol. Chem. 277, 4147–4151.

Bacchiocchi, S. and Principato, G. (2000). Mitochondrial contribution to metabolic changes in the digestive gland of *Mytilus galloprovincialis* during anaerobiosis. J. Exp. Zool. 286, 107–113.

Balaban, P., Poteryaev, D., Zakharov, I., Uvarov, P., Malyshev, A. and Belyavsky, A. (2001). Up- and down-regulation of helix command-specific 2 (HCS2) gene expression in the nervous system of terrestrial snail *Helix lucorum*. Neuroscience 103, 551–559.

Bogdanov, Y., Ovchinnikov, D., Balaban, P. and Belyavsky, A. (1994). Novel gene HCS1 is specifically expressed in the giant interneurones of the terrestrial snail. NeuroReport 5, 589–592.

Bridges, C. (1994). Ecophysiological adaptations in intertidal rockpool fishes. In: Water/Air Transition in Biology (Dalta Munshi, J., Mittal, A. et al., Eds.), pp. 59–92. Oxford IBH Publishing Co., New Delhi.

Brooks, S. and Storey, K. (1990). cGMP-Stimulated protein kinase phosphorylates pyruvate kinase in an anoxia-tolerant marine mollusc. J. Comp. Physiol. B 160, 309–316.

Brooks, S. and Storey, K. (1991). The role of protein kinases in anoxia tolerance in facultative anaerobes: purification and characterization of a protein kinase that phosphorylates pyruvate kinase. Biochim. Biophys. Acta 1073, 253–259.

Brooks, S. and Storey, K. (1997). Glycolytic controls in estivation and anoxia: a comparison of metabolic arrest in land and marine molluscs. Comp. Biochem. Physiol. A 118, 1103–1114.

Browning, D., McShane, M., Marty, C. and Ye, R. (2000). Nitric oxide activation of p38 mitogen-activated protein kinase in 293T fibroblasts requires cGMP-dependent protein kinase. J. Biol. Chem. 275, 2811–2816.

Bunn, H. and Poyton, R. (1996). Oxygen sensing and molecular adaptation to hypoxia. Physiol. Rev. 76, 839–885.

Busa, W. and Nuccitelli, R. (1984). Metabolic regulation via intracellular pH. Am. J. Physiol. 246, R409–R438.

Buttgereit, F. and Brand, M. (1995). A hierarchy of ATP-consuming processes in mammalian cells. Biochem. J. 312, 163–167.

Cai, Q. and Storey, K.B. (1996). Anoxia-induced gene expression in turtle heart: up-regulation of mitochondrial genes for NADH-ubiquinone oxidoreductase subunit 5 and cytochrome C oxidase subunit 1. Eur. J. Biochem. 241, 83–92.

Canesi, L., Ciacci, C., Betti, M. and Gallo, G. (2000). Growth factor-mediated signal transduction and redox balance in isolated digestive gland cells from *Mytilus galloprovincialis* Lam. Comp. Biochem. Physiol. C 125, 355–363.

Cohen, D. (1997a). Mitogen-activated protein kinase cascades and the signaling of hyperosmotic stress to immediate early genes. Comp. Biochem. Physiol. A 117, 291–299.

Cohen, P. (1997b). The search for physiological substrates of MAP and SAP kinases in mammalian cells. Trends Cell. Biol. 7, 353–361.

Collier, N.C., Heuser, J., Levy, M.A. and Schlesinger, M.J. (1988). Ultrastructural and biochemical analysis of the stress granule in chicken embryo fibroblasts. J. Cell Biol. 106, 1131–1139.

de Zwaan, A. (1977). Anaerobic energy metabolism in bivalve molluscs. Oceanogr. Mar. Biol. Annu. Rev. 15, 103–187.

de Zwaan, A. and Putzer, V. (1985). Metabolic adaptations of intertidal invertebrates to environmental hypoxia (a comparison of environmental anoxia to exercise anoxia). Symp. Soc. Exp. Biol. 39, 33–62.

de Zwaan, A., Cortesi, P., van den Thillart, G., Brooks, S., Storey, K.B., Roos, J., van Lieshout, G., Cattani, O. and Vitali, G. (1992). Energy metabolism of bivalves at reduced oxygen tensions. In: Marine Coastal Eutrophication (Vollenweider, R.A., Marchetti, R. and Viviani, R., Eds.) pp. 1029–1039. Elsevier, Amsterdam.

Deak, M., Clifton, A., Lucocq, L. and Alessi, D. (1998). Mitogen-activated and stress-activated protein kinase-1 (MSK1) is directly activated by MAPK and SAPK2/p38, and may mediate activation of CREB. EMBO J. 17, 4426–4441.

Derrigo, M., Cestelli, A., Savettieri, G. and Di Liegro, I. (2000). RNA-protein interactions in the control of stability and localization of messenger RNA. Int. J. Mol. Med. 5, 111–123.

Eddy, S.F. and Storey, K.B. (2002). Dynamic use of cDNA arrays: heterologous probing for gene discovery and exploration of organismal adaptation to environmental stress. In: Cell and Molecular Responses to Stress (Storey, K.B. and Storey, J.M., Eds.). Vol. 3, Chap. 22. Elsevier, Amsterdam.

Ellington, W. (1983). The extent of intracellular acidification during anoxia in the catch muscles of two bivalve molluscs. J. Exp. Zool. 227, 313–317.

Frerichs, K.U., Smith, C.B., Brenner, M., DeGracia, D.J., Krause, G.S., Marrone, L., Dever, T.E. and Hallenbeck, J.M. (1998). Suppression of protein synthesis in brain during hibernation involves inhibition of protein initiation and elongation. Proc. Nat. Acad. Sci. USA 95, 14511–14516.

Fukunaga, R. and Hunter, T. (1997). MNK1, a new MAP kinase-activated protein kinase, isolated by a novel expression screening method for identifying protein kinase substrates. EMBO J. 16, 1921–1933.

Goto, S. (2001). A novel gene that is up-regulated during recovery from cold shock in *Drosophila melanogaster*. Gene 270, 259–264.

Greenway, S. and Storey, K. (1999). The effect of prolonged anoxia on enzyme activities in oysters (*Crassostrea virginica*) at different seasons. J. Exp. Mar. Biol. Ecol. 242, 259–272.

Greenway, S. and Storey, K. (2000). Seasonal change and prolonged anoxia affect the kinetic properties of phosphofructokinase and pyruvate kinase in oysters. J. Comp. Physiol. 170, 285–293.

Greenway, S. and Storey, K. (2001). Effects of seasonal change and prolonged anoxia on metabolic enzymes of *Littorina littorea*. Can. J. Zool. 79, 907–915.

Grieshaber, M., Hardewig, I., Kreutzer, U. and Portner, H. (1994). Physiological and metabolic responses to hypoxia in invertebrates. Rev. Physiol. Biochem. Pharmacol. 125, 43–147.

Griffin, O. and Stuehr, D. (1995). Nitric oxide synthases: properties and catalytic mechanism. Annu. Rev. Physiol. 57, 707–736.

Gudi, T., Casteel, D., Vinson, C., Boss, G. and Pilz, R. (2000). NO activation of fos promoter elements requires nuclear translocation of G-kinase I and CREB phosphorylation but is independent of MAP kinase activation. Oncogene 19, 6324–6333.

Guppy, M., Fuery, C. and Flanigan, J. (1994). Biochemical principles of metabolic depression. Comp. Biochem. Physiol. 109B, 175–189.

Guppy, M. and Withers, P. (1999). Metabolic depression in animals: physiological perspectives and biochemical generalizations. Biol. Rev. 74, 1–40.

Hand, S. (1998). Quiescence in *Artemia franciscana* embryos: Reversible arrest of metabolism and gene expres-

sion at low oxygen levels. J. Exp. Biol. 201, 1233–1242.

Hand, S. and Hardewig, I. (1996). Downregulation of cellular metabolism during environmental stress: mechanisms and implications. Annu. Rev. Physiol. 58, 539–563.

Hershey, J.W.B. (1991). Translational control in mammalian cells. Annu. Rev. Biochem. 60, 717–755.

Higgins, W.J. and Greenberg, M.J. (1974). Intracellular actions of 5-hydroxytryptamine on the bivalve myocardium-II. Cyclic nucleotide-dependent protein kinase and microsomal calcium uptake. J. Exp. Zool. 190, 305–316.

Hittel, D. and Storey, K.B. (2001). Differential expression of adipose and heart type fatty acid binding proteins in hibernating ground squirrels. Biochim. Biophys. Acta 1522, 238–243.

Hittel, D. and Storey, K.B. (2002). The translation state of differentially expressed mRNAs in the hibernating thirteen-lined ground squirrel (*Spermophilus tridecemlineatus*). Arch. Biochem. Biophys. 401, 244–254.

Hochachka, P.W. and Somero, G.N. (1984). Biochemical Adaptation. Princeton University Press, Princeton.

Holwerda, D.A., Druitwagen, E.C.J. and de Bont, A.M. (1981). Regulation of pyruvate kinase and phosphoenolpyruvate carboxykinase activity during anaerobiosis in *Mytilus edulis* L. Mol. Physiol. 1, 165–171.

Ivan, M., Kondo, K., Yang, H., Kim, W., Valiando, J., Ohh, M., Salic, A., Asara, J., Lane, W. and Kaelin Jr., W. (2001). HIF-α targeted for VHL-mediated destruction by proline hydroxylation: implications for O_2 sensing. Science 292, 464–468.

Jaakkola, P., Mole, D., Tian, Y.M., Wilson, M., Gielbert, J., Gaskell, S., von Kriegsheim, A., Hebestreit, H., Mukherji, M., Schofield, C., Maxwell, P., Pugh, C. and Ratcliffe, P. (2001). Targeting of HIF-α to the von Hippel-Lindau ubiquitylation complex by O_2-regulated prolyl hydroxylation. Science 292, 468–472.

Karin, M. and Hunter, T. (1995). Transcriptional control by protein phosphorylation: signal transmission from the cell surface to the nucleus. Curr. Biol. 5, 747–757.

Kaufman, R. (1994). Control of gene expression at the level of translation initiation. Curr. Opin. Biotech. 5, 550–557.

Kedersha, N.L., Gupta, M., Li, W., Miller, I., Anderson, P. (1999). RNA-binding proteins TIA-1 and TIAR link the phosphorylation of eIF-2-α to the assembly of mammalian stress granules. J. Cell Biol. 147, 1431–1441.

Kohler, G. and Lindl, T. (1980). Effects of 5-hydroxytryptamine, dopamine and acetylcholine on accumulation of cAMP and cGMP in the anterior byssus retractor muscle of *Mytilus edulis* L. (Mollusca). Pflugers Arch. 383, 257–262.

Laine, R., Shay, N. and Kilberg, M. (1994). Nuclear retention of the induced mRNA following amino acid-dependent transcriptional regulation of mammalian ribosomal proteins L17 and S25. J. Biol. Chem. 269, 9693–9697.

Larade, K., Nimigan, A. and Storey, K. (2001). Transcription pattern of ribosomal protein L26 during anoxia exposure in *Littorina littorea*. J. Exp. Zool. 290, 759–768.

Larade, K. and Storey, K. (2002a). Reversible suppression of protein synthesis in concert with polysome disaggregation during anoxia exposure in *Littorina littorea*. Mol. Cell. Biochem. 232, 121–127.

Larade, K. and Storey, K. (2002b). Characterization of a novel gene up-regulated during anoxia exposure in the marine snail, *Littorina littorea*. Gene 283, 145–154.

Lee, J. and Horowitz, P. (1992). Sulfhydryl groups on yeast ribosomal proteins L7 and L26 are significantly more reactive in the 80S particles than in the 60S subunits. J. Biol. Chem. 267, 2502–2506.

Li, B., Nierras, C.R., Warner, J.R. (1999). Transcriptional elements involved in the repression of ribosomal protein synthesis. Mol. Cell. Biol. 19, 5393–5404.

Loomis, S. (1995). Freezing tolerance of marine invertebrates. Oceanogr. Mar. Biol. 33, 337–350.

Lucas, K., Pitari, G., Kazerounian, S., Ruiz-Stewart, I., Park, J., Schulz, S., Chepenik, K. and Waldman, S. (2000). Guanylyl cyclases and signaling by cyclic GMP. Pharmacol. Rev. 52, 375–413.

Ma, E., Xu, T. and Haddad, G. (1999). Gene regulation by O_2 deprivation: an anoxia-regulated novel gene in *Drosophila melanogaster*. Mol. Brain Res. 63, 217–224.

Mager, W. (1988). Control of ribosomal protein gene expression. Biochim. Biophys. Acta 949, 1–15.

Marsh, A., Maxson, R. and Manahan, D. (2001). High macromolecular synthesis with low metabolic cost in Antarctic sea urchin embryos. Science 291, 1950–1952.

Martin, S., Epperson, E., van Breukelen, F. (2000). Quantitative and qualitative changes in gene expression during hibernation in golden-mantled ground squirrels. In: Life in the Cold (Heldmaier, G. and Klingenspor, M., Eds.), pp. 315–324. Springer-Verlag, Berlin.

Martin-Blanco, E. (2000). p38 MAPK signaling cascades: ancient roles and new functions. BioEssays 22, 637–645.

Marzouki, A., Lavergne, J., Reboud, J. and Reboud, A. (1990). Modification of the accessibility of ribosomal proteins after elongation factor 2 binding to rat liver ribosomes and during translocation. Biochim. Biophys. Acta 1048, 238–244.

Mathews, M., Sonenberg, N. and Hershey, J. (1996). Origins and targets of translational control. In: Translational Control. (Hershey, J., Mathews, M. et al., Eds.), pp. 1–29. Cold Spring Harbor Laboratory Press, Cold Spring Harbor, NY.

Meyuhas, O. (2000). Synthesis of the translational apparatus is regulated at the translational level. Eur. J. Biochem. 267, 6321–6330.

Michaelidis, B. and Storey, K.B. (1990). Phosphofructo-

kinase from the anterior byssus retractor muscle of *Mytilus edulis*: modification of the enzyme in anoxia and by endogenous protein kinases. Int. J. Biochem. 22, 759–765.

Mitsumoto, A., Takeuchi, A., Okawa, K., Nakagawa, Y. (2002). A subset of newly synthesized polypeptides in mitochondria from human endothelial cells exposed to hydroperoxide stress. Free Radic. Biol. Med. 32, 22–37.

Moroz, L. (2000). Giant identified NO-releasing neurons and comparative histochemistry of putative nitrergic systems in gastropod molluscs. Micro. Res. Tech. 49, 557–569.

Nover, L., Scharf, K.D. and Neumann, D. (1983). Formation of cytoplasmic heat shock granules in tomato cell cultures and leaves. Mol. Cell. Biol. 3, 1648–1655.

Nover, L., Scharf, K.D. and Neumann, D. (1989). Cytoplasmic heat shock granules are formed from precursor particles and are associated with a specific set of mRNAs. Mol. Cell. Biol. 9, 1298–1308.

Nygard, O., Nilsson, L. and Westermann, P. (1987). Characterisation of the ribosomal binding site for eukaryotic elongation factor 2 by chemical cross-linking. Biochim. Biophys. Acta 910, 245–253.

Plaxton, W.C. and Storey, K.B. (1984). Purification and properties of aerobic and anoxic forms of pyruvate kinase from red muscle tissue of the channeled whelk, *Busycotypus canaliculatum*. Eur. J. Biochem. 143, 267-272.

Pyronnet, S., Imataka, H., Gingras, A., Fukunaga, R., Hunter, T. and Sonenberg, N. (1999). Human eukaryotic translation initiation factor 4G (eIF4G) recruits Mnk1 to phosphorylate eIF4E. EMBO J. 18, 270–279.

Raingeaud, J., Gupta, S., Rogers, J., Dickens, M., Han, J., Ulevitch, R. and Davis, R. (1995). Pro-inflammatory cytokines and environmental stress cause p38 mitogen-activated protein kinase activation by dual phosphorylation on tyrosine and threonine. J. Biol. Chem. 270, 7420–7426.

Rhoads, R. (1993). Regulation of eukaryotic protein synthesis by initiation factors. J. Biol. Chem. 268, 3017–3020.

Rhodes, L. and Van Beneden, R. (1997). Isolation of the cDNA and characterization of mRNA expression of ribosomal protein S19 from the soft shell clam, *Mya arenaria*. Gene 197, 295–304.

Robertson, R., El-Haj, A., Clarke, A. and Taylor, E. (2001). Effects of temperature on specific dynamic action and protein synthesis rates in the Baltic isopod crustacean, *Saduria entomon*. J. Exp. Mar. Biol. Ecol. 262, 113–129.

Rouse, J., Cohen, P., Trigon, S., Morange, M., Alonso-Llamazares, A., Zamanillo, D., Hunt, T. and Nebreda, A. (1994). A novel kinase cascade triggered by stress and heat shock that stimulates MAPKAP kinase-2 and phosphorylation of the small heat shock proteins. Cell 78, 1027–1037.

Ruan, H., Brown, C. and Morris, D. (1997). Analysis of ribosome loading onto mRNA species: implications for translational control. In: mRNA Formation and Function. (Richter, J., Ed.), pp. 305–321. Academic Press, New York.

Scharf, K.D., Heider, H., Hohfeld, I., Lyck, R., Schmidt, E. and Nover, L. (1998). The tomato Hsf system: HsfA2 needs interaction with HsfA1 for efficient nuclear import and may be localized in cytoplasmic heat stress granules. Mol. Cell. Biol. 18, 2240–2251.

Schmidt, H. and Kamp, G. (1996). The Pasteur effect in facultative anaerobic metazoa. Experientia 52, 440–448.

Semenza, G.L. (2000). HIF-1: mediator of physiological and pathophysiological responses to hypoxia. J. Appl. Physiol. 88, 1474–1480.

Snyder, M. (1999). Ribosomal proteins S27E, P2, and L37A from marine invertebrates. Mar. Biotechnol. 1, 184–190.

Sokolova, I., Bock, C. and Portner, H.-O. (2000). Resistance to freshwater exposure in White Sea *Littorina* spp. II: Acid-base regulation. J. Comp. Physiol. B 170, 105–115.

Spirin, A. (1996). Masked and translatable messenger ribonucleoproteins in higher eukaryotes. In: Translational Control. (Hershey, J., Mathews, M., et al., Eds.), pp. 319–334. Cold Spring Harbor Laboratory Press, Cold Spring Harbor, NY.

Stamler, J., Singel, D. and Loscalzo, J. (1992). Biochemistry of nitric oxide and its redox activated forms. Science 258, 1898–1902.

Storey, K. (1985). Fructose 2,6-bisphosphate and anaerobic metabolism in marine molluscs. FEBS Lett. 181, 245–248.

Storey, K. (1988). Suspended animation: the molecular basis of metabolic depression. Can. J. Zool. 66, 124–131.

Storey, K. (1992). The basis of enzymatic adaptation. In: Fundamentals of Medical Cell Biology. (Bittar, E., Ed.), pp. 137–156. JAI Press, Greenwich, CT.

Storey, K. (1993). Molecular mechanisms of metabolic arrest in mollusks. In: Surviving Hypoxia: Mechanisms of Control and Adaptation. (Hochachka, P., Lutz, P., Sick, T.J., Rosenthal, M. and van den Thillart, G., Eds.), pp. 253–269. CRC Press, Boca Raton.

Storey, K. and Storey, J. (1990). Metabolic rate depression and biochemical adaptation in anaerobiosis, hibernation and estivation. Quart. Rev. Biol. 65, 145–174.

Surks, M.I. and Berkowitz, M. (1971). Rat hepatic polysome profiles and in vitro protein synthesis during hypoxia. Am. J. Physiol. 220, 1606–1609.

Teem, J.L., Abovich, N., Kaufer, N.F., Schwindinger, W.F., Warner, J.R., Levy, A., Woolford, J., Leer, R.J., van

Raamsdonk-Duin, M.M., Mager, W.H., Planta, R.J., Schultz, L., Friesen, J.D., Fried, H., Rosbash, M. (1984). A comparison of yeast ribosomal protein gene DNA sequences. Nucleic Acids Res. 12, 8295–8312.

Tielens, A. and Van Hellemond, J. (1998). The electron transport chain in anaerobically functioning eukaryotes. Biochim. Biophys. Acta 1365, 71–78.

Truchot, J. and Duhamel-Jouve, A. (1980). Oxygen and carbon dioxide in the marine intertidal environment: diurnal and tidal changes in rockpools. Respir. Physiol. 39, 241–254.

van der Houven van Oordt, W., Diaz-Meco, M.T., Lozano, J., Krainer, A.R., Moscat, J., Caceres, J.F. (2000). The MKK(3/6)-p38-signaling cascade alters the subcellular distribution of hnRNP A1 and modulates alternative splicing regulation. J. Cell Biol. 149, 307–316.

Van Hellemond, J., Klockiewicz, M., Gaasenbeek, C., Roos, M. and Tielens, A. (1995). Rhodoquinone and complex II of the electron transport chain in anaerobically functioning eukaryotes. J. Biol. Chem. 270, 31065–31070.

Villarreal, J. and Lee, J. (1998). Yeast ribosomal protein L26 is located at the ribosomal subunit interface as determined by chemical cross-linking. Biochimie 80, 321–324.

Walsh, P., McDonald, D. and Booth, C. (1984). Acid–base balance in the sea mussel, *Mytilus edulis*. II. Effects of hypoxia and air-exposure on intracellular acid–base status. Mar. Biol. Lett. 5, 359–369.

Waskiewicz, A., Flynn, A., Proud, C. and Cooper, J. (1997). Mitogen-activated protein kinases activate the serine/threonine kinases MNK1 and MNK2. EMBO J. 16, 1909–1920.

Watanabe, T. (1998). Isolation of a cDNA encoding a homologue of ribosomal protein L26 in the decapod crustacean *Penaeus japonicus*. Mol. Mar. Biol. Biotechnol. 7, 259–262.

Willmore, W.G., English, T.E. and Storey K.B. (2001). Mitochondrial gene responses to low oxygen stress in turtle organs. Copeia 2001, 628–637.

Wingrove, J. and O'Farrell, P. (1999). Nitric oxide contributes to behavioural, cellular, and developmental responses to low oxygen in *Drosophila*. Cell 98, 105–114.

Yeh, Y., Traut, R. and Lee, J. (1986). Protein topography of the 40s ribosomal subunit from *Saccharomyces cerevisiae* as shown by chemical cross-linking. J. Biol. Chem. 261, 14148–14153.

Zhu, H. and Bunn, H. (2001). How do cells sense oxygen? Science 292, 449–451.

Sensing, Signaling and Cell Adaptation. Edited by K.B. Storey and J.M. Storey

CHAPTER 4

The Role of Adenosine in Tissue Protection During Ischemia-Reperfusion

Gail W. Sullivan and Joel Linden
Cardiovascular Research Center, University of Virginia, Charlottesville, VA 22908, USA

1. Introduction

Adenosine accumulates in ischemic or metabolically active tissues and appears to have evolved into a signaling molecule that triggers adaptations to protect tissues from injury. The goal of the present chapter is to review selected recent literature that has extended our knowledge of how adenosine evokes protection in tissues under ischemic stress. We will draw from recent studies of heart, liver, intestine, kidney, eye, central nervous system, skeletal muscle and skin.

Endogenously released adenosine has been termed a "retaliatory metabolite" because it is released from hypoxic or injured tissues and functions to prevent injury. The mechanisms of adenosine-mediated tissue protection are now being realized. In the heart, which has been intensively investigated, the tissue protective actions of adenosine include: (1) stimulation of coronary and collateral vessel vasodilation to maintain adequate blood flow; (2) reduction in myocardial oxygen demand through negative chronotropic, inotropic and dromotropic effects; (3) induction of a shift towards non-oxygen requiring glycolytic energy production; (4) promotion of purine salvage allowing ATP replenishment; (5) reduction in tissue-destructive oxygen radical release and neutrophil-mediated endothelial damage; (6) prevention of no-reflow by inhibition of neutrophil and platelet aggregation and the resulting microvascular plugging; and (7) stimulation of collateral coronary angiogenesis (reviewed by Ely and Berne, 1992).

Most of these effects have been shown to be mediated by adenosine binding to four G protein coupled adenosine receptors, A_1, A_{2A}, A_{2B} and A_3.

2. Endogenous adenosine is protective in ischemia-reperfusion injury

Evidence that endogenous adenosine can be tissue protective in ischemia-reperfusion has been observed in multiple organs by several types of experiments including methods that alter the concentrations of adenosine available for binding to adenosine receptors. These include enhancing adenosine tissue accumulation in the injured tissue by inhibiting adenosine metabolism, increasing extracellular concentrations of adenosine by blocking cellular uptake, and decreasing interstitial adenosine by blocking adenosine release from cells. Interstitial and blood adenosine concentrations are transiently increased locally from nanomolar to micromolar concentrations by ischemia, hypoxia, trauma and inflammation. In addition, tissue concentrations of endogenous adenosine can be augmented by pharmacological agents that inhibit adenosine transport and reduce adenosine metabolism.

2.1. Adenosine metabolism

Endogenous adenosine can be increased by blocking adenosine deaminase which converts adenosine to inosine, and inhibiting adenosine kinase

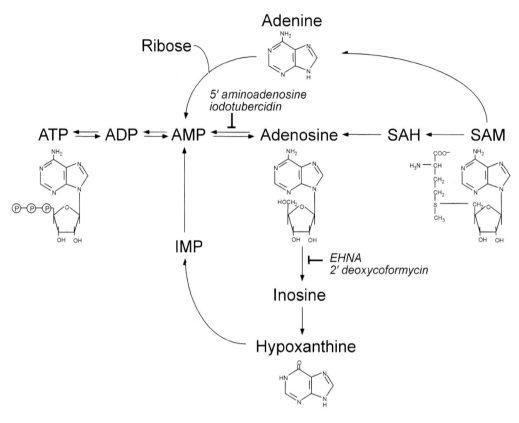

Fig. 4.1. Major pathways involved in adenosine metabolism during ischemia-reperfusion. 5′ Aminoadenosine and iodo-tubercidin are inhibitors of adenosine kinase and EHNA and 2′ deoxycoformycin are inhibitors of adenosine deaminase. Abbreviations: AMP, adenosine monophosphate; ADP, adenosine diphosphate; ATP, adenosine triphosphate; EHNA, erythro-9(2-hydroxy-3-nonyl)adenine; IMP, inosine triphosphate; SAH, S-adenosylhomocysteine; and SAM, S-adenosylmethionine.

that phosphorylates adenosine to AMP. A third possible pathway for the production of tissue adenosine is conversion of S-adenosylhomocysteine by S-adenosylhomocysteine hydrolase to adenosine. In ischemic brain tissue AMP rather than S-adenosylhomocysteine appears to be the major source of released adenosine (Fig. 4.1) (Latini et al., 1996).

The adenosine deaminase inhibitor, 2′-deoxycoformycin, protects the rat kidney from ischemia-reperfusion injury by both promoting ATP replenishment and reducing free radical-induced lipid peroxidation as evidenced by decreased formation of renal malondialdehyde (Bor et al., 1999). In the isolated perfused mouse heart, the adenosine deaminase inhibitor, erythro-9(2-hydroxy-3-nonyl)adenine (EHNA), or the adenosine kinase inhibitor, iodotubercidin, each reduce ischemia-

reperfusion injury when administered alone but the protection is lost when these agents are administered together (Peart et al., 2001). A possible explanation for this observation is that the two inhibitors together block two major pathways for purine salvage. That is, iodotubercidin inhibits the direct phosphorylation of adenosine to AMP, and EHNA, by inhibiting the formation of inosine, blocks the pathway of conversion of inosine to hypoxanthine to IMP and then to AMP (Fig. 4.1). In support of this concept, a cocktail of the adenosine metabolism inhibitors containing EHNA plus the adenosine kinase inhibitor 5′-aminoadenosine in the presence of the purine restoring nucleotide precursors, adenine and ribose, preserves heart function and prevents inflammation following 24 hours of reperfusion in a rat cardiac transplantation model (Smolenski et al., 2001).

2.2. Adenosine localization and transport

The tissue sources of endogenous adenosine are separate for blood and interstitial tissue since the vascular endothelium acts as a barrier to adenosine transport. Consequently, sources of blood adenosine include the vascular endothelium, and blood components (especially nucleotides released from stimulated platelets). Adenosine is derived from platelet-released nucleotides via the action of endothelial ecto-5′ nucleotidases such CD39 and CD73. Interstitial adenosine comes from ischemic myocytes, parenchymal cells, and metabolism of nucleotides derived from autonomic nerves and tissue-resident mast cells.

One way to increase available adenosine to membrane receptors is to block cellular uptake with adenosine transporter inhibitors. In a murine model of adult hippocampal ischemia, the transport inhibitors, dipyridamole and S-(4-nitrobenzyl)-6-thioinosine (NBMPR), not only induce an increase in extracellular adenosine (Kaku et al., 1994), but also, by an A_1 adenosine receptor-mediated mechanism, stimulate release of the protective amino acid taurine (Saransaari and Oja, 2000). In addition to other described protective effects of adenosine in the ischemic brain (reviewed by von Lubitz, 1999), adenosine stimulated taurine release prevents hippocampal hyperexcitation. However, in some instances adenosine transport inhibitors can block adenosine release and decrease receptor activation. NBMPR blocks release of adenosine from rat cardiomyocytes, and increases the incidence of ventricular fibrillation in isolated rat hearts stressed with coronary regional ischemia (Schreieck and Richardt, 1999).

3. Adenosine receptors

Four subtypes of adenosine surface G-protein coupled receptors (A_1, A_{2A}, A_{2B} and A_3) have been identified and cloned for several animal species. Pharmacological and genetic means have been used to study the specific subtypes involved in tissue protection. These include the use of receptor-selective antagonists, or selective agonists or allosteric enhancers. Genetic manipulations of adenosine receptor expression have included receptor-selective knock-out mice and transgenic gene over-expression. In cultured cells it has been possible to preclude receptor expression with antisense oligonucleotides.

3.1. Adenosine receptor antagonists

Studies with selective adenosine receptor antagonists have revealed protective roles of the separate adenosine receptor subtypes in preconditioning, ischemia and reperfusion. In a rat model of eye ischemia-reperfusion, the A_1 selective adenosine receptor antagonist, 8-cyclopentyl-1,3-dipropylxanthine (CPX), attenuates tissue recovery; whereas the A_{2A} selective adenosine receptor antagonist, 8-(3-chlorostyrl)caffeine (CSC), protects the retina (Li et al., 1999). In contrast, in other tissues, blockade of adenosine A_{2A} receptors with selective antagonists promotes tissue damage in ischemia-reperfusion models. For example, the A_{2A} adenosine receptor antagonist CSC, but not the A_1 adenosine receptor selective antagonist CPX increases the incidence of ventricular fibrillation in an isolated rat heart model of coronary regional ischemia (Schreieck and Richardt, 1999). Hence, adenosine receptor selective antagonist studies suggest that there may be divergent tissue specific actions of the separate adenosine receptor subtypes in ischemia-reperfusion injury.

3.2. Adenosine receptor agonists and allosteric enhancers

Introduction of exogenous adenosine, selective adenosine receptor agonists and adenosine receptor allosteric enhancers to ischemic and reperfused tissues have also been used to study the tissue protective effects of adenosine. For example, in the perfused isolated mouse heart studies cited above (Peart et al., 2001), the protective effects of endogenous adenosine were enhanced by the addition of 10 μM exogenous adenosine to the perfusion medium. The use of selective adenosine receptor agonists also yields data concerning the receptor subtype(s) involved in adenosine-mediated tissue protection. For example, we have observed that a selective A_{2A} adenosine receptor agonist, ATL146e (4-{3-[6-amino-9-(5-ethylcarbamoyl-3,4-di-hydroxy-tetrahydro-furan-2-yl)- 9H-purin-2-yl]-prop-2-ynyl}-cyclohexanecarboxylic acid methyl ester), decreases ischemia-reperfusion injury in a rat model of skin pressure ulcer formation (Peirce et al., 2001). In this model, the protective effects of ATL146e are counteracted by co-administration of the selective A_{2A} adenosine receptor antagonist ZM241385 indicating that the protective effects of ATL146e are from binding to A_{2A} adenosine receptors (Peirce et al., 2001).

The heart can be protected by preconditioning, i.e., short ischemic periods prior to a longer ischemic challenge (see "preconditioning" below) (Murry et al., 1986). Evidence that endogenous adenosine contributes to preconditioning by binding to A_1 adenosine receptors comes from experiments with the A_1 selective adenosine receptor allosteric enhancer PD81,723 ({2-amino-4,5-dimethyl-3-thienyl}-[3-{trifluoromethyl}phenyl] methanone). In a dog model, addition of PD81,723 during preconditioning reduces the threshold for preconditioning. Although a 5-min period of preconditioning protects the heart from a subsequent 1 h ischemic challenge (53% reduction in infarct size), neither a shorter 2.5 min preconditioning period nor treatment with the allosteric enhancer PD81,723 alone protects the heart, but the combination of 2.5 min of preconditioning in the presence of PD81,723 decreased infarct size by 44% (Mizumura et al., 1996).

3.3. Adenosine receptor expression

Adenosine receptor expression is not constant *in vivo*. For example, physiological stresses such as the release of reactive oxygen species (as occurs during reperfusion) can stimulate the expression of adenosine A_1 receptors in smooth muscle cells (Nie et al., 1998). The mechanism of enhanced A_1 adenosine receptor expression is via catalase-inhibitable H_2O_2 activation of NF-κB. Actinomycin D blocks stimulated A_1 adenosine receptor expression indicating that *de novo* receptor synthesis is occurring (Nie et al., 1998). In the mouse

isolated heart, protection from ischemia-reperfusion injury with the adenosine kinase inhibitor iodotubercidin or the adenosine deaminase inhibitor EHNA is enhanced in mice overexpressing the adenosine A_1 receptor in the heart (Peart et al., 2001).

It has been observed that the selective adenosine A_{2A} receptor antagonist, SCH58261, is neuroprotective in a rat model of cerebral ischemia (Monopoli et al., 1998). This is consistent with data from A_{2A} adenosine receptor knock-out mice. These mice are more resistant than wild-type mice to middle cerebral arterial occlusion as measured by infarct size, ischemic lesion volume and neurological function. This difference is independent of differences in genetic backgrounds, developmental changes in cerebral structure, blood flow or global A_1 adenosine receptor expression (Chen et al., 1999).

Studies using pharmacological activation with adenosine receptor agonists have suggested that adenosine binding to the A_3 subtype of receptors prior to ischemic challenge can protect the heart from ischemic challenge (reviewed by Guo et al., 2001). However, recently, mice with the A_3 adenosine receptor genetically deleted have been developed, and these mice ($A_3^{-/-}$) display a ~35% decrease in infarct size compared with wild type controls ($A_3^{+/+}$). This protective effect in $A_3^{-/-}$ mice occurs in the absence of changes in heart rate, body temperature or A_1 adenosine receptor expression. Smaller infarct size is accompanied by less cardiac inflammation as evidenced by fewer neutrophils within the infarcted regions from hearts of $A_3^{-/-}$ mice compared to the hearts of $A_3^{+/+}$ mice. These data indicate a possible stimulatory role for A_3 adenosine receptors in the inflammatory response to cardiac ischemia in mice (Guo et al., 2001). One way to reconcile these data is to propose that A_3 receptor activation provokes an inflammatory response by activating tissue-resident mast cells (Linden, 1994) which is damaging in the setting of acute inflammation, but protective in the setting of preconditioning.

Following ischemia-reperfusion challenge remodeling of the heart occurs. An overgrowth of cardiac fibroblasts contributes to pathological structural changes such as deposition of collagen and the replacement of myocytes with fibrotic scar tissue that can lead to heart failure. The potency order of adenosine receptor agonists and antagonists suggests that adenosine binding to adenosine A_{2B} receptors decreases growth factor-stimulated cardiac fibroblast proliferation. These results are consistent with results with antisense oligonucleotides (effects not seen with sense or scrambled oligonucleotides) which promote PDGF-BB stimulated cell proliferation and collagen synthesis of both basal cells and A_{2B} adenosine receptor agonist treated cells. Effects of A_{2B} antisense oligonucleotides on cell proliferation were mirrored by MAP kinase activation. That is, growth factor-stimulated MAP kinase activity is inhibited by A_{2B} agonists, and enhanced both by A_{2B} antagonists and by the presence of A_{2B} antisense oligonucleotides (Dubey et al., 2001).

3.4. Adenosine receptor down-regulation

Protection of the brain by A_{2A} receptor antagonists or A_{2A} gene deletion suggests that there are brain regions in which activation of A_{2A} receptors produce an excitatory response. Another possibility is suggested by the observation that activation of A_{2A} receptors by adenosine during ischemia down-regulates the regional expression of A_1 receptors by a protein kinase C (PKC)-dependent protein kinase A (PKA)-independent mechanism (Dixon et al., 1997). Hence, A_{2A} adenosine receptor antagonists may be tissue protective in the brain not only by blocking A_{2A} adenosine receptor stimulation of glutamate and aspartate release (O'Regan et al., 1992), but also by blocking A_{2A} adenosine receptor-mediated down-regulation of A_1 adenosine receptors.

4. Adenosine-stimulated cell pathways and effector activities

The four subtypes of adenosine receptors are all seven transmembrane spanning *N*-linked glycoprotein G-protein-coupled receptors. Glycosylation of the receptors does not affect ligand affinity, but may play a role in targeting newly formed receptors to the plasma membrane. With the exception of A_{2A} adenosine receptors, the adenosine receptors have sites for palmitoylation near the carboxyl terminus. A_1 and A_3 adenosine receptors couple to Gi and Go. The A_{2A} adenosine receptor couples to Gs and possibly Golf, and A_{2B} adenosine couple to Gs and Gq (reviewed by Linden, 2001).

4.1. Blood flow

On vascular smooth muscle cells, adenosine binds to A_{2A} and A_{2B} receptors that couple via Gs to adenylyl cyclase. This results in cAMP production that promotes vasodilation (reviewed by Ely and Berne, 1992). Potency order experiments with adenosine receptor agonists and antagonists indicate that binding to adenosine A_{2A} receptors is primarily responsible for coronary vasodilation (Belardinelli et al., 1998). Also, large A_{2A} adenosine receptor reserves within coronary arteries contribute to the high sensitivity of coronary arteries to A_{2A}-mediated adenosine-induced vasodilation (Shryock et al., 1998).

A study in hypothermic neonatal lamb isolated blood-perfused hearts indicates that more than vasodilation is involved with adenosine promotion of blood flow during reperfusion (Nomura et al., 1997). Although tissue recovery induced by adenosine in the first 30 min of reperfusion is mainly mediated by stimulation of vasodilation, there appears to be an additional, later flow-promoting effect of adenosine via binding to A_2 adenosine receptors (i.e., at 90 min reperfusion). The authors observed that improved left ventricular function occurred even when adenosine and the A_1 adenosine receptor antagonist CPX were added to reperfused hearts under conditions where the blood flow was artificially limited to equal that in the control hearts not receiving adenosine. The later action(s) of adenosine during reperfusion may be from A_{2A} adenosine receptor-mediated effects on neutrophil function (see below). Adenosine also increases blood flow during reperfusion in other tissues. For example, in rats adenosine and two selective A_{2A} adenosine receptor agonists increase hepatic tissue blood flow during reperfusion (Harada et al., 2000).

4.2. *Preconditioning*

Tissues can be protected by short preconditioning ischemic periods prior to a longer ischemic challenge that would otherwise cause tissue damage (Murry et al., 1986). In the mammalian heart, experiments with selective adenosine receptor agonists, enhancers and antagonists and studies with adenosine receptor overexpression models and with adenosine receptor knockout mice all indicate that the predominant adenosine receptor subtype contributing to adenosine stimulation of cardiac preconditioning is the A_1 receptor (reviewed by Mubagwa and Flameng, 2001). Further evidence indicates that adenosine needs to be present both during preconditioning and at the time of longer ischemic challenge to be protective (Thornton et al., 1993). In some models preconditioning stimulation of A_3 adenosine receptors may also be protective (reviewed by Cohen et al., 2000). Although stimulation of chick ventricular myocytes with A_1 or A_3 adenosine receptor agonists mimics preconditioning, stimulation of adenosine A_{2A} receptors during preconditioning attenuates preconditioning (Strickler et al., 1996). In contrast, preconditioning with the selective A_{2A} adenosine receptor agonist CGS21680 is protective to hypoxic rat hepatocytes (Carini et al., 2001). It is possible however that high concentrations of CGS21680 may activate A_3 receptors.

Adenosine is one of several agonists that stimulate preconditioning in the heart. Other physiological stimuli released during preconditioning include both receptor-mediated and receptor-independent agents such as bradykinin, opioids and free radicals (reviewed by Cohen et al., 2000). At the concentrations of the individual factors stimulated with a single cycle of preconditioning, the threshold to realize preconditioning is not reached if any of the stimuli (e.g., bradykinin) are blocked. Multiple preconditioning cycles or added exogenous stimuli can substitute for a blocked stimulus (Goto et al., 1995). These data indicate that there is additive activity of the individual stimuli during preconditioning.

There is evidence that protein kinase C (PKC) plays an important role in preconditioning

signaling stimulated by bradykinin, opioids, free radicals and adenosine. Preconditioning stimulated by any of these factors is blocked by PKC inhibitors such as staurosporine. In addition, activators of PKC such as phorbol myristate acetate can mimic preconditioning. Lck, a member of the Src family of tyrosine kinases, forms a functional signaling module with PKCε. In cardiac cells, PKCε interacts with, phosphorylates, and activates Lck. Ischemic preconditioning enhances the formation of PKCε-Lck which is required for cardioprotection (Ping et al., 2002).

Adenosine-induced preconditioning is sensitive to inhibition with pertussis toxin indicating that preconditioning is dependent on a pertussis toxin sensitive Gi/o protein (Thornton et al., 1993). Although there is evidence that in rat hepatocytes adenosine-stimulated preconditioning is phospholipase C mediated (Carini et al., 2001), adenosine receptors are not thought to couple with phospholipase C in cardiomyocytes (Cohen et al., 2000). In the rabbit heart, phospholipase D stimulation contributes to activation of PKC and tissue protection in adenosine-induced preconditioning (Cohen et al., 1996).

There is evidence that specific mitogen-activated protein kinases play a role downstream to PKC in the signaling pathway stimulated by preconditioning. In isolated rabbit hearts, the MAP kinase kinase stimulant, anisomycin, mimics preconditioning and protects the heart from infarct, and is not inhibited by the PKC inhibitor, chelerythrine (Baines et al., 1998). MAP kinase kinase can in turn activate p38 MAPK. Within this pathway MAPK-activated protein kinase 2 (MAPKAPK-2) is downstream of p38 MAPK. On activation MAPKAPK-2 activates heat shock protein 27 (HSP27). Phosphorylated HSP27 under oxidative stress stabilizes actin filaments and preserves cell viability (reviewed by Cohen et al., 2000). Hence, the cell signaling pathway (i.e., adenosine → adenosine A_1 receptor → phospholipases C/D → PKC → MAPKK → p38 MAPK → MAPKAPK-2 → HSP27 → actin polymerization) may contribute to cell survival in tissues stressed by ischemia-reperfusion (Fig. 4.2).

There is evidence that an end effector of

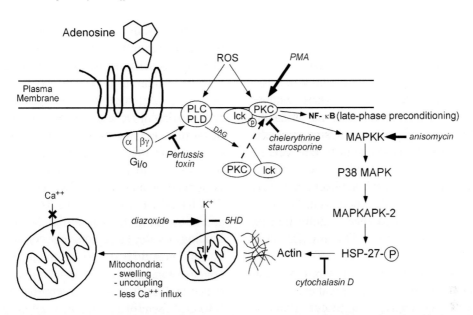

Fig. 4.2. Cell signaling pathways that are stimulated by adenosine in preconditioning. Adenosine binds to A_1 adenosine receptors and stimulates pathways that include Gi/o protein-coupled responses. Receptor coupling results in PLC and/or PLD activation to generate and DAG promote the translocation and activation of PKCε and lck-PO_4 molecules to the cell membrane. Lck-PO_4 activates NF-κB and induces late phase preconditioning. Membrane associated PKC activates MAPK. In this pathway MAPKAPK-2 phosphorylates HSP-27 which promotes actin polymerization. Polymerized actin interacts with mitochondria resulting in the opening of ATP dependent K^+ channels, mitochondrial swelling, uncoupling of electron transport and inhibition of Ca^{2+} influx. Preconditioning can be mimicked by PMA, ROS, anisomycin, diazoxide and NO and inhibited by pertussis toxin, chelerythrine, staurosporine, cytochalasin D and 5-HD. Abbreviations: 5-HD, 5-hydroxydecanoate; A_1 AR, A_1 adenosine receptor; DAG, diacylglycerol; HSP-27, heat shock protein 27; MAPK, mitogen-activated protein kinase; MAPKK, MAPK kinase; MAPKAPK-2, MAPK-activated protein kinase 2; NO, nitric oxide; PKC, protein kinase C; PLC, phospholipase C; PLD, phospholipase D; PMA, phorbol myristate acetate; and ROS, reactive oxygen species.

preconditioning may be the opening of mitochondrial K_{ATP}-channels (reviewed by Cohen et al., 2000). The mitochondrial K_{ATP}-channel blocker, 5-hydroxydecanoate (5-HD), reduces preconditioning in rabbit isolated cardiomyocytes. Conversely, the mitochondrial K_{ATP}-channel opener, diazoxide, is cardioprotective and treatment with it mimics preconditioning. In *in vitro* studies, preconditioning protects isolated rabbit cardiomyocytes from ischemic challenge, and diazoxide protects as well as preconditioning. Both preconditioning and diazoxide protection are reversed by cytochalasin D disruption of the cytoskeleton. These data indicate that in rabbit cardiomyocytes the cytoskeleton may regulate mitochondrial K_{ATP}-channel opening and consequently cell viability stimulated by preconditioning (Baines et al., 1999). It is not definite how opening of mitochondrial K_{ATP}-channels protects ischemic tissues. One

concept is that opening of mitochondrial K_{ATP}-channels results in a net influx of K^+ ions which causes mitochondrial swelling and uncoupling of electron transport thus preventing damaging ATP hydrolysis. Also mitochondrial K_{ATP}-channel opening may promote viability by reducing Ca^{2+} influx.

Another effect of adenosine released during preconditioning is stimulation of nitric oxide (NO) release from the activation of cNOS in vascular endothelium (Kuo and Chancellor, 1995; Peralta et al., 1999). NO has multiple potentially protective effects in ischemic tissue that include stimulation of vasodilation, activation of PKC, and stimulation of mitochondrial K_{ATP} channel opening (reviewed by Pagliaro et al., 2001). NO can be produced quickly in response to adenosine via activation of endothelial cNOS (Kuo and Chancellor, 1995). In addition, a delayed production of NO can come from preconditioning-stimulated PKC-mediated induction of myocardial iNOS via stimulation of tyrosine kinases and the transcription factor NF-κB (Bolli et al., 1998).

Two windows of preconditioning protection have been observed to occur. The first window comes in the first few hours following preconditioning, and then the protective effect of preconditioning wanes for the next 12–24 h. Between 12 and 24 h following preconditioning, a second window of protection is observed and lasts about 72 h (reviewed by Pagliaro et al., 2001). Evidence indicates that adenosine contributes to both the first and second windows of preconditioning (Baxter et al., 1994).

An effector of the second window of protection in preconditioning may be the antioxidant enzyme, Mn-superoxide dismutase (Mn-SOD). Prior to administration of the adenosine A_1 selective agonist, 2-chloro-N^6-cyclopentyladenosine (CCPA), treatment with an antisense oligonucleotide (but not sense or scrambled oligonucleotide) for the initiation site of rat Mn-SOD blocks the second window of preconditioning (24 h after CCPA treatment). In addition, antisense treatment prevents CCPA-stimulated Mn-SOD expression and function (Dana et al., 2000).

It has been observed that a brief period of preconditioning in one organ can protect a second "remote" organ from subsequent ischemia-reperfusion damage. Adenosine has been proposed to be a mediator in "remote preconditioning". For example in a rabbit model, renal preconditioning followed by coronary occlusion decreases cardiac tissue damage as measured by infarct size and functional assays. Renal preconditioning of cardiac muscle is counteracted by administration of the adenosine receptor antagonist, 8-sulfophenyltheophylline, prior to coronary occlusion, suggesting that adenosine is contributing to cardiac protection (Takaoka et al., 1999).

4.3. Reperfusion

Although restoration of circulation is critical for tissue recovery following ischemia, the reinstitution of blood flow during reperfusion sets off an inflammatory cascade of events that can result in further tissue damage and renewed vessel blockage due to the "no-reflow" phenomenon. Protection of previously ischemic tissue by adenosine during reperfusion appears to be primarily from adenosine binding to A_{2A} receptors. For example, in a rat model of renal ischemia-reperfusion injury an A_1 adenosine receptor agonist is protective if given prior to ischemia, but has little effect when given later. In contrast, an A_{2A} adenosine receptor agonist is protective only when given just prior to reperfusion (Lee and Emala, 2000, 2001).

Following ischemia in the heart, adenosine binding to A_{2A} receptors on coronary arteries causes vasodilation to augment blood flow (Belardinelli et al., 1998). In addition, there is evidence that binding to A_{2A} adenosine receptors on leukocytes (Jordan et al., 1999), platelets (Seligmann et al., 1998), vascular endothelium (Bouma et al., 1996) and underlying tissues (Zhao et al., 2001) contribute significantly to protection of reperfused tissue by ameliorating the inflammatory response. By binding to neutrophil A_{2A} receptors, adenosine blocks the release of reactive oxygen species (ROS) (Sullivan et al., 2001), and attenuates the upregulation of the neutrophil adhesion molecule CD11/CD18 (Wollner et al., 1993), and the release of tissue disruptive granule products including elastase (Bouma et al., 1997). Hence, by acting directly on the neutrophil, adenosine has the potential to decrease the inflammatory response during reperfusion (Fig. 4.3). Several models *in vivo* indicate that treatment with adenosine A_{2A} agonists does result in decreased neutrophil infiltration into a variety of reperfused tissues that correlate with reduced tissue injury including lung (Ross et al., 1999), kidney (Okusa et al., 2001), skin (Peirce et al., 2001), and heart (Jordan et al., 1997). Further, there is recent evidence that these effects of adenosine on neutrophil function *in vitro* and *in vivo* may be via an A_{2A} adenosine receptor-stimulated cAMP-mediated pathway that decreases phospholipase D activation (Lee and Emala, 2001; Okusa et al., 2001; Sullivan et al., 2001; Thibault et al., 2002).

There is also evidence that adenosine receptor-mediated actions during reperfusion are due to effects on several different cell types. For example,

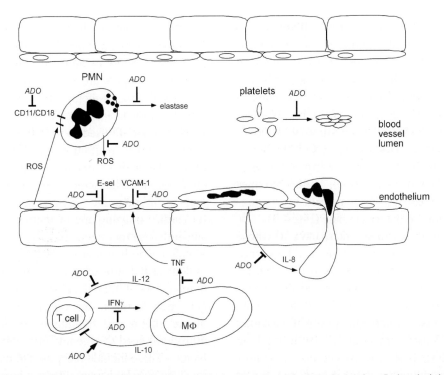

Fig. 4.3. Model of the inflammatory cascade during reperfusion and sites of adenosine action. Ischemic injury initiates an inflammatory cascade that involves blood elements, the vascular endothelium and underlying tissues. Briefly, this includes activation of the endothelium by ROS, cytokines and reactive lipids (e.g., platelet activating factor) released from resident immune cells and interstitial cells. Activation of the endothelium and the underlying tissues results in the release of cytokines, ROS, adhesion molecules and other factors which attract and recruit leukocytes, and promote platelet and leukocyte activation and aggregation. Endogenous adenosine promotes the release of the anti-inflammatory cytokine IL-10 from macrophages and inhibits release of the pro-inflammatory cytokines IFNγ, IL-8, TNFα and IL-12 from macrophages and/or T cells. In addition, adenosine decreases platelet aggregation and the neutrophil oxidative burst, PMN expression of CD11/CD18 adhesion molecules and release of elastase from PMN granules. Abbreviations: ADO, adenosine; E-sel; E-selectin; IFNγ, interferon-gamma; IL-8, interleukin-8; IL-10, interleukin-10; IL-12, interleukin-12; MФ, macrophage; PMN, neutrophil; ROS, reactive oxygen species; TNF, tumor necrosis factor-alpha; and VCAM-1, vascular cell adhesion molecule 1.

by binding to receptors on endothelial cells, adenosine can block expression of the adhesion molecules E-selectin and VCAM-1 to which recruited neutrophils and other leukocytes bind. Adenosine also inhibits endothelial release of the chemokine interleukin-8. The subtype(s) of adenosine receptors responsible for these actions on endothelial cells have not been definitively identified (Bouma et al., 1996).

It has been observed recently that adenosine, by binding to A_{2A} receptors during canine cardiac reperfusion, prevents apoptosis of myocytes by modulating the expression of Bcl-2 and Bax proteins (Zhao et al., 2001). Treatment with the selective A_{2A} adenosine receptor agonist CGS21680 during reperfusion decreases infarct size, limits neutrophil recruitment to the affected tissue and reduces myocyte apoptosis. Comparison by Western blot assay of the ratio of expression of the proteins Bax and Bcl-2 indicates that previously ischemic tissues have a higher (apoptosis favoring) ratio of Bax/Bcl-2 than non-ischemic tissues. Treatment with CGS21680 decreases the ratio of Bax to Bcl-2 within reperfusion stressed tissue, and hence may protect the cells from apoptosis.

Tissue resident immune cells may play prominent roles in initiating the inflammatory cascade during reperfusion including: Kupffer cells (macrophages) in the liver (Horie et al., 1997), lymphocytes (T cells) in the liver (Zwacka et al., 1997), lung (Frangogiannis et al., 2000), and kidney (Rabb et al., 2000) and mast cells in the small intestine (Kanwar and Kubes, 1994). There is also evidence that, via activation of A_{2A} receptors, adenosine can modulate the pro-inflammatory phenotype and viability of these immune cells. For example, by binding to A_{2A} receptors, adenosine decreases the release of tumor necrosis factor-alpha and interleukin-12 from stimulated mono-cytes/macrophages (Hasko et al., 2000; Link et al., 2000) and from liver following ischemia-reperfusion (Harada et al., 2000). These cytokines promote tissue damage during reperfusion by aug-menting the inflammatory cascade (Fig. 4.3).

By contrast, the cytokine interleukin-10 whose release from macrophages is stimulated by adeno-sine binding to A_{2A} receptors (Hasko et al., 1996) is protective during reperfusion both at the site of injury (Frangogiannis et al., 2000; Le Moine et al., 2000) and in tissues remote to the site of ischemia/reperfusion (Yoshidome et al., 1999). In the lung interleukin-10 is protective by decreasing neutro-phil recruitment and/or NF-κB-mediated tumor necrosis factor-alpha (TNFα) and macrophage inflammatory protein-2 (MIP-2) release (Yoshi-dome et al., 1999).

The activation of T cells upstream to neutrophil activation may play a crucial initiating role in ischemia-reperfusion injury in several tissues. Evi-dence comes from comparing the degree of ischemia-reperfusion injury in wild-type mice to T cell deficient mouse strains. Mice with fewer T cells display diminished ischemia-reperfusion injury and replenishment of T cells to these mice restores sensitivity to ischemia-reperfusion (Zwacka et al., 1997; Rabb et al., 2000). *In vitro* studies indicate that adenosine, by binding to lym-phocyte A_{2A} receptors, can decrease T cell activa-tion, expansion and viability (Apasov et al., 2000). It will be of great interest in the future to determine if adenosine binding to lymphocyte A_{2A} receptors affects T cell viability, activation and expansion, and consequently the downstream inflammatory response and tissue damage during reperfusion.

5.   Conclusions and clinical implications

In this review, we have outlined some recent advances in our understanding of how adenosine

affects tissue function during ischemia and reperfusion. The great potential, but also the limita-tions and dangers of applying this knowledge to therapeutic targets are now being realized. Pres-ently, adenosine is therapeutically used to correct supraventricular tachycardia and is included in solutions that are used to preserve organs prior to transplantation. In addition, pharmaceutical agents now in use may act in part through modulation of endogenous adenosine concentrations. These include dipyridamole which is used therapeutically for cardiac imaging and methotrexate which is an anti-inflammatory agent.

Adenosine is a non-specific agonist that has a biological half-life of only a few seconds within human blood. Future development of selective lon-ger acting adenosine analogs as therapeutic agents in preventing and treating ischemia-reperfusion injury will need to take into consideration several factors. These include receptor and tissue specific-ity since tissues display different receptor subtype expression and some adenosine analogs will cross-react with several subtypes of receptors and produce undesirable side effects. The choice of adenosine analog as a therapeutic agent may depend on the time of treatment since distinct receptor subtypes appear to be active at different stages of ischemia and reperfusion. There is also evidence of species differences; for example, mast cell activation is stimulated by A_3 adenosine recep-tors in rodents (Jin et al., 1997), but by A_{2B} adeno-sine receptors in man (Linden et al., 1999). In addition, chronic treatment with adenosine agonists may lead to loss of effectiveness due to downregulation and desensitization of receptors as discussed above. With this knowledge there is great potential for the development of new thera-peutic agents to treat ischemia-reperfusion injury.

References

Apasov, S.G., Chen, J.F., Smith, P.T., Schwarzschild, M.A., Fink, J.S. and Sitkovsky, M.V. (2000). Study of A(2A) adenosine receptor gene deficient mice reveals that adenosine analogue CGS 21680 possesses no A(2A) re-ceptor-unrelated lymphotoxicity. Br. J. Pharmacol. 131, 43–50.

Baines, C.P., Liu, G.S., Birincioglu, M., Critz, S.D., Cohen, M.V. and Downey, J.M. (1999). Ischemic preconditioning depends on interaction between mitochondrial K_{ATP} channels and actin cytoskeleton. Am. J. Physiol. 276, H1361–1368.

Baines, C.P., Wang, L., Cohen, M.V. and Downey, J.M. (1998). Protein tyrosine kinase is downstream of protein kinase C for ischemic preconditioning's anti-infarct effect in the rabbit heart. J. Mol. Cell. Cardiol. 30, 383–392.

Baxter, G.F., Marber, M.S., Patel, V.C. and Yellon, D.M. (1994). Adenosine receptor involvement in a delayed phase of myocardial protection 24 hours after ischemic preconditioning. Circulation 90, 2993–3000.

Belardinelli, L., Shryock, J.C., Snowdy, S., Zhang, Y., Monopoli, A., Lozza, G., Ongini, E., Olsson, R.A. and Dennis, D.M. (1998). The A2A adenosine receptor mediates coronary vasodilation. J. Pharmacol. Exp. Ther. 284, 1066–1073.

Bolli, R., Dawn, B., Tang, X.-L., Qui, Y., Ping, P., Xuan, Y.-T., Jones, W.K., Guo, Y. and Zhang, J. (1998). The nitric oxide hypothesis of late preconditioning. Basic Res. Cardiol. 93, 325–338.

Bor, M.V., Durmus, O., Bilgihan, A., Cevik, C. and Turkozkan, N. (1999). The beneficial effect of 2′-deoxycoformycin in renal ischemia-reperfusion is mediated both by preservation of tissue ATP and inhibition of lipid peroxidation. Int. J. Clin. Lab. Res. 29, 75–79.

Bouma, M.G., Jeunhomme, T.M., Boyle, D.L., Dentener, M.A., Voitenok, N.N., van den Wildenberg, F.A. and Buurman, W.A. (1997). Adenosine inhibits neutrophil degranulation in activated human whole blood: involvement of adenosine A2 and A3 receptors. J. Immunol. 158, 5400–5408.

Bouma, M.G., Vandenwildenberg, F. and Buurman, W.A. (1996). Adenosine inhibits cytokine release and expression of adhesion molecules by activated human endothelial cells. Am. J. Physiol. 39, C522–C529.

Carini, R., De Cesaris, M.G., Splendore, R., Vay, D., Domenicotti, C., Nitti, M.P., Paola, D., Pronzato, M.A. and Albano, E. (2001). Signal pathway involved in the development of hypoxic preconditioning in rat hepatocytes. Hepatology 33, 131–139.

Chen, J.F., Huang, Z., Ma, J., Zhu, J., Moratalla, R., Standaert, D., Moskowitz, M.A., Fink, J.S. and Schwarzschild, M.A. (1999). A(2A) adenosine receptor deficiency attenuates brain injury induced by transient focal ischemia in mice. J. Neurosci. 19, 9192–9200.

Cohen, M.V., Baines, C.P. and Downey, J.M. (2000). Ischemic preconditioning: from adenosine receptor of K_{ATP} channel. Ann. Rev. Physiol. 62, 79–109.

Cohen, M.V., Liu, Y., Liu, G.S., Wang, P., Weinbrenner, C., Cordis, G.A., Das, D.K. and Downey, J.M. (1996). Phospholipase D plays a role in ischemic preconditioning in rabbit heart. Circulation 94, 1713–1718.

Dana, A., Jonassen, A.K., Yamashita, N. and Yellon, D.M. (2000). Adenosine A(1) receptor activation induces delayed preconditioning in rats mediated by manganese superoxide dismutase. Circulation 101, 2841–2848.

Dixon, A.K., Widdowson, L. and Richardson, P.J. (1997). Desensisation of the adenosine A1 receptor by the A2A receptor in the rat striatum. J. Neurochem. 69, 315–321.

Dubey, R.K., Gillespie, D.G., Zacharia, L.C., Mi, Z. and Jackson, E.K. (2001). A(2b) receptors mediate the antimitogenic effects of adenosine in cardiac fibroblasts. Hypertension 37, 716–721.

Ely, S.W. and Berne, R.M. (1992). Protective effects of adenosine in myocardial ischemia. Circulation 85, 893–904.

Frangogiannis, N.G., Mendoza, L.H., Lindsey, M.L., Ballantyne, C.M., Michael, L.H., Smith, C.W. and Entman, M.L. (2000). IL-10 is induced in the reperfused myocardium and may modulate the reaction to injury. J. Immunol. 165, 2798–2808.

Goto, M., Liu, Y., Yang, X.-M., Ardell, J.L., Cohen, M.V. and Downey, J.M. (1995). Role of bradykinin in protection of ischemic preconditioning in rabbit hearts. Circ. Res. 77, 611–621.

Guo, Y., Bolli, R., Bao, W., Wu, W.J., Black, R.G., Murphree, S.S., Salvatore, C.A., Jacobson, M.A. and Auchampach, J.A. (2001). Targeted deletion of the A(3) adenosine receptor confers resistance to myocardial ischemic injury and does not prevent early preconditioning. J. Mol. Cell. Cardiol. 33, 825–830.

Harada, N., Okajima, K., Murakami, K., Usune, S., Sato, C., Ohshima, K. and Katsuragi, T. (2000). Adenosine and selective A(2A) receptor agonists reduce ischemia/reperfusion injury of rat liver mainly by inhibiting leukocyte activation. J. Pharmacol. Exp. Ther. 294, 1034–1042.

Hasko, G., Kuhel, D.G., Chen, J.F., Schwarzschild, M.A., Deitch, E.A., Mabley, J.G., Marton, A. and Szabo, C. (2000). Adenosine inhibits IL-12 and TNF-[alpha] production via adenosine A2a receptor-dependent and independent mechanisms. FASEB J. 14, 2065–2074.

Hasko, G., Szabo, C., Nemeth, Z.H., Kvetan, V., Pastores, S.M. and Vizi, E.S. (1996). Adenosine receptor agonists differentially regulate IL-10, TNF-alpha, and nitric oxide production in RAW 264.7 macrophages and in endotoxemic mice. J. Immunol. 157, 4634–4640.

Horie, Y., Wolf, R., Russell, J., Shanley, T.P. and Granger, D.N. (1997). Role of Kupffer cells in gut ischemia/reperfusion-induced hepatic microvascular dysfunction in mice. Hepatology 26, 1499–1505.

Jin, X., Shepherd, R.K., Duling, B.R. and Linden, J. (1997). Inosine binds to A3 adenosine receptors and stimulates mast cell degranulation. J. Clin. Invest. 100, 2849–2857.

Jordan, J.E., Zhao, Z.Q., Sato, H., Taft, S. and Vinten-Johansen, J. (1997). Adenosine A2 receptor activation attenuates reperfusion injury by inhibiting neutrophil accumulation, superoxide generation and coronary endothelial adherence. J. Pharmacol. Exp. Ther. 280, 301–309.

Jordan, J.E., Zhao, Z.Q. and Vinten-Johansen, J. (1999). The role of neutrophils in myocardial ischemia-reperfusion injury. Cardiovasc. Res. 43, 860–878.

Kaku, T., Hada, J. and Hayashi, Y. (1994). Endogenous adenosine exerts inhibitory effects upon the development of spreading depression and glutamate release induced by microdialysis with high K$^+$ in rat hippocampus. Brain Res. 658, 39–48.

Kanwar, S. and Kubes, P. (1994). Mast cells contribute to ischemia-reperfusion-induced granulocyte infiltration and intestinal dysfunction. Am. J. Physiol. 267, G316–G321.

Kuo, L. and Chancellor, J.D. (1995). Adenosine potentiates flow-induced dilation of coronary arterioles by activating K_{ATP} channels in endothelium. Am. J. Physiol. 269, H541–H549.

Latini, S., Corsi, C., Pedata, F. and Pepeu, G. (1996). The source of brain adenosine outflow during ischemia and electrical stimulation. Neurochem. Int. 28, 113–118.

Le Moine, O., Louis, H., Demols, A., Desalle, F., Demoor, F., Quertinmont, E., Goldman, M. and Deviere, J. (2000). Cold liver ischemia-reperfusion injury critically depends on liver T cells and is improved by donor pretreatment with interleukin 10 in mice. Hepatology 31, 1266–1274.

Lee, H.T. and Emala, C.W. (2000). Protective effects of renal ischemic preconditioning and adenosine pretreatment: role of A(1) and A(3) receptors. Am. J. Physiol. 278, F380–F387.

Lee, H.T. and Emala, C.W. (2001). Systemic adenosine given after ischemia protects renal function via A(2a) adenosine receptor activation. Am. J. Kidney Dis. 38, 610–618.

Li, B., Rosenbaum, P.S., Jennings, N.M., Maxwell, K.M. and Roth, S. (1999). Differing roles of adenosine receptor subtypes in retinal ischemia-reperfusion injury in the rat. Exp. Eye Res. 68, 9–17.

Linden, J. (1994). Cloned adenosine A3 receptors: pharmacological properties, species differences and receptor functions. Trends Pharmacol. Sci. 15, 298–306.

Linden, J. (2001). Molecular approach to adenosine receptors: receptor-mediated mechanisms of tissue protection. Ann. Rev. Pharmacol. Toxicol. 41, 775–787.

Linden, J., Thai, T., Figler, H., Jin, X. and Robeva, A.S. (1999). Characterization of human A(2B) adenosine receptors: radioligand binding, western blotting, and coupling to G(q) in human embryonic kidney 293 cells and HMC-1 mast cells. Mol. Pharmacol. 56, 705–713.

Link, A.A., Kino, T., Worth, J.A., McGuire, J.L., Crane, M.L., Chrousos, G.P., Wilder, R.L. and Elenkov, I.J. (2000). Ligand-activation of the adenosine A2a receptors inhibits IL-12 production by human monocytes. J.

Immunol. 164, 436–442.

Mizumura, T., Auchampach, J.A., Linden, J., Bruns, R.F. and Gross, G.J. (1996). PD 81,723, an allosteric enhancer of the A1 adenosine receptor, lowers the threshold for ischemic preconditioning in dogs. Circ. Res. 79, 415–423.

Monopoli, A., Lozza, G., Forlani, A., Mattavelli, A. and Ongini, E. (1998). Blockade of adenosine A$_{2A}$ receptors by SCH 58261 results in neuroprotective effects in cerebral ischaemia in rats. NeuroReport 9, 3955–3959.

Mubagwa, K. and Flameng, W. (2001). Adenosine, adenosine receptors and myocardial protection: An updated overview. Cardiovasc. Res. 52, 25–39.

Murry, C.E., Jennings, R.B. and Reimer, K.A. (1986). Preconditioning with ischemia: a delay of lethal cell injury in ischemic myocardium. Circulation 74, 1124–1136.

Nie, Z., Mei, Y., Ford, M., Rybak, L., Marcuzzi, A., Ren, H., Stiles, G.L. and Ramkumar, V. (1998). Oxidative stress increases A1 adenosine receptor expression by activating nuclear factor kappa B. Mol. Pharmacol. 53, 663–669.

Nomura, F., Forbess, J.M., Hiramatsu, T. and Mayer, J.E. (1997). Relationship of blood flow effects of adenosine during reperfusion to recovery of ventricular function after hypothermic ischemia in neonatal lambs. Circulation 96, II-227–232.

O'Regan, M.H., Simpson, R.E., Perkins, L.M. and Phillis, J.W. (1992). The selective A2 adenosine receptor agonist CGS21680 enhances excitatory transmitter amino acid release from the ischemic rat cerebral cortex. Neurosci. Lett. 138, 169–172.

Okusa, M.D., Linden, J., Huang, L., Rosin, D.L., Smith, D.F. and Sullivan, G. (2001). Enhanced protection from renal ischemia: Reperfusion injury with A2A-adenosine receptor activation and PDE 4 inhibition. Kidney Int. 59, 2114–2125.

Pagliaro, P., Gattullo, D., Rastaldo, R. and Losano, G. (2001). Ischemic preconditioning: from the first to the second window of protection. Life Sciences 69, 1-15.

Peart, J., Matherne, P.G., Cerniway, R.J. and Headrick, J.P. (2001). Cardioprotection with adenosine metabolism inhibitors in ischemic-reperfused mouse heart. Cardiovasc. Res. 52, 120–129.

Peirce, S.M., Skalak, T.C., Rieger, J.M., Macdonald, T.L. and Linden, J. (2001). Selective A(2A) adenosine receptor activation reduces skin pressure ulcer formation and inflammation. Am. J. Physiol. 281, H67–H74.

Peralta, C., Hotter, G., Closa, D., Prats, N., Xaus, C., Gelpi, E. and Rosello-Catafau, J. (1999). The protective role of adenosine in inducing nitric oxide synthesis in rat liver ischemia preconditioning is mediated by activation of adenosine A2 receptors. Hepatology 29, 126–132.

Ping, P., Song, C., Zhang, J.,Guo, Y., Cao, X., Li, R. C., Wu, W., Vondriska, T.M., Pass, J.M., Tang, X.L., Pierce, W.M. and Bolli, R. (2002). Formation of protein kinase

C(epsilon)-Lck signaling modules confers cardioprotection. J. Clin. Invest. 109, 499–507.

Rabb, H.,Daniels, F., O'Donnell, M., Haq, M.,Saba, S.R., Keane, W. and Tang, W.W. (2000). Pathophysiological role of T lymphocytes in renal ischemia-reperfusion injury in mice. Am. J. Physiol. 279, F525–F531.

Ross, S.D., Tribble, C.G., Linden, J., Gangemi, J.J., Lanpher, B.C.,Wang, A.Y. and Kron, I.L. (1999). Selective adenosine-A2A activation reduces lung reperfusion injury following transplantation. J. Heart Lung Transpl. 18, 994–1002.

Saransaari, P. and Oja, S.S. (2000). Modulation of the ischemia-induced taurine release by adenosine receptors in the developing and adult mouse hippocampus. Neuroscience 97, 425–430.

Schreieck, J. and Richardt, G. (1999). Endogenous adenosine reduces the occurrence of ischemia-induced ventricular fibrillation in rat heart. J. Mol. Cell. Cardiol. 31, 123–134.

Seligmann, C., Kupatt, C., Becker, B.F., Zahler, S. and Beblo, S. (1998). Adenosine endogenously released during early reperfusion mitigates postischemic myocardial dysfunction by inhibiting platelet adhesion. J. Cardiovasc. Pharmacol. 32, 156–163.

Shryock, J.C., Snowdy, S., Baraldi, P.G., Cacciari, B., Spalluto, G., Monopoli, A., Ongini, E., Baker, S.P. and Belardinelli, L. (1998). A2A-adenosine receptor reserve for coronary vasodilation. Circulation 98, 711–718.

Smolenski, R.T., Raisky, O., Slominska, E.M., Abunasra, H., Kalsi, K.K., Jayakumar, J., Suzuki, K. and Yacoub, M.H. (2001). Protection from reperfusion injury after cardiac transplantation by inhibition of adenosine metabolism and nucleotide precursor supply. Circulation 104, I246–I252.

Strickler, J., Jacobson, K.A. and Liang, B.T. (1996). Direct preconditioning of cultured chick ventricular myocytes: Novel functions of cardiac adenosine A2a and A3 receptors. J. Clin. Invest. 98, 1773–1779.

Sullivan, G.W., Rieger, J.M., Scheld, W.M., Macdonald, T.L. and Linden, J. (2001). Cyclic AMP-dependent inhibition of human neutrophil oxidative activity by substituted 2-propynylcyclohexyl adenosine A(2A) receptor agonists. Br. J. Pharmacol. 132, 1017–1026.

Takaoka, A., Nakae, I., Mitsunami, K., Yabe, T., Morikawa, S., Inubushi, T. and Kinoshita, M. (1999). Renal ischemia/reperfusion remotely improves myocardial energy metabolism during myocardial ischemia via adenosine receptors in rabbits: effects of "remote preconditioning". J. Am. Coll. Cardiol. 33, 556–564.

Thibault, N., Burelout, C., Harbour, D., Borgeat, P., Naccache, P.H. and Bourgoin, S.G. (2002). Occupancy of adenosine A2a receptors promotes fMLP-induced cyclic AMP accumulation in human neutrophils: impact on phospholipase D activity and recruitment of small GTPases to membranes. J. Leukocyte Biol. 71, 367–377.

Thornton, J.D., Liu, G.S. and Downey, J.M. (1993). Pretreatment with pertussis toxin blocks the protective effects of preconditioning: evidence for a G-protein mechanism. J. Mol. Cell. Cardiol. 25, 311–320.

Thornton, J.D., Thornton, C.S. and Downey, J.M. (1993). Effect of adenosine receptor blockade: preventing protective preconditioning depends on time of initiation. Am. J. Physiol. 265, H504–H508.

von Lubitz, D.K. (1999). Adenosine and cerebral ischemia: therapeutic future or death of a brave concept? Eur. J. Pharmacol. 371, 85–102.

Wollner, A., Wollner, S. and Smith, J.B. (1993). Acting via A2 receptors, adenosine inhibits the upregulation of Mac-1 (CD11b/CD18) expression on FMLP-stimulated neutrophils. Am. J. Resp. Cell Mol. Biol. 9, 179–185.

Yoshidome, H., Kato, A., Edwards, M.J. and Lentsch, A.B. (1999). Interleukin-10 inhibits pulmonary NF-κB activation and lung injury induced by hepatic ischemia-reperfusion. Am. J. Physiol. 277, L919–L923.

Zhao, Z.Q., Budde, J.M., Morris, C.,Wang, N.P., Velez, D.A., Muraki, S., Guyton, R.A. and Vinten-Johansen, J. (2001). Adenosine attenuates reperfusion-induced apoptotic cell death by modulating expression of Bcl-2 and Bax proteins. J. Mol. Cell. Cardiol. 33, 57–68.

Zwacka, R.M., Zhang, Y., Halldorson, J., Schlossberg, H., Dudus, L. and Engelhardt, J.F. (1997). CD4(+) T-lymphocytes mediate ischemia/reperfusion-induced inflammatory responses in mouse liver. J. Clin. Invest. 100, 279–289.

Sensing, Signaling and Cell Adaptation. Edited by K.B. Storey and J.M. Storey
© *2002 Elsevier Science B.V. All rights reserved.*

CHAPTER 5

NF-κB Function in Inflammation, Cellular Stress and Disease

Neil R. Chapman, Sonia Rocha, Ian M. Adcock[1] and Neil D. Perkins*
Division of Gene Regulation and Expression, School of Life Sciences, University of Dundee, Dundee, UK
[1]*Department of Thoracic Medicine, National Heart and Lung Institute, Imperial College School of*
Medicine, London, UK

1. Introduction

The cells constituting each organ system of the body are continuously challenged by an array of endocrine, paracrine and autocrine agonists which ensure the maintenance of cellular homeostasis. The endpoint of such stimulation is the alteration of gene expression within the target cell resulting in modification of cellular activity. Cytotoxic stimuli such as hypoxia, heat shock, oxidative stress, infection and associated inflammation and tissue damage are also encountered. In such cases, the cell must initiate a stress-response if it is to survive. Among the multitude of proteins that become activated during cellular stress, the Nuclear Factor κB (NF-κB) family of transcription factors, have been found to play a pivotal regulatory role.

NF-κB was first identified as a constitutively active transcription factor that bound the kappa light chain enhancer region within B-lymphocytes (Sen and Baltimore, 1986). Further work illustrated that this factor is present in virtually every other cell type within the body and as such is critically involved in a number of cellular functions including the cell cycle (Joyce et al., 2001), regulation of apoptosis (Barkett and Gilmore, 1999) and inflammation (Barnes and Karin, 1997). NF-κB itself can take a number of different forms and is composed of dimeric complexes formed from five distinct subunits, RelA (p65), RelB, c-Rel, NF-

κB1 (p105/p50) and NF-κB2 (p100/p52) (Ghosh et al., 1998; Perkins, 2000) (Fig. 5.1). These subunits all contain a region of approximately 300 amino acids within their amino termini, termed the Rel Homology Domain (RHD), which mediates DNA-binding together with dimerization and interactions with heterologous transcription factors. p105 and p100 share a high degree of homology and remain in the cytoplasm due to the masking of their nuclear localization signal (NLS) by the C-terminal ankyrin repeat domains. Both coranslational and proteolytic processing are mechanisms suggested to account for the generation of p50 (Ghosh et al., 1998), while signal-induced proteolysis resulting from NF-κB Inducing Kinase (NIK) activity is thought to account for p52 generation (Senftleben et al., 2001). In contrast, RelA, RelB and c-Rel do not require proteolytic processing and contain transactivation domains within their carboxy termini. Distinct combinations of subunits determine the specificity of transcriptional activation and gene knockouts in mice have revealed that all have distinct, non-overlapping functions (Gerondakis et al., 1999).

2. Activation of NF-κB

NF-κB activity is rapidly induced by over 150 different stimuli including ionizing radiation, cytokines, such as tumor necrosis factor α (TNF-α) and interleukin 1 (IL-1), growth factors, bacterial lipopolysaccharide (LPS) and hypoxia (Pahl, 1999). In

---

*Corresponding author

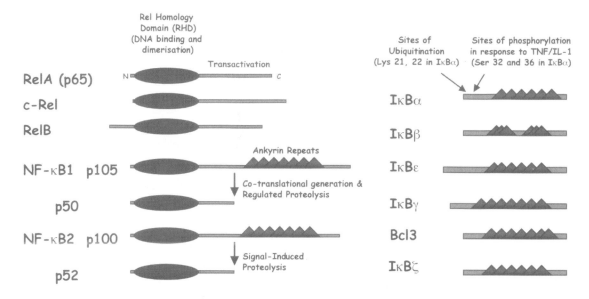

Fig. 5.1. The mammalian NF-κB family of transcription factors and inhibitors. NF-κB is composed of dimeric complexes formed from five distinct subunits, RelA (p65), RelB, c-Rel, NF-κB1 (p105/p50) and NF-κB2 (p100/p52). All contain a region of approximately 300 amino acids within their amino termini, termed the Rel Homology Domain (RHD), which mediates DNA-binding together with dimerization and interactions with heterologous transcription factors. p105 and p100 undergo proteolytic processing to yield the DNA-binding isoforms, p50 and p52. In contrast, RelA, RelB and c-Rel do not require proteolytic processing and contain transactivation domains within their carboxy termini. The IκB family of proteins serve to inhibit NF-κB function. They undergo IKK-induced phosphorylation of both conserved N-terminal serine residues. This acts as the signal for poly-ubiquitination of conserved lysine residues. Protein:protein interactions are mediated by the ankyrin repeat regions. IκBγ is an alternative splice variant of the p105 subunit and is thought to be specific for B-lymphocytes. Bcl-3 does not contain the signal responsive serine residues. Structurally, little is known about IκBζ except that its ankyrin repeat region shares the greatest homology to that in Bcl-3.

the majority of unstimulated cell types, NF-κB is retained within the cytoplasm in an inactive form, bound to its inhibitor protein, IκB. There a several members of the IκB family which are characterized by the presence of multiple ankyrin repeat motifs which mediate binding to NF-κB subunits (Huxford et al., 1998) (Fig. 5.1). Upon cellular stimulation, IκB becomes phosphorylated which in most cases results in its ubiquitination and subsequent degradation by the 26S proteasome (Karin and Ben-Neriah, 2000). NF-κB is then free to translocate to the nucleus. In most cell types, IκBα, β and ε perform this inhibitory role and all have conserved serine and lysine residues which mediate signal induced phosphorylation and ubiquitination, respectively (Ghosh et al., 1998; Karin and Delhase, 2000). While both IκBα and β share a number of similarities in their primary structures, they differ at the functional level. Regulation of NF-κB by IκBα is rapid and transient

while regulation by IκBβ is prolonged and both cell and stimulus specific (Ghosh et al., 1998; Perkins, 2000). In contrast to IκBα and β, IκBε only binds RelA/c-Rel containing complexes (Ghosh et al., 1998; Perkins, 2000). Expression levels of the different IκB isoforms can vary according to the cell type, resulting in differential NF-κB activation (Gerondakis et al., 1999).

Other IκB like proteins have also been described. IκBγ represents an alternative splice form of p105, encoding its ankyrin repeat containing C-terminus and is specific to murine B-cells (Ghosh et al., 1998). Although Bcl-3 has a similar C-terminal ankyrin repeat region to other IκB proteins, it lacks the conserved serine residues and therefore does not undergo signal induced proteolysis. Rather, Bcl-3 is nuclear and serves as a transcriptional co-activator for the p50 and p52 subunits (Ghosh et al., 1998). Recently, a further novel member of the IκB family was reported, termed IκBξ, which displays most similarity to Bcl-3 and is nuclear but appears to inhibit transcription from NF-κB complexes containing the p50 subunit (Yamazaki et al., 2001).

TNF-α-mediated induction is one of the most studied pathways of NF-κB activation and serves as a paradigm for induction of NF-κB by many other stimuli (Fig. 5.2). TNF-α binds with and

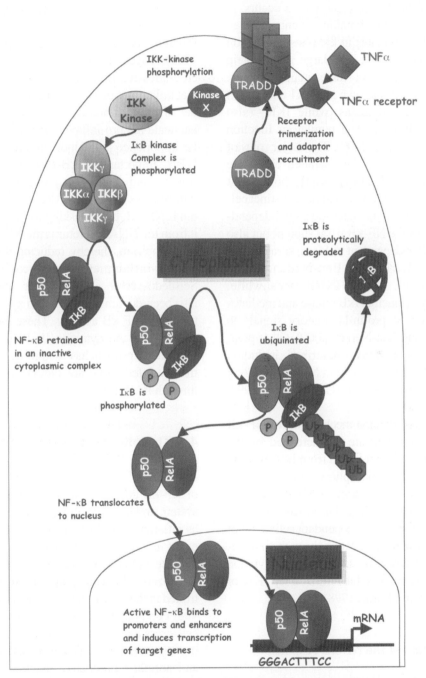

Fig. 5.2. NF-κB activation is the end point of a kinase cascade. TNF-α engages its receptor and induces trimerization. Adaptor proteins (only TRADD is illustrated for clarity) are then recruited to the cytoplasmic domains of the receptors. The signal is then transduced through a number of different IKK-kinases which serve to phosphorylate the IKKα and β subunits. Once activated, IKK then phosphorylates IκB. Phospo-IκB is then ubiquitinated and finally degraded by the 26S proteaosome. NF-κB is then free to translocate to the nucleus where it can activate gene expression.

induces TNF-receptor-1 (TNFR-1) trimerization (Baud and Karin, 2001). Recruitment of intra-cellular adaptor proteins in the order TRADD, RIP-1, FADD and finally TRAF-2, to the cytoplas-mic regions of the TNF receptors then occurs. A cascade of intracellular signalling events are then initiated which culminates in the phosphorylation of the IκB kinase (IKK). This large multiprotein complex is composed of three main subunits. IKKα and IKKβ (also known as IKK1 and IKK2) both harbor kinase domains, while IKKγ (also known as NEMO) serves a regulatory function (Karin and Ben-Neriah, 2000). The major roles of each IKK subunit have been detailed elsewhere (Karin and Ben-Neriah, 2000). Briefly, however, it is clear that both IKKα and IKKβ have distinct cel-lular functions. IKKα mediates the signal-depend-ant processing of the NF-κB p100 subunit but also has an NF-κB-independent role in keratinocyte development (Karin and Ben-Neriah, 2000; Senftleben et al., 2001). IKKβ, meanwhile, appears to be the principal IκB kinase and mediates the transduction of pro-inflammatory signals to NF-κB (Karin and Ben-Neriah, 2000). Other puta-tive IκB kinases have been described, including IKKε and pp90^{rsk} but their role in NF-κB signaling has not been clearly defined (Ghoda et al., 1997; Peters et al., 2000).

Other, less well defined mechanisms of NF-κB activation do exist and are characterized by the way in which NF-κB is released from IκB. Briefly, hypoxic injury induces phosphorylation of IκBα at tyrosine-42. This in turn promotes the dissociation of IκBα rather than its degradation (Imbert et al., 1996). Furthermore, UV-C dependant activation of NF-κB causes IκBα degradation through an unde-fined pathway that does not utilize phosphorylation of the critical serine residues utilized by IKK or tyrosine-42 (Li and Karin, 1998).

3. Attenuation of the NF-κB response

In non-diseased tissue, NF-κB activity is tightly regulated. Upon stimulation, the effects of NF-κB are typically transient and its signaling ability is rapidly attenuated by a number of negative

feedback loops (Perkins, 1997). Once activated by an appropriate stimulus, NF-κB will induce the expression of a specific subset of genes, which, in the context of immune regulation, includes immuno-receptors, cell adhesion molecules, pro-teins involved with antigen presentation, pro-inflammatory cytokines and chemokines (Pahl, 1999). Together, these gene products co-ordinate an aggressive inflammatory response. To ensure that inflammation is transient and localized, a num-ber of proteins are also upregulated that serve to attenuate the pro-inflammatory signal. Of particu-lar importance is the induction and resynthesis of IκBα. Importantly, IκBα contains both nuclear import and export sequences. Upon NF-κB induced resynthesis, IκBα enters the nucleus, binds NF-κB, both masking its NLS and removing it from the DNA and returns the NF-κB complex to the cytoplasm, thereby reducing the transcription of pro-inflammatory molecules (Arenzana-Seisdedos et al., 1997). Other compounds have also been documented to act in a feedback loop to terminate NF-κB activity. These include the A20 protein and the cytokine IL-10 (Perkins, 2000). This latter protein has a number of anti-inflam-matory effects that are thought to arise, in part, through its ability to suppress NF-κB function.

4. The inflammatory stress response

As briefly discussed above, one of the major roles of NF-κB concerns the regulation of the immune system and its associated downstream effects such as inflammation. Inflammation is a multi-factorial event involving a coordinated response from host defence systems and can be viewed as a highly organized cellular response to stress. When con-trolled in the appropriate fashion, inflammation is a vital component of the physiological response to cell stress and is associated with processes includ-ing regeneration of damaged and infected tissue and normal reproductive function. Dysregulation of this process however, ultimately leads to self-destruction of bodily tissues and occasionally to death. Chronic inflammation underlies disease occurring at a number of locations within the body.

Consistent with its regulation of cytokines and immunoreceptors, aberrant activation of NF-κB is often associated with inflammatory conditions including rheumatoid arthritis, asthma and inflammatory bowel disease (Barnes and Adcock, 1997; Bondeson et al., 1999; Jobin and Sartor, 2000; Neurath et al., 1998).

5. Rheumatoid arthritis

Rheumatoid arthritis (RA) is the most disabling form of rheumatic disease. It is an autoimmune condition characterized by synovial tissue hyperplasia and joint tissue destruction (Feldmann, 2001). Both TNF-α and IL-1 are pro-inflammatory cytokines whose expression is upregulated in the RA synovium. These cytokines are regulated, in part, through the activity of NF-κB and, in addition, induce NF-κB nuclear translocation and transcriptional activity (Pahl, 1999). There is strong evidence that NF-κB plays a critical role in the progression of RA. For example, an increased level of NF-κB activity has been observed in synovia from patients with RA (Yamasaki et al., 2001). This increased expression was correlated with both an increased proliferation and decreased level of apoptosis of such synovial cells. Significantly, inhibition of NF-κB activity in the rheumatoid synovium, using an adenovirus expressed dominant negative IκB super repressor, results in both a reduction in the level of inflammation seen and the severity of tissue destruction observed (Bondeson et al., 1999). Targeting NF-κB function directly, possibly through inhibiting IKK, might therefore prove to be a useful method of treating this disease, providing possible side effects caused by inhibition of normal NF-κB function can be avoided.

6. NF-κB and the airway

Asthma is a complex, multi-factorial condition characterized by chronic airway inflammation (Wills-Karp, 1999), thought to result, in part, from inappropriate activity of NF-κB and its downstream effectors (Barnes and Adcock, 1997). The majority of asthma-associated allergens, including oxidants and other environmental factors, induce NF-κB. Once activated, NF-κB co-ordinates the induction of multiple pro-inflammatory genes. Levels of a number of cytokines, including IL-1β, IL-8, macrophage inflammatory protein-1 (MIP-1) and TNF-α are seen to be elevated in secretions from macrophages obtained from bronchoalveolar lavage from patients with mild asthma (John et al., 1998). Other pro-inflammatory mediators, including adhesion molecules such as E-selectin, VCAM-1 and ICAM-1, which mediate the initial recruitment of inflammatory cells from the plasma into the epithelium, are also activated (Barnes and Adcock, 1997).

There are a number of possible therapeutic strategies for the treatment of asthma and these have been extensively reviewed elsewhere (Barnes, 1999). To date, however, the most efficacious therapy for the long-term management of asthmatic symptoms is the use of glucocorticoids. Glucocorticoids, once bound to the glucocorticoid receptor (GR), exert their effect, in part, through inhibiting inappropriate pro-inflammatory gene expression mediated by NF-κB (Barnes, 1998). It currently remains unresolved how GR exerts all its anti-inflammatory effects. It is clear that RelA and GR associate to form part of a large complex that does not inhibit RelA-DNA binding (Barnes, 1998; Nissen and Yamamoto, 2000). Furthermore, GR can indirectly modify both the phosphorylation of RNA-polymerase II CTD (Nissen and Yamamoto, 2000) and the acetylation pattern of histone H4 (Ito et al., 2000). There is, however, still some contention regarding whether glucocorticoids affect IκBα expression (Adcock et al., 1999).

A second major disease of the airway is cystic fibrosis (CF). While CF airway inflammation was originally perceived to arise through chronic bacterial infections, it became clear that airway epithelial cells harboring CF mutations also possessed increased NF-κB activity. This elevated NF-κB transactivation was concomitant with increases in pro-inflammatory cytokines such as IL-8 which in turn recruited increased numbers of polymorphonuclear leukocytes (Blackwell et al., 2001). The mechanism through which NF-κB is dysregulated in CF epithelia is now beginning to emerge. Weber et al. (2001) demonstrated that the proteins with the major CF mutation (Δ508) are mis-trafficked within the endoplasmic reticulum causing cell stress and the resultant activation of NF-κB. A second mutant form of CFTR, G551D, which does not function appropriately as a Cl⁻ channel, was trafficked normally, however, and did not cause ER-overload or NF-κB activation. This observation illustrates clearly that, while NF-κB activation does not influence the primary deficits of CF, the secondary, inflammatory effects associated with this factor greatly exacerbate the disease already present and warrant clinical intervention to prevent further tissue damage occurring. Interestingly, persistently nuclear NF-κB:IκBβ complexes, capable of binding DNA and stimulating transactivation of NF-κB responsive genes, have been observed in CF cells (Venkatakrishnan et al., 2000), suggesting that novel mechanisms of NF-κB regulation might be utilized in these diseased tissues.

7.   NF-κB and gastro-intestinal disease

The cells of the airway and intestines share a common embryonic origin resulting in many similar structural features such as modes of neural innervation and smooth muscle patterning. Importantly, they also share some similar mechanistic dysfunctions, salient examples being cystic fibrosis and epithelial inflammation. Inflammatory diseases affect all parts of the gastrointestinal tract and transcriptionally active NF-κB has been described in many of these cases. For example, the RelA subunit was identified in the antral region of the human stomach in G-(acid secreting) cells in *Helicobacter pylori*-associated gastritis. In these cells, which were also the major source of TNF-α, the presence of RelA was positively correlated with the severity of the disease state (van den Brink et al., 2000). Conversely, activation of NF-κB in rat gastric fibroblasts was vital for ulcer healing since pharmacological impairment of its function with indomethacin prolonged the healing time required (Takahashi et al., 2001). The reasons for these differences in function largely remain

undefined but may reflect the different spatial expression of NF-κB within the stomach wall. The fibroblasts form the part of the basement layer, while the G-cells are located within the gastric epithelium. More importantly, such differences illustrate the importance of maintaining stringent control of NF-κB activity within the cell. Chronic activation underlies disease while inhibition of function prevents an efficient healing process occurring.

Inflammation of other bowel sections, such as that observed in Crohn's disease and ulcerative colitis, are characterized by having highly inflamed mucosal layers concomitant with the up-regulation of a number of NF-κB regulated pro-inflammatory genes, including TNF-α, IL-1β, IL-6, and IL-8 (Jobin and Sartor, 2000; Neurath et al., 1998). Both animal models and human studies demonstrate the critical role played by NF-κB in these forms of inflammatory bowel disease (IBD). In mice, RelA antisense oligonucleotides have been shown to reverse chemically-induced chronic experimental colitis (Neurath et al., 1996). Moreover, further studies, using human biopsy specimens, have clearly identified activated NF-κB in both macrophages and epithelial cells of the inflamed mucosa (Rogler et al., 1998; Schreiber et al., 1998).

While these clinical studies clearly demonstrate that activated NF-κB plays a central role in the inflammatory disease process, little is known about how the pro-inflammatory signals are perpetuated and not attenuated once their function is served. An exception to this however, is the role of IL-10 in IBD. At the molecular level, IL-10 is thought to suppress a number of anti-inflammatory effects through the inhibition of both IKK activity and the DNA-binding activity of NF-κB (Schottelius et al., 1999). Interestingly, IL-10 deficient mice have been reported to have severe inflammatory bowel disease (Papadakis and Targan, 2000). Furthermore, topical IL-10 enema treatment of patients with ulcerative colitis caused a down regulation of pro-inflammatory cytokines (Papadakis and Targan, 2000). This illustrates the critical importance of anti-inflammatory cytokines, such as IL-10, being able to attenuate the pro-inflammatory signal once its function has been served.

Failure to do so facilitates chronic NF-κB activation with subsequent pathological consequences.

8. NF-κB and reproductive function

When controlled appropriately, inflammation is a vital physiological process. In the mammalian female reproductive tract, highly controlled, localized inflammatory reactions occur regularly. While such processes are not instigated as part of a typical stress response, they remain vital for the normal function of this system and are involved in ovulation, menstruation, implantation and labor.

8.1. The uterine cycle

The human menstrual cycle has been associated with the increased expression of a number of pro-inflammatory agents including IL-8, TNF-α and prostaglandins (Critchley et al., 1999) all of which are, at the level of transcription, under the control of NF-κB-responsive promoters. Recently, it was demonstrated that expression of the RelA subunit of NF-κB in human endometrial epithelial cells followed a distinct cycle (Laird et al., 2000). Epithelial cell staining for RelA was weak during the proliferative phase (roughly days 5–14) of the cycle but increased during the secretory phase (roughly days 14–28), being maximal at the time of implantation (Laird et al., 2000). During a normal menstrual cycle, serum progesterone levels start to decrease during the peri-menstrual phase (just prior to menstruation) when levels of NF-κB and other NF-κB-inducing agents are increased (King et al., 2001; Laird et al., 2000). Progesterone is viewed as the major hormone of pregnancy and a mutual antagonism between both progesterone and NF-κB has been previously reported (Kalkhoven et al., 1996). Thus, one can speculate that the decreasing progesterone level functionally relieves inhibition of NF-κB. At the molecular level, this would facilitate the expression of NF-κB responsive genes needed for menstruation (such as COX-2; a regulator of prostaglandin synthesis) and the progression of the normal cycle. If pregnancy is successful, progesterone levels do not decline but remain elevated due to the presence of a functional corpus luteum. It is possible that this elevation of progesterone is required to repress NF-κB and facilitate both the maintenance of the decidua and "maternal tolerance" of the developing fetus (Parham, 1996).

8.2. Implantation and labor

Approximately six days after successful fertilization, the embryo must bury itself into the endometrial wall to initiate placentation. Around the time of embryo implantation, various NF-κB-regulated proteins, including inflammatory mediators such as IL-1, IL-6, Leukaemia Inhibitory Factor (LIF) (Critchley et al., 1999) and certain matrix metalloproteinases such as gelatinase-B show increased expression (Campbell and Lees, 2000). Such observations suggest a role for NF-κB in this localized inflammatory reaction. Interestingly, NF-κB has also been proposed to play a pivotal role in the onset of labor through its ability to up-regulate pro-inflammatory genes including COX-2 (Allport et al., 2001) and IL-8 (Kunsch and Rosen, 1993) both of which are essential for the initiation of parturition. Interestingly, it has recently been reported that IKKα and NIK levels can be found elevated in early pregnancy while IKKβ levels are reduced (King et al., 2001). As detailed above, it is thought IKKα and IKKβ mediate differing arms of the NF-κB response with the former regulating signal-induced processing of the NF-κB p100 subunit while the latter mediates pro-inflammatory signaling through IκB (Karin and Delhase, 2000; Senftleben et al., 2001). Such selective activation of NF-κB could provide a mechanism for maintaining a limited NF-κB response in the presence of progesterone. Complete activation of NF-κB dependent immune and inflammatory response could result in the destruction of the "foreign" trophoblast and embryo *in utero*.

8.3. Inflammatory diseases of the reproductive tract

Certain inflammatory diseases of the reproductive tract such as endometriosis, (defined as the ectopic growth of endometriotic-like tissue outside the uterine cavity), are now believed to have an auto-immune component. Patients with endometriosis are seen to have elevated levels of certain pro-inflammatory cytokines and chemokines (e.g. RANTES) (Lebovic et al., 2001), T- and B-lymphocyte abnormalities and decreased cell apoptosis (Nothnick, 2001). These are all factors associated with other inflammatory disorders such as IBD and, as such, it is tempting to speculate on a role for chronically activated NF-κB in such gynaecological diseases.

9. Other NF-κB associated diseases

9.1. NF-κB, cancer and stress

Over the last few years, it has become apparent that NF-κB can frequently be found in an aberrantly active form in many forms of cancer. These include breast cancer, prostate cancer, pancreatic adenocarcinoma, melanoma, head and neck squamous cell carcinoma, acute lymphoblastic leu-kemia and hepatocellular carcinoma (Baldwin, 2001; Rayet and Gelinas, 1999). NF-κB has also been found to be either induced by or required for the function of a number of oncogenes (Mayo et al., 2001). The stressful conditions found within the tumor are also likely to result in the activation of NF-κB. For example, both serum starvation and hypoxia are known inducers of NF-κB activity (Pahl, 1999). Moreover, NF-κB is also activated by, and reduces the effectiveness of, many forms of cancer therapy (Baldwin, 2001). Although muta-tions and translocations of some NF-κB subunits have been reported in tumor cells, these appear to be relatively rare occurrences. Instead, similar to the situation in inflammatory diseases, active NF-κB in cancer is associated with the constitu-tively nuclear localization of "normal" NF-κB, often resulting from constitutively active IKK (Baldwin, 2001; Rayet and Gelinas, 1999). For these reasons, it is possible that drugs developed to inhibit NF-κB function in inflammatory diseases might also have useful applications as anti-cancer therapies. The association of NF-κB with cancer

also helped reveal the critical role that NF-κB plays as a regulator of apoptosis and the cell cycle. For example, NF-κB can both induce proliferation, through upregulation of Cyclin D1 and D2 (Joyce et al., 2001) and, under some circumstances, inhibit apoptosis (Barkett and Gilmore, 1999; Perkins, 2000). Both of these effects are characteristics acquired by transformed cells during tumor devel-opment. The contribution made by NF-κB to the progression of cancer is almost certainly more complex than this, however. For example, NF-κB activity has been associated with the growth, angiogenesis and metastasis of human melanoma cells in nude mice (Huang et al., 2000). It also is probable that the pro-inflammatory actions of NF-κB also contribute to tumor development under some circumstances (Fitzpatrick, 2001).

Two important physiological activators of NF-κB, relevant to our understanding of its role in cancer, are reactive oxygen species (ROS) and hypoxia/re-oxygenation. Several studies have shown direct activation of NF-κB following treat-ment with H_2O_2 (Li and Karin, 1999). However, most of the evidence for the role of ROS and reac-tive oxygen intermediates (ROI) comes from the observation that a variety of anti-oxidants can inhibit NF-κB activation (Li and Karin, 1999). The generation of ROS is a normal event during the life of a cell, and can be induced by external agents. In fact, most anti-cancer and anti-inflammatory agents rely on the induction of ROS for their thera-peutic effects (Hockel and Vaupel, 2001). In the context of cancer research, it is known that most solid tumors present extensive hypoxic or low oxy-genated areas, increasing the resistance to these therapeutic agents (Hockel and Vaupel, 2001). In addition to cancer, oxidative stress is a major factor in many other pathological conditions including several inflammatory diseases, neurodegenerative disorders and sunburns, as well as more normal events such as aging (Li and Karin, 1999). Interest-ingly, in most of these cases, NF-κB has been described to have an active role (Helenius, 2001).

In contrast to the situation with pro-inflam-matory cytokines, however, our knowledge of the pathways involved in NF-κB activation by oxida-tive stress is less advanced. Some observations indicate that alternative signaling pathways might be employed under these circumstances, suggesting that drugs developed to inhibit the IKK pathway may not prove useful for all pathological conditions involving NF-κB function. For example, activation of NF-κB by hypoxia identified a different site of IκBα phosphorylation to that seen with activators such as TNF-α. Instead of phosphorylation at serines 32 and 36, hypoxia induced phosphorylation at tyrosine 42 (Koong et al., 1994). Tyrosine phosphorylation of IκBα has also been demonstrated following oxidative stress (Schoonbroodt et al., 2000). This type of post-translational modification of IκBα, generally does not result in its degradation, but causes dissociation, thus freeing NF-κB and allowing it to translocate to the nucleus (Imbert et al., 1996).

Hypoxic conditions also activate the transcription factor HIF-1, the major mediator of both physiological and pathological responses to hypoxia (Semenza, 2000). As such, HIF-1 regulates the neo-vascularization process (mainly through direct induction of VEGF) and has been shown to prevent cell death in hypoxic conditions. Since HIF-1 is activated after hypoxia, most solid tumors over express this transcription factor (Semenza, 2000). Indeed, HIF-1 is considered to be a new target for anti-cancer therapy and recently it has been shown that disruption of hypoxia-induced transcription, suppresses tumor growth (Kung et al., 2000). Despite the observations that both HIF-1 and NF-κB are activated under hypoxic conditions and often co-operate towards the final cellular response, sharing some common responsive genes and co-activators, an interaction between these transcription factors has yet to be investigated. NF-κB function has been shown to be influenced both by both direct and indirect associations with many heterologous transcription factors (Perkins, 1997; Perkins, 2000) and it would be of great interest to determine whether a more direct interaction exists between these proteins.

9.2. Viral infection

The NF-κB pathway is a major target for a number of common eukaryotic viruses including Human Immunodeficiency Virus (HIV-1), Human T-cell Leukaemia Virus (HTLV-1), Hepatitis-B and -C, Epstein Bar Virus and Influenza Virus (Hiscott et al., 2001). All these viruses have evolved different mechanisms to commandeer the host NF-κB signaling pathway for their own ends (Hiscott et al., 2001). The (mis)use of the NF-κB pathway in viral-infected cells is important in an evolutionary context. From this perspective it would give the virus a very strong foothold within the host to utilize those critical genes which modulate the host's immune response, growth rate and survival.

9.3. Cardiovascular disease

NF-κB has been described in human atherosclerotic plaques while being absent in normal vessels (Collins and Cybulsky, 2001). This suggested a role for NF-κB in the induction of atherogenic lesions underlying cardiovascular failure. Interestingly, a number of NF-κB regulated gene products associated with the inflammatory response, including adhesion molecules (VCAM-1, ICAM-1 and E-selectin), chemokines and interleukins such as IL-8 are implicated in the genesis of atherosclerotic disease (Collins and Cybulsky, 2001). The induction of NF-κB-dependent gene expression appears to be a multi-factorial process with a number of stress-stimuli, including hypercholesterolaemia and hypoxia, initiating intracellular signaling pathways which are then integrated by NF-κB leading to characteristic patterns of gene expression observed in the diseased state (Collins and Cybulsky, 2001).

9.4. Neurology

NF-κB activity has been linked to normal neurological development, and both neuroprotective and neuropathological states (Mattson and Camandola, 2001). Neuronal cell agonists capable of activating NF-κB include nerve growth factor and erythropoietin, the latter stimulating NF-κB to act in an anti-apoptotic fashion and protect hypoxic neurones (Carter et al., 1996; Digicaylioglu and Lipton, 2001). Furthermore, activation of NF-κB by β-amyloid protein may underlie the function of

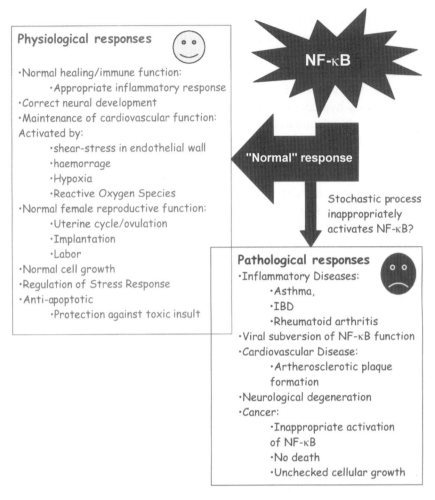

Fig. 5.3. NF-κB plays a critical role in regulating the cellular response to stress. Activated NF-κB is vital for a range of physiological processes including normal immune function and cell growth. Inappropriate activation of NF-κB however, is associated with a number of pathological conditions including inflammatory diseases, cardiovascular diseases and cancer. The dividing line between an appropriate cellular response to stress and a disregulated response that leads to disease is not clearly defined and there will be overlap between both cases.

NF-κB in the progression of Alzheimer's disease (Kaltschmidt et al., 1997).

10. Conclusions

The critical importance of NF-κB in mediating the cellular response to stress is illustrated in Fig. 5.3. While NF-κB is vital for a wide range of normal cellular functions, its inappropriate activation can either be a primary cause or contributing factor to many human diseases. This has now been clearly documented for many inflammatory and malignant diseases. While it is clear that the range of

pathological conditions associated with inappropriate NF-κB activation is increasing, the mechanism by which such conditions are manifest are often generally similar, however. For example, inflammatory diseases are characterized by having chronic over expression of a key set of pro-inflammatory genes and/or a deficit in the activity of anti-inflammatory mediators.

The involvement of the NF-κB pathway in human disease has resulted in it becoming an attractive therapeutic target. The effectiveness of such drugs will be of considerable interest. It is important to recognize that NF-κB does not serve as a lone regulator of the stress response, however. As discussed above, the HIF-1 transcription factor also has a critical function mediating the response to hypoxia. Indeed, how NF-κB functions co-ordinately with other stress activated transcription factors or signaling pathways is likely to be of considerable importance. Interestingly, treatment with SP600125, an inhibitor of c-Jun N-terminal kinase (JNK) activity, which regulates the AP-1 transcription factor, virtually abolished joint destruction in a rat model of inflammatory arthritis (Han et al., 2001). Such experiments suggest that small molecule inhibitors of stress activated kinases may have real therapeutic potential and it will be interesting to see if anti-IKK compounds will have a similar effect. Inhibitors of other pathways might also work well in combination with inhibition of NF-κB. Since NF-κB is an important mediator of normal cellular functions, it is possible that high doses of compounds that inhibit this pathway alone might result in a number of deleterious side effects, such as immune deficiency. An understanding of the other pathways that function co-operatively with NF-κB might allow combination therapies to be employed, either with new or existing treatments, with subsequently reduced dosages of the anti-NF-κB reagents. Such an approach might result in more effective therapy with more limited side effects. Nonetheless, advances in our understanding of NF-κB function offers real hope for new and novel treatments for many diseases where current therapies are of limited effect.

Acknowledgements

We would like to thank Lisa Anderson, Donna Bumpass, Kirsteen Campbell, David Gregory, Elisa Garcia, Kevin Roche, Omar Sharif, Alison Thain and the members of the Division of Gene Regulation and Expression at the University of Dundee for their help and assistance on this project. NDP is funded by a Royal Society University Fellowship, NRC is supported by a component grant from the Medical Research Council Dundee Gene Expression Co-operative Group, SR is funded by a grant from the Cancer Research Campaign and IMA is supported by GlaxoSmithKline.

References

Adcock, I.M., Nasuhara, Y., Stevens, D.A. and Barnes, P.J. (1999). Ligand-induced differentiation of glucocorticoid receptor (GR) trans-repression and transactivation: preferential targetting of NF-κB and lack of IκB involvement. Brit. J. Pharmacol. 127, 1003–1011.

Allport, V.C., Pieber, D., Slater, D.M., Newton, R., White, J.O. and Bennett, P.R. (2001). Human labour is associated with nuclear factor-κB activity which mediates cyclo-oxygenase-2 expression and is involved with the 'functional progesterone withdrawal'. Mol. Hum. Reprod. 7, 581–586.

Arenzana-Seisdedos, F., Turpin, P., Rodriguez, M., Thomas, D., Hay, R.T., Virelizier, J. and Dargemont, C. (1997). Nuclear localization of IκBα promotes active transport of NF-κB from the nucleus to the cytoplasm. J. Cell Sci. 110, 369–378.

Baldwin, A.S. (2001). Control of oncogenesis and cancer therapy resistance by the transcription factor NF-κB. J. Clin. Invest. 107, 241–246.

Barkett, M. and Gilmore, T.D. (1999). Control of apoptosis by Rel/NF-κB transcription factors. Oncogene 18, 6910–6924.

Barnes, P.J. (1998). Anti-inflammatory actions of glucocorticoids: molecular mechanisms. Clin. Sci. 94, 557–572.

Barnes, P.J. (1999). Therapeutic strategies for allergic diseases. Nature 402, B31–B38.

Barnes, P.J. and Adcock, I.M. (1997). NF-κB: A pivotal role in asthma and a new target for therapy. Trends Pharmacol. Sci. 18, 46–50.

Barnes, P.J. and Karin, M. (1997). Mechanisms of disease—Nuclear factor-κB—A pivotal transcription factor in chronic inflammatory diseases. New Eng. J. Med. 336, 1066–1071.

Baud, V. and Karin, M. (2001). Signal transduction by tumor necrosis factor and its relatives. Trends Cell Biol. 11, 372–377.

Blackwell, T.S., Stecenko, A.A. and Christman, J.W. (2001). Dysregulated NF-κB activation in cystic fibrosis: evidence for a primary inflammatory disorder. Am. J. Physiol.-Lung C. 281, L69–L70.

Bondeson, J., Foxwell, B., Brennan, F. and Feldmann, M. (1999). Defining therapeutic targets by using adenovirus: Blocking NF-κB inhibits both inflammatory and destructive mechanisms in rheumatoid synovium but spares anti-inflammatory mediators. Proc. Natl. Acad. Sci. U.S.A. 96, 5668–5673.

Campbell, S. and Lees, C. (2000). Obstetrics by Ten Teachers, 17 Edition, Arnold, London.

Carter, B.D., Kaltschmidt, C., Kaltschmidt, B., Offenhauser, N., BohmMatthaei, R., Baeuerle, P.A., and Barde, Y.A. (1996). Selective activation of NF-κB by nerve growth factor through the neurotrophin receptor p75. Science 272, 542–545.

Collins, T. and Cybulsky, M.I. (2001). NF-κB: pivotal mediator or innocent bystander in atherogenesis? J. Clin. Invest. 107, 255–264.

Critchley, H.O.D., Jones, R.L., Lea, R.G., Drudy, T.A., Kelly, R.W., Williams, A.R.W. and Baird, D.T. (1999). Role of inflammatory mediators in human endometrium during progesterone withdrawal and early pregnancy. J. Clin. Endocr. Metab. 84, 240–248.

Digicaylioglu, M., and Lipton, S.A. (2001). Erythropoietin-mediated neuroprotection involves cross-talk between Jak2 and NF-κB signalling cascades. Nature 412, 641–647.

Feldmann, M. (2001). Pathogenesis of arthritis: recent research progress. Nat. Immunol. 2, 771–773.

Fitzpatrick, F.A. (2001). Inflammation, carcinogenesis and cancer. Int. Immunopharmacol. 1, 1651–1667.

Gerondakis, S., Grossmann, M., Nakamura, Y., Pohl, T. and Grumont, R. (1999). Genetic approaches in mice to understand Rel/NF-κB and IκB function: transgenics and knockouts. Oncogene 18, 6888–6895.

Ghoda, L., Lin, X. and Greene, W.C. (1997). The 90-kDa ribosomal S6 kinase (pp90(rsk)) phosphorylates the N-terminal regulatory domain of IκB alpha and stimulates its degradation *in vitro*. J. Biol. Chem. 272, 21281–21288.

Ghosh, S., May, M.J. and Kopp, E.B. (1998). NF-κB and rel proteins: Evolutionarily conserved mediators of immune responses. Annu. Rev. Immunol. 16, 225–260.

Han, Z.N., Boyle, D.L., Chang, L.F., Bennett, B., Karin, M., Yang, L., Manning, A.M. and Firestein, G.S. (2001). C-Jun N-terminal kinase is required for metalloproteinase expression, and joint destruction in inflammatory arthritis. J. Clin. Invest. 108, 73–81.

Helenius, M., Kyrylenko, S., Vehvilainen, P. and Salminen, A. (2001). Characterization of aging-associated up-regulation of constitutive NF-κB binding activity. Antioxid. Redox Signal. 3, 147–156.

Hiscott, J., Kwon, H. and Genin, P. (2001). Hostile takeovers: viral appropriation of the NF-κB pathway. J. Clin. Invest. 107, 143–151.

Hockel, M. and Vaupel, P. (2001). Tumor hypoxia: Definitions and current clinical, biologic, and molecular aspects. J. Natl. Cancer I. 93, 266–276.

Huang, S., DeGuzman, A., Bucana, C.D. and Fidler, I.J. (2000). Nuclear factor-κB activity correlates with growth, angiogenesis, and metastasis of human melanoma cells in nude mice. Clin. Cancer Res. 6, 2573–2581.

Huxford, T., Huang, D.B., Malek, S. and Ghosh, G. (1998). The crystal structure of the IκB α/NF-κB complex reveals mechanisms of NF-κB inactivation. Cell 95, 759–770.

Imbert, V., Rupec, R.A., Livolsi, A., Pahl, H.L., Traenckner, E.B., Mueller-Dieckmann, C., Farahifar, D., Rossi, B., Auberger, P., Baeuerle, P.A. and Peyron, J. (1996). Tyrosine phosphorylation of IκB-α activates NF-κB without proteolytic degradation of IκB-α. Cell 86, 787–798.

Ito, K., Barnes, P.J. and Adcock, I.M. (2000). Glucocorticoid receptor recruitment of histone deacetylase 2 inhibits interleukin-1 beta-induced histone H4 acetylation on lysines 8 and 12. Mol. Cell. Biol. 20, 6891–6903.

Jobin, C. and Sartor, R.B. (2000). The IκB/NF-κB system: a key determinant of mucosal inflammation and protection. Am. J. Physiol. 278, C451–C462.

John, M., Lim, S., Seybold, J., Jose, P., Robichaud, A., O'Connor, B., Barnes, P.J. and Chung, K.F. (1998). Inhaled corticosteroids increase interleukin-10 but reduce macrophage inflammatory protein-1 alpha, granulocyte-macrophage colony-stimulating factor, and interferon-gamma release from alveolar macrophages in asthma. Am. J. Resp. Crit. Care 157, 256–262.

Joyce, D., Albanese, C., Steer, J., Fu, M.F., Bouzahzah, B. and Pestell, R.G. (2001). NF-κB and cell-cycle regulation: the cyclin connection. Cytokine Growth F. R. 12, 73–90.

Kalkhoven, E., Wissink, S., van der Saag, P T. and van der Burg, B. (1996). Negative interaction between the RelA(p65) subunit of NF-κB and the progesterone receptor. J. Biol. Chem. 271, 6217–6224.

Kaltschmidt, B., Uherek, M., Volk, B., Baeuerle, P.A. and Kaltschmidt, C. (1997). Transcription factor NF-κB is activated in primary neurons by amyloid beta peptides and in neurons surrounding early plaques from patients with Alzheimer disease. Proc. Natl. Acad. Sci. U.S.A. 94, 2642–2647.

Karin, M. and Ben-Neriah, Y. (2000). Phosphorylation meets ubiquitination: The control of NF-κB activity. Annu. Rev. Immunol. 18, 621–663.

Karin, M. and Delhase, M. (2000). The IκB kinase (IKK) and NF-κB: key elements of proinflammatory signalling. Semin. Immunol. 12, 85–98.

King, A.E., Critchley, H.O.D. and Kelly, R.W. (2001). The NF-κB pathway in human endometrium and first trimester decidua. Mol. Human Reprod.7, 175–183.

Koong, A.C., Chen, E.Y. and Giaccia, A.J. (1994). Hypoxia causes the activation of nuclear factor κB through the phosphorylation of IκBα on tyrosine residues. Cancer Res. 54, 1425–1430.

Kung, A.L., Wang, S., Klco, J.M., Kaelin, W.G. and Livingston, D.M. (2000). Suppression of tumor growth through disruption of hypoxia-inducible transcription. Nat. Med. 6, 1335–1340.

Kunsch, C. and Rosen, C. A. (1993). NF-κB subunit-specific regulation of the interleukin-8 promoter. Mol. Cell. Biol. 13, 6137–6146.

Laird, S.M., Tuckerman, E.M., Cork, B.A. and Li, T.C. (2000). Expression of nuclear factor κB in human endometrium; role in the control of interleukin 6 and leukaemia inhibitory factor production. Mol. Hum. Reprod. 6, 34–40.

Lebovic, D.I., Chao, V.A., Martini, J.F. and Taylor, R.N. (2001). IL-1 beta induction of RANTES (regulated upon activation, normal T cell expressed and secreted) chemokine gene expression in endometriotic stromal cells depends on a Nuclear Factor-κB site in the proximal promoter. J. Clin. Endocr. Metab. 86, 4759–4764.

Li, N. and Karin, M. (1999). Is NF-κB the sensor of oxidative stress? FASEB J. 13, 1137–1143.

Li, N.X. and Karin, M. (1998). Ionizing radiation and short wavelength UV activate NF-κB through two distinct mechanisms. Proc. Natl. Acad. Sci. USA 95, 13012–13017.

Mattson, M.P. and Camandola, S. (2001). NF-κB in neuronal plasticity and neurodegenerative disorders. J. Clin. Invest. 107, 247–254.

Mayo, M.W., Norris, J.L. and Baldwin, A.S. (2001). Ras regulation of NF-κB and apoptosis. Meth. Enzymol. 333, 73–87.

Neurath, M.F., Becker, C. and Barbulescu, K. (1998). Role of NF-κB in immune and inflammatory responses in the gut. Gut 43, 856–860.

Neurath, M.F., Pettersson, S., zum Buschenfelde, K.H.M. and Strober, W. (1996). Local administration of antisense phosphorothioate oligonucleotides to the p65 subunit of NF-κB abrogates established experimental colitis in mice. Nat. Med. 2, 998–1004.

Nissen, R.M. and Yamamoto, K.R. (2000). The glucocorticoid receptor inhibits NF-κB by interfering with serine-2 phosphorylation of the RNA polymerase II carboxy-terminal domain. Genes Dev. 14, 2314–2329.

Nothnick, W.B. (2001). Treating endometriosis as an autoimmune disease. Fertil. Steril. 76, 223–231.

Pahl, H.L. (1999). Activators and target genes of Rel/NF-κB transcription factors. Oncogene 18, 6853–6866.

Papadakis, K.A. and Targan, S.R. (2000). Role of cytokines in the pathogenesis of inflammatory bowel disease. Annu. Rev. Med. 51, 289–298.

Parham, P. (1996). Immunology: Keeping mother at bay. Curr. Biol. 6, 638–641.

Perkins, N.D. (1997). Achieving transcriptional specificity with NF-κB. Int. J. Biochem. Cell Biol. 29, 1433–1448.

Perkins, N.D. (2000). The Rel/NF-κB family: friend and foe. Trends Biochem. Sci. 25, 434–440.

Peters, R.T., Liao, S.M. and Maniatis, T. (2000). IKKε is part of a novel PMA-inducible IκB kinase complex. Mol. Cell 5, 513–522.

Rayet, B. and Gelinas, C. (1999). Aberrant rel/nfkb genes and activity in human cancer. Oncogene 18, 6938–6947.

Rogler, G., Brand, K., Vogl, D., Page, S., Hofmeister, R., Andus, T., Knuechel, R., Baeuerle, P. A., Scholmerich, J. and Gross, V. (1998). Nuclear factor κB is activated in macrophages and epithelial cells of inflamed intestinal mucosa. Gastroenterology 115, 357–369.

Schoonbroodt, S., Ferreira, V., Best-Belpomme, M., Boelaert, J.R., Legrand-Poels, S., Korner, M. and Piette, J. (2000). Crucial role of the amino-terminal tyrosine residue 42 and the carboxyl-terminal PEST domain of IκBα in NF-κB activation by an oxidative stress. J. Immunol. 164, 4292–4300.

Schottelius, A.J.G., Mayo, M.W., Sartor, R.B. and Baldwin, A.S. (1999). Interleukin-10 signaling blocks Inhibitor of κB kinase activity and Nuclear Factor κB DNA binding. J. Biol. Chem. 274, 31868–31874.

Schreiber, S., Nikolaus, S. and Hampe, J. (1998). Activation of nuclear factor κB in inflammatory bowel disease. Gut 42, 477–484.

Semenza, G.L. (2000). HIF-1: mediator of physiological and pathophysiological responses to hypoxia. J. Appl. Physiol. 88, 1474–1480.

Sen, R. and Baltimore, D. (1986). Inducibility of kappa immunoglobulin enhancer-binding protein NF-κB by a posttranslational mechanism. Cell 47, 921–928.

Senftleben, U., Cao, Y.X., Xiao, G.T., Greten, F.R., Krahn, G., Bonizzi, G., Chen, Y., Hu, Y.L., Fong, A., Sun, S.C. and Karin, M. (2001). Activation by IKK α of a second, evolutionary conserved, NF-κB signaling pathway. Science 293, 1495–1499.

Takahashi, S., Fujita, T. and Yamamoto, A. (2001). Role of nuclear factor-κB in gastric ulcer healing in rats. Am. J. Physiol. 280, G1296–G1304.

van den Brink, G.R., ten Kate, F.J., Ponsioen, C.Y., Rive, M.M., Tytgat, G.N., van Deventer, S.J.H. and Peppelenbosch, M.P. (2000). Expression and activation of NF-κB in the antrum of the human stomach. J. Immunol. 164, 3353–3359.

Venkatakrishnan, A., Stecenko, A.A., King, G., Blackwell, T.R., Brigham, K.L., Christman, J.W. and Blackwell, T.S. (2000). Exaggerated activation of Nuclear Factor-κB and altered IκB-β processing in cystic fibrosis bronchial epithelial cells. Am. J. Resp. Cell Mol. 23, 396–403.

Weber, A.J., Soong, G., Bryan, R., Saba, S. and Prince, A. (2001). Activation of NF-κB in airway epithelial cells is dependent on CFTR trafficking and Cl⁻ channel function. Am. J. Physiol. 281, L71–L78.

Wills-Karp, M. (1999). Immunologic basis of antigen-induced airway hyperresponsiveness. Annu. Rev. Immunol. 17, 255–281.

Yamasaki, S., Kawakami, A., Nakashima, T., Nakamura, H., Kamachi, M., Honda, S., Hirai, Y., Hida, A., Ida, H., Migita, K., Kawabe, Y., Koji, T., Furuichi, I., Aoyagi, T. and Eguchi, K. (2001). Importance of NF-κB in rheumatoid synovial tissues: in situ NF-κB expression and in vitro study using cultured synovial cells. Ann. Rheum. Dis. 60, 678–684.

Yamazaki, S., Muta, T. and Takeshige, K. (2001). A novel IκB protein, IκB-ξ, induced by proinflammatory stimuli, negatively regulates Nuclear Factor-κB in the nuclei. J. Biol. Chem. 276, 27657–27662.

Sensing, Signaling and Cell Adaptation. Edited by K.B. Storey and J.M. Storey

CHAPTER 6

MAPping Stress Survival in Yeasts: From the Cell Surface to the Nucleus

Aaron Ngocky Nguyen and Kazuhiro Shiozaki*
Section of Microbiology, University of California, Davis, CA 95616, USA

1. Introduction

Mitogen-activated protein (MAP) kinase cascades represent an evolutionarily conserved signaling mechanism used by eukaryotic cells to respond to alterations in their environment (Schaeffer and Weber, 1999). As such, they regulate a wide range of cellular processes, including gene expression, cellular homeostasis, and differentiation (Cobb and Goldsmith, 1995; Herskowitz, 1995). The basic mechanism of signal transduction through these MAP kinase (MAPK) modules appears to be very similar. A MAPK is activated by the phosphorylation of neighboring threonine and tyrosine residues within its kinase activation loop, a step carried out by a dual-specificity MAPK kinase (MAPKK or MEK). The MAPKK itself is activated by the phosphorylation of specific serine/threonine residues in its catalytic domain, a reaction mediated by a MAPKK kinase (MAPKKK or MEKK). An activated MAPK then executes effector functions through the phosphorylation of substrates, such as nuclear transcription factors and cytosolic proteins (Marshall, 1994; Hill and Treisman, 1995). Thus, a MAPK cascade involves three sequentially acting kinases, MAPKKK → MAPKK → MAPK.

Over the last decade, MAPK homologs that are responsive to environmental stress stimuli have been identified in various organisms and are referred to as stress-activated protein kinases

(SAPKs). These SAPKs include the highly homologous Hog1 family of MAPKs consisting of Hog1 in the budding yeast *Saccharomyces cerevisiae* (Brewster et al., 1993), Spc1 (also known as Sty1 or Phh1) in the fission yeast *Schizosaccharomyces pombe* (Millar et al., 1995; Shiozaki and Russell, 1995; Kato et al., 1996), and p38 in mammalian cells (Han et al., 1994; Lee et al., 1994). Studies in higher eukaryotes have shown that the p38 MAPK cascade is activated by a variety of stress conditions (e.g., osmotic stress, heat shock, oxidative stress), bacterial endotoxins and inflammatory cytokines (Han et al., 1994; Lee et al., 1994; Rouse et al., 1994; Raingeaud et al., 1995); however, the detailed mechanisms of how mammalian cells perceive these environmental cues are not fully understood.

In this chapter, we review the current understanding of stress-activated MAPK cascades in budding and fission yeasts. Reviews of these pathways with more historical perspectives are available (Gustin et al., 1998; Posas et al., 1998a; Millar, 1999). Here, we describe the components of the Hog1 and Spc1 cascades and the functions these proteins perform. For each pathway, we first focus on the upstream elements that are important for relaying stress signals to the MAPK cascades. Subsequently, we discuss how stress signals are conveyed to the nucleus, where the induction of stress-response genes occurs. Because of the wealth of genetic and biochemical data as well as the conservation of the stress MAPK cascades through evolution, budding and fission yeasts

---

*Corresponding author

provide excellent model systems for understanding how eukaryotic cells detect and respond to diverse environmental stimuli.

2. The Hog1 MAPK cascade in budding yeast

Saccharomyces cerevisiae adapts to growth under increased external osmolarity through activation of the high osmolarity glycerol response (HOG) pathway (Gustin et al., 1998; Posas et al., 1998a). Upon a high osmolarity stimulus, Hog1 MAPK is activated by Pbs2 MAPKK through phosphorylation at conserved threonine and tyrosine residues (Brewster et al., 1993). Pbs2, on the other hand, is activated by two independent upstream mechanisms that involve either the Ste11 MAPKKK or one of two partially redundant MAPKKKs, Ssk2

and Ssk22 (Fig. 6.1) (Maeda et al., 1994; Maeda et al., 1995; Posas and Saito, 1997). Activated Hog1 translocates from the cytoplasm to the nucleus to initiate a series of adaptive responses, such as the transcription of stress-protective genes and the intracellular accumulation of glycerol, a compatible osmolyte to counterbalance the high osmolarity in the extracellular environment (Blomberg and Adler, 1992; Gustin et al., 1998).

2.1. Regulation of the Hog1 pathway by two independent osmosensing pathways

Budding yeast cells detect extracellular hyperosmolarity by two independent pathways that utilize different transmembrane proteins, Sln1 and Sho1, to activate Hog1 MAPK cascade (Fig. 6.1) (Maeda et al., 1994; Maeda et al., 1995). Sln1 is a homolog of bacterial two-component signal

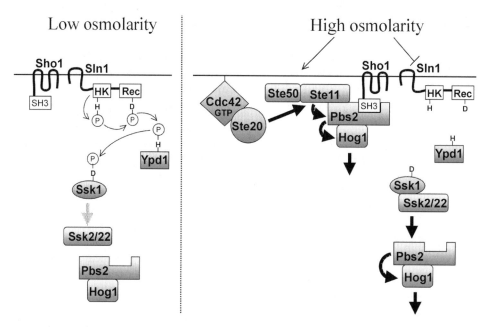

Fig. 6.1. Activation of the Hog1 MAPK cascade by the Sln1 and Sho1 pathways. Under low osmolarity conditions (left), the Sln1 histidine kinase is active, and a phosphoryl group is transferred to the Ssk1 response regulator via Ypd1. Phosphorylated Ssk1 cannot activate the Hog1 cascade. In contrast, high osmolarity stress inhibits Sln1 (right), and the non-phosphorylated form of Ssk1 activates Ssk2/22 MAPKKKs, which is followed by sequential activation of Pbs2 MAPKK and Hog1 MAPK. High osmolarity stress may also result in the recruitment of Pbs2 MAPKK to the plasma membrane through the interaction between a proline-rich sequence within its N-terminus and an SH3 domain of Sho1. Because Pbs2 also binds Hog1 and Ste11 MAPKKK, and Ste11 constitutively binds Ste50, it is possible that a macromolecular assembly comprised of Ste50, Ste11, Pbs2 and Hog1 is recruited to the plasma membrane. Under the stress condition, active (GTP-bound) Cdc42 may localize Ste20 to the plasma membrane, where it can activate Ste11 MAPKKK. Ste11 then activates Pbs2 to transmit the high osmolarity signal to Hog1 MAPK.

transducers; it is composed of an N-terminal extracellular domain, a cytoplasmic histidine kinase domain, and a C-terminal receiver domain (Ota and Varshavsky, 1993). Genetic and biochemical investigations indicate that under low osmolarity conditions, the Sln1 histidine kinase is active and phosphorylates a histidine residue within the kinase domain (Maeda et al., 1994; Posas et al., 1996). Next, the phosphate is transferred to an aspartate residue of the Sln1 receiver domain. Finally, the phosphate is transferred, via a phosphorelay mechanism involving the Ypd1 histidine phosphotransfer protein, to an aspartate residue within the receiver domain of the Ssk1 response regulator (Fig. 6.1). Under high osmolarity conditions, however, the activity of Sln1 is suppressed, resulting in an abundance of non-phosphorylated Ssk1 (Maeda et al., 1994; Posas et al., 1996). It is believed that the unphosphorylated form of Ssk1 binds to the N-terminal autoinhibitory region of Ssk2 (and perhaps Ssk22) MAPKKK, causing partial activation of the kinase (Maeda et al., 1995; Posas and Saito, 1998). Ssk2 then autophosphorylates serine and threonine residues, which may increase its catalytic activity towards Pbs2 MAPKK (Posas and Saito, 1998).

Transmission of the high osmolarity signal to the Hog1 MAPK cascade by the Sho1 pathway is, however, less straightforward (Fig. 6.1). The Sho1 protein contains four N-terminal transmembrane domains and a cytoplasmic SH3 domain (Maeda et al., 1995). Sho1 appears to be localized at the plasma membrane, specifically at sites of polarized growth (Raitt et al., 2000; Reiser et al., 2000). It is postulated that Pbs2 MAPKK is recruited to these sites in response to osmotic stress through the interaction between the Sho1 SH3 domain and a proline-rich motif near the N-terminus of Pbs2 (Maeda et al., 1995; Raitt et al., 2000; Reiser et al., 2000). Because Pbs2 can also physically associate with Hog1 MAPK (Posas and Saito, 1997) as well as Ste11 MAPKKK, which forms a complex with an essential subunit, Ste50 (Posas et al., 1998b), it is likely that a macromolecular assembly comprised of Ste50, Ste11, Pbs2, and Hog1 is localized to the plasma membrane by Sho1. Recent genetic studies indicate that the Cdc42 GTPase and a

p21-activated kinase (PAK) homolog, Ste20, are also involved in the Sho1 branch of the HOG pathway (O'Rourke and Herskowitz, 1998; Raitt et al., 2000; Reiser et al., 2000). Interestingly, Cdc42 and Ste20 have been implicated in the establishment of cell polarity during the cell cycle as well as in the cellular responses to mating pheromone and nutritional limitation (Pringle et al., 1995; Johnson, 1999). It was demonstrated that binding of the activated, GTP-bound form of Cdc42 to Ste20 localizes Ste20 to the plasma membrane, at sites of polarized growth (Peter et al., 1996). By inference to these situations, a model has been proposed such that in response to high osmolarity stress, the binding of activated Cdc42 to Ste20 juxtaposes Ste20 to the Sho1-maintained macromolecular assembly in the plasma membrane, where Ste20 activates the Ste11 MAPKKK by direct phosphorylation (Fig. 6.1) (Wu et al., 1995; Mosch et al., 1996; Raitt et al., 2000). Subsequently, Ste11 activates Pbs2 to transmit the stress signal to Hog1 MAPK (Posas and Saito, 1997). However, all Pbs2 molecules do not need to be localized to the plasma membrane, since in Δ*sho1* cells Hog1 is efficiently activated through the Sln1 pathway (Maeda et al., 1995). It is possible that one population of Pbs2 remains in the cytoplasm, where it can be activated by Ssk2 and Ssk22 MAPKKKs, while another population exists close to the plasma membrane, where it can be stimulated by Ste11 MAPKKK (Fig. 6.1).

2.2. Nuclear translocation of Hog1 upon osmotic stress

In order for the high osmolarity signal to be converted to the expression of stress-response genes, components of the Hog1 pathway, thought to reside in the cytoplasm, must cross the nuclear envelope. Silver and colleagues showed that while Pbs2 MAPKK and Ste11 MAPKKK are mostly cytoplasmic, Hog1 MAPK transiently accumulates in the nucleus following osmotic stress (Ferrigno et al., 1998). Notably, Hog1 phosphorylation, rather than its catalytic activity, is necessary for nuclear import of the MAPK (Ferrigno et al., 1998; Reiser et al., 1999). Furthermore, Hog1 import appears to be an active process since it depends on the small

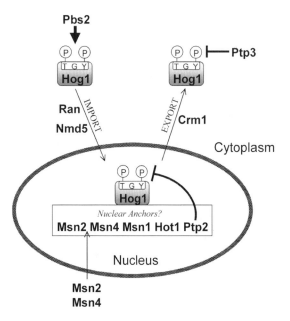

Fig. 6.2. Nuclear translocation of Hog1 MAPK in response to high osmolarity stress. Following activation by Pbs2 MAPKK, Hog1 transiently accumulates in the nucleus. Import of Hog1 into the nucleus requires Hog1 phosphorylation but not its catalytic activity. Hog1 nuclear translocation appears to be an active process since Ran (Gsp1) and Nmd5 (β-importin homolog) are required. Once in the nucleus, retention of Hog1 is dependent on the Msn1, Msn2, Msn4 and Hot1 transcription factors as well as on Ptp2, a protein tyrosine phosphatase (PTP) that negatively regulates Hog1. Msn2 and Msn4 are known to move into the nucleus upon stress. Hog1 transport out of the nucleus requires the exportin Crm1. Another PTP enzyme, Ptp3, also contributes to Hog1 dephosphorylation in the cytoplasm.

GTP binding protein, Ran (Gsp1), and on the β-importin homolog, Nmd5 (Fig. 6.2) (Ferrigno et al., 1998). Although it remains to be determined whether Nmd5 directly binds Hog1, it was found that Nmd5 translocates into the nucleus upon osmotic stress with similar kinetics as those of Hog1. Therefore, Nmd5 may be a specific nuclear import receptor for active Hog1 (Ferrigno et al., 1998).

Once in the nucleus, Hog1 is thought to regulate transcription factors to allow for the expression of various genes required for stress survival. Studies have shown that Hog1 nuclear staining remains high for up to 15 minutes, then gradually declines to a steady-state level within 1 hour after osmotic stress (Reiser et al., 1999). How is the retention of

Hog1 in the nucleus accomplished? As detailed in the next section, experiments in fission yeast suggest that Spc1 MAPK remains in the nucleus by associating with its nuclear target, the Atf1 transcription factor (Gaits et al., 1998). In a similar scenario, Hog1 is postulated to reside transiently in the nucleus by its interaction with various nuclear proteins (Fig. 6.2). These include the Msn1 and Hot1 transcription factors that play a role in osmotic stress-induced gene expression (Rep et al., 1999) as well as the general stress response transcription factors, Msn2 and Msn4, which undergo nuclear localization upon stress (Gorner et al., 1998; Reiser et al., 1999). It is important to note, however, that no clear physical interaction between Hog1 and these factors have been detected. Rather, this nuclear anchor model is based on localization of Hog1 in mutants lacking these transcription factors (Reiser et al., 1999; Rep et al., 1999). In addition, nuclear tyrosine phosphatase, Ptp2, which negatively regulates Hog1 by dephosphorylation, has been proposed as an anchor for Hog1 in the nucleus, since strains lacking Ptp2 show decreased Hog1 nuclear retention (Mattison and Ota, 2000).

As mentioned above, nuclear accumulation of Hog1 in response to high osmolarity stress is only transient (Ferrigno et al., 1998; Reiser et al., 1999). Rapid nuclear export of Hog1 after stress adaptation may be essential for cell survival since hyperactivation of the pathway, for example in strains expressing the constitutively active form of Ssk2 MAPKKK, is lethal (Maeda et al., 1995). Because nuclear import and export of Hog1 correlate with its phosphorylation and dephosphorylation, respectively (Ferrigno et al., 1998; Reiser et al., 1999), it appears that dephosphorylation of the MAPK drives its export. However, normal nuclear export of Hog1 is observed in a strain lacking the protein tyrosine phosphatases (PTPs) that dephosphorylate Hog1, indicating that tyrosine dephosphorylation of Hog1 is not essential for nuclear export (Mattison and Ota, 2000). On the other hand, export of Hog1 from the nucleus requires Hog1 catalytic activity (Reiser et al., 1999), the reason for which is uncertain. Hog1 activity might be required to stimulate an export system or to induce the release of Hog1 from its nuclear anchors

(Reiser et al., 1999). Like the export of other proteins (Stade et al., 1997), the transport of Hog1 out of the nucleus depends on the exportin, Crm1 (Xpo1) (Fig. 6.2), although it is unclear whether Crm1 interacts with Hog1 directly (Ferrigno et al., 1998).

2.3. Induction of adaptive responses to osmostress through Hog1 targets

As stated earlier, budding yeast cells respond to a hyperosmolarity environment by up-regulating a set of stress-defensive genes, a number of which are known to be regulated by Hog1 MAPK (Posas et al., 2000; Rep et al., 2000). These include ones that are important for increases in the intracellular osmolarity, such as genes encoding enzymes involved in glycerol synthesis (i.e., *GPD1*, *GPP2*, and *GLO1*), and ones that are critical for the general stress response, like genes encoding heat shock proteins (*HSP12*, *HSP70*, and *HSP104*) and cytosolic catalase (*CTT1*) (Gustin et al., 1998). Interestingly, it was recently shown that Hog1 associates with the promoters of some of these genes, a process that depends on various transcription factors. Thus, it was postulated that Hog1 constitutes a part of transcription activation complexes (Alepuz et al., 2001). However, the mechanisms of gene regulation by activated Hog1 have not been completely understood, mainly due to the elusiveness of the direct substrates of Hog1. While the transcription factors Msn1, Msn2, Msn4 and Hot1 have been implicated in osmotic stress-induced gene expression (Rep et al., 1999), direct interaction between these factors and Hog1 remains to be demonstrated.

Recently, by yeast two-hybrid screens for proteins that interact with Hog1, a protein kinase, Rck2, was identified (Bilsland-Marchesan et al., 2000; Teige et al., 2001), which is structurally related to yeast and mammalian Ca^{2+}/calmodulin-dependent kinases (Melcher and Thorner, 1996). Various experimental results are consistent with Rck2 being the direct target of Hog1 MAPK (Bilsland-Marchesan et al., 2000; Teige et al., 2001); biochemically, Rck2 physically associates with Hog1 and is phosphorylated in response to

osmotic stress in a Hog1-dependent manner. In addition, in-vitro experiments have demonstrated that Hog1 phosphorylates Rck2 in the C-terminal autoinhibitory domain to induce Rck2 activation. Genetically, overexpression of *RCK2* suppresses the osmotic sensitivity of Δ*hog1* and Δ*pbs2* mutants, and deletion of *RCK2* suppresses cell lethality caused by hyperactivation of the HOG pathway. Thus, it is likely that upon osmotic stress, Hog1 phosphorylates the Rck2 kinase, and stimulated Rck2 may then phosphorylate substrates to control cellular responses to osmotic stress, such as the inhibition of protein synthesis (Teige et al., 2001). Unexpectedly, unlike Δ*hog1* and Δ*pbs2* mutants, cells deleted for *RCK2* do not display osmosensitivity, indicating that Hog1 has another downstream target which has an overlapping function with Rck2 (Bilsland-Marchesan et al., 2000).

Another well-characterized protein that appears to be directly regulated by Hog1 is the Sko1 transcriptional repressor, which binds cAMP responsive element (CRE) sequences in the promoter regions of various osmotic stress-inducible genes, such as *ENA1* (Na^+ extrusion ATPase) (Proft and Serrano, 1999), *HAL1* (ion homeostasis protein) (Pascual-Ahuir et al., 2001b), and *GRE2* (homolog of plant isoflavonoid reductase) (Garcia-Gimeno and Struhl, 2000; Proft et al., 2001; Rep et al., 2001). It is thought that Sko1 recruits the Ssn6-Tup1 corepressor complex to the promoters of those genes, and thereby inhibits their expression under normal growth conditions (Proft and Serrano, 1999; Garcia-Gimeno and Struhl, 2000; Pascual-Ahuir et al., 2001b; Proft et al., 2001). In response to high osmolarity stress, Sko1 is phosphorylated by Hog1, resulting in transcriptional derepression via the disruption of the interaction between Sko1 and the Ssn6-Tup1 complex (Pascual-Ahuir et al., 2001a; Proft et al., 2001). Remarkably, Sko1 is also regulated by the cAMP-activated protein kinase A (PKA) pathway, and PKA phosphorylation sites have been mapped in Sko1 (Proft et al., 2001). It is believed that high PKA activity under normal growth conditions increases the repression activity of Sko1, whereas low PKA activity upon osmotic stress leads to derepression by Sko1 (Pascual-Ahuir et al., 2001a;

Proft et al., 2001). Transcriptional repression by Sko1 requires nuclear accumulation of Sko1, a process that appears to be dependent on PKA activity (Pascual-Ahuir et al., 2001a).

3.   The Spc1 MAPK cascade in fission yeast

Whereas Hog1 is believed to be responsive mainly to high osmolarity stress (Schüller et al., 1994), SAPKs in mammals and the fission yeast *S. pombe* are activated by a multitude of stress signals (reviewed in Banuett, 1998). Studies in fission yeast demonstrate that the Spc1 MAPK is stimulated by high osmolarity, oxidative stress, heat shock, UV irradiation, and nutritional limitation. In fact, *spc1* null (Δ*spc1*) mutants are hypersensitive to these conditions, indicating that the function of Spc1 is crucial for the cellular resistance to multiple forms of stress (Millar et al., 1995; Shiozaki and Russell, 1995; Degols et al., 1996; Kato et al., 1996; Shiozaki and Russell, 1996; Degols and Russell, 1997). In response to harsh environmental conditions, Spc1 is activated by the phosphorylation of conserved residues, Thr-171 and Tyr-173, mediated by the Wis1 MAPKK (Millar et al., 1995; Shiozaki and Russell, 1995; Degols et al., 1996; Degols and Russell, 1997). Activators of Wis1 include the Wis4 and Win1 MAPKKKs (Samejima et al., 1997; Shieh et al., 1997; Shiozaki et al., 1997; Samejima et al., 1998). Importantly, Wis1 activity is counteracted by PTPs, Pyp1 and Pyp2, with Pyp1 having the major cellular activity that dephosphorylates Tyr-173 of Spc1 (Millar et al., 1995; Shiozaki and Russell, 1995; Degols et al., 1996).

Once activated, Spc1 induces the expression of stress-protective genes such as *gpd1+* and *ctt1+* (Degols et al., 1996; Wilkinson et al., 1996; Degols and Russell, 1997). The gene *gpd1+* encodes glycerol-3-phosphate dehydrogenase (Pidoux et al., 1990), a key enzyme in glycerol synthesis that is important for increasing the intracellular osmolarity in a hyperosmolarity environment (Ohmiya et al., 1995). Cytosolic catalase, encoded by *ctt1+*, decomposes hydrogen peroxide to protect cells from oxidative damage (Nakagawa et al., 1995).

The regulation of these genes is under the control of a bZIP transcription factor, Atf1 (also known as Gad7) (Takeda et al., 1995; Kanoh et al., 1996). Both Δ*spc1* and Δ*atf1* mutants are defective in stress-induced expression of the same set of genes and are sensitive to various stress treatments (Shiozaki and Russell, 1996; Wilkinson et al., 1996). Moreover, Atf1 is phosphorylated by Spc1 both *in vitro* and *in vivo* (Shiozaki and Russell, 1996). These findings strongly suggest that Atf1 is a downstream target of Spc1. Because Atf1 is most homologous to ATF2, one of the key substrates of human p38 and JNK SAPKs (Gupta et al., 1995; Livingstone et al., 1995; van Dam et al., 1995; Raingeaud et al., 1996), the discovery of the link between Spc1 and Atf1 underscores the high conservation of SAPK pathways in *S. pombe* and mammalian cells.

3.1.   Multiple mechanisms of Spc1 activation

As described, fission yeast and human SAPK cascades are very similar, with regards to the stress signals that activate the pathways as well as the target transcription factors that are activated by them. Due to its genetic amenability, fission yeast is therefore an excellent model system to address key issues in eukaryotic stress signaling. One obvious question deals with how cells are able to detect various stress stimuli and transmit them to SAPKs. Is there a main sensor that can perceive all the different environmental signals? Or, are there multiple sensors, each with the capacity to specifically detect individual types of stress?

The first hint at the possibility of multiple sensors conveying different stress signals to the fission yeast Spc1 MAPK came from studies examining whether various stress stimuli are transmitted to Wis1 MAPKK through MAPKKKs (Shieh et al., 1998; Shiozaki et al., 1998). These experiments utilized Wis1 proteins having MAPKKK phosphorylation sites, Ser-469 and Thr-473, substituted with alanine or aspartic acid residues to create the *wis1AA* and *wis1DD* mutants, respectively. As in the Δ*wis1* strain, it was found that Spc1 phosphorylation is not detectable in the *wis1AA* mutant under any stress condition, indicating that Ser-469 and Thr-473 are essential for Wis1 activation. In the absence of stress, Spc1 phosphorylation is higher in *wis1DD* mutant cells than in wild-type cells, which suggests that the *wisDD* mutation partially stimulates Wis1 activity by mimicking phosphorylation of Ser-469 and Thr-473. However, high osmolarity and oxidative stress do not induce further activation of Spc1 in *wis1DD* cells, suggesting that these stress signals are mediated by MAPKKKs through the phosphorylation of Ser-469 and Thr-473 and that substitution of these sites with unphosphorylatable residues abolishes signaling to Spc1. Surprisingly, heat shock induces strong activation of Spc1 in both wild-type and *wis1DD* cells, indicating that heat shock signals can be transmitted to Spc1 independently of MAPKKKs (Shiozaki et al., 1998). Hence, these results provide the first indication that stress signals can be transmitted to Spc1 MAPK via different mechanisms.

3.2. Regulation of Spc1 in response to heat shock

How does heat shock activate Spc1 in *wis1DD* mutant cells? Two possible scenarios can be envisioned. First, heat shock may somehow increase the activity of the Wis1DD protein, independently of the MAPKKK phosphorylation sites. Alternatively, in the presence of constitutive Wis1DD activity, heat shock may inhibit the phosphatases that negatively regulate Spc1. By assaying Wis1 activity *in vitro*, it was found that Wis1 activation upon heat shock is relatively weak and transient, which contrasts with strong Wis1 activation induced by osmotic stress and oxidative stress. Significantly, the activity of Wis1DD was shown to be constitutive and not affected by heat shock, suggesting that heat shock-induced activation of Spc1 in *wis1DD* cells is not mediated by an increase in Wis1DD activity (Nguyen and Shiozaki, 1999). Consistent with the second hypothesis, Pyp1 and perhaps Pyp2, the PTPs that dephosphorylate Tyr-173 of Spc1, are significantly inhibited in heat-shocked cells; heat shock disrupts the interaction between Pyp1 and Spc1 and causes Pyp1 to become insoluble (Nguyen and Shiozaki, 1999).

These results strongly suggest that heat shock activates Spc1 by simultaneously increasing Wis1 kinase activity and inhibiting Pyp1/Pyp2 phosphatase activity (Fig. 6.3). Importantly, the increase in Spc1 activity in heat-shocked cells is rapidly attenuated by the dephosphorylation of Thr-171 of Spc1, carried out by the type 2C protein phosphatase (PP2C) enzymes, Ptc1 and Ptc3 (Fig. 6.3) (Nguyen and Shiozaki, 1999). Other threonine phosphatases may also contribute to Spc1 Thr-171 dephosphorylation, since Thr-171 dephosphorylation is not completely inhibited in $\Delta ptc1$ $\Delta ptc3$ mutants. Because the expression of $ptc1^+$ and $pyp2^+$ are known to be induced by the Spc1-Atf1 pathway in response to stress stimuli (Shiozaki et al., 1994; Degols et al., 1996; Wilkinson et al., 1996; Gaits et al., 1997), these phosphatases constitute dual loops of negative feedback to regulated Spc1 activity (Fig. 6.3). Thus, transient activation of Spc1 upon heat shock is ensured by differential regulation of Thr-171 and Tyr-173 phosphorylation.

Accumulating evidence suggest that regulation of stress MAPKs by PP2C enzymes may be a common phenomenon in eukaryotes. First, PP2Cα has been proposed to be a negative regulator of mammalian p38 and JNK pathways and has been shown to dephosphorylate the conserved threonine residue of p38 *in vitro* (Takekawa et al., 1998). Second, like fission yeast $ptc1^+$, the expression of human *WIP1*, encoding a PP2C enzyme, is induced upon stress to attenuate p38 activity (Takekawa et al., 2000). Lastly, a PP2C in budding yeast, Ptc1, has been demonstrated to specifically dephosphorylate Hog1 MAPK (Warmka et al., 2001).

Down-regulation of phosphatases in response to heat shock may not be unique to fission yeast. The inhibition of a JNK phosphatase has recently been proposed as a mechanism for JNK activation in response to heat shock (Meriin et al., 1999). Additionally, M3/6, the dual-specificity phosphatase that negatively regulates JNK, was also shown to be inhibited in heat-shocked cells (Palacios et al., 2001). Interestingly, like fission yeast Pyp1, M3/6 also becomes insoluble upon heat shock. It is possible that this mechanism of SAPK activation in

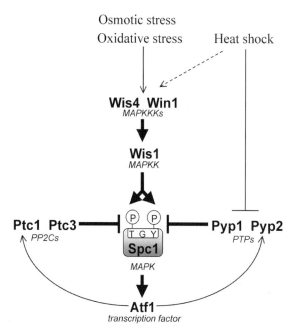

Fig. 6.3. Regulation of Spc1 MAPK by threonine- and tyrosine-specific phosphatases. Upon stress stimuli such as osmotic stress, oxidative stress, and heat shock, Wis1 MAPKK activates Spc1 by phosphorylating Thr-171 and Tyr-173. Osmotic stress and oxidative stress activate Wis1 through the phosphorylation of Ser-469 and Thr-473, which is carried out by Wis4 and Win1 MAPKKKs. Heat shock weakly activates Wis1 in a MAPKKK-dependent manner and simultaneously inhibits Pyp1 and Pyp2, the Spc1 Tyr-173 phosphatases, resulting in strong activation of Spc1. Although Tyr-173 remains phosphorylated in heat-shocked cells because of inhibition of the tyrosine phosphatases, Spc1 activity is attenuated by Thr-171 dephosphorylation, which is carried out by Ptc1, Ptc3, and other threonine phosphatases. Transcription of $pyp2^+$ and $ptc1^+$ is induced by the Spc1-Atf1 pathway in response to stress stimuli, which constitutes dual loops of negative feedback.

response to heat shock (i.e., down-regulation of inhibitory phosphatases) may be conserved from yeast to human. Regulation by PP2C and PTP enzymes thus provides another layer of conservation between fission yeast and mammalian SAPK pathways.

3.3. Oxidative stress signaling in fission yeast

Homologs of the *S. cerevisiae* Sln1-Ypd1-Ssk1 multistep phosphorelay proteins have been identified in *S. pombe*. These include the Mak2 and

Mak3 histidine kinases, Mpr1 (also known as Spy1) histidine phosphotransfer protein (HPt), and Mcs4 response regulator (Cottarel, 1997; Shieh et al., 1997; Shiozaki et al., 1997; Aoyama et al., 2000; Nguyen et al., 2000; Buck et al., 2001). Fission yeast Mpr1 and Mcs4 are homologous to budding yeast Ypd1 and Ssk1, respectively. In addition to a histidine kinase domain and a receiver domain, Mak2 and Mak3 also harbor a PAS/PAC domain, a GAF domain, and a domain with homology to a family of atypical serine/threonine kinases found in prokaryotes (Buck et al., 2001). PAS/PAC domains are present in proteins that regulate cellular responses to light and redox (Crews and Fan, 1999) while GAF domains have been identified in phytochromes, redox-responsive transcription factors, and plant ethylene receptors (Aravind and Ponting, 1997). Unlike the homologous signaling module in budding yeast, the fission yeast phosphorelay system does not appear to be responsive to high osmolarity stress. Instead, the Mak2/3-Mpr1-Mcs4 system is specific and essential for transmitting oxidative stress signals to the Spc1 MAPK cascade (Fig. 6.4) (Nguyen et al., 2000; Buck et al., 2001).

Genetic and biochemical data suggest that the fission yeast phosphorelay proteins function in the same order as do those in budding yeast and prokaryotes, histidine kinase → HPt → response regulator (reviewed by Santos and Shiozaki, 2001). Various lines of evidence are consistent with the idea that phosphotransfer through the *S. pombe* phosphorelay system is critical for transmitting oxidative stress signals to Spc1. First, Spc1 is not activated by oxidative stress in cells expressing Mpr1HQ, a mutant Mpr1 protein having the putative histidine phosphorylation site (His-221) substituted with glutamine (Nguyen et al., 2000). Likewise, oxidative stress does not activate Spc1 in cells expressing Mcs4(D412N), with the putative phospho-accepting residue (Asp-412) of Mcs4 substituted with asparagine (Buck et al., 2001). Finally, the Mpr1HQ protein cannot bind Mcs4 whereas wild-type Mpr1 shows a significant increase in binding to Mcs4 in response to oxidative stress (Nguyen et al., 2000). Gathered together, these results suggest that oxidative stress

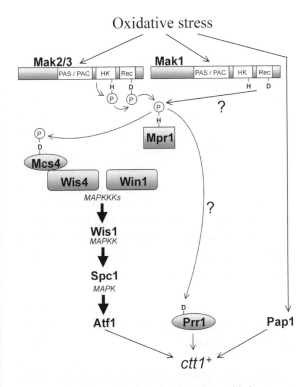

Fig. 6.4. Multiple mechanisms for detecting oxidative stress. A multistep phosphorelay system, composed of Mak2 and Mak3 histidine kinases, Mpr1 histidine phosphotransfer protein, and Mcs4 response regulator, transmits oxidative stress signals to the Spc1 MAPK cascade. Activated Spc1 induces the expression of *ctt1+* and other stress-response genes through the Atf1 transcription factor. The Pap1 transcription factor, Mak1 histidine kinase, and Prr1 response regulator can also mediate *ctt1+* expression independently of the Spc1 pathway in response to oxidative stress. It is unknown whether Mak1, Mpr1, and Prr1 constitute another multistep phosphorelay system or whether the PAS/PAC domains in Mak1, Mak2, and Mak3 are responsible for sensing the oxidative stress signals.

signals may be sensed by the PAS/PAC motifs in Mak2 and Mak3, leading to the stimulation of histidine kinase activity in these proteins. By inference to the budding yeast phosphorelay mechanism, a phosphate moiety may be transferred sequentially from a histidine residue within the Mak2/Mak3 histidine kinase domain to an aspartate residue of the Mak2/Mak3 receiver domain (Fig. 6.4). The phosphate may then be relayed to His-221 of Mpr1, which may result in an increase in the affinity of Mpr1 for Mcs4. It is conceivable that the enhanced interaction between

Mpr1 and Mcs4 facilitates the final acceptance of the phosphate at Asp-412 of Mcs4. Phosphotransfer from Mpr1 to Mcs4 has been detected *in vitro* (Aoyama et al., 2000). Because Mcs4 constitutively binds the Wis4 MAPKKK (Buck et al., 2001), phosphorylation of Mcs4 may ultimately lead to the activation of the Spc1 cascade. Interestingly, Spc1 is still weakly activated upon oxidative stress in the Δ*mpr1* Δ*mcs4* double mutant, suggesting that fission yeast has an Mcs4-independent pathway that can also convey oxidative stress signals to Spc1 (Nguyen et al., 2000).

The expression of *ctt1+*, encoding cytoplasmic catalase (Nakagawa et al., 1995), is highly induced in response to oxidative stress, a process that is dependent on the Spc1-Atf1 pathway (Shieh et al., 1997; Nguyen et al., 2000). Surprisingly, oxidative stress-induced transcription of *ctt1+* is still evident in Δ*mpr1* cells, in which no Spc1 activation is observed (Nguyen et al., 2000). Consistent with this, Δ*mpr1* mutants are resistant to oxidative stress (Nguyen et al., 2000). It was discovered that the Spc1-independent expression of *ctt1+* is mediated by Pap1 (Fig. 6.4) (Nguyen et al., 2000), a bZIP transcription factor with significant homology and similar DNA-binding specificity to mammalian c-Jun (Toda et al., 1991; Toda et al., 1992). Other studies indicate that *ctt1+* expression is also regulated by the Mak1 histidine kinase (Buck et al., 2001) and the Prr1 response regulator (Fig. 6.4) (Ohmiya et al., 1999). However, it remains to be determined whether Mak1 and Prr1 constitute part of a phosphorelay system and whether there is any interaction among Mak1, Prr1 and Pap1. Because oxidative stress is known to be damaging to many cellular components, it may not be surprising that fission yeast cells possess a repertoire of mechanisms for sensing this particular stress.

3.4. Activation of Spc1 in response to high osmolarity stress

As described in the previous two sections, transmission of the heat shock and oxidative stress signals in *S. pombe* is beginning to be deciphered. In contrast, osmotic stress signaling in this organism continues to be elusive. Proteins important for mediating the high osmolarity signal include the Mcs4 response regulator (Shieh et al., 1997; Shiozaki et al., 1997), Wis4 and Win1 MAPKKKs (Shieh et al., 1997; Shiozaki et al., 1997; Samejima et al., 1998), Wis1 MAPKK, and Spc1 MAPK (Millar et al., 1995; Shiozaki and Russell, 1995). Although Mcs4 is known to be involved, it appears that a phosphorelay mechanism through Mcs4 is not essential (Buck et al., 2001). Additionally, unlike the situation upon oxidative stress, osmotic stress does not induce the interaction between Mpr1 and Mcs4 (Nguyen et al., 2000). In fact, Spc1 activation in response to osmotic stress in Δ*mpr1* cells is similar to that in wild-type cells, suggesting that Mpr1 is not essential for this process (Nguyen et al., 2000). It is important to note that Spc1 activation is not completely inhibited in Δ*mcs4* cells (Shieh et al., 1997; Shiozaki et al., 1997; Nguyen et al., 2000), implying that other factors acting in parallel with Mcs4 may be important for mediating high osmolarity signals to Spc1. Extensive studies are therefore being conducted to identify these components.

3.5. Recognition of Spc1 by its cognate MAPKK

In fission yeast, three MAPKs have been identified, which share certain structural similarities (Hughes, 1995; Millar et al., 1995; Shiozaki and Russell, 1995; Kato et al., 1996; Toda et al., 1996; Zaitsevskaya-Carter and Cooper, 1997). However, Wis1 MAPKK phosphorylates specifically Spc1 MAPK in response to different stress stimuli. Aiming to understand the mechanism of this signaling specificity, we carried out extensive analyses of Wis1 (Nguyen et al., 2002). Wis1 is a 605-amino acid protein, with the C-terminal half making up the kinase domain (Warbrick and Fantes, 1991); no function has been attributed to the N-terminal, non-catalytic domain of Wis1. By constructing a series of N-terminal truncations of the Wis1 protein, we discovered a region (residues 201-300) adjacent to the Wis1 catalytic domain that is important for phosphorylation of its substrate, Spc1. Two clusters of residues within this

region, [234]RRAPPGKLDL[243] and [260]RRGLNI[265], resemble the proposed consensus sequence for a MAPK docking site in MAPKKs (at least two basic amino acid residues separated by a spacer of 2-6 residues from a hydrophobic-X-hydrophobic sequence) (Bardwell and Thorner, 1996; Bardwell et al., 2001). Indeed, mutating the basic residues within both putative MAPK docking sites in Wis1 significantly reduced the affinity of Wis1 for Spc1. Furthermore, cells expressing Wis1 with mutated MAPK docking sites show reduced activation of Spc1 in response to mild stress conditions. These data support the idea that the MAPK docking motifs in Wis1 are important for stress signaling to Spc1 MAPK (Nguyen et al., 2002).

3.6. Regulated nuclear translocation of Spc1 upon stress stimuli

Like *S. cerevisiae* Hog1 (Ferrigno et al., 1998; Reiser et al., 1999) and mammalian JNK (Cavigelli et al., 1995), Spc1 MAPK transiently accumulates in the nucleus in response to environmental stress (Gaits et al., 1998; Gaits and Russell, 1999). Remarkably, the mechanism of Spc1 translocation into and out of the nucleus resembles that of Hog1 translocation, although there are notable differences. Like the situation in budding yeast, stress-induced nuclear import of Spc1 requires Spc1 phosphorylation but not its catalytic activity (Gaits and Russell, 1999). Import of Spc1 also appears to be an active process since nuclear accumulation of the MAPK is defective in *pim1-d1* cells (Gaits and Russell, 1999); *pim1-d1* is a temperature-sensitive mutation in the *pim1*+ gene, which encodes a homolog of mammalian RCC1 (Matsumoto and Beach, 1991; Sazer and Nurse, 1994), a guanine nucleotide exchange factor that is essential for nucleocytoplasmic trafficking (Nigg, 1997). Once in the nucleus, Spc1 is believed to interact with and phosphorylate the Atf1 transcription factor (Shiozaki and Russell, 1996; Wilkinson et al., 1996); the interaction with Atf1 is postulated to be crucial for maintaining Spc1 in the nucleus (Gaits et al., 1998).

As in the nuclear export of Hog1, the transport

of Spc1 out of the nucleus requires the exportin Crm1 (Xpo1) (Gaits et al., 1998), and a direct interaction between Spc1 and Crm1 was detected (Gaits and Russell, 1999). It is unlikely that phosphorylation of Spc1 regulates the interaction with Crm1, since even unphosphorylatable Spc1 mutant proteins can bind Crm1 *in vitro* (Gaits and Russell, 1999). Rather, Spc1 may bind Crm1 when it is in the nucleus, and Spc1-Crm1 complex is exported to the cytoplasm, where PTPs and PP2Cs dephosphorylate and inactivate Spc1 (Gaits and Russell, 1999 and our unpublished results).

During the characterization of the N-terminus of Wis1 MAPKK, we identified a sequence that resembles a nuclear export signal (NES) (Fischer et al., 1995; Wen et al., 1995). Various results are consistent with this stretch of amino acids constituting a functional NES that is important for maintaining Wis1 in the cytoplasm (Nguyen et al., 2002). First, mutations in this region disturbed the cytoplasmic localization of Wis1. Second, Wis1 was found throughout the cell, both in the nucleus and in the cytoplasm, in the *crm1-809* strain, a cold-sensitive mutant defective in NES-dependent protein export from the nucleus (Fukuda et al., 1997). Lastly, the putative NES sequence of Wis1, when fused to GST (glutathione *S*-transferase), conferred nuclear exclusion of GST.

The aforementioned results prompted studies to determine the physiological role of the Wis1 NES (Nguyen et al., 2002). We targeted Wis1 into the nucleus by substituting its NES with a nuclear localization signal (NLS). As expected, it was found that cytoplasmic localization of Wis1 is important for the activation of Spc1 MAPK, a mainly cytoplasmic protein in the absence of stress (Gaits et al., 1998; Gaits and Russell, 1999). Surprisingly, the maintenance of Wis1 in the cytoplasm is crucial for the translocation of Spc1 MAPK into the nucleus in response to stress stimuli. The exact role of cytoplasmic Wis1 in nuclear transport of Spc1 is currently unclear. Because a population of Wis1 also translocates into the nucleus upon stress (Nguyen et al., 2002), one attractive hypothesis is that Wis1 acts as a carrier for Spc1 into the nucleus.

3.7. Crucial link between mitosis/meiosis and changes in the extracellular environment

In addition to its significance in stress survival, Spc1 MAPK is also involved in coordinating cellular proliferation and differentiation with changes in the extracellular environment. It was observed that loss of functional Spc1 leads to a G_2 cell cycle delay, a defect that is greatly exacerbated under stress conditions and in minimal medium (Millar et al., 1995; Shiozaki and Russell, 1995). Interestingly, cells lacking Atf1, a substrate of Spc1, do not share this cell cycle delay phenotype, implying that Spc1 regulates mitosis in response to environmental cues through different target(s) (Shiozaki and Russell, 1996).

Spc1 also has an important role in regulating sexual development (Shiozaki and Russell, 1996; Wilkinson et al., 1996). This conclusion is based on the observation that Spc1 is activated in response to nitrogen limitation (Shiozaki and Russell, 1996), a condition known to promote a program of sexual development in fission yeast. In addition, *spc1* and *atf1* mutants are sterile, possibly due to the inability of these cells to arrest in G_1 and to induce the expression of *ste11+*, which encodes a transcription factor that regulates the commitment to meiosis (Sugimoto et al., 1991). Thus, the Spc1-Atf1 pathway links the nutritional status of the extracellular environment with meiosis.

Besides linking cellular processes with extracellular conditions, Spc1 MAPK also plays a crucial role in controlling mitosis in response to perturbations within the cell. It was recently demonstrated that Spc1 is important for establishing a mitotic delay in response to disturbed actin cytoskeleton (Gachet et al., 2001). This delay appears to be a checkpoint mechanism that prevents sister chromatid separation when the mitotic spindle is misoriented in the absence of proper actin organization.

4. Concluding remarks

In this chapter, we have reviewed the current knowledge of the stress-activated MAPK cascades in two distantly related yeasts, S. *cerevisiae* and *S. pombe*. Studies in these organisms have provided an invaluable source of information for understanding MAPK activation and regulation in eukaryotic cells. In fact, it is becoming increasingly clear that mechanisms for responding to environmental stress are conserved from yeast to human. This conservation is highlighted by three observations. First, SAPKs are stimulated by a similar range of environmental insults, including high osmolarity stress, oxidative stress, and heat shock. Second, SAPKs regulate similar targets such as the bZIP transcription factors. Lastly, SAPKs appear to be negatively regulated by PTPs and PP2C enzymes.

Multistep phosphorelay systems play crucial roles in mediating extracellular signals to the stress MAPK cascades in budding and fission yeasts. Available data from the human genome project suggest that homologs of yeast phosphorelay proteins are absent in humans. This finding raises the perplexing question of how human SAPKs are activated in response to environmental stress. At least with regard to heat shock, the mechanism for activating mammalian SAPKs appears to be similar to that for fission yeast SAPK; these SAPKs are mainly activated through the down-regulation of the SAPK inhibitory phosphatases.

Upon stress stimuli, both Hog1 and Spc1 MAPKs transiently accumulate in the nucleus, where they regulate the expression of stress-response genes. However, the mechanisms for these translocation processes are only beginning to be addressed. Various proteins have been identified as mediators of MAPK transport into and out of the nucleus but their exact functions remain unclear. Other issues that need to be investigated include mechanisms for maintaining SAPKs in the nucleus during stress, determinants for SAPK release from nuclear anchors, and ways for transporting SAPKs back to the cytoplasm after stress adaptation.

Within a single cell, multiple MAPK cascades act in parallel in response to different extracellular stimuli. Thus, another area of limited knowledge concerns signaling specificity. This situation is especially complex in mammalian cells, in which the JNK and p38 SAPKs are activated by an almost identical set of stress signals. Specific MAPK docking sites found in regulators and substrates of MAPKs may play a key role in establishing the observed signaling specificities in different MAPK pathways.

In summary, it is evident that mechanisms of SAPK regulation are conserved among eukaryotes. Studies in budding and fission yeasts will continue to provide valuable insights into the molecular responses to stress in human cells.

Acknowledgements

We thank members of the Shiozaki lab for critical reading of the manuscript. A.N.N. was supported by the National Institutes of Health (NIH) Molecular and Cellular Biology Training Program at University of California, Davis (T32 GM07377) and the George G. Lee Fellowship. Studies of the Spc1 MAPK cascade in this lab are supported by an NIH grant awarded to K.S. (GM59788).

References

Alepuz, P.M., Jovanovic, A., Reiser, V. and Ammerer, G. (2001). Stress-induced map kinase Hog1 is part of transcription activation complexes. Mol. Cell. 7, 767–777.

Aoyama, K., Mitsubayashi, Y., Aiba, H. and Mizuno, T. (2000). Spy1, a histidine-containing phosphotransfer signaling protein, regulates the fission yeast cell cycle through the Mcs4 response regulator. J. Bacteriol. 182, 4868–4874.

Aravind, L. and Ponting, C.P. (1997). The GAF domain: an evolutionary link between diverse phototransducing proteins. Trends Biochem. Sci. 22, 458–459.

Banuett, F. (1998). Signalling in the yeasts: an informational cascade with links to the filamentous fungi. Microbiol. Mol. Biol. Rev. 62, 249–274.

Bardwell, A.J., Flatauer, L.J., Matsukuma, K., Thorner, J. and Bardwell, L. (2001). A conserved docking site in MEKS mediates high-affinity binding to MAP kinases and cooperates with a scaffold protein to enhance signal transmission. J. Biol. Chem. 276, 10374–10386.

Bardwell, L. and Thorner, J. (1996). A conserved motif at the amino termini of MEKs might mediate high-affinity interaction with the cognate MAPKs. Trends Biochem. Sci. 21, 373–374.

Bilsland-Marchesan, E., Arino, J., Saito, H., Sunnerhagen, P. and Posas, F. (2000). Rck2 kinase is a substrate for the osmotic stress-activated mitogen- activated protein kinase Hog1. Mol. Cell. Biol. 20, 3887–3895.

Blomberg, A. and Adler, L. (1992). Physiology of osmotolerance in fungi. Adv. Microb. Physiol. 33, 145–212.

Brewster, J.L., de Valoir, T., Dwyer, N.D., Winter, E. and Gustin, M.C. (1993). An osmosensing signal transduction pathway in yeast. Science 259, 1760–1763.

Buck, V., Quinn, J., Soto Pino, T., Martin, H., Saldanha, J., Makino, K., Morgan, B.A. and Millar, J.B. (2001). Peroxide sensors for the fission yeast stress-activated mitogen-activated protein kinase pathway. Mol. Biol. Cell. 12, 407–419.

Cavigelli, M., Dolfi, F., Claret, F.-X. and Karin, M. (1995). Induction of c-fos expression through JNK-mediated TCF/Elk1 phosphorylation. EMBO J. 14, 5957–5964.

Cobb, M.H. and Goldsmith, E.J. (1995). How MAP kinases are regulated. J. Biol. Chem. 270, 14843–14846.

Cottarel, G. (1997). Mcs4, a two-component system response regulator homologue, regulates the *Schizosaccharomyces pombe* cell cycle control. Genetics 147, 1043–1051.

Crews, S.T. and Fan, C.M. (1999). Remembrance of things PAS: regulation of development by bHLH-PAS proteins. Curr. Opin. Genet. Dev. 9, 580–587.

Degols, G. and Russell, P. (1997). Discrete roles of the Spc1 kinase and the Atf1 transcription factor in the UV response of *Schizosaccharomyces pombe*. Mol. Cell. Biol. 17, 3356–3363.

Degols, G., Shiozaki, K. and Russell, P. (1996). Activation and regulation of the Spc1 stress-activated protein kinase in *Schizosaccharomyces pombe*. Mol. Cell. Biol. 16, 2870–2877.

Ferrigno, P., Posas, F., Koepp, D., Saito, H. and Silver, P.A. (1998). Regulated nucleo/cytoplasmic exchange of HOG1 MAPK requires the importin beta homologs NMD5 and XPO1. Embo J. 17, 5606–5614.

Fischer, U., Huber, J., Boelens, W.C., Mattaj, I.W. and Luhrmann, R. (1995). The HIV-1 Rev activation domain is a nuclear export signal that accesses an export pathway used by specific cellular RNAs. Cell 82, 475–483.

Fukuda, M., Asano, S., Nakamura, T., Adachi, M., Yoshida, M., Yanagida, M. and Nishida, E. (1997). CRM1 is responsible for intracellular transport mediated by the nuclear export signal. Nature 390, 308–311.

Gachet, Y., Tournier, S., Millar, J.B. and Hyams, J.S. (2001). A MAP kinase-dependent actin checkpoint ensures proper spindle orientation in fission yeast. Nature 412, 352–355.

Gaits, F., Degols, G., Shiozaki, K. and Russell, P. (1998). Phosphorylation and association with the transcription factor Atf1 regulate localization of Spc1/Sty1 stress-activated kinase in fission yeast. Genes Dev. 12, 1464–1473.

Gaits, F. and Russell, P. (1999). Active nucleocytoplasmic shuttling required for function and regulation of stress-activated kinase Spc1/StyI in fission yeast. Mol. Biol. Cell 10, 1395–1407.

Gaits, F., Shiozaki, K. and Russell, P. (1997). Protein phosphatase 2C acts independently of stress-activated kinase cascade to regulate the stress response in fission yeast. J. Biol. Chem. 272, 17873–17879.

Garcia-Gimeno, M.A. and Struhl, K. (2000). Aca1 and Aca2, ATF/CREB activators in *Saccharomyces cerevisiae*, are important for carbon source utilization but not the response to stress. Mol. Cell. Biol. 20, 4340–4349.

Gorner, W., Durchschlag, E., Martinez-Pastor, M.T., Estruch, F., Ammerer, G., Hamilton, B., Ruis, H. and Schuller, C. (1998). Nuclear localization of the C2H2 zinc finger protein Msn2p is regulated by stress and protein kinase A activity. Genes Dev. 12, 586–597.

Gupta, S., Campbell, D., Dérijard, B. and Davis, R.J. (1995). Transcription factor ATF2 regulation by the JNK signal transduction pathway. Science 267, 389–393.

Gustin, M.C., Albertyn, J., Alexander, M. and Davenport, K. (1998). MAP kinase pathways in the yeast *Saccharomyces cerevisiae*. Microbiol. Mol. Biol. Rev. 62, 1264–1300.

Han, J., Lee, J.-D., Bibbs, L. and Ulevitch, R.J. (1994). A MAP kinase targeted by endotoxin and hyperosmolarity in mammalian cells. Science 265, 808–811.

Herskowitz, I. (1995). MAP kinase pathways in yeast: for mating and more. Cell 80, 187–197.

Hill, C.S. and Treisman, R. (1995). Transcriptional regulation by extracellular signals: mechanisms and specificity. Cell 80, 199–211.

Hughes, D.A. (1995). Control of signal transduction and morphogenesis by Ras. Semin. Cell. Biol. 6, 89–94.

Johnson, D.I. (1999). Cdc42: An essential Rho-type GTPase controlling eukaryotic cell polarity. Microbiol. Mol. Biol. Rev. 63, 54–105.

Kanoh, J., Watanabe, Y., Ohsugi, M., Iino, Y. and Yamamoto, M. (1996). *Schizosaccharomyces pombe* gad7[+] encodes a phosphoprotein with a bZIP domain, which is required for proper G1 arrest and gene expression under nitrogen starvation. Genes Cells 1, 391–408.

Kato, T.J., Okazaki, K., Murakami, H., Stettler, S., Fantes, P.A. and Okayama, H. (1996). Stress signal, mediated by a Hog1-like MAP kinase, controls sexual development in fission yeast. FEBS Lett. 378, 207–212.

Lee, J.C., Laydon, J.T., McDonnell, P.C., Gallagher, T.F., Kumar, S., Green, D., McNulty, D., Blumenthal, M.J., Heys, J.R., Landvatter, S.W., Stickler, J.E., McLaughlin, M.M., Siemens, I.R., Fisher, S.M., Livi, G.P., White, J.R., Adams, J.L. and Young, P.R. (1994). A protein kinase involved in the regulation of inflammatory cytokine biosynthesis. Nature 372, 739–746.

Livingstone, C., Patel, G. and Jones, N. (1995). ATF-2 contains a phosphorylation-dependent transcriptional activation domain. EMBO J. 14, 1785–1797.

Maeda, T., Takekawa, M. and Saito, H. (1995). Activation of yeast PBS2 MAPKK by MAPKKKs or by binding of an SH3-containing osmosensor. Science 269, 554–558.

Maeda, T., Wurgler-Murphy, S.M. and Saito, H. (1994). A two-component system that regulates an osmosensing MAP kinase cascade in yeast. Nature 369, 242–245.

Marshall, C.J. (1994). MAP kinase kinase kinase, MAP kinase kinase and MAP kinase. Curr. Opin. Gen. Dev. 4, 82–89.

Matsumoto, T. and Beach, D. (1991). Premature initiation of mitosis in yeast lacking RCC1 or an Interacting GTPase. Cell 66, 347–360.

Mattison, C.P. and Ota, I.M. (2000). Two protein tyrosine phosphatases, Ptp2 and Ptp3, modulate the subcellular localization of the Hog1 MAP kinase in yeast. Genes Dev. 14, 1229–1235.

Melcher, M.L. and Thorner, J. (1996). Identification and characterization of the CLK1 gene product, a novel CaM kinase-like protein kinase from the yeast *Saccharomyces cerevisiae*. J. Biol. Chem. 271, 29958–29968.

Meriin, A.B., Yaglom, J.A., Gabai, V.L., Zon, L., Ganiatsas, S., Mosser, D.D. and Sherman, M.Y. (1999). Protein-damaging stresses activate c-Jun N-terminal kinase via inhibition of its dephosphorylation: a novel pathway controlled by HSP72. Mol. Cell. Biol. 19, 2547–2555.

Millar, J.B. (1999). Stress-activated MAP kinase (mitogen-activated protein kinase) pathways of budding and fission yeasts. Biochem. Soc. Symp. 64, 49–62.

Millar, J.B.A., Buck, V. and Wilkinson, M.G. (1995). Pyp1 and Pyp2 PTPases dephosphorylate an osmosensing MAP kinase controlling cell size at division in fission yeast. Genes Dev. 9, 2117–2130.

Mosch, H.U., Roberts, R.L. and Fink, G.R. (1996). Ras2 signals via the Cdc42/Ste20/mitogen-activated protein kinase module to induce filamentous growth in *Saccharomyces cerevisiae*. Proc. Natl. Acad. Sci. USA 93, 5352–5356.

Nakagawa, C.W., Mutoh, N. and Hayashi, Y. (1995). Transcriptional regulation of catalase gene in the fission yeast *Schizosaccharomyces pombe*: molecular cloning of the catalase gene and northern blot analyses of the transcript. J. Biochem. 118, 109–116.

Nguyen, A.N., Ikner, A.D., Shiozaki, M., Warren, S.M. and Shiozaki, K. (2002). Cytoplasmic localization of Wis1 MAPKK by nuclear export signal (NES) is important for nuclear targeting of Spc1/Sty1 MAPK in fission yeast. Mol. Biol. Cell. In press.

Nguyen, A.N., Lee, A., Place, W. and Shiozaki, K. (2000). Multistep phosphorelay proteins transmit oxidative stress signals to the fission yeast stress-activated protein kinase. Mol. Biol. Cell 11, 1169–1181.

Nguyen, A.N. and Shiozaki, K. (1999). Heat shock-induced activation of stress MAP kinase is regulated by threonine- and tyrosine-specific phosphatases. Genes Dev. 13, 1653–1663.

Nigg, E.A. (1997). Nucleocytoplasmic transport: signals, mechanisms and regulation. Nature 386, 779–787.

O'Rourke, S.M. and Herskowitz, I. (1998). The Hog1 MAPK prevents cross talk between the HOG and pheromone response MAPK pathways in *Saccharomyces cerevisiae*. Genes Dev. 12, 2874–2886.

Ohmiya, R., Kato, C., Yamada, H., Aiba, H. and Mizuno, T. (1999). A fission yeast gene (*prr1⁺*) that encodes a response regulator implicated in oxidative stress response. J. Biochem. (Tokyo) 125, 1061–1066.

Ohmiya, R., Yamada, H., Nakashima, K., Aiba, H. and Mizuno, T. (1995). Osmoregulation of fission yeast: cloning of two distinct genes encoding glycerol-3-phosphate dehydrogenase, one of which is responsible for osmotolerance for growth. Mol. Microbiol. 18, 963–973.

Ota, I.M. and Varshavsky, A. (1993). A yeast protein similar to bacterial two-component regulators. Science 262, 566–569.

Palacios, C., Collins, M.K. and Perkins, G.R. (2001). The JNK phosphatase M3/6 is inhibited by protein-damaging stress. Curr. Biol. 11, 1439–1443.

Pascual-Ahuir, A., Posas, F., Serrano, R. and Proft, M. (2001a). Multiple levels of control regulate the yeast cAMP-response element-binding protein repressor Sko1p in response to stress. J. Biol. Chem. 276, 37373–37378.

Pascual-Ahuir, A., Serrano, R. and Proft, M. (2001b). The Sko1p repressor and Gcn4p activator antagonistically modulate stress-regulated transcription in *Saccharomyces cerevisiae*. Mol. Cell. Biol. 21, 16–25.

Peter, M., Neiman, A.M., Park, H.O., van Lohuizen, M. and Herskowitz, I. (1996). Functional analysis of the interaction between the small GTP binding protein Cdc42 and the Ste20 protein kinase in yeast. EMBO J. 15, 7046-7059.

Pidoux, A.L., Fawell, E.H. and Armstrong, J. (1990). Glyerol-3-phosphate dehydrogenase homologue from *Schizosaccharomyces pombe*. Nuc. Acids Res. 18, 7145.

Posas, F., Chambers, J.R., Heyman, J.A., Hoeffler, J.P., de Nadal, E. and Arino, J. (2000). The transcriptional response of yeast to saline stress. J. Biol. Chem. 275, 17249–17255.

Posas, F. and Saito, H. (1997). Osmotic activation of the HOG MAPK pathway via Ste11p MAPKKK: Scaffold role of Pbs2p MAPKK. Science 276, 1702–1705.

Posas, F. and Saito, H. (1998). Activation of the yeast SSK2 MAP kinase kinase kinase by the SSK1 two-component response regulator. EMBO J. 17, 1385–1394.

Posas, F., Takekawa, M. and Saito, H. (1998a). Signal transduction by MAP kinase cascades in budding yeast.

Curr. Opin. Microbiol. 1, 175–182.

Posas, F., Witten, E.A. and Saito, H. (1998b). Requirement of STE50 for osmostress-induced activation of the STE11 mitogen-activated protein kinase kinase kinase in the high-osmolarity glycerol response pathway. Mol. Cell. Biol. 18, 5788–5796.

Posas, F., Wurgler-Murphy, S.M., Maeda, T., Witten, E.A., Thai, T.C. and Saito, H. (1996). Yeast HOG1 MAP kinase cascade is regulated by a multiple phosphorelay mechanism in the SLN1-YPD1-SSK1 'two-component' osmosensor. Cell 86, 865–875.

Pringle, J.R., Bi, E., Harkins, H.A., Zahner, J.E., De Virgilio, C., Chant, J., Corrado, K. and Fares, H. (1995). Establishment of cell polarity in yeast. Cold Spring Harbour Symp. Quant. Biol. 60, 729–744.

Proft, M., Pascual-Ahuir, A., de Nadal, E., Arino, J., Serrano, R. and Posas, F. (2001). Regulation of the Sko1 transcriptional repressor by the Hog1 MAP kinase in response to osmotic stress. EMBO J. 20, 1123–1133.

Proft, M. and Serrano, R. (1999). Repressors and upstream repressing sequences of the stress-regulated ENA1 gene in *Saccharomyces cerevisiae*: bZIP protein Sko1p confers HOG-dependent osmotic regulation. Mol. Cell. Biol. 19, 537–546.

Raingeaud, J., Gupta, S., Rogers, J., Dickens, M., Han, J., Ulevitch, R. and Davis, R. (1995). Pro-inflammatory cytokines and environmental stress cause p38 mitogen-activated protein kinase activation by dual phosphorylation on tyrosine and threonine. J. Biol. Chem. 270, 7420–7426.

Raingeaud, J., Whitmarsh, A. J., Barrett, T., Dérijard, B. and Davis, R.J. (1996). MKK3- and MKK6-regulated gene expression is mediated by the p38 mitogen-activated protein kinase signal transduction pathway. Mol. Cell. Biol. 16, 1247–1255.

Raitt, D.C., Posas, F. and Saito, H. (2000). Yeast Cdc42 GTPase and Ste20 PAK-like kinase regulate Sho1-dependent activation of the Hog1 MAPK pathway. EMBO J. 19, 4623–4631.

Reiser, V., Ruis, H. and Ammerer, G. (1999). Kinase activity-dependent nuclear export opposes stress-induced nuclear accumulation and retention of Hog1 mitogen-activated protein kinase in the budding yeast *Saccharomyces cerevisiae*. Mol. Biol. Cell. 10, 1147–1161.

Reiser, V., Salah, S.M. and Ammerer, G. (2000). Polarized localization of yeast Pbs2 depends on osmostress, the membrane protein Sho1 and Cdc42. Nature Cell. Biol. 2, 620–627.

Rep, M., Krantz, M., Thevelein, J.M. and Hohmann, S. (2000). The transcriptional response of *Saccharomyces cerevisiae* to osmotic shock. Hot1p and Msn2p/Msn4p are required for the induction of subsets of high osmolarity glycerol pathway-dependent genes. J. Biol. Chem. 275, 8290–8300.

Rep, M., Proft, M., Remize, F., Tamas, M., Serrano, R.,

Thevelein, J.M. and Hohmann, S. (2001). The *Saccharomyces cerevisiae* Sko1p transcription factor mediates HOG pathway-dependent osmotic regulation of a set of genes encoding enzymes implicated in protection from oxidative damage. Mol. Microbiol. 40, 1067–1083.

Rep, M., Reiser, V., Gartner, U., Thevelein, J. M., Hohmann, S., Ammerer, G. and Ruis, H. (1999). Osmotic stress-induced gene expression in *Saccharomyces cerevisiae* requires Msn1p and the novel nuclear factor Hot1p. Mol. Cell. Biol. 19, 5474–5485.

Rouse, J., Cohen, P., Trigon, S., Morange, M.A.-L., A., Zamanillo, D., Hunt, T. and Nebreda, A. (1994). A novel kinase cascade triggered by stress and heat shock that stimulates MAPKAP kinase-2 and phosphorylation of the small heat shock proteins. Cell 78, 1027–1037.

Samejima, I., Mackie, S. and Fantes, P.A. (1997). Multiple modes of activation of the stress-responsive MAP kinase pathway in fission yeast. EMBO J. 16, 6162–6170.

Samejima, I., Mackie, S., Warbrick, E., Weisman, R. and Fantes, P.A. (1998). The fission yeast mitotic regulator *win1*⁺ encodes a MAP kinase kinase kinase that phosphorylates and activates Wis1 MAP kinase kinase in response to high osmolarity. Mol. Biol. Cell. 9, 2325–2335.

Santos, J.L. and Shiozaki, K. (2001). Fungal histidine kinases. Sci STKE 2001, RE1.

Sazer, S. and Nurse, P. (1994). A fission yeast RCC1-related protein is required for the mitosis to interphase transition. EMBO J. 13, 606–615.

Schaeffer, H.J. and Weber, M.J. (1999). Mitogen-activated protein kinases: specific messages from ubiquitous messengers. Mol. Cell. Biol. 19, 2435–2444.

Schüller, C., Brewster, J.L., Alexander, M.R., Gustin, M.C. and Ruis, H. (1994). The HOG pathway controls osmotic regulation of transcription via the stress response element (STRE) of the *Saccharomyces cerevisiae CTT1* gene. EMBO J. 13, 4382–4389.

Shieh, J.-C., Wilkinson, M.G., Buck, V., Morgan, B.A., Makino, K. and Millar, J.B.A. (1997). The Mcs4 response regulator coordinately controls the stress-activated Wak1-Wis1-Sty1 MAP kinase pathway and fission yeast cell cycle. Genes Dev. 11, 1008–1022.

Shieh, J.C., Martin, H. and Millar, J.B.A. (1998). Evidence for a novel MAPKKK-independent pathway controlling the stress activated Sty1/Spc1 MAP kinase in fission yeast. J. Cell. Sci. 111, 2799–2807.

Shiozaki, K., Akhavan-Niaki, H., McGowan, C.H. and Russell, P. (1994). Protein phosphatase 2C encoded by *ptc1*⁺ is important in the heat shock response of fission yeast. Mol. Cell. Biol. 14, 3743–3751.

Shiozaki, K. and Russell, P. (1995). Cell-cycle control linked to the extracellular environment by MAP kinase pathway in fission yeast. Nature 378, 739–743.

Shiozaki, K. and Russell, P. (1996). Conjugation, meiosis and the osmotic stress response are regulated by Spc1 kinase through Atf1 transcription factor in fission yeast. Genes Dev. 10, 2276–2288.

Shiozaki, K., Shiozaki, M. and Russell, P. (1997). Mcs4 mitotic catastrophe suppressor regulates the fission yeast cell cycle through the Wik1-Wis1-Spc1 kinase cascade. Mol. Biol. Cell. 8, 409–419.

Shiozaki, K., Shiozaki, M. and Russell, P. (1998). Heat stress activates fission yeast Spc1/Sty1 MAPK by a MEKK-independent mechanism. Mol. Biol. Cell. 9, 1339–1349.

Stade, K., Ford, C.S., Guthrie, C. and Weis, K. (1997). Exportin 1 (Crm1p) is an essential nuclear export factor. Cell 90, 1041–1050.

Sugimoto, A., Iino, Y., Maeda, T., Watanabe, Y. and Yamamoto, M. (1991). *Schizosaccharomyces pombe* ste11$^+$ encodes a transcription factor with an HMG motif that is a critical regulator of sexual development. Genes Dev. 5, 1990–1999.

Takeda, T., Toda, T., Kominami, K., Kohnosu, A., Yanagida, M. and Jones, N. (1995). *Schizosaccharomyces pombe atf1$^+$* encodes a transcription factor required for sexual development and entry into stationary phase. EMBO J. 14, 6193–6208.

Takekawa, M., Adachi, M., Nakahata, A., Nakayama, I., Itoh, F., Tsukuda, H., Taya, Y. and Imai, K. (2000). p53-inducible wip1 phosphatase mediates a negative feedback regulation of p38 MAPK-p53 signaling in response to UV radiation. EMBO J. 19, 6517–6526.

Takekawa, M., Maeda, T. and Saito, H. (1998). Protein phosphatase 2Cα inhibits the human stress-responsive p38 and JNK MAPK pathways. EMBO J. 17, 4744–4752.

Teige, M., Scheikl, E., Reiser, V., Ruis, H. and Ammerer, G. (2001). Rck2, a member of the calmodulin-protein kinase family, links protein synthesis to high osmolarity MAP kinase signaling in budding yeast. Proc. Natl. Acad. Sci. USA 98, 5625–5630.

Toda, T., Dhut, S., Superti-Furga, G., Gotoh, Y., Nishida, E., Sugiura, R. and Kuno, T. (1996). The fission yeast pmk1+ gene encodes a novel mitogen-activated protein kinase homolog which regulates cell integrity and func-

tions coordinately with the protein kinase C pathway. Mol. Cell. Biol. 16, 6752–6764.

Toda, T., Shimanuki, M., Saka, Y., Yamano, H., Adachi, Y., Shirakawa, M., Kyogoku, Y. and Yanagida, M. (1992). Fission yeast pap1-dependent transcription is negatively regulated by an essential nuclear protein, crm1. Mol. Cell. Biol. 12, 5474–5484.

Toda, T., Shimanuki, M. and Yanagida, M. (1991). Fission yeast genes that confer resistance to staurosporine encode an AP-1-like transcription factor and a protein kinase related to the mammalian ERK1/MAP2 and budding yeast *FUS3* and *KSS1* kinases. Genes Dev. 5, 60–73.

van Dam, H., Wilhelm, D., Herr, I., Steffen, A., Herrlich, P. and Angel, P. (1995). ATF-2 is preferentially activated by stress-activated protein kinases to mediate c-jun induction in response to genotoxic agents. EMBO J. 14, 1798–1811.

Warbrick, E. and Fantes, P.A. (1991). The wis1 protein is a dosage-dependent regulator of mitosis in *Schizosaccharomyces pombe*. EMBO J. 10, 4291–4299.

Warmka, J., Hanneman, J., Lee, J., Amin, D. and Ota, I. (2001). Ptc1, a type 2C Ser/Thr phosphatase, inactivates the HOG pathway by dephosphorylating the mitogen-activated protein kinase hog1. Mol. Cell. Biol. 21, 51–60.

Wen, W., Meinkoth, J.L., Tsien, R.Y. and Taylor, S.S. (1995). Identification of a signal for rapid export of proteins from the nucleus. Cell 82, 463–473.

Wilkinson, M.G., Samuels, M., Takeda, T., Toone, W.M., Shieh, J.-C., Toda, T., Millar, J.B.A. and Jones, N. (1996). The Atf1 transcription factor is a target for the Sty1 stress-activated MAP kinase pathway in fission yeast. Genes Dev. 10, 2289–2301.

Wu, C., Whiteway, M., Thomas, D.Y. and Leberer, E. (1995). Molecular characterization of Ste20p, a potential mitogen-activated protein or extracellular signal-regulated kinase kinase (MEK) kinase kinase from *Saccharomyces cerevisiae*. J. Biol. Chem. 270, 15984–15992.

Zaitsevskaya-Carter, T. and Cooper, J.A. (1997). Spm1, a stress-activated MAP kinase that regulates morphogenesis in *S. pombe*. EMBO J. 16, 1318–1331.

Sensing, Signaling and Cell Adaptation. Edited by K.B. Storey and J.M. Storey

CHAPTER 7

Calcium Signaling Mediated by Cyclic ADP-Ribose and NAADP: Roles in Cellular Response to Stress

Hon Cheung Lee

Department of Pharmacology, University of Minnesota, Minneapolis, MN 55455, USA

1. Introduction

Cells are surrounded by a plasma membrane which isolates them from the environment and, in so doing, defines their existence. Internally, cells are likewise compartmentalized by endo-membranes into organelles with diverse and specific functions. Information transfer, or signal transduction, across these membranes is fundamental to coordinating and integrating all cellular activities. Cells have developed various strategies to bypass the isolation imposed by the plasma membrane so that they can detect and adjust to environmental changes. Chemical signals are generally sensed by surface receptors, which upon binding specific ligands, activate intracellular production of second messengers. Cyclic AMP was the first of these second messengers identified (Rall and Sutherland, 1958). In many cases, surface receptor activation can also lead to mobilization of intracellular Ca^{2+} stores. Inositol trisphosphate (IP_3) is the best-known second messenger for this process (Streb et al., 1983). More recently, two novel nucleotides, cyclic ADP-ribose (cADPR) and nicotinic acid adenine dinucleotide phosphate (NAADP) have been shown to be as effective as IP_3 in mobilizing internal Ca^{2+} stores (Lee et al., 1989; Lee et al., 1994; Lee and Aarhus, 1995). So far, over 45 different cell types from three kingdoms—protist, plant and animal—have been reported to be responsive to these two nucleotide messengers (reviewed in Lee, 1997; Lee, 2001).

Since its identification more than a decade ago, much has been learned about the functions of cADPR. Remarkably, not only is it involved in sensing chemical signals, like IP_3, but it also functions in mediating cellular responses to physical stimuli such as heat and drought. Indeed, even functions that are totally controlled internally, such as cell cycle regulation, have been found to involve cADPR. The pervasive presence of cADPR in eukaryotic signaling may in fact be taken advantage of by some bacteria to aid infection. In this article, I shall review the current knowledge of the signaling pathways mediated by cADPR and NAADP and then focus specifically on some of their novel roles in stress-related functions.

2. Calcium signaling mediated by cyclic ADP-ribose and NAADP

2.1. Structure of cyclic ADP-ribose

The Ca^{2+} mobilizing activities of cADPR and NAADP were first reported in sea urchin eggs (Clapper et al., 1987). It was found that two unknown metabolites derived respectively from NAD and NADP can activate Ca^{2+} release from the egg microsomes in a manner totally independent of the IP_3-receptor. Both are more effective than IP_3 and are insensitive to heparin, an antagonist of the IP_3-receptor. Subsequent structural determination by X-ray crystallography established that the

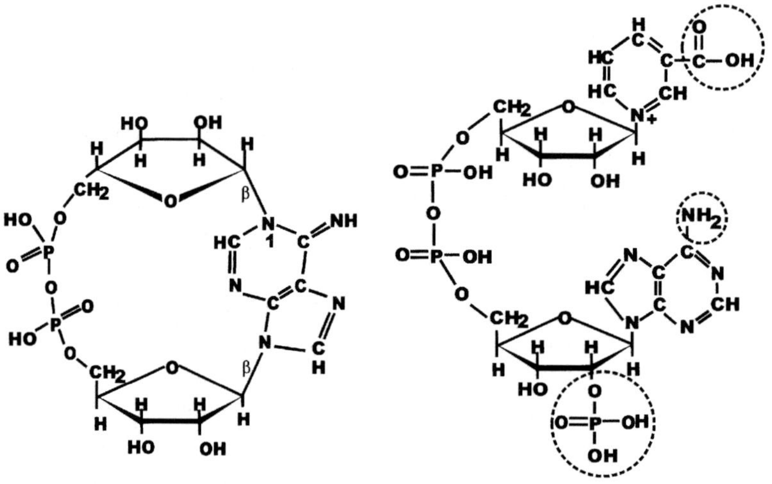

Fig. 7.1. Structures of cADPR and NAADP. The site of cyclization of cADPR is at the N1-position and the linkages to the two anomeric carbons are both in the β-configuration. Dashed circles in the NAADP formula indicate the three structural determinants important for its Ca^{2+} releasing activity.

active metabolite from NAD is cADPR (Lee et al., 1989; Lee et al., 1994). The structure of cADPR is unique. The long linear molecule of NAD is completely circularized, with the adenine group linked back to the distal ribose unit in a head-to-tail manner (Fig. 7.1). The sites of linkage are the N1-nitrogen of the adenine ring and the anomeric carbon of the distal ribose (Lee et al., 1994). The nicotinamide group, which attaches to the anomeric carbon of the distal ribose, is released during the cyclization process (Lee et al., 1989). Of all the known metabolites of NAD, cADPR is the only one that has this unique cyclic structure. The cyclic linkage is remarkably stable with a half-life of hydrolysis in days, even under acidic conditions (Lee and Aarhus, 1993). The uniqueness of cADPR is also evidenced by the fact that it is completely resistant to many common hydrolytic enzymes, such as NADase, nucleotidases and phosphatases (Takahashi et al., 1995; Graeff et al., 1997; Graeff and Lee, 2001). As will be described later, the only known enzyme that can hydrolyze the cyclic linkage of cADPR is CD38 (Jackson and Bell, 1990; Howard et al., 1993; Kim et al., 1993; Lee et al., 1993).

Cyclic ADP-ribose is endogenously present in cells and tissues and its levels are modulated by various stimuli (reviewed in Lee, 1997; Lee, 2001). The first assay for cADPR was a bioassay based on its Ca^{2+} release activity in sea urchin egg microsomes (Walseth et al., 1991; Walseth et al., 1997b). A more sensitive radio-immunoassay is also available (Takahashi et al., 1995; Graeff et al., 1997). Most recently, a novel coupled-enzyme cycling assay with nanomolar sensitivity has been developed (Graeff and Lee, 2001). Not only is the assay as sensitive and specific as the radio-immunoassay, but more importantly, all the reagents it employs are commercially available. It is also a single-step fluorimetric assay that can be run in a multi-well plate reader, allowing ready automation and high throughput screening. It is likely to become the method of choice in the field.

A large number of analogs of cADPR have been produced (reviewed in (Lee et al., 1999; Lee, 2001)). Total synthesis by chemical means alone has also been achieved (Shuto et al., 2001). Some of the more useful analogs include 8-amino-cADPR, 8-Br-cADPR (Walseth and Lee, 1993; Walseth et al., 1997a) and 7-deaza-8-Br-cADPR (Sethi et al., 1997); all are specific antagonists. The latter two, although less potent than 8-amino-cADPR, are more useful since both are cell permeant (Sethi et al., 1997). 3-deaza-cADPR

(Wong et al., 1999) and cyclic ADP-carbocylic-ribose (Shuto et al., 2001) are agonists of cADPR. Both are non-hydrolyzable and are even more potent than cADPR itself. They should be particularly useful when used in systems that have high cADPR-hydrolyzing activity or in situations that require long-term incubation with the agonist. An inactive caged analog of cADPR has also been synthesized, which upon UV-irradiation can regenerate active cADPR (Aarhus et al., 1995a). It is useful when localized release of cADPR in a cell is required. It has also been used to functionally visualize the distribution of the cADPR-sensitive Ca^{2+} stores in live sea urchin eggs (Lee and Aarhus, 2000).

2.2. Structure of NAADP

The structure of the active metabolite of NADP was determined in 1995 (Lee and Aarhus, 1995). It is not a cyclic compound as cADPR, but a simple derivative of NADP (Fig. 7.1). The only difference between NAADP and NADP is the replacement of the $-NH_2$ group of the nicotinamide in NADP with a $-OH$ group, turning the amide group into a carboxyl group. In other words, the nicotinamide group is replaced by nicotinic acid. NAADP is thus only one mass unit higher than its precursor (Lee and Aarhus, 1995). Indeed, regular proton NMR spectra of the two are indistinguishable (Lee and Aarhus, 1995; Chini et al., 1995). NAADP is thus a common nucleotide and, similar to NADP, it is also readily hydrolyzed by many phosphatases as well as pyrophosphatases (Lee et al., 1997). This is in sharp contrast to the unique stability observed with cADPR.

A series of analogs of NAADP has been synthesized to probe the structural determinants important for its biological activity (Lee and Aarhus, 1997). Expectedly, the carboxyl group of the nicotinic acid moiety is absolutely required for its Ca^{2+} release activity. The position of the 2'-phosphate is less critical. Only moderate decrease in potency is seen in analogs with the 3'-phosphate or 2',3'-cyclic phosphate. Attachment of a caged group on the 2'-phosphate, however, eliminates its activity and the caged analog has proved to be a versatile and

widely used tool for investigating the Ca^{2+} signaling functions of NAADP (Lee et al., 1997; Lee and Aarhus, 2000; Santella et al., 2000; Churchill and Galione, 2001b). Modifications of the adenine ring such as in the analogs, etheno-NAADP and etheno-aza-NAADP, reduce but do not eliminate the Ca^{2+} releasing activity (Lee and Aarhus, 1998). The modifications render the analogs fluorescent and can potentially be used to visualize the receptor. These specific structural requirements indicate the biological effect of NAADP is mediated by a receptor capable of recognizing small structural changes in the molecule.

Currently, there is no assay for the endogenous NAADP in cells. In principle, the sea urchin egg microsomes, which show remarkable sensitivity to NAADP (Lee and Aarhus, 1995), can be used as a bioassay. The egg microsomes also bind NAADP specifically and with high affinity (Aarhus et al., 1996), allowing the possibility for developing a radio-receptor assay. Perhaps, the most desirable assay would be a cycling assay analogous to that recently developed for cADPR (Graeff and Lee, 2001). An assay with sufficient sensitivity and employing readily available reagents would greatly advance our knowledge of the regulation of the endogenous levels of NAADP in cells and how they may change in response to stimuli.

2.3. Calcium mobilization

Not only are cADPR and NAADP structurally distinct, they also activate totally separate mechanisms of Ca^{2+} mobilization. 8-Amino-cADPR and 8-Br-cADPR were synthesized as specific and potent antagonists of cADPR and, as expected, neither the NAADP- nor the IP_3-sensitive Ca^{2+} release is affected by them (Walseth and Lee, 1993; Lee and Aarhus, 1995). Conversely, heparin, an antagonist of the IP_3-receptor inhibits neither the cADPR nor the NAADP mechanisms (Clapper et al., 1987; Dargie et al., 1990; Lee and Aarhus, 1995). On the other hand, the cADPR-mechanism is respectively stimulated and blocked by agonists and antagonists of ryanodine receptors (Galione et al., 1991; Lee, 1993; Lee et al., 1995). Indeed, the cADPR-sensitive Ca^{2+} channels reconstituted into lipid bilayers exhibit characteristics very similar to the ryanodine-sensitive Ca^{2+} channels (Lokuta et al., 1998; Perez et al., 1998). Also, ryanodine receptors isolated from muscles are found to be sensitive to cADPR (Sonnleitner et al., 1998; Li et al., 2001) and binding of ryanodine to the receptor can likewise be modulated by cADPR (Guse et al., 1999; Zhang et al., 1999). These results provide convincing evidence that the target of cADPR is the ryanodine receptor.

Less is known about the identity of the NAADP-sensitive Ca^{2+} channel. As described above, structure-function studies of the analogs of NAADP suggest that its action is mediated by a unique receptor. Specific binding sites with high affinity have been detected in sea urchin egg (Aarhus et al., 1996), brain (Patel et al., 2000) and cardiac microsomes (Bak et al., 2001). So far, only L-type Ca^{2+} channel antagonists, such as diltazem, have been reported to be moderately effective in blocking the NAADP-mechanism (Genazzani et al., 1997). Not only is the NAADP-mechanism pharmacologically different from those of cADPR and IP_3, the NAADP-sensitive Ca^{2+}-stores can also be separated from those sensitive to cADPR by fractionation of cell homogenates as reported in sea urchin eggs (Lee and Aarhus, 1995) and plants (Navazio et al., 2000). Physical separation of the two types of Ca^{2+} stores have also been achieved in live sea urchin eggs, using centrifugation to stratify their organelles (Lee and Aarhus, 2000). It was observed that the cADPR- and IP_3-sensitive Ca^{2+} stores are segregated to one pole of the centrifuged eggs together with the endoplasmic reticulum and the cell nucleus, while the NAADP-sensitive stores are moved to the opposite pole. These results strongly suggest that not only is the NAADP-mechanism unique, but even the organelles that NAADP targets are different from the endoplasmic reticulum, a known Ca^{2+} store sensitive to IP_3 and cADPR. Indeed, the NAADP-stores may represent a hitherto unknown organelle in cells.

2.4. Multiple Ca^{2+} stores in cells

The question of why cells need multiple Ca^{2+} stores was first addressed in a study using *Ascidian*

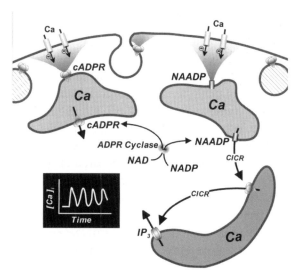

Fig. 7.2. Spatial segregation of Ca^{2+} stores. In some cells, the cADPR- and NAADP-sensitive Ca^{2+} stores are mainly present in the cortex and their mobilization results in rapid modulation of plasma membrane Ca^{2+} channel (long cylinders) activity. Tetrameric globules and short cylinders respectively represent the cADPR and the NAADP mechanisms. Exocytic vesicles in the cell cortex may also contain the cADPR-channels, where activation induces membrane fusion and exocytosis. The IP_3-receptors (tetrameric ellipsoids) are mainly in the endoplasmic reticulum distributed in the cytoplasm and separated from the cell cortex. The cADPR-mechanism is also found to be present in the endoplasmic reticulum in some cells. Ca^{2+} released from the NAADP-stores can be amplified to become propagative Ca^{2+} waves by the Ca^{2+}-induced Ca^{2+} release (CICR) mechanism that involves both the cADPR and IP_3 mechanisms. Interaction between the NAADP and the cADPR stores can also lead to the generation of prolonged Ca^{2+} oscillations in some cells. A single enzyme, ADP-ribosyl cyclase (ADPR cyclase) is responsible for synthesizing cADPR from NAD and NAADP from NADP.

oocytes (Albrieux et al., 1998). The large size of the oocytes provides the necessary advantage for investigating the spatial arrangement of Ca^{2+} stores. Analyses indicate a hierarchical organization, with the NAADP-sensitive stores situated closest to the plasma membrane, providing fast and immediate modulation of the activity of the membrane Ca^{2+} channels (Fig. 7.2). Almost exclusive localization of the NAADP-sensitive stores in the cortex has also been observed in starfish oocytes (Lim et al., 2001).

The cADPR-sensitive stores are located slightly removed from the plasma membrane. Activation by cADPR results in fusion and insertion of the cortical vesicles into the plasma membrane (Albrieux et al., 1998). This fusion event can not be activated by NAADP. Similar to NAADP, the cADPR-mediated Ca^{2+} release is effective in modulating the membrane Ca^{2+} channels. The cADPR-mechanism may in fact be part of the cortical vesicles as has been found in zymogen granules (Gerasimenko et al., 1996) and the insulin containing secretory vesicles (Mitchell et al., 2001) in pancreatic acinar and β-cells, respectively.

The IP_3-sensitive Ca^{2+} stores are mainly distributed throughout the cytoplasm. Activation of these stores, although readily elevating cytoplasmic Ca^{2+} levels, is ineffective in inducing vesicular fusion or modulating plasma membrane Ca^{2+} channels (Albrieux et al., 1998). The cortical region of the oocytes thus represents a restricted zone, in which the Ca^{2+} concentration is highly buffered and insulated from cytoplasmic changes, perhaps by the abundance of microsomal ATPase.

Segregation of Ca^{2+} stores can thus provide a means to effect local regulation of selective functions without causing global activation of the cell. This spatial organization, however, does not preclude interaction between stores. By strategically positioning different types of Ca^{2+} stores in close proximity, Ca^{2+} signals can be transmitted, amplified or propagated between them (Fig. 7.2). Indeed, results obtained from pancreatic acinar cells suggest that the action of cholecystokinin is mediated by a triggering Ca^{2+} signal from the NAADP-sensitive stores, which is then amplified and propagated as Ca^{2+} waves by the cADPR-and IP_3-sensitive stores through, perhaps, the Ca^{2+}-induced Ca^{2+} release mechanism (Cancela et al., 1999).

Interaction between stores can also lead to Ca^{2+} oscillation as is observed following activation of the NAADP-sensitive stores by photolyzing caged NAADP inside live sea urchin eggs (Aarhus et al., 1996). It is proposed that Ca^{2+} released from the NAADP-stores overloads the cADPR-dependent stores, which is known to be sensitive to Ca^{2+} priming (Galione et al., 1993a), and results in spontaneous and cyclic release (Lee, 1997). This two-store mechanism for Ca^{2+} oscillation has recently been verified experimentally (Churchill and Galione, 2001a). It is clear, therefore, that the presence of multiple Ca^{2+} stores provides cells with a versatile means of localizing as well as integrating diverse Ca^{2+} signals.

2.5. Enzymatic synthesis and degradation

Despite the fact that cADPR and NAADP differ in structure, precursor, mechanism of action and Ca^{2+} stores targeted, both messengers are synthesized by a single enzyme, ADP-ribosyl cyclase (Aarhus et al., 1995b). This novel enzyme, a 30 kDa soluble protein, was first identified in *Aplysia* ovotestis (Lee and Aarhus, 1991). Amino acid sequence comparison shows that CD38, a membrane antigen first identified in lymphocytes (Jackson and Bell, 1990), is not only a structural homologue but also possesses similar enzymatic activity as the cyclase (States et al., 1992; Howard et al., 1993; Lee et al., 1993). Both can cyclize NAD to produce cADPR. Additionally, in the presence of nicotinic acid, both can catalyze a base exchange reaction, exchanging the nicotinamide group of NADP with nicotinic acid to produce NAADP (Aarhus et al., 1995b). CD38 is now known to be a ubiquitous protein not only expressed on the cell surface but also intracellularly and present in lymphocytes as well as a wide range of other cells and tissues including brain, pancreas, liver, etc. (reviewed in Lee, 2000).

Another antigen, CD157 (previously designated as BST-1), a glycosylphosphatidyl–inositol anchored protein, is also a homologue of the *Aplysia* cyclase and CD38, and has been shown to possess similar enzyme activities (Hirata et al., 1994). The sequence identity of the three proteins is about 25–30%.

The 3-D structure of the *Aplysia* cyclase has been solved. It is a homodimer both in solution (Munshi et al., 1998) and when crystallized (Prasad et al., 1996). The α-helices and β-structures are respectively segregated between the amino and carboxyl domains. The ten cysteine residues are paired in five disulfide linkages, three in the amino and two in the carboxyl domain (Prasad et al., 1996) (Fig. 7.3). Glu179 is identified as the catalytic residue by site-directed mutagenesis

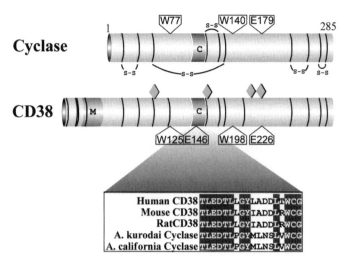

Fig. 7.3. Homology between the *Aplysia* ADP-ribosyl cyclase and mammalian CD38. The cyclase contains 285 amino acids. The ten cysteine residues are paired in disulfide linkages (S-S). The catalytic residue is E179. W77 and W140 are also critical for catalysis. CD38 has a transmembrane domain (M), a segment (C) highly conserved with the cyclase, and four glycosylation sites (diamonds). Ten of its cysteines (represented as lines) can be aligned with those of the cyclase. E226 is the catalytic residue, while E146 controls its cyclase activity. Two additional tryptophans, corresponding to those in the cyclase are also critical for catalysis.

(Munshi et al., 1999). Two other critical residues are Trp77 and Trp140. Although these three residues span over 100 residues in the primary sequence, the 3D structure shows that they all cluster in the active site pocket near the central cleft of the molecule (Munshi et al., 1999). A segment of highly conserved residues, found in both the cyclase and CD38 as indicated in Fig. 7.3, forms part of the base of the active pocket. It appears that the two tryptophans may be responsible for positioning the substrate NAD into a folded conformation such that the two ends of the linear molecule can cyclize.

The catalytic residue of CD38 was identified as Glu226 by site-directed mutagenesis (Munshi et al., 2000). Similar to the cyclase, two tryptophans, Trp125 and Trp198, are critical for enzymatic activities. Glu226 is equivalent to Glu179 in the cyclase and it has also been found to be covalently ribosylated during catalysis (Sauve et al., 2000), providing further support that it is the catalytic residue. Homology modeling shows that all three critical residues are clustered in a pocket similar to that observed in the cyclase (Munshi et al., 2000). In addition to catalyzing the synthesis of cADPR and NAADP, CD38 also hydrolyzes cADPR efficiently to ADP-ribose (Howard et al., 1993; Kim et

al., 1993; Lee et al., 1993). CD38 is thus a novel enzyme that can catalyze a multitude of enzyme activities. In fact, CD38 is the only enzyme so far reported to be capable of hydrolyzing cADPR. The critical residue that controls the cyclization and hydrolysis reaction is Glu146, which is part of the highly conserved segment that forms part of the base of the active site pocket (Graeff et al., 2001). Mutagenizing Glu146 depressed the hydrolase activity but greatly stimulates the cyclase activity. A unified model emphasizing a single enzyme intermediate and the accessibility of water to the active site has been proposed to account for the novel multiplicity of catalysis of the cyclase and CD38 (Lee, 1999; Lee, 2000; Sauve et al., 2000).

A crucial factor that controls the catalysis of the cyclase and CD38 is pH. Under acidic conditions, both enzymes mainly catalyze the base-exchange reaction and produce NAADP from NADP and nicotinic acid, whereas at neutral or alkaline pH, the cyclization reaction is dominant (Aarhus et al., 1995b). The cyclase activity in some cells, such as sea urchin eggs, can also be stimulated by cGMP-dependent phosphorylation (Galione et al., 1993b). This appears to be a general regulatory mechanism as it has also been documented in such widely different systems as plants (described below) and hippocampus (Reyes-Harde et al., 1999). In fact, it has been observed that nitric oxide, a gaseous messenger that is known to elevate cGMP levels in a wide variety of cells, can indeed activate Ca^{2+} mobilization by stimulating cADPR production (Willmott et al., 1996; Reyes-Harde et al., 1999). As will be described later, this cross-talk between the nitric oxide- and cADPR-signaling pathways plays an important role in plant defense against pathogen infection.

As mentioned above, CD38 is the only known enzyme that can effectively degrade cADPR. In contrast to this remarkable stability of cADPR, NAADP is highly sensitive toward common hydrolytic enzymes and can be easily hydrolyzed by enzymes such as alkaline phosphatases and pyrophosphatases (Lee, 1997; Lee et al., 1997). This heightened susceptibility could ensure rapid removal of this extraordinarily effective Ca^{2+} messenger after its signaling function is completed.

3. cADPR and plant response to environmental stress

Unlike animals, plants are immobile and cannot actively avoid adverse conditions. Various strategies were evolved in plants to handle environmental stress. A particular robust response is the regulation of stomata on leaves, pores through which gas exchange takes place. By closing the stomata, plants can conserve water during periods of drought by reducing loss through evaporation. Stomatal aperture is made up of and controlled by two parallel guard cells which are elongated in shape and are physically linked at the two ends, but not longitudinally. When water is plentiful, swelling of the guard cells induced by solute and water intake results in opening of the pore. Closure of the stomata is regulated by a plant hormone, abscisic acid (ABA), which triggers solute efflux from the guard cells and causes cell shrinkage.

A complex series of ion movements in guard cells is activated by ABA (Schroeder et al., 2001). Principal among them is cytoplasmic elevation of Ca^{2+} concentration, which in turn activates anion channels in the plasma membrane of the cell.

Efflux of anions depolarizes the cell and opens the outward-rectifying K^+-channels present also in the plasma membrane. The net effect is the efflux of KCl from the guard cell leading to reduction in turgor pressure, cell shrinkage and closure of the stomatal aperture.

A series of recent studies shows that cADPR plays a crucial role in initiating the Ca^{2+} signal induced by ABA. A main source of this Ca^{2+} signal is by release from the large vacuoles present in plant cells. Isolated vacuolar microsomes can respond to either cADPR or IP_3 and release Ca^{2+} (Allen et al., 1995). Patch-clamp studies of individual vacuoles show the presence of cADPR-gated Ca^{2+} channels in the vacuolar membrane (Allen et al., 1995; Leckie et al., 1998). Microinjection of cADPR into guard cells likewise induces Ca^{2+} elevation and effects loss of turgor pressure in the injected cell (Leckie et al., 1998). Conversely, treatment of guard cells with nicotinamide, an inhibitor of the ADP-ribosyl cyclase (described above), significantly inhibits stomatal closure induced by ABA.

The ABA-induced Ca^{2+} signal occurs in the form of a prolonged oscillation, which appears to be important for stomatal closure (Leckie et al., 1998; Allen et al., 2000). A plant mutant which lacks the ability to produce a Ca^{2+} oscillation in response to certain stimuli also fails to close the stomata in response to the same stimuli (Allen et al., 2000). The exact mechanism of how ABA activates this Ca^{2+} oscillation has not been worked out. Evidence obtained in sea urchin eggs indicates that specific interactions between the cADPR- and NAADP-sensitive Ca^{2+} stores can lead to prolonged Ca^{2+} oscillation (Aarhus et al., 1996; Lee, 1997; Churchill and Galione, 2001a). Indeed, plants, like the eggs, also possess both types of Ca^{2+} stores. Fractionation studies show that the NAADP-sensitive Ca^{2+} stores are mainly co-purified with the endoplasmic reticulum (Navazio et al., 2000). The cADPR-sensitive stores, on the other hand, are associated with both the vacuolar membranes and the endoplasmic reticulum (Allen et al., 1995; Navazio et al., 2001). The pharmacology of the NAADP-sensitive mechanism in plants is also similar to that observed in animal cells, being insensitive to 8-amino-cADPR and heparin and showing potent self-desensitization (Aarhus et al., 1996; Genazzani et al., 1996; Navazio et al., 2001). It has also been shown that plant homogenates are capable of enzymatic synthesis of NAADP from NADP and nicotinic acid, as observed in animal cells (Aarhus et al., 1995b; Navazio et al., 2001). It is thus clear that both Ca^{2+} signaling mechanisms, like the IP_3 mechanism, are highly conserved in evolution, testifying to their fundamental importance.

Stomatal closure represents a fast and immediate response to ABA, which is also capable of inducing long-term adaptation through expression of specific genes. This process is likewise mediated by cADPR. Microinjection of cADPR into plant cells activates expression of ABA-specific genes, such as *rd29A*, a desiccation-responsive gene (Wu et al., 1997), as well as *kin2* and *BN115*, which are cold-inducible genes (Wu et al., 1997; Sangwan et al., 2001). Indeed, cADPR-induced expression of *BN115* effects significant protection of the leaves to damage by freezing (Sangwan et al., 2001). The protective effect of cADPR is more specific than simply elevating cytoplasmic Ca^{2+}, since A23187, a non-specific Ca^{2+} ionophore, can produce much less protection, indicating a high degree of specificity of the cADPR-pathway in mediating cold-acclimatization in plants (Sangwan et al., 2001).

Expression of both cold-induced and desiccation-induced genes, *kin2* and *rd29A*, can also be activated by injection of the *Aplysia* ADP-ribosyl cyclase, which presumably produces cADPR using the cytoplasmic NAD as substrate (Wu et al., 1997). Indeed, elevation of endogenous cADPR levels is observed to follow ABA application, preceding the expression of *kin2* and *rd29A*. Conversely, the induction of gene expression by ABA is blocked by microinjection of an antagonist of cADPR, 8-amino-cADPR (Wu et al., 1997). Interestingly, heparin, an antagonist of the IP_3 receptor, does not block the ABA induction. These results indicate that the cADPR-signaling pathway is primarily responsible for mediating the induction of gene expression by ABA. Similar to that observed in *Ascidian* oocytes described above, the plant cell

thus represents yet another system where different Ca^{2+}-signaling pathways coexist within the same cell and each subserves separate functions.

In addition to the robust response to environmental stress, effective defense against pathogen infection is of equal importance for plant survival. Emerging evidence indicates that the cADPR-signaling pathway also plays an important role in this crucial process. Infection of resistant, but not susceptible, tobacco with tobacco mosaic virus results in enhanced nitric oxide synthase activity (Durner et al., 1998). As described above, work in sea urchin eggs has shown that nitric oxide can activate ADP-ribosyl cyclase through a cGMP-dependent protein kinase mechanism (Galione et al., 1993b; Willmott et al., 1996; Graeff et al., 1998), suggesting a role for cADPR in plant defense. This is supported by the observation that nitric oxide, cGMP and cADPR all can activate the expression of the same defense gene (Durner et al., 1998). Furthermore, ruthenium red, a known blocker of the cADPR-sensitive Ca^{2+} channel, can inhibit the gene expression activated by both cGMP and cADPR. If nitric oxide can indeed turn on the cADPR-pathway, one would expect it to also activate stomatal closure. This has been shown to be the case in three different plant species (Mata and Lamattina, 2001). In all three cases, nitric oxide is found to be effective in promoting water retention during drought.

The results described above clearly show that in plants, as in animal cells, nitric oxide elevates cGMP levels and through a cGMP-dependent kinase, stimulates the synthesis of cADPR, leading eventually to mobilization of intracellular Ca^{2+} stores. Plants, in fact, make ample use of this evolutionarily conserved signaling pathway to regulate defense and stress responses, two of most important processes for their survival.

4. cADPR and sponge response to heat stress

Marine organisms, lacking the internal temperature regulation system prevalent in warm-blooded animals, are particularly susceptible to heat. A massive die-off of sponges was found in the Mediterranean in 1999, following an unusual warming of the sea by several degrees. The biochemical responses of sponges to heat stress has recently been investigated (Zocchi et al., 2001a). The initial interest was focused on the ADP-ribosyl cyclase in sponges. Marine organisms are special in the field of cADPR-signaling. The cyclase was first described in sea urchin eggs (Clapper et al., 1987). Also, the most abundant source of the cyclase ever reported is from a marine organism, *Aplysia*, and it was from this source that the cyclase was first purified (Lee and Aarhus, 1991). Like other marine organisms, sponges do possess high levels of the cyclase and its catalytic properties are most similar to the *Aplysia* cyclase, having mainly NAD cyclization activity and low cADPR hydrolytic activity (Zocchi et al., 2001a). This is in contrast to the mammalian CD38 as described above. More interestingly, the activity of the sponge cyclase is highly dependent on the temperature at which the sponge cells are maintained. Even a few minutes of exposure of the cells to elevated temperature, 26°C instead of the normal sea temperature of 14°C, is sufficient to activate the sponge cyclase activity by 3–5 fold. This remarkable sensitivity suggests the cyclase is intimately associated with the temperature sensor and is likely involved in mediating the organism's response to heat stress.

Accompanying the increase in cyclase activity, cADPR levels in sponge cells are elevated, resulting in a progressive increase in cytoplasmic Ca^{2+} concentration (Zocchi et al., 2001a). The latter can be blocked by preincubating the cells with a cell permeant antagonist of cADPR, 8-Br-cADPR. Similar to that observed in plant and animal cells, the activation of the cyclase by heat stress turns out to be also due to protein phosphorylation. The activation is, however, mediated by a cAMP-dependent process instead of cGMP (Zocchi et al., 2001a).

The upstream event that leads to the stimulation of the cyclase by phosphorylation is found to be the activation of a cation channel in the plasma membrane of sponge cells. This ion channel shows a strong outward rectification with a single-channel conductance of about 13 pS, is permeable to Rb^+, blocked by bupivacaine and trivalent cations such as Gd^{+3}, and activated by heat, mechanical stretch, as well as by arachidonic acid (Zocchi et al., 2001a). All these characteristics indicate the sponge channel is very similar to the stretch-receptor channel identified in mammalian cells, which has been shown to be responsible for thermosensing also.

The most unexpected finding concerning the heat-stress response in sponge, perhaps, is how the activation of the cation channel is linked to stimulation of the cyclase. The link is found to be ABA, the same hormone responsible for mediating stress response in plants. Thus, heat stress rapidly increases the production of endogenous ABA in sponge, which is completely blocked by Gd^{+3}, an inhibitor of the cation channel (Zocchi et al., 2001a). Treatment of sponge cells with exogenous ABA, likewise activates phosphorylation and stimulation of the cyclase and the consequential elevation of intracellular Ca^{2+} concentration.

Thus, sponges, phylogenetically the oldest metazoan animals, are the only organisms outside the plant kingdom that have been reported to be responsive to ABA. It is truly remarkable that the salient features of the cADPR-signaling pathway are so evolutionarily conserved among organisms from different kingdoms, from plants to animals.

5.   cADPR in bacterial infection and immune responses

How far down the phylogenetic tree can one go and still find the presence of the cADPR-signaling pathway? The protozoan, *Euglena*, has been found to possess a membrane bound cyclase with catalytic properties similar to the soluble *Aplysia* cyclase. Both the activity of the *Euglena* cyclase and the cellular cADPR levels show periodic elevations correlating with cell cycles (Masuda et al., 1997), suggesting that in protozoa, as in human HeLa cells and mouse fibroblasts (Zocchi et al., 1998; Franco et al., 2001), the cADPR-pathway is also involved in regulating cell proliferation.

Even bacteria, such as *S. pyrogenes*, possess the cyclase (Karasawa et al., 1995). Prokaryotes do not have internal Ca^{2+} stores. It is curious what role the cyclase plays in *S. pyrogenes*. A recent study proposes that it is involved in mediating bacterial cytotoxicity (Madden et al., 2001).

A strategy used by *S. pyrogenes* during infection is the secretion of streptolysin O, which forms transmembrane pores in the host cells and allows transfer of specific bacterial factors into the cells. Analyses show that only a 52 kDa protein is transferred by *S. pyrogenes* to host cells and this protein is identified by N-terminal sequencing as the bacterial cyclase (Madden et al., 2001). Mutant bacteria deficient in this 52 kDa protein are much less cytotoxic to the infected cells, indicating that the cyclase, once transferred to the host cell, is active in mediating cytolysis. Although the exact mechanism of this process remains to be worked out, it is clear that introducing an exogenous cyclase into cells would lead to unregulated conversion of endogenous NAD to cADPR and ADP-ribose. The consequential Ca^{2+} elevation could be part of the cytotoxic process.

The cADPR-pathway is utilized not only by bacteria but also by the host's immune system to combat the infection. A variety of cells in the immune system are known to be responsive to cADPR (reviewed in Lee, 2001). For examples, activation of the T-cell receptor in lymphocytes leads to cADPR production, Ca^{2+} mobilization and stimulation of the capacitative Ca^{2+} influx, culminating in cell proliferation (Guse et al., 1999). Likewise, proliferation of human hemopoietic progenitors also involves the cADPR-pathway (Podesta et al., 2000; Zocchi et al., 2001b).

More relevant against bacterial infection is the neutrophil, which functions to confine and combat the bacteria at the infection site and prevent systemic spread of the bacteria. Migration of neutrophils to the infection site is thus crucial in this process and is chemotactically directed by bacterial products, mainly N-formylpeptides. In response to the peptide, the Ca^{2+} stores in neutrophils are mobilized and this is followed by capacitative Ca^{2+} influx due to store depletion (Partida-Sanchez et al., 2001). This Ca^{2+} influx is critical for the directed migration of neutrophils. 8-Br-cADPR, a permeant antagonist of cADPR, effectively inhibits the store-depletion stimulated

Ca^{2+} influx and blocks the peptide-induced neutrophil chemotaxis (Partida-Sanchez et al., 2001). The chemotactic response to host-produced cytokines, such as IL-8, is not associated with Ca^{2+} influx and is not affected by 8-Br-cADPR, showing exquisite specificity.

That the cADPR-dependent chemotaxis indeed is crucial for the host to combat bacterial infection is shown using transgenic mice having the CD38 gene ablated (Partida-Sanchez et al., 2001). Neutrophils isolated from the knockout mice lack CD38 and can not produce cADPR. They exhibit normal chemotactic response to IL-8 but not to N-formylpeptides. The latter also fails to elicit the Ca^{2+}-influx associated with chemotaxis in the mutant neutrophils.

The CD38 knockout mice are particularly susceptible to bacterial infection (Partida-Sanchez et al., 2001). The bacterial dose required for 50% fatality is ten times lower than in the wild-type mice. Similar susceptibility is seen in irradiated wild-type mice if they are reconstituted with bone-marrow cells derived from the mutant mice but not from wild-type mice (Partida-Sanchez et al., 2001). The cause of the susceptibility in the mutant mice apparently is the failure of neutrophils to accumulate in the lungs, the site of infection. Consequently, the infecting bacteria dissimulate rapidly and their titers in the blood of the mutant mice are much higher than in the wild-type mice (Partida-Sanchez et al., 2001). Thus, the lack of CD38 in the mutant neutrophils prevents them from producing cADPR in response to the bacterial peptide, impairing their ability to move toward the site of infection.

6. Conclusion

Results described in this article clearly indicate that the cADPR-signaling pathway is not just a curiosity observed *in vitro*, but, in fact, is present and fully functional in living organisms. Indeed, as shown in the CD38 knockout mice, its correct operation is critical for the survival of the mice during bacterial infection. This is also the case for plants in surviving freezing cold. It is truly remarkable that the pathway and various components of it are conserved throughout evolution and are utilized by various organisms, from prokaryotes to mouse, to serve diverse functions crucial for their survival.

Results described also establish firmly the presence of multiple Ca^{2+} signaling pathways in cells. The multiplicity allows cells to commit specific pathways for specific functions. This is directly shown in *Ascidian* oocytes and is also the case in neutrophils, where the cADPR-pathway is specifically used in chemotaxis directed by bacterial peptides but not by endogenous cytokines. In addition to having multiple Ca^{2+} signaling pathways, spatial segregation of different types of Ca^{2+} stores can provide further hierarchical complexity and versatility necessary for cells to respond to the myriad of stimuli and stresses of an ever-changing environment.

References

Aarhus, R., Dickey, D.M., Graeff, R.M., Gee, K.R., Walseth, T.F. and Lee, H.C. (1996). Activation and inactivation of Ca^{2+} release by NAADP⁺. J. Biol. Chem. 271, 8513–8516.

Aarhus, R., Gee, K. and Lee, H.C. (1995a). Caged cyclic ADP-ribose—synthesis and use. J. Biol. Chem. 270, 7745–7749.

Aarhus, R., Graeff, R.M., Dickey, D.M., Walseth, T.F. and Lee, H.C. (1995b). ADP-ribosyl cyclase and CD38 catalyze the synthesis of a calcium-mobilizing metabolite from NADP. J. Biol. Chem. 270, 30327–30333.

Albrieux, M., Lee, H.C. and Villaz, M. (1998). Calcium signaling by cyclic ADP-ribose, NAADP, and inositol trisphosphate are involved in distinct functions in Ascidian oocytes. J. Biol. Chem. 273, 14566–14574.

Allen, G.J., Chu, S.P., Schumacher, K., Shimazaki, C.T., Vafeados, D., Kemper, A., Hawke, S.D., Tallman, G., Tsien, R.Y., Harper, J.F., Chory, J. and Schroeder, J.I. (2000). Alteration of stimulus-specific guard cell calcium oscillations and stomatal closing in *Arabidopsis* det3 mutant. Science 289, 2338–2342.

Allen, G.J., Muir, S.R. and Sanders, D. (1995). Release of Ca^{2+} from individual plant vacuoles by both InsP₃ and cyclic ADP-ribose. Science 268, 735–737.

Bak, J., Billington, R.A., Timar, G., Dutton, A.C. and Genazzani, A.A. (2001). NAADP receptors are present and functional in the heart. Curr. Biol. 11, 987–990.

Cancela, J.M., Churchill, G.C. and Galione, A. (1999). Co-ordination of agonist-induced Ca^{2+}-signalling patterns by NAADP in pancreatic acinar cells. Nature. 398, 74–76.

Chini, E.N., Beers, K.W. and Dousa, T.P. (1995). Nicotinate adenine dinucleotide phosphate (NAADP) triggers a specific calcium release system in sea urchin eggs. J. Biol. Chem. 270, 3216–3223.

Churchill, G.C. and Galione, A. (2001a). NAADP induces Ca^{2+} oscillations via a two-pool mechanism by priming IP₃- and cADPR-sensitive Ca^{2+} stores. EMBO J. 20, 2666–2671.

Churchill, G.C. and Galione, A. (2001b). Prolonged inactivation of nicotinic acid adenine dinucleotide phosphate-induced Ca2+ release mediates a spatiotemporal Ca^{2+} memory. J. Biol. Chem. 276, 11223–11225.

Clapper, D.L., Walseth, T.F., Dargie, P.J. and Lee, H.C. (1987). Pyridine nucleotide metabolites stimulate calcium release from sea urchin egg microsomes desensitized to inositol trisphosphate. J. Biol. Chem. 262, 9561–9568.

Dargie, P.J., Agre, M.C. and Lee, H.C. (1990). Comparison of Ca^{2+} mobilizing activities of cyclic ADP-ribose and inositol trisphosphate. Cell Regul. 1, 279–290.

Durner, J., Wendehenne, D. and Klessig, D.F. (1998). Defense gene induction in tobacco by nitric oxide, cyclic GMP, and cyclic ADP-ribose. Proc. Natl. Acad. Sci. USA. 95, 10328–10333.

Franco, L., Zocchi, E., Usai, C., Guida, L., Bruzzone, S., Costa, A. and De Flora, A. (2001). Paracrine roles of NAD⁺ and cyclic ADP-ribose in increasing intracellular calcium and enhancing cell proliferation of 3T3 fibroblasts. J. Biol. Chem. 276, 21642–21648.

Galione, A., Lee, H.C. and Busa, W.B. (1991). Ca^{2+}-induced Ca^{2+} release in sea urchin egg homogenates: modulation by cyclic ADP-ribose. Science 253, 1143–1146.

Galione, A., McDougall, A., Busa, W.B., Willmott, N., Gillot, I. and M., W. (1993a). Redundant mechanisms of calcium-induced calcium release underlying calcium waves during fertilization of sea urchin eggs. Science 261, 348–352.

Galione, A., White, A., Willmott, N., Turner, M., Potter, B.V. and Watson, S.P. (1993b). cGMP mobilizes intracellular Ca^{2+} in sea urchin eggs by stimulating cyclic ADP-ribose synthesis. Nature 365, 456–459.

Genazzani, A.A., Empson, R.M. and Galione, A. (1996). Unique inactivation properties of NAADP-sensitive Ca^{2+} release. J. Biol. Chem. 271, 11599–11602.

Genazzani, A.A., Mezna, M., Dickey, D.M., Michelangeli, F., Walseth, T.F. and Galione, A. (1997). Pharmacological properties of the Ca^{2+}-release mechanism sensitive to NAADP in the sea urchin egg. Br. J. Pharm. 121, 1489–1495.

Gerasimenko, O.V., Gerasimenko, J.V., Belan, P.V. and Petersen, O.H. (1996). Inositol trisphosphate and cyclic ADP-ribose-mediated release of Ca^{2+} from single isolated pancreatic zymogen granules. Cell. 84, 473–480.

Graeff, R. and Lee, H.C. (2002). A novel cycling assay for cellular cyclic ADP-ribose with nanomolar sensitivity. Biochem. J. 361, 379–381.

Graeff, R., Munshi, C., Aarhus, R., Johns, M. and Lee, H.C. (2001). A single residue at the active site of CD38 determines its NAD cyclizing and hydrolyzing activities. J. Biol. Chem. 276, 12169–12173.

Graeff, R.M., Franco, L., Deflora, A. and Lee, H.C. (1998). Cyclic GMP-dependent and -independent effects on the synthesis of the calcium messengers cyclic ADP-ribose and nicotinic acid adenine dinucleotide phosphate. J. Biol. Chem. 273, 118–125.

Graeff, R.M., Walseth, T.F. and Lee, H.C. (1997). A radio-immunoassay for measuring endogenous levels of cyclic ADP-ribose in tissues. Meth. Enzymol. 280, 230–241.

Guse, A.H., da Silva, C.P., Berg, I., Skapenko, A.L., Weber, K., Heyer, P., Hohenegger, M., Ashamu, G.A., Schulze-Koops, H., Potter, B.V.L. and Mayr, G.W. (1999). Regulation of calcium signalling in T lymphocytes by the second messenger cyclic ADP-ribose. Nature 398, 70–73.

Hirata, Y., Kimura, N., Sato, K., Ohsugi, Y., Takasawa, S., Okamoto, H., Ishikawa, J., Kaisho, T., Ishihara, K. and Hirano, T. (1994). ADP ribosyl cyclase activity of a novel bone marrow stromal cell surface molecule, BST-1. FEBS Lett. 356, 244–248.

Howard, M., Grimaldi, J.C., Bazan, J.F., Lund, F.E., Santos-Argumedo, L., Parkhouse, R.M., Walseth, T.F. and Lee, H.C. (1993). Formation and hydrolysis of cyclic ADP-ribose catalyzed by lymphocyte antigen CD38. Science 262, 1056–1059.

Jackson, D.G. and Bell, J.I. (1990). Isolation of a cDNA encoding the human CD38 (T10) molecule, a cell surface glycoprotein with an unusual discontinuous pattern of expression during lymphocyte differentiation. J. Immunol. 144, 2811–2815.

Karasawa, T., Takasawa, S., Yamakawa, K., Yonekura, H., Okamoto, H. and Nakamura, S. (1995). NAD⁺-glycohydrolase from *Streptococcus pyogenes* shows cyclic ADP-ribose forming activity. FEMS Microbiol. Lett. 130, 201–204.

Kim, H., Jacobson, E.L. and Jacobson, M.K. (1993). Synthesis and degradation of cyclic ADP-ribose by NAD glycohydrolases. Science 261, 1330–1333.

Leckie, C.P., McAinsh, M.R., Allen, G.J., Sanders, D. and Hetherington, A.M. (1998). Abscisic acid-induced stomatal closure mediated by cyclic ADP-ribose. Proc. Natl. Acad. Sci. USA. 95, 15837–15842.

Lee, H.C. (1993). Potentiation of calcium- and caffeine-induced calcium release by cyclic ADP-ribose. J. Biol. Chem. 268, 293–299.

Lee, H.C. (1997). Mechanisms of calcium signaling by cyclic ADP-ribose and NAADP. Physiol. Rev. 77, 1133–1164.

Lee, H.C. (1999). A unified mechanism for enzymatic synthesis of two calcium messengers, cyclic ADP-ribose and NAADP. Biol. Chem. 380, 785–793.

Lee, H.C. (2000). Enzymatic functions and structures of CD38 and homologs. Chem. Immunol. 75, 39–59.

Lee, H.C. (2001). Physiological functions of cyclic ADP-ribose and NAADP as calcium messengers. Ann. Rev. Pharmacol. Toxicol. 41, 317–345.

Lee, H.C. and Aarhus, R. (1991). ADP-ribosyl cyclase: an enzyme that cyclizes NAD+ into a calcium-mobilizing metabolite. Cell Regul. 2, 203–209.

Lee, H.C. and Aarhus, R. (1993). Wide distribution of an enzyme that catalyzes the hydrolysis of cyclic ADP-ribose. Biochim. Biophys. Acta 1164, 68–74.

Lee, H.C. and Aarhus, R. (1995). A derivative of NADP mobilizes calcium stores insensitive to inositol trisphosphate and cyclic ADP-ribose. J. Biol. Chem. 270, 2152–2157.

Lee, H.C. and Aarhus, R. (1997). Structural determinants of nicotinic acid adenine dinucleotide phosphate important for its calcium-mobilizing activity. J. Biol. Chem. 272, 20378–20383.

Lee, H.C. and Aarhus, R. (1998). Fluorescent analogs of NAADP with calcium mobilizing activity. Biochim. Biophys. Acta. 1425, 263–271.

Lee, H.C. and Aarhus, R. (2000). Functional visualization of the separate but interacting calcium stores sensitive to NAADP and cyclic ADP-ribose. J. Cell Sci. 113, 4413–4420.

Lee, H.C., Aarhus, R., Gee, K.R. and Kestner, T. (1997). Caged nicotinic acid adenine dinucleotide phosphate—Synthesis and use. J. Biol. Chem. 272, 4172–4178.

Lee, H.C., Aarhus, R. and Graeff, R.M. (1995). Sensitization of calcium-induced calcium release by cyclic ADP-ribose and calmodulin. J. Biol. Chem. 270, 9060–9066.

Lee, H.C., Aarhus, R. and Levitt, D. (1994). The crystal structure of cyclic ADP-ribose. Nature Struct. Biol. 1, 143–144.

Lee, H.C., Munshi, C. and Graeff, R. (1999). Structures and activities of cyclic ADP-ribose, NAADP and their metabolic enzymes. Mol. Cell. Biochem. 193, 89–98.

Lee, H.C., Walseth, T.F., Bratt, G.T., Hayes, R.N. and Clapper, D.L. (1989). Structural determination of a cyclic metabolite of NAD⁺ with intracellular Ca²⁺-mobilizing activity. J. Biol. Chem. 264, 1608–1615.

Lee, H.C., Zocchi, E., Guida, L., Franco, L., Benatti, U. and De Flora, A. (1993). Production and hydrolysis of cyclic ADP-ribose at the outer surface of human erythrocytes. Biochem. Biophys. Res. Commun. 191, 639–645.

Li, P.-L., Tang, W.-X., Valdivia, H.H., Zou, A.-P. and Campbell, W.B. (2001). cADP-ribose activates reconstituted ryanodine receptors from coronary arterial smooth muscle. Am. J. Physiol. 280, H208–215.

Lim, D., Kyozuka, K., Gragnaniello, G., Carafoli, E. and Santella, L. (2001). NAADP⁺ initiates the Ca²⁺ response during fertilization of starfish oocytes. FASEB J. 15, 2257–2267.

Lokuta, A.J., Darszon, A., Beltran, C. and Valdivia, H.H. (1998). Detection and functional characterization of ryanodine receptors from sea urchin eggs. J. Physiol. 510.1, 155–164.

Madden, J.C., Ruiz, N. and Caparon, M. (2001). Cytolysin-mediated translocation (CMT): A functional equivalent of type III secretion in Gram-positive bacteria. Cell 104, 143–152.

Masuda, W., Takenaka, S., Inageda, K., Nishina, H., Takahashi, K., Katada, T., Tsuyama, S., Inui, H., Miyatake, K. and Nakano, Y. (1997). Oscillation of ADP-ribosyl cyclase activity during the cell cycle and function of cyclic ADP-ribose in a unicellular organism, *Euglena gracilis*. FEBS Lett. 405, 104–106.

Mata, C.G. and Lamattina, L. (2001). Nitric oxide induces stomatal closure and enhances the adaptive plant responses against drought stress. Plant Physiol. 126, 1196–1204.

Mitchell, K.J., Pinton, P., Varadi, A., Tacchetti, C., Ainscow, E.K., Pozzan, T., Rizzuto, R. and Rutter, G.A. (2001). Dense core secretory vesicles revealed as a dynamic Ca^{2+} store in neuroendocrine cells with a vesicle-associated membrane protein aequorin chimaera. J. Cell Biol. 155, 41–51.

Munshi, C., Aarhus, R., Graeff, R., Walseth, T.F., Levitt, D. and Lee, H.C. (2000). Identification of the enzymatic active site of CD38 by site-directed mutagenesis. J. Biol. Chem. 275, 21566–21571.

Munshi, C., Baumann, C., Levitt, D., Bloomfield, V.A. and Lee, H.C. (1998). The homo-dimeric form of ADP-ribosyl cyclase in solution. Biochim. Biophys. Acta 1388, 428–436.

Munshi, C., Thiel, D.J., Mathews, I.I., Aarhus, R., Walseth, T.F. and Lee, H.C. (1999). Characterization of the active site of ADP-ribosyl cyclase. J. Biol. Chem. 274, 30770–30777.

Navazio, L., Bewell, M.A., Siddiqua, A., Dickinson, G.D., Galione, A. and Sanders, D. (2000). Calcium release from the endoplasmic reticulum of higher plants elicited by the NADP metabolite nicotinic acid adenine dinucleotide phosphate. Proc. Natl. Acad. Sci. USA. 97, 8693–8698.

Navazio, L., Mariani, P. and Sanders, D. (2001). Mobilization of Ca^{2+} by cyclic ADP-ribose from the endoplasmic reticulum of cauliflower florets. Plant Physiol. 125, 2129–2138.

Partida-Sanchez, S., Cockayne, D., Monard, S., Jacobson, E.L., Oppenheimer, N., Garvy, B., Kusser, K., Goodricj, S., Howard, M., Harmsen, A., Randall, T. and Lund, F.E. (2001). Cyclic ADP-ribose production by CD38 regulates intracellular calcium release, extracellular calcium influx and chemotaxis in neutrophils and is required for bacterial clearance in vivo. Nature Med. 7, 1209–1216.

Patel, S., Churchill, G.C., Sharp, T. and Galione, A. (2000). Widespread distribution of binding sites for the novel Ca^{2+}-mobilizing messenger, nicotinic acid adenine dinucleotide phosphate, in the brain. J. Biol. Chem. 275, 36495–36497.

Perez, C.F., Marengo, J.J., Bull, R. and Hidalgo, C. (1998). Cyclic ADP-ribose activates caffeine-sensitive calcium channels from sea urchin egg microsomes. Am. J. Physiol. 274, C 430–C 439.

Podesta, M., Zocchi, E., Pitto, A., Usai, C., Franco, L., Bruzzone, S., Guida, L., Bacigalupo, A., Scadden, D.T., Walseth, T.F., De Flora, A. and Daga, A. (2000). Extracellular cyclic ADP-ribose increases intracellular free calcium concentration and stimulates proliferation of human hemopoietic progenitors. FASEB J. 14, 680–690.

Prasad, G.S., McRee, D.E., Stura, E.A., Levitt, D.G., Lee, H.C. and Stout, C.D. (1996). Crystal structure of *Aplysia* ADP ribosyl cyclase, a homologue of the bifunctional ectozyme CD38. Nature Struct. Biol. 3, 957–964.

Rall, T.W. and Sutherland, E.W. (1958). Formation of a cyclic adenine ribonucleotide by tissue particles. J. Biol. Chem. 232, 1065–1076.

Reyes-Harde, M., Potter, B.V.L., Galione, A. and Stanton, P.K. (1999). Induction of hippocampal LTD requires nitric-oxide-stimulated PKG activity and Ca^{2+} release from cyclic ADP-ribose-sensitive stores. J. Neurophysiol. 82, 1569–1576.

Sangwan, V., Foulds, I., Singh, J. and Dhindsa, R.S. (2001). Cold-activation of *Brassica napus* BN115 promoter is mediated by structural changes in membranes and cytoskeleton, and requires Ca^{2+} influx. Plant J. 27, 1–12.

Santella, L., Kyozuka, K., Genazzani, A.A., De Riso, L. and Carafoli, E. (2000). Nicotinic acid adenine dinucleotide phosphate-induced Ca^{2+} release. Interactions among distinct Ca^{2+} mobilizing mechanisms in starfish oocytes. J. Biol. Chem. 275, 8301–8306.

Sauve, A.A., Deng, H.T., Angeletti, R.H. and Schramm, V.L. (2000). A covalent intermediate in CD38 is responsible for ADP-ribosylation and cyclization reactions. J. Am Chem. Soc. 122, 7855–7859.

Schroeder, J.I., Kwak, J.M. and Allen, G.J. (2001). Guard cell abscisic acid signalling and engineering drought hardiness in plants. Nature 410, 327–330.

Sethi, J.K., Empson, R.M., Bailey, V.C., Potter, B.V.L. and Galione, A. (1997). 7-Deaza-8-bromo-cyclic ADP-ribose, the first membrane-permeant, hydrolysis-resistant cyclic ADP-ribose antagonist. J. Biol. Chem. 272, 16358–16363.

Shuto, S., Fukuoka, M., Manikowsky, A., Ueno, Y., Nakano, T., Kuroda, R., Kuroda, H. and Matsuda, A. (2001). Total synthesis of cyclic ADP-carbocyclic-ribose, a stable mimic of Ca^{2+}-mobilizing second messenger cyclic ADP-ribose. J. Am. Chem. Soc. 123, 8750–8759.

Sonnleitner, A., Conti, A., Bertocchini, F., Schindler, H.

and Sorrentino, V. (1998). Functional properties of the ryanodine receptor type 3 (Ryr3) Ca^{2+} release channel. EMBO J. 17, 2790–2798.

States, D.J., Walseth, T.F. and Lee, H.C. (1992). Similarities in amino acid sequences of *Aplysia* ADP-ribosyl cyclase and human lymphocyte antigen CD38. Trends Biochem. Sci. 17, 495.

Streb, H., Irvine, R.F., Berridge, M.J. and Schulz, I. (1983). Release of Ca^{2+} from a nonmitochondrial intracellular store in pancreatic acinar cells by inositol-1,4,5-trisphosphate. Nature 306, 67–69.

Takahashi, K., Kukimoto, I., Tokita, K., Inageda, K., Inoue, S., Kontani, K., Hoshino, S., Nishina, H., Kanaho, Y. and Katada, T. (1995). Accumulation of cyclic ADP-ribose measured by a specific radioimmunoassay in differentiated human leukemic HL-60 cells with all-trans-retinoic acid. FEBS Lett. 371, 204–208.

Walseth, T.F., Aarhus, R., Gurnack, M.E., Wong, L., Breitinger, H.-G.A., Gee, K.R. and Lee, H.C. (1997a). Preparation of cyclic ADP-ribose antagonists and caged cyclic ADP-ribose. Meth. Enzymol. 280, 294–305.

Walseth, T.F., Aarhus, R., Zeleznikar, R.J. and Lee, H.C. (1991). Determination of endogenous levels of cyclic ADP-ribose in rat tissues. Biochim. Biophys. Acta. 1094, 113–120.

Walseth, T.F. and Lee, H.C. (1993). Synthesis and characterization of antagonists of cyclic-ADP-ribose-induced Ca^{2+} release. Biochim. Biophys. Acta. 1178, 235–242.

Walseth, T.F., Wong, L., Graeff, R.M. and Lee, H.C. (1997b). A bioassay for determining endogenous levels of cyclic ADP-ribose. Meth. Enzymol. 280, 287–294.

Willmott, N., Sethi, J.K., Walseth, T.F., Lee, H.C., White, A.M. and Galione, A. (1996). Nitric oxide-induced mo-

bilization of intracellular calcium via the cyclic ADP-ribose signaling pathway. J. Biol. Chem. 271, 3699–3705.

Wong, L., Aarhus, R., Lee, H.C. and Walseth, T.F. (1999). Cyclic 3-deaza-adenosine diphosphoribose: a potent and stable analog of cyclic ADP-ribose. Biochim. Biophys. Acta. 1472, 555–564.

Wu, Y., Kuzma, J., Marechal, E., Graeff, R., Lee, H.C., Foster, R. and Chua, N.H. (1997). Abscisic acid signaling through cyclic ADP-ribose in plants. Science. 278, 2126-2130.

Zhang, X., Wen, J., Bidasee, K.R., Besch Jr, H.R., Wojcikiewicz, R.J., Lee, B. and Rubin, R.P. (1999). Ryanodine and inositol trisphosphate receptors are differentially distributed and expressed in rat parotid gland. Biochem. J. 340, 519–527.

Zocchi, E., Carpaneto, A., Cerrano, C., Bavestrello, G., Giovine, M., Brozzone, S., Guida, L., Franco, L. and Usai, C. (2001a). The temperature-signaling cascade in sponges involves a heat-gated cation channel, abscisic acid and cyclic ADP-ribose. Proc. Natl. Acad. Sci. USA. 98, 14859–14864.

Zocchi, E., Daga, A., Usai, C., Franco, L., Guida, L., Bruzzone, S., Costa, A., Marchetti, C. and Deflora, A. (1998). Expression of CD38 increases intracellular calcium concentration and reduces doubling time in HeLa and 3T3 cells. J. Biol. Chem. 273, 8017–8024.

Zocchi, E., Podesta, M., Pitto, A., Usai, C., Bruzzone, S., Franco, L., Guida, L., Bacigalupo, A. and De Flora, A. (2001b). Paracrinally stimulated expansion of early human hemopoietic progenitors by stroma-generated cyclic ADP-ribose. FASEB J. 15, 1610–1612.

Sensing, Signaling and Cell Adaptation. Edited by K.B. Storey and J.M. Storey
© 2002 Elsevier Science B.V. All rights reserved.

CHAPTER 8

The Cellular and Molecular Basis of the Detection of Pain

Jennifer K. Bonnington, David R. Robinson, Vittorio Vellani and Peter A. McNaughton*
Department of Pharmacology, University of Cambridge, Cambridge, UK

1. Introduction

A wide range of environmental stimuli can cause damage, and an animal must possess a correspondingly wide range of mechanisms by which these stimuli can be detected and the mechanisms for avoiding them put in train. Nerve terminals specialized for the detection of noxious stimuli, or nociceptors (a term coined by Sherrington some one hundred years ago) form the first stage of the defensive interface between the body and the many dangers in the external world. The terminals of nociceptors innervate almost all areas of the skin surface and internal organs, and the cell bodies of most of these reside within the dorsal root ganglia (DRG), immediately outside the spinal cord. Nociceptive afferents fall into two main categories: the rapidly conducting, myelinated Aδ-fibres, and the much more slowly conducting unmyelinated C fibres. Both classes synapse in peripheral dorsal layers of the spinal cord, from which information is relayed to higher structures of the central nervous system.

One can deduce much about the properties of nociceptors by reflecting on the parameters of their tasks. The organism will be more concerned with detecting a noxious stimulus, and with reacting appropriately in order to minimize damage, than with analyzing the precise nature of the damaging stimulus. Thus, we might expect nociceptors to respond to a wide range of noxious stimuli rather

than to the very specific stimuli detected by other sensory systems (for instance, the photoreceptors of the eye detect only light, and over a restricted band of wavelengths at that). And so it proves to be: nociceptors are typically *polymodal*, i.e. a single nerve terminal typically responds to a large number of potentially damaging stimuli. These stimuli can broadly be divided into those originating in the external world, and those released within the body, often as a result of inflammation or tissue damage. Damaging external stimuli include strong mechanical stimuli, heat above about 43°C or cold below 8°C, and chemical stimuli such as acid or capsaicin, the active ingredient of chilli peppers. Internal stimuli include mechanical stimuli resulting from the overextension of a joint or excess distension of the viscera. Internal heat stimuli might be supposed to be uncommon, at least in the homothermic internal environment of a mammal, but as we show below, the threshold for eliciting a response from the mechanism responsible for detecting noxious heat is lowered in inflammation, such that ordinary bodily heat may be sufficient to evoke a sensation of heat pain. Finally, the range of internal chemical stimuli capable either of directly exciting nociceptors, or of lowering their threshold such that other stimuli can excite them more readily, is large and almost certainly remains to be fully characterized.

Substances that directly *excite* nociceptors include acid, which is released by anaerobic glycolysis in over-exercised muscle or in tissues with an inadequate blood supply; potassium ions, released during electrical activity or as a result of

---

*Corresponding author

cell rupture; and ATP, which is present at high concentrations inside cells but not normally in the extracellular milieu. Each of these stimuli excites nociceptors by depolarizing the nerve terminal to the threshold for initiating an action potential. H^+ ions and ATP do this by a direct action on specific proton and ATP-gated ion channels, whereas K^+ ions act by altering the Nernst equilibrium potential for potassium which is the principal contributor to the nerve resting potential. Other excitatory stimuli such as bradykinin bind to G-protein coupled receptors, which activate intracellular signaling pathways and in turn open ion channels.

A much larger class of internally generated substances *sensitize* the nociceptive nerve terminals without directly exciting them. Sensitisation results in a diminished threshold for activation and an enhanced responsiveness to a noxious stimulus. These phenomena are not observed in isolated neurons, and are mediated by endogenous chemicals released following stress or damage to adjacent tissues. Prostaglandins potentiate the action of algogenic stimuli such as heat by lowering the threshold for the generation of action potentials. Bradykinin and nerve growth factor, amongst other sensitizing agents, activate intracellular signalling pathways which have more specific and direct actions on the heat-gated ion channel, leading to a lowering of the temperature threshold for activation of this ion channel. Some of these diverse processes are discussed below.

Many studies of nociception, particularly those on intact animals, have for experimental convenience used stimuli impinging on the skin, and it is sometimes implicitly assumed that these studies of somatic nociceptors are of "typical" nociceptors. There is increasing evidence, however, that nociceptive terminals in internal organs such as the heart and intestine have different properties than those in the skin. In the final section of this chapter we discuss pain transduction in viscera, and the differences between visceral and somatic nociceptors.

2. Ion channels involved in nociception

Nociceptive nerve terminals are almost always located far (in electrical terms) from their central connections, and action potentials are therefore essential in order to transmit information over such distances. Excitation of a nociceptive nerve terminal must therefore involve the activation of an inward current—a *generator current*—in order to depolarise the terminal to the threshold for activating voltage-sensitive Na channels, and consequently for initiating an action potential. An example, of the action of heat on an isolated nociceptor, is shown in Fig. 8.1. The left-hand panel is a recording of the neuronal membrane potential, showing the depolarisation from the resting potential of around -65 mV, followed, once the action potential threshold of -40 mV is reached, by the rapid upstroke of the nerve action potential (Fig. 8.1A). The intensity of pain is coded for by the frequency of action potentials, which is determined by the rate of rise of membrane potential in the interval between action potentials, in turn determined principally by the magnitude of the generator current elicited by the noxious heat stimulus. The generator current can be recorded directly by voltage-clamping the neuronal membrane potential at its resting level, as shown in the centre panel (Fig. 8.1B). In heat-sensitive neurons a large inward current is elicited with a slight delay after a rapid elevation in temperature (lower trace in center panel). Many sensory neurons in the dorsal root ganglion (DRG) detect non-noxious sensory modalities, and are therefore not heat-sensitive. In these neurons only a small and probably non-specific current is observed on application of a heat stimulus (upper trace).

2.1. Ion channels gated by heat

When the skin is heated the sensation is first perceived as pleasant warmth, but above about $\sim43°C$ the sensation changes to one of pain, mild at first but growing in intensity at an exponential rate as the temperature is elevated (Belmonte and Giraldez, 1981). An inward membrane current elicited by noxious heat in DRG neurons has been characterised (Cesare and McNaughton, 1996). The current is carried by a relatively non-selective cation channel, is activated above $\sim43°C$, and, like the sensation of pain and the action potential

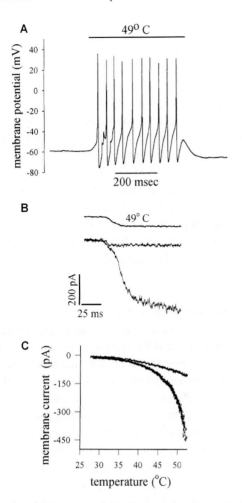

Fig. 8.1. Action potentials and generator currents elicited in a nociceptive neuron by a noxious heat stimulus. (A) Application of a 49°C heat stimulus depolarises a nociceptive neuron to threshold and elicits a train of action potentials. This temperature did not damage the neuron as repeated application of the stimulus gave a similar result. (B) Generator current elicited by a heat stimulus, recorded by maintaining the cell membrane potential at –70 mV using the whole-cell patch clamp technique. Currents recorded from two neurons are shown, one heat-insensitive (upper trace) and one heat-sensitive (lower trace). (C) Dependence of the membrane current on temperature in a heat-insensitive neuron (upper trace) and in a heat-sensitive neuron (lower trace). The current recorded from the heat-insensitive neuron is much smaller and is probably due to a non-specific dependence of membrane resistance on temperature. Modified from Cesare et al. (1999b).

frequency recorded in intact preparations, increases exponentially as the temperature is raised (see Fig. 8.1Cl). An ion channel with these properties was subsequently cloned by expressing a DRG library and searching for responsiveness to capsaicin, the active ingredient of chilli peppers. The channel was named the vanilloid receptor 1 (VR1) (Caterina et al., 1997). VR1 exhibits many of the properties of the heat-gated channel: it is a non-selective cation channel that opens in response to temperatures greater than 43°C, and is also activated by low pH. Its predicted topological organisation consists of five trans-membrane domains with a hydrophobic loop between the fifth and sixth domain, and recent experiments have indicated that it forms a functional tetramer in the plasma membranes of DRG neurons (Kuzhikandathil et al., 2001). The channel is a member of the TRP ion channel family thought to be responsible for a diverse range of functions, including generating the receptor potential in invertebrate photoreceptors and refilling intracellular stores with calcium (Clapham et al., 2001).

Studies using membrane permeable and impermeable analogues of capsaicin and the competitive antagonist capsazepine have shown that capsaicin gates the channel by binding to an internal site (Jung et al., 1999), and more recent studies using domain swaps between mammalian VR1, which is activated by capsaicin, and chicken VR1, which is not, have localised the capsaicin-binding domain to a short region of amino acids in the cytoplasmic linker between transmembrane domains 2 and 3 (Jordt and Julius, 2002). Following capsaicin binding VR1 undergoes a conformational change that it does not undergo in response to activation by protons or thermal stimuli. These structural rearrangements include the putative pore domain, and reveal the location of a second intracellular domain that contributes to the positive cooperativity seen for capsaicin activation (Welch et al., 2000). Mild external acidity (pH 6–7) potentiates the capsaicin and heat responses, while more acid solutions directly activate VR1 (Tominaga et al., 1998). These two processes involve protonation of distinct residues, with the glutamic acid residue E648 being involved in direct gating, and E600 being involved in sensitisation (Jordt et al., 2000). VR1 is also directly activated by several hydrophobic substances, including the endogenous cannabinoid anandamide (Zygmunt et al., 1999) and products of the lipoxygenase pathway (Sun et al., 2000).

The importance of VR1 in noxious heat trans-duction *in vivo* was thrown into doubt, however, when two papers were published on studies of VR1 knock-out mice. Disruption of the VR1 gene com-pletely abolished sensitivity to capsaicin and greatly diminished inflammatory hyperalgesia, but heat sensitivity remained largely unaltered (Caterina et al., 2000; Davis et al., 2000). VR1, therefore, appeared to be the receptor responsible for capsaicin binding and sensitisation of the heat response on exposure to inflammatory mediators, but not the only heat-gated channel. A possible alternative heat-activated channel has been cloned and named the vanilloid receptor–like channel (VRL-1) (Caterina et al., 1999). This channel does not bind capsaicin, but is found in DRG neurons and has been proposed to open when temperatures reach 52°C, though other groups have found that VRL-1 is not gated by heat (Garcia et al, 1999).

Several recent reports have characterised a channel gated by cold and menthol (Reid and Flonta, 2001; McKemy et al., 2002; Peier et al., 2002). Moderate, non-painful cold, below 25°C, activates the channel, and the activation threshold is shifted to higher temperatures by exposure to menthol. Like VR1, the channel is a member of the TRP family of ion channels, which may point to a more general role for TRP channels in thermo-sensation. We know, of course, that several tem-perature-sensitive ion channels remain to be discovered: the mechanisms for sensing damaging cold (below around 8°C) or for moderate warmth are still mysterious at the molecular level, and it seems clear from the VR1 knockout studies that further mechanisms sensing noxious heat remain to be characterised.

2.2. ATP-gated ion channels

ATP was first proposed to have an algogenic role in 1977 when it was shown that ATP application to a blister base evoked pain (Bleehen and Keele, 1977). High concentrations of ATP are present in all cells and virtually absent from extracellular fluid, and it follows that ATP release is likely to be a sensitive indicator of cell damage. Purinergic

receptors are widely distributed throughout most tissues and have been broadly classified into the ionotropic receptors (P2X) and metabotropic receptors (P2Y). The first P2X receptors were cloned in 1994 (Brake et al., 1994; Valera et al., 1994) and over the following years the family expanded to currently include seven known mem-bers (North and Surpranant, 2000). The discovery of P2X3, a member of the ATP-gated ion channel family that is exclusively expressed in nociceptive neurons, suggested a mechanism for the activation of nociceptors by ATP (Chen et al., 1995). Upon activation, however, heterologously expressed P2X3 was found to desensitise in less than 100 ms and to recover only in longer than 20 minutes. These are not the characteristics of a channel that must maintain a sustained signal in the presence of continuing damage. Heteromeric receptors of P2X3 and the widely expressed P2X2 have been observed in isolated nociceptors, albeit from the nodose ganglion, where most sensory neurons are of visceral origin, and these do display the slowly desensitising characteristics required for sustained signalling (Lewis et al., 1995). The P2X2/P2X3 heteromer is sensitised when the pH drops below 7.0, a regular occurrence during anoxia and inflam-mation (Stoop et al., 1997), although a possible sensitisation of P2X receptors by other endoge-nous compounds remains to be tested. Deletion of the gene for P2X3 caused little evidence of impair-ment in nociception (Souslova et al., 2000). The main deficit was in a failure of bladder emptying, probably because ATP acts as a signalling mole-cule between distended epithelial cells and the nociceptive neurons that trigger detrusor muscle contraction (Ferguson et al., 1997; Cockayne et al., 2000). The current view, therefore, is that activa-tion of P2X3 ion channels by ATP is probably an important mechanism for pain signalling only in visceral afferents such as those innervating the bladder.

2.3. Proton-gated ion channels

There can be little doubt that acid causes pain, as anyone who has splashed lemon juice on a cut can testify. Acidification occurs in inflamed or ischaemic tissue, where the pH may fall as low as 5.4 (Steen et al., 1992). The pain of ischaemic muscle is driven by this local acidosis, and usually precludes further exercise by the subject. In pathological circumstances such as arthritis, acidosis contributes to the sensation of prolonged pain from the affected area. Both small diameter (unmyelinated) and larger-diameter (myelinated) nerve fibres are involved in acid detection.

There have recently been significant advances in our understanding of the molecular mechanism of acid sensation. The first Acid Sensing Ion Channel (ASIC, now called ASIC1) was cloned by Laszdunski's group (Waldmann et al., 1997b). ASIC1 remains closed at pH 7.4, opening upon acidosis to pass a sodium selective, amiloride sensitive current. ASIC1 is a member of a family which includes the amiloride-sensitive epithelial Na channel, the FMRF-amide gated Na channel and the mammalian homologues (MDEG 1 and 2) of degenerins first identified in *C. elegans* (Waldmann et al., 1999). Four genes of the ASIC family have now been identified, two of which are alternatively spliced to generate a total of six major transcripts (Chen et al., 1998; Waldmann et al., 1999; Akopian et al., 2000). Only one member of the ASIC family is expressed exclusively in sensory neurons: Dorsal Root ganglion specific ASIC, or DRASIC —now called ASIC3 (Waldmann et al., 1999). The ASIC family typically respond to step acidification with a transient inward current, and in all cases adapt rapidly to pH change, although coexpression of two or more ASICs to form heteromeric channels may produce a more sustained response (Babinski et al., 2000). The transient nature of the current in expression systems (Waldmann et al., 1999) and in at least some DRG neurons (Petruska et al., 2000) presents a difficulty in explaining how pain-sensitive neurons might use members of the ASIC family to signal low pH in a sustained manner.

A second potential acid-sensing mechanism is the capsaicin and heat-gated ion channel, VR1 (Caterina et al., 1997; Tominaga et al., 1998). As discussed above, this ion channel is also activated by protons, and moreover generates a sustained inward current in response to acidification, which makes it a possible candidate for the sustained signalling of low pH *in vivo*. Sensory neurons and skin-nerve preparations from VR1 KO mice have reduced responsiveness to acid, particularly in unmyelinated fibres (Caterina et al., 2000), but the role of VR1 in the behavioural responses to acid has not so far been determined.

2.4. *Voltage-gated sodium channels*

The voltage-gated sodium currents (VGSCs) are essential for the initiation and propagation of action potentials. There are two distinct classes of neuronal VGSCs, identified by their sensitivity to tetrodotoxin (TTX). The TTX-sensitive current is blocked by nanomolar concentrations of TTX and is found widely distributed throughout the sensory nervous system; the TTX-resistant current needs concentrations of TTX greater than 10 μM before inhibition can be achieved, and exists exclusively in a subpopulation of neurons with the characteristics of nociceptors (Baker and Wood, 2001). Two channels responsible for the TTX-resistant current in DRG neurons have been cloned and named Nav1.8 (formerly SNS/PN3) (Akopian et al., 1996) and Nav1.9 (formerly NaN/SNS2) (Tate et al., 1998; Dib-Hajj et al., 1998). Nav1.8 activates and inactivates at much more positive voltages than the normal TTX-sensitive channel, and therefore the ratio of expression of these two types of VGSCs can profoundly alter the excitability of a neuron. Endogenous mediators released upon injury can also modulate Nav1.8 by shifting its threshold in the negative direction, leading to an enhanced excitability of the neuron. Many hyperalgesic inflammatory mediators which activate the cAMP/PKA pathway, such as PGE_2 and serotonin, increase the number of action potentials evoked from a sensitised nociceptor in response to a given stimulus current (Gold et al., 1996; England et al., 1996a). This topic is discussed in more detail below.

3. Sensitisation of nociceptors

All sensory receptors adapt to a constant stimulus. This useful property reduces the gain of sensory transduction when the background level of stimulation increases, thereby enabling us to detect low-intensity stimuli with high gain when there is a low level of background, but to reduce the amplification and thereby avoid saturation in an environment with a high level of background. Nociceptors, like all other sensory receptors, adapt to just-threshold levels of stimulation, sufficient to activate the nociceptor but not to cause tissue damage. It is obviously not in the interests of the organism to adapt to damaging stimulation, however, and at higher levels of stimulus intensity, sufficient to cause tissue damage, the nociceptor *sensitises*, that is to say the gain of sensory transduction increases. The processes underlying adaptation have been little characterised, but they are known to be intrinsic to the nociceptor, because adaptation is observed in isolated nociceptive neurons. Sensitization, on the other hand, is not observed in isolated nociceptors, and is caused *in vivo* by the release of pro-inflammatory mediators from surrounding damaged or inflamed tissues. The number of potential mediators released by cell damage is large, and as it is vital for nociceptors to be able to detect any type of damage we would expect to find a correspondingly large range of surface membrane receptors able to detect these mediators and activate the intracellular pathways leading to sensitization. The number of intracellular signalling pathways is much smaller than the number of mediators, however, as many receptors converge onto and activate a few intracellular signalling pathways. Two main pathways have been characterised to date: the cAMP/PKA pathway and the PLC/PKC pathway, but others no doubt await discovery.

3.1. Sensitisation by PGE_2: the PKA pathway

The algogenic compounds PGE_2, adenosine and serotonin sensitise nociceptors by increasing the magnitude of the Na current carried by the TTX-resistant VGSC Nav 1.8. PGE_2 was shown to double the current amplitude and to shift the activation threshold in the negative direction by 14 mV in cell attached recordings (Gold et al., 1996). This effect of PGE_2 can be mimicked by the addition of

cAMP analogues or by direct activation of adenylate cyclase by forskolin (England et al., 1996b). Sensitisation can also be abolished through the use of selective PKA inhibitors or by selective activation of the μ-opioid receptor, which activates G_i and results in diminished cAMP levels (Gold and Levine, 1996). These observations highlight cAMP and PKA as key members of the PGE_2 sensitisation pathway. In subsequent studies, however, PKC has also been implicated and may be the enzyme which actually phosphorylates Nav1.8, as inhibition of PKC blocks sensitisation produced by PKA activation, while block of PKA does not block sensitisation caused by PKC activation (Gold et al., 1998).

In a similar manner, serotonin and adenosine also couple to G_s and activate PKA to sensitise the TTX-resistant VGSC (Gold et al., 1996). The receptor mediating the serotonin induced increase in the TTX-resistant VGSC belongs to the $5HT_4$ category, which in other tissues has been shown to couple to G_s, resulting in the activation of adenylate cyclase (Cardenas et al., 1997). PKA involvement was identified through phosphodiesterase inhibition, addition of cAMP analogues, or either forskolin or 3-isobutyl-L-methylxanthine (IBMX) treatments. All mimicked the action of serotonin and therefore suggest that adenylyl cyclase activity mediates sensitisation (Cardenas et al., 2001).

The key residues phosphorylated by PKA have now been identified on the intracellular linker region between domains 1 and 2 of the α pore-forming subunit of Nav1.8 (Fitzgerald et al., 1999). Studies on Nav1.8 knockout mice have indicated a role for this channel in inflammatory hyperalgesia, with a delayed onset of hyperalgesia in response to the inflammatory compound carrageenan being observed in the knockout mice (Akopian et al., 1999).

3.2. Sensitisation by bradykinin: the PKC pathway

Bradykinin (BK) has been described as the most potent algogenic compound known to man. It is produced when proteolytic enzymes liberated by cell damage or inflammation cause cleavage of a

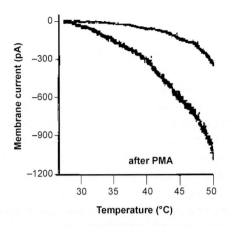

Fig. 8.2. The heat-gated membrane current in a nociceptive neuron is sensitised by activation of PKC. The heat-gated membrane current is activated at ~43°C in control neurons, but after activation of PKC by phorbol myristate acetate (PMA), a PKC-specific activator, the threshold shifts to lower temperatures and the magnitude of the current increases. Later work showed that one PKC isoform, PKCε, was responsible for the effect (see text). Modified from Cesare and McNaughton (1996).

precursor protein, kininogen. Studies in DRG neurons have shown that BK can significantly enhance the current evoked by a noxious heat stimulus, and that a crucial step in this process of sensitization is the activation of protein kinase C (Cesare and McNaughton, 1996). As with the PKA pathway, more than one inflammatory mediator can activate the PKC pathway and cause sensitization. ATP, acting on the metabotropic P2Y1 receptor, has recently been shown to cause sensitisation of the heat-gated current by a PKC-dependent mechanism (Tominaga et al., 2001), and other mediators using this pathway will no doubt be identified in future. Figure 8.2 illustrates the shift in the dependence of the heat-gated membrane current on temperature following PKC activation. This shift is strikingly similar to the change in dependence on temperature of both the subjective sensation of pain and the frequency of action potentials in primary nerve fibres following damage to the tissue surrounding a primary nociceptor *in vivo* (Belmonte and Giraldez, 1981; Treede et al., 1992).

A similar enhancement of the current gated by capsaicin identifies VR1 as the ion channel modulated by bradykinin (Premkumar and Ahern, 2000;

Vellani et al., 2001). Of the many PKC isoforms which could mediate this effect, PKCε has been shown to be specifically involved. PKCε alone is translocated to the plasma membrane by bradykinin stimulation, and infusion of a constitutively active PKCε into the DRG neuron resulted in sensitisation, whilst inhibition with specific peptides abolished the effect (Cesare et al., 1999a). Direct phosphorylation of VR1 by PKCε has recently been demonstrated, and the crucial serine residues whose phosphorylation mediates the effect have been identified (Tominaga et al., 2001). It has also been suggested that the TTX-resistant VGSC may be sensitised by activation of PKCε (Khasar et al., 1999). This alternative pathway may act in parallel with VR1 sensitisation and further experiments will be needed to clarify the contribution of each mechanism to sensitization *in vivo*.

Figure 8.3 summarises mechanisms causing activation and sensitization of VR1 (Vellani et al., 2001). VR1 is a non-selective cation channel. The principal current carrier under physiological conditions is Na$^+$ ions, and an influx of Ca^{2+} ions has important implications for desensitisation (see below). The channel is activated by a range of stimuli, as noted above. Capsaicin binds at the internal membrane surface, and anandamide and

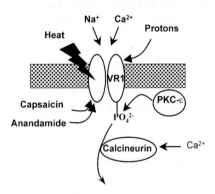

Fig. 8.3. Gating and sensitisation of the heat-sensitive ion channel, VR1. The heat-sensitive ion channel is gated by heat, capsaicin, anandamide and protons. The ion channel is cation-selective, and the current is carried principally by Na$^+$ and Ca^{2+} ions in normal circumstances. Phosphorylation by PKCε potentiates gating, and this potentiation is antagonized by dephosphorylation by the calcium-dependent phosphatase, calcineurin. Modified from Vellani et al. (2001).

other hydrophobic ligands presumably act at the same site. Protons activate at an external site. The site of action of heat has not been determined.

The state of sensitisation of the channel—and therefore the temperature threshold for its activation—is influenced by the state of internal phosphorylation, which in turn is determined by the "yin and yang" balance between phosphorylation by PKCε and dephosphorylation by the calcium-dependent phosphatase calcineurin (Vellani et al., 2001). Activation of PKCε, by the binding of bradykinin to B2 receptors or of ATP to P2Y1 receptors, causes enhanced phosphorylation and a shift in the activation threshold to lower temperatures (see Fig. 8.2). Activation of calcineurin causes dephosphorylation, a decline in heat- or capsaicin-gated current, and an elevation in temperature threshold for activation. The principal mechanism by which calcineurin is activated appears to be the influx of calcium through VR1 itself. The activation of VR1 is therefore self-limiting, unless of course calcium influx is prevented by activating VR1 in the absence of external calcium.

3.3. Sensitisation by nerve growth factor

Nerve growth factor (NGF) levels increase during injury and inflammation. NGF can act on the nociceptors directly, and can also activate mast cells to cause degranulation and further NGF release, resulting in higher NGF concentrations (Mendell, 1999). NGF was shown to have both a long-term (hours to days) and a short-term (minutes) hyperalgesic effect following injection into a rat's paw (Lewin et al., 1993). An initial, rapid sensitisation was only seen for thermal stimulation, and a long-term sensitisation was observed to both mechanical and thermal treatments. When hyperalgesia is induced by injection of complete Freund's adjuvant, the removal of NGF from tissues by injection of anti-NGF antibodies abolished both thermal and mechanical hyperalgesia (Woolf et al., 1994). NGF is therefore a critical hyperalgesic mediator, and is responsible for both short-term and long-term hyperalgesia. As the mechanisms underlying the short-term and long-term

effects are probably distinct it is useful to consider them separately.

Rapid sensitisation occurs *in vivo* within 10 minutes or less of NGF treatment. The rapid effect was originally thought to be caused by mast cell degranulation and the release of a "soup" of inflammatory mediators, but experiments on isolated nociceptors have shown that NGF acts directly on the nociceptors themselves to sensitise the current gated by both heat and capsaicin (Shu and Mendell, 1999). Such rapid effects must involve local signalling events in nociceptor terminals rather than an upregulation of gene transcription. Experiments with cloned VR1 expressed in a HEK283 cell line suggest that rapid enhancement of the VR1 current by the binding of NGF to the TrkA receptor may be caused by the activation of phospholipase C, metabolism of PIP_2, and a consequent release of VR1 from a PIP_2 mediated receptor block (Chuang et al., 2001). Whether the same mechanism is active in sensory neurones has yet to be clearly established, but it must be borne in mind that pathways active in a heterologous expression system, where levels of expressed proteins are vastly above those in primary cells, may be different from those in real sensory neurons. The only study to date implicates the activation of PKA in the signalling pathway between activation of TrkA by NGF and sensitisation of VR1, but further details of how PKA is activated by TrkA, and of whether its action on VR1 is direct or indirect, remain to be elucidated (Shu and Mendell, 2001).

The binding of NGF to TrkA receptors has been shown to up- or down-regulate the expression of many proteins involved in hyperalgesia, via both transcriptional and post-translational mechanisms. Current thinking is that the long-term hyperalgesic effects of NGF are at least partly explained by changes in expression of ion channels and receptors in the terminals of nociceptive neurons. Responses to capsaicin and noxious heat are potentiated by NGF treatment, indicating that VR1 is upregulated by NGF (Bevan and Winter, 1995), a result which has been confirmed by immunocytochemistry (Michael and Priestley, 1999). Nav1.8 is down-regulated by axotomy, which causes deprivation of target derived NGF, and is upregulated by NGF treatment (Waxman et al., 1999). The lack of thermal hyperalgesia seen in Nav1.8 knock out mice in response to systemic NGF addition suggests that Nav1.8 may represent another potential target of NGF downstream signalling (Kerr et al., 2001). Finally, expression of the B2 receptor for bradykinin is upregulated by NGF (Lee et al., 2002), which may explain the hypersensitivity to bradykinin in NGF-treated animals (Bennett et al., 1998). Expression of the B2 receptor is also upregulated by nerve crush, an effect that is due to infiltration of the distal stump by activated macrophages, which release significant quantities of NGF (Lee et al., 2002).

In summary, NGF is commonly thought of as a factor that promotes axonal growth in sensory neurones, so its recently-emerging role as an important factor in promoting hyperalgesia is something of a surprise. NGF appears to act by at least two mechanisms: a short-term action that enhances the sensitivity of VR1 to heat and capsaicin by an as yet undetermined intracellular signalling pathway; and a longer-term effect that appears to involve the upregulation of the expression of a number of genes crucial for the transduction of algogenic stimuli. The relative importance of these effects, and the pathways involved in each, have yet to be fully established and will be exciting areas for future investigation.

4. Pain detection in the viscera

Visceral pain has different characteristics from somatic pain, upon which the vast majority of experimental studies concentrate. This final section will outline our current understanding of pain transduction in visceral neurons. Visceral pain itself has five main features that differentiate it from somatic pain (Cervero, 1995). These are that visceral pain:

a) is not evoked from all viscera;
b) is not always linked to visceral injury;
c) is frequently referred to other locations;
d) is diffuse and poorly localised; and
e) is accompanied by motor and autonomic reflexes.

Points (a) and (b) can be explained, not as previously thought by the fact that some viscera lacked afferent innervation, but because much of the viscera is innervated by receptors whose activation does not evoke conscious perception (Cervero, 2000). The remaining three characteristics result from the fact that visceral nociceptive pathways converge onto somatic pathways (Cervero, 2000). Interestingly, many forms of noxious stimulation do not produce a conscious sensation of pain in the viscera, and thus it is difficult to accurately describe a "visceral noxious stimulus". The liver and kidneys, for example, are insensitive to any form of stimulation, and the hollow organs, including the colon, whilst being highly sensitive to distension or inflammation, are insensitive to burning or cutting stimuli (Sengupta and Gebhart, 1998). The viscera principally provide afferent input, albeit rarely consciously perceived, encoding only discomfort and pain, and as such are unlike skin from which a much wider range of both nociceptive and non-nociceptive afferent input originates.

This section will focus on gastrointestinal (GI), specifically colonic, and cardiac visceral nociception in order to highlight some examples of how pain can be perceived from the viscera, and how this perception may be enhanced by hyperalgaesia.

4.1. Visceral afferent neurons

Many visceral sensory fibres terminate in the spinal cord (spinal afferents) with their cell soma in the dorsal root ganglia (DRG), after travelling through (but not synapsing in) prevertebral sympathetic ganglia. Vagal afferents, by contrast, have their cell soma in the nodose ganglia and synapse in the brainstem. Those sensory neurons that innervate the colon, amongst other viscera, are almost exclusively small diameter myelinated Aδ fibres or unmyelinated C fibres. Activity in most of these does not reach a conscious level, but provides regular CNS inputs to help regulation and control of the visceral environment. Electrophysiological data from cats has suggested that a difference exists between the C and Aδ afferent mechanoreceptors innervating the colon in that the slowly adapting C fibres encode tonic stimuli, whilst primary phasic stimuli are encoded by rapidly adapting Aδ fibres (Janig and Koltzenburg, 1991).

The most common experimental method for studying GI tract pain in both humans and non-human animals is balloon distension (Ness and Gebhart, 1990). Such techniques have enabled the identification of two populations of mechano-sensory afferent neurons in the hollow viscera (Sengupta and Gebhart, 1994; Sengupta and Gebhart, 1998), the first of which account for 20–30% of the total population and have a high (e.g. >30 mmHg distending pressure) threshold of response that is very close to the noxious range (ca. >40 mmHg). These are often C fibre afferents and are probably only activated by noxious stimuli. The remaining mechanosensory population have low (e.g. <5 mmHg) thresholds of response, well below the range of pressures causing damage. Cervero and Janig (1992) proposed a convergence model in which visceral pain perception is mediated via these high- and low-threshold mechano-sensory afferents. In this model "wide-dynamic range" (also known as "multireceptive") dorsal horn neurons receive converging inputs from low-threshold mechanoreceptors. Normally, the low threshold (intensity-encoding) mechanosensory afferents activate these wide-dynamic range neurons, and consequently elicit non-painful sensation and activate regulatory reflex pathways. These low levels of activity are not sufficient to activate spinothalamic pain pathways, which are excited either directly by the high-threshold mechano-receptors, or by both threshold mechanoreceptor subtypes together, in response to transient high-intensity stimuli.

4.2. Visceral hyperalgesia

Some irritable bowel syndrome (IBS) patients show a significantly lower tolerance for colorectal balloon distension than healthy subjects (Whitehead et al., 1990). It is now commonly thought that the discomfort experienced by many functional bowel disorder patients is due to afferent neuron sensitisation, and since IBS patients have normal, or even higher, thresholds for somatic nociception

(Cook et al., 1987), any hypersensitivity exhibited is due to specific visceral mechanisms, i.e. to visceral hyperalgaesia. In the case of IBS, it is thought that chronic sensitisation of primary afferent sensory neurons in the GI tract account for its symptoms (Mayer and Gebhart, 1994).

Nociceptor sensitisation, or primary hyperalgesia, can be caused by factors such as inflammation and neuropathies, including those resulting from viral attack or ischemia. During inflammation and tissue injury, spinal afferent C fibre receptive fields that lie predominantly in the muscle layer, serosa and mesentery, increase in size and become detectable in the mucosa (Mayer and Gebhart, 1994). Perhaps more crucially, however, inflammation releases endogenous inflammatory mediators, which in the GI tract include cytokines, bradykinin, tachykinins, CGRP (calcitonin gene related peptide), prostaglandins, histamine, catecholamines, prostanoids, H^+ and K^+ ions, adenosine, 5-HT (5-hydroxytryptamine) and neurotrophins (Sengupta and Gebhart, 1998). As discussed in preceding sections, these inflammatory mediators act to sensitise visceral afferent receptors by modulating the properties of ion channels involved in nociception, or by altering receptor expression within the afferents themselves. It has, for example, been shown electrophysiologically that the introduction of acetic acid or mustard oil leads to sensitisation of pelvic nerve colonic afferents (Su et al., 1997). An increased magnitude of action potential firing and decreased threshold of the response was observed in all of the high-threshold afferents, whereas only approximately half of the low-threshold afferents showed increased activity in response to colorectal distension.

Acute visceral inflammation or tissue injury can induce mechanically insensitive ("silent" or "sleeping") afferent neurons to become spontaneously active, as shown in response to colorectal distension by Sengupta and Gebhart (1998), resulting in the manifestation of chronic visceral pain. Up to half of the unmyelinated fibres innervating the colon may be these so-called silent afferents, which do not respond even to stimuli intensities as high as 100 mmHg. They may only become active following the release of inflammatory mediators.

An increased peripheral afferent input to the spinal cord can increase the release of neurochemicals such as substance P (SP) and glutamate in the dorsal horn of the spinal cord to produce an increased excitability of second order spinal neurons and the development of secondary hyperalgesia (or allodynia) and prolonged reflex maintenance. Since the *N*-Methyl-D-Aspartate (NMDA) glutamate receptor subtype antagonist MK-801 prevents the development of this central sensitization (Woolf and Thompson, 1991), it is thought that increased neuronal excitability within the spinal cord may result from primary afferent terminal glutamate release and activation of NMDA receptors. The current hypothesis for the involvement of these receptors in central sensitisation is that strong responses, after acute noxious visceral stimulation, recruit NMDA receptor-mediated neural mechanisms (Cervero, 2000).

Central sensitisation cannot, however, account for the longer-term alterations in visceral sensory thresholds associated with functional bowel disorders, and alterations in the level of expression of genes involved in nociception may account, for example, for the symptoms of IBS. Experimentally, changes in the levels of various peptides, such as SP, CGRP and VIP, have been observed as a consequence of inflammation (Weihe et al., 1988).

4.3. Cardiac nociception

Like GI afferents, sensory cardiac afferents mediate pain, which in the case of angina occurs when the amount of blood reaching the heart is insufficient for its metabolic requirements, thus causing a drop in blood pH. It has been shown by whole-cell patch-clamp electrophysiology of fluorescently labelled rat cardiac sensory neurons that these neurons contain large amplitude acid-evoked depolarising currents (Benson et al., 1999) that are evoked at pH 7 or below. The molecular identity of this current is proposed to be the ASIC3, or DRASIC, a member of the ASIC family of proton-activated ion channels (Sutherland et al., 2001) which, unlike any of the other ASIC channels, is found only in nociceptive sensory neurons and

reproduces the functions of the native channel (Waldmann et al., 1997a). As discussed above, VR1 is another possible contributor to acid sensation, but this possibility was ruled out since the cardiac currents are blocked by amiloride and are selective for Na^+ over K^+, both characteristics of ASIC3 but not of VRI (Sutherland et al., 2001). These findings suggest that ASIC3 is a pain sensor in cardiac neurons and may be a useful pharmaceutical target for the treatment of ischemic pain such as angina.

4.4. The peptide content of primary visceral neurons

The neuropeptides substance P and CGRP are expressed in a much higher fraction of visceral than somatic afferents, with 82% of rat visceral afferents expressing substance P, compared to only 21% of somatic afferents (Perry and Lawson, 1998). Initial studies using a retrograde labelling protocol suggest that the visceral spinal afferents innervating the descending colon of the mouse are also mostly of the CGRP-containing type (D.R. Robinson, unpublished observations), and similar findings have been previously reported for other visceral targets including rat bladder (Bennett et al. 1996) and rat renal afferents (Zheng and Lawson, 1994). Data from NK1 receptor knockout mice (De Felipe et al., 1998) have suggested that both SP and the NK1 receptor have a more important role in pain of visceral than of somatic origin, and SP antagonists reduce the sensitisation of spinothalamic cells caused by intradermal capsaicin injection (Rees et al., 1998).

The C fibre DRG nociceptor population seems to be divided into two subpopulations, those that express Ret receptors and are sensitive to GDNF, and those that express trkA receptors and are sensitive to NGF (Snider and McMahon, 1998). The first subpopulation of nociceptors project to lamina II of the spinal cord and do not express CGRP or SP, but instead express GFRα and P2X3, and bind the plant lectin from *Griffonia simplicifolia*, isolectin B_4 (IB_4). The second subpopulation expresses the inflammatory mediator peptides CGRP and SP, and projects to both laminae I and

II. As discussed above, visceral afferent neurons contain a greater proportion of trkA/CGRP-expressing neurons than do cutaneous afferents. VR1 is expressed in a subfraction of both of these subpopulations (Michael and Priestley, 1999), and both populations are believed to respond to a wide range of thermal, chemical and mechanical noxious stimuli. The CGRP-containing neurons are thought to play a direct role in inflammation through the release of neuropeptides on stimulation, which may be particularly important in inflammatory bowel disease in view of the preponderance of peptide-containing neurons in the viscera, while the IB$_4$-positive neurons may be more involved in chronic neuropathic pain (see Caterina and Julius, 1999).

Finally, agonists of the proteinase-activated receptor PAR2 have recently been shown to have proinflammatory effects, and interestingly PAR2 is co-expressed with CGRP and SP in DRG neurons (Steinhoff et al., 2000). PAR receptors, which are expressed on sensory neurons and whose stimulation can result in hyperalgaesia, may therefore be another interesting target for future study of both somatic and visceral pain.

Acknowledgements

Research in the authors' lab was supported by grants from the BBSRC, MRC and Wellcome Trust to P.McN. and by PhD studentships from the BBSRC to J.K.B. and from Glaxo SmithKline to D.R.R.

References

Akopian, A.N., Chen, C.-C., Ding, Y., Cesare, P. and Wood, J.N. (2000). A new member of the acid-sensing ion channel family. NeuroReport 11, 2217–2222.

Akopian, A.N., Sivilotti, L. and Wood, J.N. (1996). A tetrodotoxin-resistant voltage-gated sodium channel expressed by sensory neurons. Nature 379, 257–262.

Akopian, A.N., Souslova, V., England, S., Okuse, K., Ogata, N., Ure, J., Smith, A., Kerr, B.J., McMahon, S.B., Boyce, S., Hill, R., Stanfa, L.C., Dickenson, A.H. and Wood, J.N. (1999). The tetrodotoxin-resistant sodium channel SNS has a specialized function in pain pathways. Nature Neurosci. 2, 541–548.

Babinski, K., Catarsi, S., Biagini, G. and Seguela, P. (2000). Mammalian ASIC2a and ASIC3 subunits co-assemble into heteromeric proton-gated channels sensitive to Gd^{3+}. J. Biol. Chem. 275, 28519–28525.

Baker, M.D. and Wood, J.N. (2001). Involvement of Na$^+$ channels in pain pathways. Trends Pharmacol. Sci. 22, 27–31.

Belmonte, C. and Giraldez, F. (1981). Responses of cat corneal sensory receptors to mechanical and thermal stimulation. J. Physiol. 321, 355–368.

Bennett, D.L.H., Dmietrieva, N., Priestley, J.V., Clary, D. and McMahon, S.B. (1996). trkA, CGRP and IB4 expression in retrogradely labelled cutaneous and visceral primary sensory neurones in the rat. Neurosci. Lett. 206, 33–36.

Bennett, D.L.H., Koltzenburg, M., Priestley, J.V., Shelton, D.L. and McMahon, S.B. (1998). Endogenous nerve growth factor regulates the sensitivity of nociceptors in the adult rat. Eur. J. Neurosci. 10, 1282–1291.

Benson, C.J., Eckert, S.P. and McCleskey, E.W. (1999). Acid-evoked currents in cardiac sensory neurons: A possible mediator of myocardial ischemic sensation. Circ. Res. 84, 921–928.

Bevan, S. and Winter, J. (1995). Nerve growth factor (NGF) differentially regulates the chemosensitivity of adult rat cultured sensory neurons. J. Neurosci. 15, 4918–4926.

Bleehen, T. and Keele, C.A. (1977). Observations on the algogenic actions of adenosine compounds on the human blister base preparation. Pain 3, 367–377.

Brake, A.J., Wagenbach, M.J. and Julius, D. (1994). New structural motif for ligand-gated ion channels defined by an ionotropic ATP receptor. Nature 371, 519–523.

Cardenas, C.G., Del Mar, L.P., Cooper, B.Y. and Scroggs, R.S. (1997). 5HT4 receptors couple positively to tetrodotoxin-insensitive sodium channels in a subpopulation of capsaicin-sensitive rat sensory neurons. J. Neurosci. 17, 7181–7189.

Cardenas, L.M., Cardenas, C.G. and Scroggs, R.S. (2001). 5HT increases excitability of nociceptor-like rat dorsal root ganglion neurons via cAMP-coupled TTX-resistant Na(+) channels. J. Neurophysiol. 86, 241–248.

Caterina, M.J. and Julius, D. (1999). Sense and specificity: A molecular identity for nociceptors. Curr. Opin. Neurobiol. 9, 525–530.

Caterina, M.J., Leffler, A., Malmberg, A.B., Martin, W.J., Trafton, J., Petersen-Zeitz, K.R., Koltzenburg, M., Basbaum, A.I. and Julius, D. (2000). Impaired nociception and pain sensation in mice lacking the capsaicin receptor. Science 288, 306–313.

Caterina, M.J., Rosen, T.A., Tominaga, M., Brake, A.J. and Julius, D. (1999). A capsaicin-receptor homologue with a high threshold for noxious heat. Nature 398, 436–441.

Caterina, M.J., Schumacher, M.A., Tominaga, M., Rosen, T.A., Levine, J. D. and Julius, D. (1997a). The capsaicin receptor: a heat-activated ion channel in the pain pathway. Nature 389, 816–824.

Cervero, F. (1995). Visceral pain: mechanisms of peripheral and central sensitization. Ann. Med. 27, 235–239.

Cervero, F. (2000). Visceral pain-central sensitisation. Gut 47 Suppl 4, iv56–iv57.

Cervero, F. and Janig, W. (1992). Visceral nociceptors: a new world order? Trends Neurosci. 15, 374–378.

Cesare, P., Dekker, L.V., Sardini, A., Parker, P.J. and McNaughton, P.A. (1999a). Specific involvement of PKC-epsilon in sensitization of the neuronal response to painful heat. Neuron 23, 617–624.

Cesare, P. and McNaughton, P. (1996). A novel heat-activated current in nociceptive neurons and its sensitization by bradykinin. Proc. Natl. Acad. Sci. USA 93, 15435–15439.

Cesare, P., Moriondo, A., Vellani, V. and McNaughton, P.A. (1999b). Ion channels gated by heat. Proc. Natl. Acad. Sci. USA 96, 7658–7663.

Chen, C.-C., England, S., Akopian, A.N. and Wood, J.N. (1998). A sensory neuron-specific, proton-gated ion channel. Proc. Natl. Acad. Sci. USA 95, 10240–10245.

Chen, C.C., Akopian, A.N., Sivilotti, L., Colquhoun, D., Burnstock, G. and Wood, J.N. (1995). A P2X purinoceptor expressed by a subset of sensory neurons. Nature 377, 428–431.

Chuang, H.H., Prescott, E.D., Kong, H., Shields, S., Jordt, S.E., Basbaum, A.I., Chao, M.V. and Julius, D. (2001). Bradykinin and nerve growth factor release the capsaicin receptor from PtdIns(4,5)P_2-mediated inhibition. Nature 411, 957–962.

Clapham, D.E., Runnels, L.W. and Strubing, C. (2001). The TRP ion channel family. Nat. Rev. Neurosci. 2, 387–396.

Cockayne, D.A., Hamilton, S.G., Zhu, Q.M., Dunn, P.M., Zhong, Y., Novakovic, S., Malmberg, A.B., Cain, G., Berson, A., Kassotakis, L., Hedley, L., Lachnit, W.G., Burnstock, G., McMahon, S.B. and Ford, A.P. (2000). Urinary bladder hyporeflexia and reduced pain-related behaviour in P2X3-deficient mice. Nature 407, 1011–1015.

Cook, I.J., van Eeden, A. and Collins, S.M. (1987). Patients with irritable bowel syndrome have greater pain tolerance than normal subjects. Gastroenterology 93, 727–733.

Davis, J.B., Gray, J., Gunthorpe, M.J., Hatcher, J.P., Davey, P.T., Overend, P., Harries, M.H., Latcham, J., Clapham, C., Atkinson, K., Hughes, S.A., Rance, K., Grau, E., Harper, A.J., Pugh, P.L., Rogers, D.C., Bingham, S., Randall, A. and Sheardown, S.A. (2000). Vanilloid receptor-1 is essential for inflammatory thermal hyperalgesia. Nature 405, 183–187.

De Felipe, C., Herrero, J.F., O'Brien, J.A., Palmer, J.A., Doyle, C.A., Smith, A.J.H., Laird, J.M.A., Belmonte, C., Cervero, F. and Hunt, S.P. (1998). Altered nociception, analgesia and aggression in mice lacking the receptor for substance P. Nature 392, 394–397.

Dib-Hajj, S.D., Tyrrell, L., Black, J.A. and Waxman, S.G. (1998). NaN, a novel voltage-gated Na channel, is expressed preferentially in peripheral sensory neurons and down-regulated after axotomy. Proc. Natl. Acad. Sci. USA 95, 8963–8968.

England, S., Bevan, S. and Docherty, R.J. (1996). PGE2 modulates the tetrodotoxin-resistant sodium current in neonatal rat dorsal root ganglion neurones via the cyclic AMP-protein kinase A cascade. J. Physiol. 495, 429–440.

Ferguson, D.R., Kennedy, I. and Burton, T.J. (1997). ATP is released from rabbit urinary bladder epithelial cells by hydrostatic pressure changes—a possible sensory mechanism? J. Physiol. 505, 503–511.

Fitzgerald, E.M., Okuse, K., Wood, J.N., Dolphin, A.C. and Moss, S.J. (1999). cAMP-dependent phosphorylation of the tetrodotoxin-resistant voltage-dependent sodium channel SNS. J. Physiol. 516, 433–446.

Garcia, R., Liapi, A., Cesare, P., Bonnert, T., Wafford, K., Clark, S., Young, J., Delmas, P., Whiting, P., McNaughton, P.A. and Wood, J.N. (1999). VR-L, a vanilloid receptor-like orphan receptor, is expressed in T cells and sensory neurons. J. Physiol. 518, 126P.

Gold, M.S. and Levine, J.D. (1996). DAMGO inhibits prostaglandin E2-induced potentiation of a TTX-resistant Na^+ current in rat sensory neurons in vitro. Neurosci. Lett 212, 83–86.

Gold, M.S., Levine, J.D. and Correa, A.M. (1998). Modulation of TTX-R I(Na) by PKC and PKA and their role in PGE_2-induced sensitization of rat sensory neurons in vitro. J. Neurosci. 18, 10345–10355.

Gold, M.S., Reichling, D.B., Shuster, M.J. and Levine, J.D. (1996). Hyperalgesic agents increase a tetrodotoxin-resistant Na+ current in nociceptors. Proc. Natl. Acad. Sci. USA 93, 1108–1112.

Janig, W. and Koltzenburg, M. (1991). Receptive properties of sacral primary afferent neurons supplying the colon. J. Neurophysiol. 65, 1067–1077.

Jordt, S.E. and Julius, D. (2002). Molecular basis for species-specific sensitivity to "hot" chili peppers. Cell 108, 421–430.

Jordt, S.E., Tominaga, M. and Julius, D. (2000). Acid potentiation of the capsaicin receptor determined by a key extracellular site. Proc. Natl. Acad. Sci. USA 97, 8134–8139.

Jung, J., Hwang, S.W., Kwak, J., Lee, S.Y., Kang, C.J., Kim, W.B., Kim, D. and Oh, U. (1999). Capsaicin binds to the intracellular domain of the capsaicin-activated ion channel. J. Neurosci. 19, 529–538.

Kerr, B.J., Souslova, V., McMahon, S.B. and Wood, J.N. (2001). A role for the TTX-resistant sodium channel Nav 1.8 in NGF-induced hyperalgesia, but not neuropathic pain. Neuroreport 12, 3077–3080.

Khasar, S.G., Lin, Y.H., Martin, A., Dadgar, J., McMahon, T., Wang, D., Hundle, B., Aley, K.O., Isenberg, W.,

McCarter, G., Green, P.G., Hodge, C.W., Levine, J.D. and Messing, R.O. (1999). A novel nociceptor signaling pathway revealed in protein kinase C epsilon mutant mice. Neuron 24, 253–260.

Kuzhikandathil, E.V., Wang, H., Szabo, T., Morozova, N., Blumberg, P.M. and Oxford, G.S. (2001). Functional analysis of capsaicin receptor (vanilloid receptor subtype 1) multimerization and agonist responsiveness using a dominant negative mutation. J. Neurosci. 21, 8697–8706.

Lee, Y.-J., Zachrisson, O., Tonge, D.A. and McNaughton, P.A. (2002). Upregulation of bradykinin b2 receptor expression by neurotrophic factors and nerve injury in mouse sensory neurons. Mol. Cell. Neurosci. 19, 186–200.

Lewin, G.R., Ritter, A.M. and Mendell, L.M. (1993). Nerve growth factor-induced hyperalgesia in the neonatal and adult rat. J. Neurosci. 13, 2136–2148.

Lewis, C., Neidhart, S., Holy, C., North, R.A., Buell, G. and Surprenant, A. (1995). Coexpression of P2X2 and P2X3 receptor subunits can account for ATP-gated currents in sensory neurons. Nature 377, 432–435.

Mayer, E.A. and Gebhart, G.F. (1994). Basic and clinical aspects of visceral hyperalgesia. Gastroenterology 107, 271–293.

McKemy, D.D., Neuhausser, W.M. and Julius, D. (2002). Identification of a cold receptor reveals a general role for TRP channels in thermosensation. Nature 416, 52–58.

Mendell, L.M. (1999). Neurotrophin action on sensory neurons in adults: An extension of the neurotrophic hypothesis. Pain 82, S127–S132.

Michael, G.J. and Priestley, J.V. (1999). Differential expression of the mRNA for the vanilloid receptor subtype I in cells of the adult rat dorsal root and nodose ganglia and its downregulation by axotomy. J. Neurosci. 19, 1844–1854.

Ness, T.J. and Gebhart, G.F. (1990). Visceral pain: a review of experimental studies. Pain 41, 167–234.

North, R.A. and Surprenant, A. (2000). Pharmacology of cloned P2X receptors. Annu. Rev. Pharmacol Toxicol. 40, 563–580.

Peier, A.M., Moqrich, A., Hergarden, A.C., Reeve, A.J., Andersson, D.A., Story, G.M., Earley, T.J., Dragoni, I., McIntyre, P., Bevan, S. and Patapoutian, A. (2002). A TRP channel that senses cold stimuli and menthol. Cell 108, 705–715.

Perry, M.J. and Lawson, S.N. (1998). Differences in expression of oligosaccharides neuropeptides, carbonic anhydrase and neurofilament in rat primary afferent neurons retrogradely labelled via skin, muscle or visceral nerves. Neuroscience 85, 293–310.

Petruska, J.C., Napaporn, J., Johnson, R.D., Gu, J.G. and Cooper, B.Y. (2000). Subclassified acutely dissociated cells of rat DRG: histochemistry and patterns of capsaicin-, proton-, and ATP-activated currents. J.

Neurophysiol. 84, 2365–2379.

Premkumar, L.S. and Ahern, G.P. (2000). Induction of vanilloid receptor channel activity by protein kinase C. Nature 408, 985–990.

Rees, H., Sluka, K.A., Urban, L., Walpole, C.J. and Willis, W.D. (1998). The effects of SDZ NKT 343, a potent NK1 receptor antagonist, on cutaneous responses of primate spinothalamic tract neurones sensitized by intradermal capsaicin injection. Exp. Brain Res. 121, 355–358.

Reid, G. and Flonta, M.L. (2001). Cold current in thermoreceptive neurons. Nature 413, 480.

Sengupta, J.N. and Gebhart, G.F. (1994). Characterization of mechanosensitive pelvic nerve afferent fibers innervating the colon of the rat. J. Neurophysiol. 71, 2046–2060.

Sengupta, J.N. and Gebhart, G.F. (1998). The sensory innervation of the colon and its modulation. Curr. Opin. Gastroenter. 14, 15–20.

Shu, X. and Mendell, L.M. (1999). Nerve growth factor acutely sensitizes the response of adult rat sensory neurons to capsaicin. Neurosci. Lett 274, 159–162.

Shu, X. and Mendell, L.M. (2001). Acute sensitization by NGF of the response of small-diameter sensory neurons to capsaicin. J. Neurophysiol. 86, 2931–2938.

Snider, W.D. and McMahon, S.B. (1998). Tackling pain at the source: New ideas about nociceptors. Neuron 20, 629–632.

Souslova, V., Cesare, P., Ding, Y., Akopian, A.N., Stanfa, L., Suzuki, R., Carpenter, K., Dickenson, A., Boyce, S., Hill, R., Nebenuis-Oosthuizen, D., Smith, A.J., Kidd, E.J. and Wood, J.N. (2000). Warm-coding deficits and aberrant inflammatory pain in mice lacking P2X3 receptors. Nature 407, 1015–1017.

Steen, K.H., Reeh, P.W., Anton, F. and Handwerker, H.O. (1992). Protons selectively induce lasting excitation and sensitization to mechanical stimulation of nociceptors in rat skin, *in vitro*. J. Neurosci. 12, 86–95.

Steinhoff, M., Vergnolle, N., Young, S.H., Tognetto, M., Amadesi, S., Ennes, H.S., Trevisani, M., Hollenberg, M.D., Wallace, J.L., Caughey, G.H., Mitchell, S.E., Williams, L.M., Geppetti, P., Mayer, E.A. and Bunnett, N.W. (2000). Agonists of proteinase-activated receptor 2 induce inflammation by a neurogenic mechanism. Nature Med 6, 151–158.

Stoop, R., Surprenant, A. and North, R.A. (1997). Different sensitivities to pH of ATP-induced currents at four cloned P2X receptors. J. Neurophysiol. 78, 1837–1840.

Su, X., Sengupta, J.N. and Gebhart, G.F. (1997). Effects of kappa opioid receptor-selective agonists on responses of pelvic nerve afferents to noxious colorectal distension. J. Neurophysiol. 78, 1003–1012.

Sun, W.H., Cho, H., Kwak, J., Lee, S.-Y., Kang, C.-J., Jung, J., Cho, S., Kyung, Hoon, M., Suh, Y.-G., Kim, D. and Oh, U. (2000). Direct activation of capsaicin receptors by products of lipoxygenases: Endogenous capsaicin-like substances. Proc. Natl. Acad. Sci. USA 97, 6155–6160.

Sutherland, S.P., Benson, C.J., Adelman, J.P. and McCleskey, E.W. (2001). Acid-sensing ion channel 3 matches the acid-gated current in cardiac ischemia-sensing neurons. Proc. Natl. Acad. Sci. USA 98, 711–716.

Tate, S., Benn, S., Hick, C., Trezise, D., John, V., Mannion, R.J., Costigan, M., Plumpton, C., Grose, D., Gladwell, Z., Kendall, G., Dale, K., Bountra, C. and Woolf, C.J. (1998). Two sodium channels contribute to the TTX-R sodium current in primary sensory neurons. Nature Neurosci. 1, 653–655.

Tominaga, M., Caterina, M.J., Malmberg, A.B., Rosen, T.A., Gilbert, H., Skinner, K., Raumann, B.E., Basbaum, A.I. and Julius, D. (1998). The cloned capsaicin receptor integrates multiple pain-producing stimuli. Neuron 21, 531–543.

Tominaga, M., Wada, M. and Masu, M. (2001). Potentiation of capsaicin receptor activity by metabotropic ATP receptors as a possible mechanism for ATP-evoked pain and hyperalgesia. Proc. Natl. Acad. Sci. USA 98, 6951–6956.

Treede, R.D., Meyer, R.A., Raja, S.N. and Campbell, J.N. (1992). Peripheral and central mechanisms of cutaneous hyperalgesia. Prog. Neurobiol. 38, 397–421.

Valera, S., Hussy, N., Evans, R.J., Adami, N., North, R.A., Surprenant, A. and Buell, G. (1994). A new class of ligand-gated ion channel defined by P2x receptor for extracellular ATP. Nature 371, 516–519.

Vellani, V., Mapplebeck, S., Moriondo, A., Davis, J.B. and McNaughton, P.A. (2001). Protein kinase C activation potentiates gating of the vanilloid receptor, VR1, by capsaicin, protons, heat and anandamide. J. Physiol. 534, 813–825.

Waldmann, R., Bassilana, F., De Weille, J., Champigny, G., Heurteaux, C. and Lazdunski, M. (1997a). Molecular cloning of a non-inactivating proton-gated Na+ channel specific for sensory neurons. J. Biol. Chem. 272, 20975–20978.

Waldmann, R., Champigny, G., Bassilana, F., Heurteaux, C. and Lazdunski, M. (1997b). A proton-gated cation channel involved in acid sensing. Nature 386, 173–177.

Waldmann, R., Champigny, G., Lingueglia, E., De, W.J., Heurteaux, C. and Lazdunski, M. (1999). H+-gated cation channels. Ann. NY Acad. Sci. 868, 67–76.

Waxman, S.G., Dib-Hajj, S., Cummins, T.R. and Black, J.A. (1999). Sodium channels and pain. Proc. Natl. Acad. Sci. USA 96, 7635–7639.

Weihe, E., Nohr, D., Millan, M.J., Stein, C., Muller, S., Gramsch, C. and Herz, A. (1988). Peptide neuroanatomy of adjuvant-induced arthritic inflammation in rat. Agents Actions 25, 255–259.

Welch, J.M., Simon, S.A. and Reinhart, P.H. (2000). The activation mechanism of rat vanilloid receptor 1 by capsaicin involves the pore domain and differs from the activation by either acid or heat. Proc. Natl. Acad. Sci. USA 97, 13889–13894.

Whitehead, W.E., Holtkotter, B., Enck, P., Hoelzl, R., Holmes, K.D., Anthony, J., Shabsin, H.S. and Schuster, M.M. (1990). Tolerance for rectosigmoid distention in irritable bowel syndrome. Gastroenterology 98, 1187–1192.

Woolf, C.J., Safieh-Garabedian, B., Ma, Q.P., Crilly, P. and Winter, J. (1994). Nerve growth factor contributes to the generation of inflammatory sensory hypersensitivity. Neuroscience 62, 327–331.

Woolf, C.J. and Thompson, S.W. (1991). The induction and maintenance of central sensitization is dependent on N-methyl-D-aspartic acid receptor activation; implications for the treatment of post-injury pain hypersensitivity states. Pain 44, 293–299.

Zheng, F. and Lawson, S.N. (1994). Immunocytochemical properties of rat renal afferent neurons in dorsal root ganglia: A quantitative study. Neuroscience 63, 295–306.

Zygmunt, P.M., Petersson, J., Andersson, D.A., Chuang, H.-H., Sorgard, M., Di Marzo, V., Julius, D. and Hogestatt, E.D. (1999). Vanilloid receptors on sensory nerves mediate the vasodilator action of anandamide. Nature 400, 452–457.

Sensing, Signaling and Cell Adaptation. Edited by K.B. Storey and J.M. Storey

CHAPTER 9

Acquired Freezing Tolerance in Higher Plants: The Sensing and Molecular Responses to Low Nonfreezing Temperatures

Zhanguo Xin

Plant Stress and Germplasm Development Unit, Plant Stress and Water Conservation Laboratory, USDA, Agricultural Research Service, Lubbock, TX 79415, USA

Abbreviations used:
ABA: abscisic acid; CBF: C-repeat-binding factor; *cft*: constitutively freezing tolerant; *COR*: cold-regulated; *cos*: constitutive expression of osmotically responsive genes; CRT: C-repeat; DRE: dehydration-responsive element; DREB: DRE-binding protein; *hos*: high expression of osmotically responsive genes; *los*: low expression of osmotically responsive genes; RD: response to dehydration; 35S: the cauliflower mosaic virus ^{35}S promoter; *sfr*: sensitive to freezing.

1. Introduction

Freezing temperatures represent a major environmental constraint for many living organisms. Organisms must either face, respond, and make adjustments to endure both daily and seasonal changes in the ambient temperatures or die. Many animals use the option of migration or movement into a thermally-buffered microenvironment to protect themselves from subzero exposure but plants have no such options. Plants are sessile organisms with little ability to directly regulate their body temperature and virtually no chance to avoid freezing when the temperatures drop below the freezing point of their internal fluids. To survive, they must be capable of perceiving changes in ambient temperature and adjust their physiology to avoid permanent damage or death due to freezing. One of the better characterized temperature

responses in plants adapted to temperate regions is their ability to increase their tolerance to freezing in response to a period of low nonfreezing temperatures, a process called cold acclimation (Levitt, 1980). The increased freezing tolerance that results from cold acclimation is not constitutive; rather it is a transiently acquired phenomenon. In nature, low non-freezing temperatures in late fall or early winter serve as the main triggers of cold acclimation, although light quality and photoperiod may also be involved. In the laboratory, cold acclimation is induced by exposing plants to low nonfreezing temperatures (2–6°C) under appropriate light conditions for certain period of time. Depending on the plant species, it may take a few days to several weeks to reach maximum levels of freezing tolerance and the tolerance produced ranges from −10°C to below −30°C in annual and biennial plants (Gilmour et al., 1988; Webb et al., 1994). In extreme cases, certain clones of dogwood trees can survive immersion in liquid nitrogen (−196°C) when fully acclimated (Sakai, 1960).

The ability of most temperate plants to enhance their freezing tolerance in the face of low non-freezing temperatures has inspired many researchers to explore the biochemical and molecular determinants of freezing tolerance. Biologically, cold acclimation in plants is very complex, involving numerous changes in gene expression, metabolism, and morphology (Xin and Browse, 1998). These changes include: increased expression of many genes (Thomashow, 1999); reduction or cessation of growth; transient increases in abscisic acid (ABA) levels; changes in membrane lipid composition; accumulation of compatible osmolytes such as proline, betaine, polyols and soluble sugars; and increased levels of antioxidants (Xin and Browse, 2000). In short, almost every cellular process is altered during cold acclimation. However, it has been a great challenge to separate the processes that are adaptive to freezing from those that are responses to injury due to exposure to low non-freezing temperatures.

In the last few years, exciting progress has been made with the introduction of new molecular and genetic technologies, and especially, the adoption of *Arabidopsis* as a model plant in study of cold acclimation. This review focuses on recent progress and discusses the major challenges still to be met in this field. For more detailed discussion on the mechanisms of cold acclimation the reader is referred to reviews by Guy (1990), Steponkus (1984), and Thomashow (1994; 1999).

2.   Freezing injury in plants

Most plants have little ability to regulate their temperature; therefore, plant tissues are directly influenced by changes in ambient temperature, making it nearly impossible for plants to avoid freezing events. In general, there are two types of freezing experienced by plant tissues, intracellular freezing and intercellular freezing. Intracellular freezing, the formation of ice inside the cytoplasm, is considered lethal to most plants. In nature, intercellular freezing is the most common type of freezing, since the intercellular fluid is more dilute than the cytoplasm, thus, has a higher freezing point. When plants are subjected to freezing temperatures the intercellular fluid in the apoplast compartment usually freezes first. Upon freezing, the ice-state of water has a much lower water potential than liquid solution and this difference increases as temperature decreases (Guy, 1990). Therefore, when ice forms intercellularly, there is a sudden drop in water potential outside the cell, and water moves out of cytoplasm by osmosis, leading to cellular dehydration. The net amount of water removed from the cell depends on both the initial solute

concentration of the cytoplasm and the actual temperature, which determines the water potential of the intercellular ice. For example, freezing at $-10°C$ results in a water potential of -11.6 MPa. This will remove 90% of the osmotically active water from the cell, assuming the initial osmotic potential of cell is -1 MPa. If the initial osmotic potential of the cell is -2 MPa due to accumulation of solutes during cold acclimation, the same freezing only removes 80% of cellular water. Thus, cellular dehydration is a major component of freezing injury. This is one explanation for the extensive overlap between cold acclimation responses and dehydration responses in both biochemical changes and in transcriptional regulation of gene expression (Shinozaki and Yamaguchi-Shinozaki, 2000).

Dehydration may disrupt cellular functions in a variety of ways. However, in the case of freezing stress, injury normally involves disruption of membrane structure and function (Uemura et al., 1995; Webb et al., 1994). Since 1912, the plasma membrane has been considered as the primary site of freezing injury (Levitt, 1980). The work of Steponkus and colleagues (Uemura et al., 1995; Webb et al., 1994) provided evidence that freezing-induced destabilization of plasma membrane involves three different types of lesions. In protoplasts from non-acclimated rye leaves, the reduction in cell volume at temperatures down to $-5°C$ is accompanied by a loss of surface area as a result of invagination of the plasma membrane and the subsequent budding off of endocytotic vesicles. Upon rewarming, the melted water is drawn back into the cells. As a consequence of the loss of plasma membrane surface area, the protoplast cannot regain its original volume and thus it may burst because of the increase in hydrostatic pressure created by an influx of water. This type of lesion is known as "expansion-induced lysis" (Uemura et al., 1995; Webb et al., 1994). In contrast, protoplasts prepared from cold-acclimated leaves do not form endocytotic vesicles. Instead, the plasma membrane is retained as exocytotic extrusions that allow re-expansion of the cell upon thawing.

At colder temperatures (and greater dehydration), plasma membranes are brought into close apposition with endomembranes, such as the chloroplast envelope. Membrane lipids in non-acclimated tissues undergo lateral phase separations. Certain lipids aggregate to form an inverted structure with hexagonal packing symmetry, called HexII phase. The lipid molecules are arranged in cylinders with the head groups oriented toward an aqueous core of 20 Å in diameter, which disrupts the membrane bilayer. The plasma membrane becomes permeable to both water and solutes and thus loses osmotic responsiveness upon rewarming. Formation of the HexII phase is considered as an interbilayer event and may involve the participation of two or more bilayers since the HexII phase is only observed in multilamellar or stacked bilayer regions. The HexII phase is more frequently observed in regions where the plasma membrane is brought into close apposition with the chloroplast outer membrane. As with expansion-induced lysis, the HexII phase is associated with protoplasts isolated from non-acclimated tissues and is largely precluded by cold acclimation (Uemura et al., 1995; Webb et al., 1994).

In cold acclimated protoplasts, freezing injury is associated with the formation of a "fracture-jump lesion". This type of membrane lesion is exemplified by a localized deviation in the freeze-fracture plane of the plasma membrane during cryo-electron microscopy, probably due to the localized fusion of plasma membrane with other cellular membranes, especially chloroplast envelopes (Uemura et al., 1995; Webb et al., 1994). It is believed that both HexII and fracture-jump lesions are formed from a common structural membrane intermediate (Uemura et al., 1995; Webb et al., 1994). However, it is unclear why HexII lesions are observed only in protoplasts isolated from non-acclimated tissues whereas fracture-jump lesions are observed exclusively in protoplasts from cold acclimated tissues. Furthermore, the temperatures at which fracture-jump lesions are observed vary greatly among plant species (Uemura et al., 1995; Webb et al., 1994). Little is known about the physiological, biochemical, and molecular basis of this variation, but it is considered a measure of the relative freezing tolerance achieved between different species. It is worth noting that these freezing-induced membrane lesions are primarily observed in isolated protoplasts, and may not reflect what occurs in intact plant cells.

3. Biochemical changes associated with cold acclimation

Although it has been known for over 100 years that most temperate plants can acquire a greater ability to withstand freezing in response to a period of low non-freezing temperatures, we still do not know what changes in metabolism during cold acclimation are critical to freezing tolerance. Numerous physiological and biochemical changes are known to occur during cold acclimation, and are shown schematically in Fig. 9.1. The most notable changes include a reduction or cessation of growth; reduction of tissue water content (Levitt, 1980); transient increase in abscisic acid (ABA) levels (Chen et al., 1983); changes in membrane lipid composition (Lynch and Steponkus, 1987; Uemura and Steponkus, 1994); the accumulation of compatible osmolytes such as proline, betaine, polyols and soluble sugars; and increased levels of antioxidants (Dörffling et al., 1997; Kishitani et al., 1994; Koster and Lynch, 1992; Murelli et al., 1995; Nomura et al., 1995; Tao et al., 1998). These complicated responses have made it very difficult to separate the processes responsible for freezing tolerance from those that represent non-protective

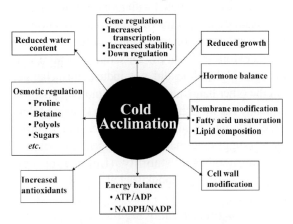

Fig. 9.1. Changes commonly observed when plants are subjected to low, non-freezing temperatures.

responses to low, non-freezing temperatures. However, it is clear from a wide range of plant responses to low temperature that cold acclimation requires many changes in cell biology and metabolism. Here, we will briefly review evidence for some of the most critical changes that have been implicated as important in cold acclimation.

3.1. Alteration in membrane structure and lipid composition

Since cell membranes are thought to be the primary sites of freezing injury, changes in membrane behavior during cold acclimation are likely to be critical to the development of freezing tolerance. Ultrastructural changes in the plasma membrane can be observed within six hours of the start of cold acclimation in *Arabidopsis* (Ristic and Ashworth, 1993). The underlying biochemical basis for these changes in membrane behavior is not completely known. Alterations in membrane lipid composition are correlated with membrane cryostability as they have been observed during cold acclimation in all species examined (Steponkus, 1984; Uemura et al., 1995; Uemura and Steponkus, 1994). There is also direct evidence for the involvement of membrane lipids in freezing and dehydration tolerance. For example, when non-acclimated rye protoplasts are placed in hyperosmotic medium, their plasma membrane buds off endocytotic vesicles as they do during freezing. However, Steponkus et al. (1988) demonstrated that when non-acclimated protoplasts were preincubated with mono-unsaturated or di-unsaturated species of phosphatidylcholine, hyperosmotic treatment resulted in the formation of exocytotic extrusions. Di-saturated species of phosphatidylcholine did not induce this change. On the other hand, alterations in lipid composition typically lag far behind increases in freezing tolerance (Uemura et al., 1995; Wanner and Junttila, 1999), suggesting that other changes also contribute to membrane stability (Lineberger and Steponkus, 1980; Steponkus et al., 1998).

3.2. Changes in soluble sugars

Accumulation of soluble sugars during cold

acclimation is well documented for many plants including *Arabidopsis*, and the time course of sugar accumulation correlates well with development of freezing tolerance during cold acclimation (Ristic and Ashworth, 1993; Wanner and Junttila, 1999). In addition, genetic evidence is available to support the role of soluble sugars in freezing tolerance. For example, one of the sensitive to freezing (*sfr*) mutants of *Arabidopsis* (*sfr4*; see below), that is impaired in its ability to cold acclimate, does not accumulate soluble sugars in response to low temperature (McKown et al., 1996). In contrast, *esk1*, a constitutively freezing tolerant mutant, accumulates sugars at warm temperatures (Xin and Browse, 1998). Several roles for sugars in protecting cells from freezing injury have been proposed, including functioning as cryoprotectants for specific enzymes (Carpenter et al., 1986), as molecules promoting membrane stability (Lineberger and Steponkus, 1980), and as osmolytes to prevent excessive dehydration during freezing (Steponkus, 1984). However, soluble sugars alone are insufficient for full freezing tolerance. Several *sfr* mutants that accumulate soluble sugars normally during cold acclimation are, nevertheless, defective in freezing tolerance. Transformation of tobacco with a bacterial pyrophosphatase or invertase gene increased the levels of soluble sugars but did not provide a reproducible increase in freezing tolerance (Hincha et al., 1996).

3.3. Changes in betaines

Accumulation of betaines has been reported in several plant species in response to low temperature (Kishitani et al., 1994; Nomura et al., 1995). *Arabidopsis* does not accumulate betaines. However, over-expression of a single bacterial gene, choline oxidase, in *Arabidopsis* results in accumulation of betaine to between 0.1 and 1 μmol g^{-1} fresh weight (Hayashi et al., 1997). Although low compared to the levels found in plants that naturally accumulate betaines (20–300 μmol g^{-1} fresh weight), this increase significantly improved the tolerance to various stresses including cold and freezing (Hayashi et al., 1997; Sakamoto et al., 2000). Even though the accumulation of the limited amount of betaines improved freezing tolerance in *Arabidopsis*, many plants that accumulate high levels of betaine are, nevertheless, chilling- or freezing-sensitive species (Wood et al., 1996; Yang et al., 1995).

3.4. Accumulation of proline

Free proline accumulates in plants in response to many stresses (Delauney and Verma, 1993), however, its role in stress tolerance remains equivocal. Examination of somatic mutants that accumulate proline in potato has demonstrated a correlation between elevated proline concentrations and enhanced freezing tolerance (Van Swaaij et al., 1986) and similarly in winter wheat (Dörffling et al., 1997). *Arabidopsis* accumulates proline during cold acclimation but the increase in proline content lags behind the development of freezing tolerance (Wanner and Junttila, 1999). The strongest genetic evidence that proline may contribute to the increased freezing tolerance comes from the isolation of several constitutively freezing tolerant mutants of *Arabidopsis* that accumulate proline in the absence of a low-temperature treatment (Xin and Browse, 1998). In wild-type *Arabidopsis* proline content increases 10-fold during two days of cold acclimation at 4°C. In the absence of acclimation, the *esk1-1* mutant contains proline at levels 30-fold higher than non-acclimated wild-type plants. At least two other freezing tolerant mutants also contain high levels of proline (Xin and Browse, unpublished) suggesting that proline does play an important role in freezing tolerance. However, proline accumulation is not required for freezing tolerance since some constitutively freezing tolerant mutants contain the same low levels of proline as non-acclimated wild-type plants.

Apparently, plants employ multiple mechanisms to ensure freezing tolerance. At present, our knowledge of the metabolic changes that contribute to freezing tolerance is incomplete, but information about the biochemical processes contributing to freezing tolerance is essential for successful engineering of freezing tolerance in crop plants.

4. Alteration of gene expression associated with cold acclimation

4.1. Diversity of cold induced genes

It has been well established that cold acclimation is associated with complex changes in gene expression (Guy, 1990; Thomashow, 1999). Some cold acclimation induced genes encode proteins with known enzymatic functions, such as alcohol dehydrogenase (Jarillo et al., 1993), phenylalanine ammonia lyase, chalcone synthase (Leyva et al., 1995), fatty acid desaturase, *FAD8* (Gibson et al., 1994), lipid transfer protein (Dunn et al., 1998), translation initiation factor (Dunn et al., 1993), a thiol protease (Shaffer and Fischer, 1988), catalases (Prasad et al., 1994), Δ-pyrroline-5-carboxylate synthase (the first enzyme committed to proline biosynthesis) (Yoshiba et al., 1995), and mRNA stabilization proteins (Phillips et al., 1997). Some encode proteins that are similar to antifreeze proteins (Kurkela and Franck, 1990), heat shock proteins or molecular chaperones (Anderson et al., 1994; Ukaji et al., 1999). Others encode various signal transduction or regulatory proteins, such as MAP kinases (Jonak et al., 1996; Mizoguchi et al., 1996), calcium dependent protein kinases (Tahtiharju et al., 1997), and actin depolymerization factor (Danyluk et al., 1996; Ouellet et al., 2001). The diverse array of genes induced by low temperature clearly exemplifies the complexity of cold acclimation in plants. The presumed functions of low temperature induced proteins may shed light on the mechanisms by which plants increase their freezing tolerance; however, the importance of many of the identified enzymes and proteins to the development of freezing tolerance remains questionable. For example, phenylalanine ammonia lyase and chalcone synthase, which are involved in biosynthesis of anthocyanin, are induced to high levels of expression during cold acclimation. *Arabidopsis* mutants deficient in either of these genes are not measurably affected in their ability to develop full levels of freezing tolerance (Leyva et al., 1995). Similarly, alcohol dehydrogenase is induced to high levels of expression during cold acclimation, but a null mutant of

Arabidopsis, which lacks alcohol dehydrogenase, can develop full level of freezing tolerance (Jarillo et al., 1993). Thus, one should be cautious when pursuing a gene based solely on demonstrated cold inducibility.

4.2. COR genes

Most of the enzyme-encoding cold regulated genes are expressed at normal growth temperatures and undergo 2–5-fold induction during cold acclimation. By contrast, one class of cold-regulated genes is strongly induced by cold acclimation (typically 50–100 fold). The encoded proteins are not enzymes and their precise physiological function remains a matter of debate although at least some appear to act as cryoprotective proteins or dehydrins (Thomashow, 1999). These genes could be relatively easily identified by differential screening and thus have been studied by several groups. Thus, diverse nomenclatures have been used to describe this class of genes, including *COR* for cold regulated; *LTI* for low temperature induced; *CAS* for cold acclimation specific; and *RD* for responsive to desiccation genes. We shall refer all of these genes as *COR* in the following discussion. The COR proteins have no known enzymatic activity, but many are very hydrophilic and remain stable upon boiling in dilute aqueous solution (Thomashow, 1999). These unusual characteristics suggest that they may function as cryoprotectants. Over-expression of a single *COR* gene, *COR15a*, has been demonstrated to offer some modest protection to chloroplasts and protoplasts derived from non-acclimated transgenic plants (Artus et al., 1996). Localized in the chloroplast stroma, the mature protein, COR15am, is proposed to function by deferring the formation of freezing-induced hexagonal II lipid phase to a lower temperature through the alteration of the intrinsic curvature of the inner membrane of the chloroplast envelope (Steponkus et al., 1998). However, no measurable improvement in freezing tolerance was observed in the *COR15a* over-expression lines at the whole plant level (Artus et al., 1996; Jaglo-Ottosen et al., 1998). Over-expression of other cold-induced genes in transgenic plants has shown little or no

enhancement in freezing tolerance (Kaye et al., 1998; Zhu et al., 1996). In all likelihood, individual components of freezing tolerance can only work within the context of a broader cold acclimation response.

5. CBF transcription factors define a major cold response pathway in flowering plants

COR genes have been particularly useful in investigating the signal pathways associated with cold acclimation using molecular-genetic approaches in *Arabidopsis*. Gene fusion studies have uncovered *cis*-acting elements in the promoter region of *COR* genes (Shinozaki and Yamaguchi-Shinozaki, 2000; Thomashow, 1999). These elements are designated drought responsive elements (DREs) or low temperature responsive element (LTREs). Both elements contain a core sequence of CCGAC referred as the C-repeat (CRT). The DRE/CRT elements have been used as "baits" to isolate two groups of DRE/CRT binding proteins using the yeast one-hybrid system (Liu et al., 1998; Stockinger et al., 1997). These proteins are called *DREB* (for DRE-binding factor) or CBF (for *CRT*-repeat binding factors). Five DRE/CRT-binding proteins have been isolated (Liu et al., 1998) and are classified into two groups, *DREB1* and *DREB2*. There are three *DREB1* proteins that are encoded by genes tandemly repeated on chromosome 4 in the order of *DREB1B* (*CBF1*), *DREB1A* (*CBF3*), and *DREB1C* (*CBF2*). There are two *DREB2* proteins, *DREB2A* and *DREB2B*. All five proteins bind specifically to DRE/CRT and function as transcription factors to activate the expression of *COR* genes.

Over-expression of *DREB1B/CBF1* in transgenic *Arabidopsis* induced the expression of the entire battery of *COR* genes that have the common DNA-elements in their promoter region at warm temperatures (Jaglo-Ottosen et al., 1998; Liu et al., 1998). When freezing tolerance was assayed by an ion-leakage test, transgenic tissues showed a 3.3°C improvement in freezing tolerance over non-transgenic control plants in the absence of cold acclimation. A much greater freezing tolerance is achieved by over-expression of another transcription factor,

DREB1A (*CBF3*) (Kasuga et al., 1999; Liu et al., 1998). Transcripts of *DREB1A/CBF3* are at very low levels in wild-type *Arabidopsis* plants even after seven days of cold acclimation at 5°C. Transgenic plants containing a *35S::CBF3* construct displayed very high levels of *CBF3* transcript, and an over-induction of *COR* polypeptides to levels 3–5 fold higher than those found in fully acclimated wild-type plants (Gilmour et al., 2000). In this study, non-acclimated wild-type plants exhibited an EL_{50} (temperature that causes a 50% leakage of electrolytes) of approximately –4.5°C, which was improved to –6°C after a seven-day acclimation at 5°C. Three independent *CBF3*-over-expressing lines had EL_{50} values of approximately –8°C without any cold treatment. However, after seven days at 5°C, the freezing tolerance of *CBF3*-over-expressing plants increased dramatically to values of –11°C or lower. One of the cold acclimated CBF3-over-expressing lines was so cold tolerant that tissues did not suffer 50% leakage of electrolytes even at the lowest temperature (–16°C) used in these experiments. Thus, by over-expressing a single regulator, it is possible to improve freezing tolerance well beyond that which can be achieved naturally by the normal mechanisms of cold acclimation. This raises the question of why cold acclimation at 5°C does not lead directly to the maximum frost tolerance that is physiologically possible. The answer may be that cold acclimation signaling has evolved to modulate the level of freezing tolerance so that it is sufficient for the conditions found within the geographic range of a given ecotype without unnecessarily limiting the competitiveness of the plants (Xin and Browse, 1998). Consistent with this hypothesis, transgenic plants over-expressing *DREB1A/CBF3* are severely compromised in growth and development even under the benign growth conditions of controlled environments (Gilmour et al., 2000; Liu et al., 1998). A first step toward overcoming (or at least ameliorating) such deleterious effects was taken in Shinozaki's group by placing the *DREB1A/CBF3* coding sequence under control of a *COR* gene promoter (*RD29A: DREB1A*). By comparison with *35S:DREB1A* transgenics, these plants were greatly improved in appearance although they retained a slight growth retardation relative to wild-type controls; again under controlled-environment conditions (Kasuga et al., 1999).

Recently, it has been shown that constitutive over-expression of CBF transcription factors in canola induces the expression of canola homologs of *Arabidopsis* CBF-mediated genes and enhanced the freezing tolerance of the transgenic plants in the absence of cold acclimation (Jaglo et al., 2001). Moreover, homologs of CBF genes have been identified in a range of plant species including tomato, which is a chilling sensitive plant and does not cold acclimate. Thus, the CBF cold response pathway may represent a highly conserved process in flowering plants and may serve other functions in addition to the regulation of cold acclimation.

6. Mutational analysis of freezing tolerance

Recently, two mutational screens have been used in *Arabidopsis* to dissect the mechanisms of freezing tolerance. The first approach involved isolating mutants that are defective in developing freezing tolerance after cold acclimation. This approach offers the potential to identify a wide range of signaling components mediating cold acclimation. Warren et al. (1996) have isolated seven *Arabidopsis* mutants that fail to develop full freezing tolerance even after two weeks of cold acclimation. These mutants are named *sfr* for sensitive to freezing. Some of the *sfr* mutants are deficient in accumulation of soluble sugars during cold acclimation (McKown et al., 1996). One of these freezing-sensitive mutants, *sfr6*, was shown to be deficient in *CBF1*-mediated induction of *COR* genes (Knight et al., 1999), confirming the importance of the *CBF1* pathway in cold acclimation. However, with the exception of *sfr6*, the remaining classes of *sfr* mutants all show strong induction of the *COR* genes. Thus, *COR* gene induction is only one component of freezing tolerance.

The second approach involved the isolation of constitutively freezing tolerant (*cft*) mutants, i.e., mutants that, in the absence of cold acclimation, are more freezing tolerant than wild-type plants. The rationale for this screen is that a single gene mutation that results in significant increases in freezing tolerance is most likely a mutation in a signaling component of cold acclimation, which mediates to the expression of a suite of terminal genes to provide increased freezing tolerance. Twenty-six constitutive freezing tolerant mutants were isolated from 800,000 EMS-mutagenized M2 seedlings of *Arabidopsis* (Xin and Browse, 1998). One of the best characterized *cft* mutants, *eskimo1*, tolerates freezing to −10.6°C without cold acclimation. This improvement in freezing tolerance (5.1°C over non-acclimated wild type) represents 70% of the freezing tolerance found in fully acclimated wild-type plants. The *esk1* mutants contain high levels of proline and soluble sugars but have similar levels of *COR* genes as in non-acclimated wild-type plants (Xin and Browse, 1998). This implies that *CBF1*-regulated genes are not required for the constitutive freezing tolerance observed in *esk1* mutant. Interestingly, cold acclimation of *esk1* produces plants that are more than 2°C more freezing tolerant than fully-acclimated wild type, suggesting that the *esk1* mutation may hyper-activate some aspect(s) of freezing tolerance.

Since the *cft* mutants are freezing tolerant when grown at warm temperatures, genes that are constitutively activated in these mutants are likely to contribute to the observed constitutive freezing tolerance. Genes which are merely responsive to low temperature but play no role in freezing tolerance should be at levels similar to wild-type plants in the *cft* mutants at warm temperature, unless they are regulated by the corresponding *cft* locus. Thus, by comparison of the gene expression profile of non-acclimated *cft* mutants with acclimated and non-acclimated wild-type plants, it should be possible to identify genes that are both induced by cold acclimation in wild-type plants and are constitutively activated in one or more non-acclimated *cft* mutants. These genes can be tested in reverse genetics or over-expression experiments to determine their importance to freezing tolerance, and thus provide a strategy to separate genes that are essential to freezing tolerance from those that are merely responsive to low temperature exposure. The biochemical and molecular basis for the constitutive freezing tolerance is unknown in any of the *cft* mutants. Molecular cloning of the genes defined by the *cft* mutations and extensive analysis of the biochemical and molecular basis of the freezing tolerance in each mutant will likely result in new insights into the mechanisms of plant acclimation to freezing temperatures.

7. Emerging signaling processes regulating cold acclimation

We will not understand cold acclimation and freezing tolerance until we have a complete picture of the signaling processes involved. Tremendous progress has been made in recent years in our understanding of the signaling pathways involved in cold acclimation, from perception of low non-freezing temperatures to activation of genes and increased freezing tolerance. It has become increasingly clear that plants employ multiple interacting pathways to regulate cold acclimation. The traditional linear signaling pathways we have been studying constitute only a part of a complex signaling network (Knight and Knight, 2001).

7.1. Abscisic acid (ABA) as a plant hormone mediating cold acclimation

ABA has also been shown to mediate the development of freezing tolerance. Four lines of evidence suggest that ABA may play a central role in the signal transduction of cold acclimation:

1. ABA treatment at normal growth temperatures can increase freezing tolerance of a wide range of plants including *Arabidopsis* (Lang et al., 1994);

2. endogenous ABA levels increases in certain plants in response to low temperatures (Chen et al., 1983);

3. ABA-deficient mutants are impaired in developing freezing tolerance during cold acclimation; however, the freezing tolerance can be restored to the wild-type level by adding ABA into the culture medium (Gilmour and Thomashow, 1991; Heino et al., 1990);

4. ABA treatment can induce all of the *COR* genes (Gilmour and Thomashow, 1991; Heino et al., 1990).

In contrast, all of the ABA-insensitive mutants examined so far can cold acclimate to the same degree as wild-type *Arabidopsis* (Gilmour and Thomashow, 1991). It is unlikely that the ability of all these mutants to acclimate is due to "leakiness" of the mutations, as they all show insensitivity to ABA during germination. One possible explanation for these observations is that cold acclimation is mediated by a different ABA receptor or pathway from that regulating germination. *COR* genes can be also induced by low temperature in ABA-deficient and ABA-insensitive mutants (Gilmour and Thomashow, 1991; Nordin et al., 1991), indicating that there are low-temperature signaling pathways independent of ABA. Indeed, the CRT/DRE elements present in the promoter regions of *COR* genes are not responsive to ABA treatment in promoter analyses (Yamaguchi-Shinozaki and Shinozaki, 1994). Rather, a different DNA element, ABRE, is responsible for the ABA inducibility of these *COR* genes.

7.2. Involvement of calcium in temperature sensing

Several lines of evidence indicate that changes in cellular calcium levels may be involved in temperature sensing. A transient increase in cytosolic calcium occurs almost immediately when plants are exposed to low temperatures (Knight et al., 1996; Monroy and Dhindsa, 1995; Sheen, 1996), implying that calcium may be involved in the transduction of the low temperature signal. In addition, the temperature that induces a calcium influx coincides with the temperature that induces cold acclimation (Monroy and Dhindsa, 1995). Furthermore, calcium channel blockers or chelators inhibit the expression of *COR* (*CAS*) genes at low temperature while a calcium ionophore causes calcium influx and induces the expression of *COR* (*CAS*) genes in warm temperatures (Knight et al., 1996; Monroy and Dhindsa, 1995). Other cellular processes may be required to relay the low temperature signal since similar calcium increases are also observed in chilling sensitive plants which are injured by exposure to low temperature (Knight et al., 1996; Sheen, 1996).

7.3. Protein kinases and phosphatases

There is increasing evidence to indicate that protein kinases and phosphatases are also involved in transduction of low temperature signals during cold acclimation in plants. Several protein kinases responsive to low temperature have been isolated from higher plants based on sequence homology or cross-reaction of antibodies (Anderberg and Walker-Simmons, 1992; Jonak et al., 1996; Mizoguchi et al., 1996; Tahtiharju and Palva, 2001; Tahtiharju et al., 1997). Using specific inhibitors, Monroy et al. (1998) demonstrated that a protein kinase inhibitor, staurosporine, could prevent induction of *CAS15* by low temperature whereas a protein phosphatase inhibitor, okadaic acid, could induce expression of *CAS15* at 25°C. These researchers further identified a protein phosphatase 2A that may serve as an early target for cold-inactivation. These experiments demonstrate that plants may use several components similar to those associated with other signaling pathways to control cold acclimation.

8. Molecular and genetic analysis of signaling transduction of stress responses

Forward genetic approaches have proved to be particularly useful in dissecting the signaling network of stress responses in plants. *COR78* (also known as *RD29A*) is one of the many genes induced to high levels in response to cold, drought, salt, and ABA (Horvath et al., 1993; Yamaguchi-Shinozaki and Shinozaki, 1994). Ishitani et al. (Ishitani et al., 1997) have engineered a transgenic *Arabidopsis* with the promoter of the *COR 78* (*RD29*) gene fused to a firefly luciferase reporter gene. After mutagenesis of the transgenic plants, mutants with aberrant expression of the reporter gene were isolated. Since the promoter of *COR78* has cold-, drought-, and ABA-responsive elements, this approach allows simultaneous dissection of the signal pathways for cold, drought, and ABA. Hundreds of mutants with altered luciferase activity have been identified. These mutants fall into three major categories: the *cos* mutants which show constitutive expression of osmotically responsive genes; the *los* mutants which show loss of expression of these genes; and *hos* mutants which show hyper-expression after induction by drought, cold or ABA. These mutants were further grouped into 14 classes according to the defect in their responses to one or a combination of stresses. Phenotypic analyses of these mutants suggest that multiple signaling pathways cross-talk and converge to activate the *COR78* gene. Several of these mutants have been characterized in detail, including *hos1* (Ishitani et al., 1998; Lee et al., 2001), *hos2* (Lee et al., 1999), *hos5* (Xiong et al., 1999), *fry1* (also a *hos* mutant) (Xiong et al., 2001b), and *los5* (Xiong et al., 2001a). In addition to the expression of *COR* genes, these mutants also display aberrance in freezing and osmotic stress tolerance. *FRY1* and *LOS5* genes have been identified by map-based cloning and *FRY1* encodes an inositol polyphosphate 1-phosphatase. Consistent with its function, *fry1* mutant plants accumulate more inositol-1,4,5-trisphosphate than wild-type plants in response to ABA treatment. The *los5* mutant is allelic to a previously known mutant *aba3* which is defective in the biosynthesis of ABA. *LOS5* encodes a molybdenum cofactor sulfurase required by the aldehyde oxidase that catalyzes the last step in the biosynthesis of ABA. Further characterization and cloning of the remaining mutants will yield valuable information about the stress signaling network in plant. It is worth noting that all the *hos* mutants characterized so far are either compromised in their inherent freezing tolerance or defective in developing freezing tolerance upon cold acclimation (Ishitani et al., 1998; Lee et al., 1999; Xiong et al., 1999; Xiong et al., 2001b). This is intriguing because these mutants accumulate higher levels of all *COR* gene transcripts than wild-type plants upon exposure to low non-freezing temperatures.

9. Multiple signal pathways mediate cold acclimation

Many lines of evidence indicate that parallel or branched signaling pathways activate distinct suites of cold acclimation responses. For example, studies of *COR* gene expression indicate that both

ABA-dependent and ABA-independent pathways are involved (Gilmour and Thomashow, 1991; Ishitani et al., 1997; Nordin et al., 1993), while the *esk1* mutant of *Arabidopsis* revealed that considerable freezing tolerance could be achieved in the absence of *COR* gene expression (Xin and Browse, 1998). Support for a model with multiple pathways also comes from the analysis of *sfr* mutants that are not able to fully acclimate (Knight et al., 1999; Warren et al., 1996). Eight *sfr* mutants representing seven complementation groups have been isolated in *Arabidopsis* and all are compromised in their ability to develop full levels of freezing tolerance during cold acclimation. None of these mutants have completely lost their ability to increase freezing tolerance during cold acclimation; most of them retain over 50% of the wild-type capacity to cold acclimate (Warren et al., 1996). The simplest explanation for this observation is that individual *sfr* mutations each block a single signaling pathway, so that each mutant is still able to partially cold acclimate through signaling pathways that are not disrupted in the mutant plant. Map-based cloning of the *sfr* loci and genes corresponding to constitutively freezing tolerant mutants such as *esk1* (Xin and Browse, 1998) will undoubtedly contribute to understanding the complexities of cold acclimation signaling.

Of course, the number of pathways an investigator sees depends on what level in the signaling cascade he or she is observing. The *esk1* mutant contains elevated levels of proline and soluble sugars that are believed to contribute to the constitutive freezing tolerance of this mutant (Xin and Browse, 1998). Plants over-expressing *DREB1A/CBF3* contain increased proline and soluble sugars as well as exhibiting constitutive expression of the *COR* genes (Gilmour et al., 2000). The expectation then is that *DREB1A/CBF3* acts in a common pathway that leads into both the *COR* pathway and the *ESK1* pathway. Based on these results, it is proposed that *CBF3* acts as a master regulator that integrates multiple pathways working in concert to enhance freezing tolerance. However, for all the *CBF3* over-expressing lines, freezing tolerance can be significantly further enhanced following cold acclimation at 5°C for seven days (Gilmour et al., 2000). The best *CBF3*-over-expressing line, A30, reaches 50% ion leakage when the plants are frozen to –9°C. After cold acclimation, the plant never reaches 50% ion leakage even after freezing to –16°C. A simple alternative explanation for these observations is that *DREB1A/CBF3* activates a subset of the total cold acclimation responses and that a very high degree of freezing tolerance can be induced because the *DREB1A/CBF3*-responsive components are hyperactivated in the transgenic plants at warm temperature. The further increase in freezing tolerance during acclimation can be brought about by other signaling pathways. It is still a matter of debate whether cold acclimation is controlled by a single temperature sensor and a trunk pathway on which *DREB1A/CBF3* acts as a master switch or by multiple temperature sensors and parallel pathways.

10. Hunting for the temperature sensors

A central issue in considering single *versus* multiple signaling pathways for cold acclimation is the question of whether there is one temperature sensor in plant cells or several sensors, which may respond to the same or different physical manifestations of a drop in temperature? Experiments from *Synechocystis* PCC6803 indicate that cyanobacteria employ two or more temperature sensors to monitor the changes in the ambient temperatures (Suzuki et al., 2000). Up to now, the nature and identity of the temperature sensors in higher plants is not known.

10.1. Changes in membrane fluidity may serve as the basis for temperature sensing in higher plants

Recently, it was proposed that the reduction in membrane fluidity occurring during cold acclimation as a mechanism for temperature sensing. Orvar et al. (2000) employed pharmacological agents to artificially alter the membrane fluidity to study their effect on *CAS30* gene expression and freezing tolerance in alfalfa cell cultures. When DMSO was used to reduce membrane fluidity of

cells at 25°C, there was a modest increase in *CAS30* transcript levels and in the freezing tolerance of cells. Conversely, pretreatment of cells with benzyl alcohol (a reagent that increases membrane fluidity) before cold acclimation at 4°C reduced the expression of *CAS30* and the ability of cells to fully cold acclimate. In this model membrane fluidity leads to cold acclimation through changes in cytoskeleton organization and the induction of Ca^{2+} influx into the cells. Other evidence also indicates that Ca^{2+} fluxes are an important early step in low temperature signaling (Kiegle et al., 2000; Plieth et al., 1999). This model of cold acclimation is also supported by *in situ* assays using transgenic *Brassica napus* carrying a β-glucuronidase (GUS) gene under the control of promoter from *BN115*, an ortholog of the *Arabidopsis COR15* gene (Sangwan et al., 2001). Both GUS activity and accumulation of *BN115* transcript are activated at 25°C by the membrane rigidifier, DMSO, the microfilament destabilizer, latrunculin B, and the calcium ionophore, A23187. The induction of *BN115* at 25°C is associated with increases in freezing tolerance. Consistent with this model, a cold regulated gene, *Wcor719*, has been show to encode an actin-depolymerization factor, TaADF (Ouellet et al., 2001). Accumulation of TaADF protein correlates closely with freezing tolerance during cold acclimation and across cultivars that vary in their capacity to develop freezing tolerance. However, accumulation of TaADF is detected only after two days of cold acclimation, while microfilament depolymerization is a much earlier event (Orvar et al., 2000). These experiments suggest that membrane fluidity and microfilament depolymerization play an important role in temperature sensing; however, it is a great challenge to establish mechanistically how the low temperature signal is actually transmitted.

10.2. Histidine kinases serve as one type of temperature sensors in Synechocystis

Fortunately, a far better-developed model for temperature signaling has emerged from the work of Norio Murata's group on the cyanobacterium *Synechocystis* PCC6803. This model provides a useful basis for discussing concepts for temperature sensors in the more complex system found in plants. When *Synechocystis* is shifted from a growth temperature of 34°C to 22°C, transcripts of three fatty acid desaturase genes, *desA* (Δ12), *desB* (ω3), and *desD* (Δ6), are induced approximately ten-fold while a fourth desaturase gene, *desC* (Δ9), remains constant (Los et al., 1997). The increased expression of fatty acid desaturase genes is believed to be an adaptive response of cyanobacterium to low temperature in order to modulate the degree of phospholipid desaturation, and ultimately membrane fluidity. Several years ago, Vigh and colleagues showed that transcription of *desA* can be induced at 34°C by reducing membrane fluidity through Palladium-catalyzed hydrogenation of fatty acids in the plasma membrane of *Synechocystis* PCC6803 (Vigh et al., 1993).

In hunting for temperature sensors, Murata's group focused on the possibility that the sensors would be two-component regulators and took advantage of the *Synechocystis* system for its convenience in targeted gene replacement (Suzuki et al., 2000). The genome of *Synechocystis* has been completely sequenced and is shown to contain 43 putative histidine kinase genes (Mizuno et al., 1996). To identify which could be involved in temperature sensing, the histidine kinases were systematically inactivated by targeted insertional mutagenesis (Suzuki et al., 2000). To monitor the changes in the induction of the *desB* gene in response to low temperature in the knockout mutants, Suzuki et al. used a reporter construct containing the bacterial *luxAB* gene fused to the *desB* promoter. In these experiments, two histidine kinases, *Hik33* and *Hik19*, were identified as necessary for the low-temperature induction of *desB*. *Synechocystis* is multiploid, containing at least ten copies of the chromosome sets per cell so that isolation of knockout mutants required extended selection for the antibiotic marker of the insertion cassette. After six months in culture on 20 μg/ml spectinomycin, PCR analysis revealed that wild-type alleles of *Hik33* and *Hik19* were still present at low levels in the mutants. However, levels of the corresponding proteins were apparently reduced substantially so that low-temperature

signaling was blocked. To identify other components of the temperature-sensing pathway, Suzuki et al. (2000) performed random insertional mutagenesis using a spectinomycin-resistance cassette. From approximately 20,000 spectinomycin-resistant lines, 18 mutants that exhibit reduced luciferase activity in response to a down-shift in temperature were isolated. One of the genes identified in this way, *Rer1*, encodes a putative protein with domains typical of a response element in a two-component regulator. The working-model presented by Suzuki and colleagues is that *Hik33* is activated by reduced membrane fluidity allowing autophosphorylation of *Hik33* and subsequent transfer of a phosphate group to *Hik19* and finally *Rer1* (Fig. 9.2). Traditional approaches (Los and Murata, 1999) and now microarray analysis using the *Hik*⁻ mutants (Suzuki et al., 2001) demonstrate that many genes besides the desaturases are regulated by this thermosensor system. Many of these genes encode proteins involved in RNA stability and translation efficiency in the cyanobacterium.

Further experiments with the *Synechocystis* mutants provide evidence for at least one additional temperature sensor in this cyanobacterium. Mutations in *Hik33* and *Hik19* genes greatly reduced (but did not eliminate) the cold induction

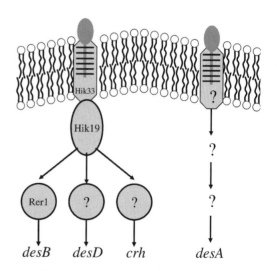

Fig. 9.2. Molecular and mutational analysis of temperature signaling in *Synechocystis* PCC6803 demonstrates the existence of at least two temperature sensors in this prokaryote. See text for details. Redrawn from Suzuki et al. (2000).

of *desB* and *desD* genes, reduced the cold induction of *Crh* (a cold-inducible RNA helicase), but had no effect on the cold induction of the *desA* gene. Thus, a separate sensor apparently operates to regulate the cold induction of the *desA* gene and to partially control *desB*, *desD* and *Crh* expression (Fig. 9.2). It is also worth noting that the response regulator, *Rer1*, which functions down-stream of *Hik33* and *Hik19*, only regulates the cold induction of *desB* and not *desD* or *Crh*, which presumably are regulated by additional response elements (Suzuki et al., 2000). A genome-wide microarray analysis (Suzuki et al., 2001) identified 70 genes in *Synechocystis* that are induced (24 genes) or repressed (46 genes) by a temperature shift from 34°C to 22°C. Transcriptional changes for 14 genes were eliminated in the ΔHik33 mutant, changes for 35 genes were reduced, while 21 genes showed the same level of induction or repression in the ΔHik33 mutant as in wild-type cells. These results further support the model for two temperature sensors (Fig. 9.2). Apparently each sensor regulates a unique set of genes and share responsibility for the regulation of a third set of genes.

Of course, the temperature-sensing process described for *Synechocystis* involves a temperature shift from 34°C to 22°C while plants undergoing cold acclimation are responding to temperatures below 10°C. Nevertheless, the evidence for (at least) two independent temperature sensors in this prokaryote establishes a *prima facie* case for considering the possibility of multiple sensors in higher plants with their more complex cellular organization and differentiated tissues and organs.

10.3. Search for temperature sensors in higher plants

Two-component regulators are present in plants and it is likely that such a system is the basis of an osmotic sensor in *Arabidopsis* (Urao et al., 1999; Urao et al., 2000). However, the *Hik33* and *Hik19* genes do not have recognizable homologs in higher plants, although putative homologs of *Hik33* have been identified in the chloroplast genomes of *Porphyra purpurea* and *Cyanidium caldarium* (Suzuki et al., 2000). For this reason, it is not useful

to speculate on whether higher plants employ histidine kinases or two-component response regulators as temperature sensors and signal transducers. Equally plausible arguments could be made for fluidity affecting Ca^{2+} channels in the membrane (Orvar et al., 2000) and inducing Ca^{2+} fluxes that activate a MAP kinase (or other) signaling cascade(s) (Monroy et al., 1998; Plieth et al., 1999). It is worth noting, however, that of the approximately 1,000 protein kinases identified in the *Arabidopsis* genome, only about a dozen are considered to be potential histidine kinases of two-component systems (The Arabidopsis Genome Initiative, 2000). Projects to systematically characterize whole families of *Arabidopsis* genes (e.g., the *Arabidopsis* 2010 Project) will soon make it possible to test knockout mutants corresponding to candidate genes in cold-acclimation signaling.

11. Conclusions and perspectives

Research on cold acclimation has reached an exciting stage. It has been established that cold acclimation is associated with complex changes in metabolism and gene expression. The *DREB1/ CBF* transcription factors have been shown to be major regulators of increased freezing tolerance in *Arabidopsis* and other plant species. Many molecular and genetic tools are now available that promise even greater discoveries in cold acclimation. For example, the completion of the *Arabidopsis* genome sequence allows for the construction of microarray chips containing sequences representing the complete set of genes from this plant. Thus, it will soon be possible to examine the regulation of every cold responsive gene during cold acclimation. The effort to construct T-DNA and transposon knockout lines that saturate the *Arabidopsis* genome will make it possible to establish the relative importance of every cold responsive gene to the establishment of freezing tolerance. At the same time, facile map-based cloning of the *cft, sfr, hos, los, cos* loci, using information from the *Arabidopsis* genome sequence, will allow characterization of important steps in the signaling cascade and uncover novel mechanisms of freezing tolerance. The most compelling issue is the identification of the sensors of low non-freezing temperature in higher plants. The molecular and mutational analysis of the two-component temperature sensor in *Synechocystis* points towards new and exciting prospects for understanding the initiation and execution of cold acclimation in plants.

Acknowledgements

I would like to thank Dr. Jeff Velten, Dr. Mel Oliver, Dr. John Burke, and Dr. Hong Zhang for their suggestions to improve the manuscript.

Disclaimer

Mention of trade names or commercial products in this article is solely for the purpose of providing specific information and does not imply recommendation or endorsement by the U.S. Department of Agriculture.

References

Anderberg, R.J. and Walker-Simmons, M.K. (1992). Isolation of a wheat cDNA clone for an abscisic acid-inducible transcript with homology to protein kinases. Proc. Natl. Acad. Sci. USA 89, 10183–10187.

Anderson, J.V., Haskell, D.W. and Guy, C.L. (1994). Differential influence of ATP on native spinach 70-kilodalton heat-shock cognates. Plant Physiol. 104, 1371–1380.

Artus, N.N., Uemura, M., Steponkus, P.L., Gilmour, S.J., Lin, C. and Thomashow, M.F. (1996). Constitutive expression of the cold-regulated *Arabidopsis thaliana* *COR15a* gene affects both chloroplast and protoplast freezing tolerance. Proc. Natl. Acad. Sci. USA 93, 13404–13409.

Carpenter, J.F., Hand, S.C., Crowe, L.M. and Crowe, J.H. (1986). Cryoprotection of phosphofructokinase with organic solutes: characterization of enhanced protection in the presence divalent cations. Arch. Biochem. Biophys. 250, 505–512.

Chen, H.-H., Brenner, M.L. and Li, P.H. (1983). Involvement of abscisic acid in potato cold acclimation. Plant Physiol. 71, 362–365.

Danyluk, J., Carpentier, E. and Sarhan, F. (1996). Identification and characterization of a low temperature regulated gene encoding an actin-binding protein from wheat. FEBS Lett. 389, 324–327.

Delauney, A.J. and Verma, D.P.S. (1993). Proline biosynthesis and osmoregulation in plants. Plant J. 4, 215–223.

Dörffling, K., Dörffling, H., Lesselich, G., Luck, E., Zimmermann, C., Melz, G. and Jürgens, H.U. (1997). Heritable improvement of frost tolerance in winter wheat by in-vitro-selection of hydroxyproline-resistant proline overproducing mutants. Plant Mol. Biol. 23, 221–225.

Dunn, M.A., Morris, A., Jack, P.L. and Hughes, M.A. (1993). A low-temperature-responsive translation elongation factor 1 alpha from barley (*Hordeum vulgare* L.). Plant Mol. Biol. 23, 221–225.

Dunn, M.A., White, A.J., Vural, S. and Hughes, M.A. (1998). Identification of promoter elements in a low-temperature-responsive gene (blt4.9) from barley (*Hordeum vulgare* L.). Plant Mol Biol 38, 551–564.

Gibson, S., Arondel, V., Iba, K. and Somerville, C. (1994). Cloning of a temperature-regulated gene encoding a chloroplast *omega*-3 desaturase from *Arabidopsis thaliana*. Plant Physiol. 106, 1615–1621.

Gilmour, S.J., Hajela, R.K. and Thomashow, M.F. (1988). Cold acclimation in *Arabidopsis thaliana*. Plant Physiol. 87, 745–750.

Gilmour, S.J., Sebolt, A.M., Salazar, M.P., Everard, J.D. and Thomashow, M.F. (2000). Overexpression of the *Arabidopsis* CBF3 transcriptional activator mimics multiple biochemical changes associated with cold acclimation. Plant Physiol. 124, 1854–1865.

Gilmour, S.J. and Thomashow, M.F. (1991). Cold acclimation and cold-regulated gene expression in ABA mutants of *Arabidopsis thaliana*. Plant Mol. Biol. 17, 1233–1240.

Guy, C.L. (1990). Cold acclimation and freezing stress tolerance: role of protein metabolism. Annu. Rev. Plant Physiol. Plant Mol. Biol. 41, 187–223.

Hayashi, H., Alia, Mustardy, L., Deshnium, P., Ida, M. and Murata, N. (1997). Transformation of *Arabidopsis thaliana* with the codA gene for choline oxidase; accumulation of glycinebetaine and enhanced tolerance to salt and cold stress. Plant J. 12, 133–142.

Heino, P., Sandman, G., Lang, V., Nordin, K. and Palva, E.T. (1990). Abscisic acid deficiency prevents development of freezing tolerance in *Arabidopsis thaliana* (L.) Heynh. Theor. Appl. Genet. 79, 801–806.

Hincha, D.K., Sonnewald, U., Willmitzer, L. and Schmitt, J.M. (1996). The role of sugar accumulation in leaf frost hardiness: Investigations with transgenic tobacco expressing a bacterial pyrophosphatase or a yeast invertase gene. J. Plant Physiol. 147, 604–610.

Horvath, D.P., McLarney, B.K. and Thomashow, M.F. (1993). Regulation of *Arabidopsis thaliana* L. (Heyn) cor78 in response to low temperature. Plant Physiol. 103, 1047–1053.

Initiative, T.A.G. (2000). Analysis of the genome sequence of the flowering plant *Arabidopsis thaliana*. Nature 408, 796–815.

Ishitani, M., Xiong, L., Lee, H., Stevenson, B. and Zhu, J.K. (1998). HOS1, a genetic locus involved in cold-responsive gene expression in *Arabidopsis*. Plant Cell 10, 1151–1162.

Ishitani, M., Xiong, L., Stevenson, B. and Zhu, J.K. (1997). Genetic analysis of osmotic and cold stress signal transduction in *Arabidopsis*: interactions and convergence of abscisic acid-dependent and abscisic acid-independent pathways. Plant Cell 9, 1935–1949.

Jaglo, K.R., Kleff, S., Amundsen, K.L., Zhang, X., Haake, V., Zhang, J.Z., Deits, T. and Thomashow, M.F. (2001). Components of the *Arabidopsis* C-repeat/dehydration-responsive element binding factor cold-response pathway are conserved in *Brassica napus* and other plant species. Plant Physiol. 127, 910–917.

Jaglo-Ottosen, K.R., Gilmour, S.J., Zarka, D.G., Schabenberger, O. and Thomashow, M.F. (1998). *Arabidopsis* CBF1 overexpression induces COR genes and enhances freezing tolerance. Science 280, 104–106.

Jarillo, J.A., Leyva, A., Salinas, J. and Martinez Zapater, J.M. (1993). Low temperature induces the accumulation of alcohol dehydrogenase mRNA in *Arabidopsis thaliana*, a chilling-tolerant plant. Plant Physiol. 101, 833–837.

Jonak, C., Kiegerl, S., Ligterink, W., Barker, P.J., Huskisson, N.S. and Hirt, H. (1996). Stress signaling in plants: a mitogen-activated protein kinase pathway is activated by cold and drought. Proc. Natl. Acad. Sci. USA 93, 11274–11129.

Kasuga, M., Liu, Q., Miura, S., Yamaguchi-Shinozaki, K. and Shinozaki, K. (1999). Improving plant drought, salt, and freezing tolerance by gene transfer of a single stress-inducible transcription factor. Nature Biotechnol. 17, 287–291.

Kaye, C., Neven, L., Hofig, A., Li, Q.B., Haskell, D. and Guy, C. (1998). Characterization of a gene for spinach CAP160 and expression of two spinach cold-acclimation proteins in tobacco. Plant Physiol. 116, 1367–1377.

Kiegle, E., Moore, C.A., Haseloff, J., Tester, M.A. and Knight, M.R. (2000). Cell-type-specific calcium responses to drought, salt and cold in the *Arabidopsis* root. Plant J. 23, 267–278.

Kishitani, S., Watanabe, K., Yasuda, S., Arakawa, K. and Takabe, T. (1994). Accumulation of glycinebetaine during cold acclimation and freezing tolerance in leaves of winter and spring barley plants. Plant Cell Environ. 17, 89–95.

Knight, H. and Knight, M.R. (2001). Abiotic stress signalling pathways: specificity and cross-talk. Trends Plant Sci 6, 262–267.

Knight, H., Trewavas, A.J. and Knight, M.R. (1996). Cold calcium signaling in *Arabidopsis* involves two cellular pools and a change in calcium signature after acclimation. Plant Cell 8, 489–503.

Knight, H., Veale, E.L., Warren, G.J. and Knight, M.R. (1999). The sfr6 mutation in *Arabidopsis* suppresses low-temperature induction of genes dependent on the CRT/DRE sequence motif. Plant Cell 11, 875–886.

Koster, K.L. and Lynch, D.V. (1992). Solute accumulation and compartmentation during the cold acclimation of Puma rye. Plant Physiol. 98, 108–113.

Kurkela, S. and Franck, M. (1990). Cloning and characterization of a cold- and ABA-inducible *Arabidopsis* gene. Plant Mol. Biol. 15, 137–144.

Lang, V., Mantyla, E., Welin, B., Sundberg, B. and Palva, E.T. (1994). Alterations in water status, endogenous abscisic acid content, and expression of rab18 gene during the development of freezing tolerance in *Arabidopsis thaliana*. Plant Physiol. 104, 1341–1349.

Lee, H., Xiong, L., Gong, Z., Ishitani, M., Stevenson, B. and Zhu, J.K. (2001). The *Arabidopsis* HOS1 gene negatively regulates cold signal transduction and encodes a RING finger protein that displays cold-regulated nucleo-cytoplasmic partitioning. Genes Dev. 15, 912–924.

Lee, H., Xiong, L., Ishitani, M., Stevenson, B. and Zhu, J.K. (1999). Cold-regulated gene expression and freezing tolerance in an *Arabidopsis thaliana* mutant. Plant J. 17, 301–308.

Levitt, J. (1980) Responses of Plants to Environmental Stresses. Academic Press, Orlando, Florida.

Leyva, A., Jarillo, J.A., Salinas, J. and Martinez Zapater, J.M. (1995). Low temperature induces the accumulation of phenylalanine ammonia-lyase and chalcone synthase mRNAs of *Arabidopsis thaliana* in a light-dependent Manner. Plant Physiol. 108, 39–46.

Lineberger, R.D. and Steponkus, P.L. (1980). Cryoprotection by glucose, sucrose, and raffinose to chloroplast thylakoids. Plant Physiol. 65, 298–304.

Liu, Q., Kasuga, M., Sakuma, Y., Abe, H., Miura, S., Yamaguchi-Shinozaki, K. and Shinozaki, K. (1998). Two transcription factors, DREB1 and DREB2, with an EREBP/AP2 DNA binding domain separate two cellular signal transduction pathways in drought- and low-temperature-responsive gene expression, respectively, in *Arabidopsis*. Plant Cell 10, 1391–1406.

Los, D.A. and Murata, N. (1999). Responses to cold shock in cyanobacteria. J. Mol. Microbiol. Biotechnol. 1, 221–230.

Los, D.A., Ray, M.K. and Murata, N. (1997). Differences in the control of the temperature-dependent expression of four genes for desaturases in *Synechocystis* sp. PCC 6803. Mol. Microbiol. 25, 1167–1175.

Lynch, D.V. and Steponkus, P.L. (1987). Plasma membrane lipid alterations associated with cold acclimation of winter rye seedlings (*Secale cereale* L. cv Puma). Plant Physiol. 83, 761–767.

McKown, R., Kuroki, G. and Warren, G. (1996). Cold responses of *Arabidopsis* mutants impaired in freezing tolerance. J. Exp. Bot. 47, 1919–1925.

Mizoguchi, T., Irie, K., Hirayama, T., Hayashida, N., Yamaguchi-Shinozaki, K., Matsumoto, K. and Shinozaki, K. (1996). A gene encoding a mitogen-activated protein kinase kinase kinase is induced simultaneously with genes for a mitogen-activated protein kinase and an S6 ribosomal protein kinase by touch, cold, and water stress in *Arabidopsis thaliana*. Proc. Natl. Acad. Sci. USA 93, 765–769.

Mizuno, T., Kaneko, T. and Tabata, S. (1996). Compilation of all genes encoding bacterial two-component signal transducers in the genome of the cyanobacterium, *Synechocystis* sp. strain PCC 6803. DNA Res. 3, 407–414.

Monroy, A.F. and Dhindsa, R.S. (1995). Low-temperature signal transduction: induction of cold acclimation-specific genes of alfalfa by calcium at 25°C. Plant Cell 7, 321–331.

Monroy, A.F., Sangwan, V. and Dhindsa, R.S. (1998). Low temperature signal transduction during cold acclimation: protein phosphatase 2A as an early target for cold-inactivation. Plant J. 13, 653–660.

Murelli, C., Rizza, F., Albini, F.M., Dulio, A., Terzi, V. and Cattivelli, L. (1995). Metabolic changes associated with cold-acclimation in contrasting cultivars of barley. Physiol. Plant. 94, 87–93.

Nomura, M., Muramoto, Y., Yasuda, S., Takabe, T. and Kishitani, S. (1995). The accumulation of glycine-betaine during cold acclimation in early and late cultivars of barley. Euphytica 83, 247–250.

Nordin, K., Heino, P. and Palva, E.T. (1991). Separate signal pathways regulate the expression of a low-temperature-induced gene in *Arabidopsis thaliana* (L.) Heynh. Plant Mol. Biol. 16, 1061–1071.

Nordin, K., Vahala, T. and Palva, E.T. (1993). Differential expression of two related, low-temperature-induced genes in *Arabidopsis thaliana* (L.) Heynh. Plant Mol. Biol. 21, 641–653.

Orvar, B.L., Sangwan, V., Omann, F. and Dhindsa, R.S. (2000). Early steps in cold sensing by plant cells: the role of actin cytoskeleton and membrane fluidity. Plant J. 23, 785–794.

Ouellet, F., Carpentier, E., Cope, M.J., Monroy, A.F. and Sarhan, F. (2001). Regulation of a wheat actin-depolymerizing factor during cold acclimation. Plant Physiol. 125, 360–368.

Phillips, J.R., Dunn, M.A. and Hughes, M.A. (1997). mRNA stability and localisation of the low-temperature-responsive barley gene family blt14. Plant Mol. Biol. 33, 1013–1023.

Plieth, C., Hansen, U.P., Knight, H. and Knight, M.R. (1999). Temperature sensing by plants: the primary characteristics of signal perception and calcium response. Plant J. 18, 491–497.

Prasad, T.K., Anderson, M.D., Martin, B.A. and Stewart, C.R. (1994). Evidence for chilling-induced oxidative stress in maize seedlings and a regulatory role for hydro-gen peroxide. Plant Cell 6, 65–74.

Ristic, Z. and Ashworth, E.N. (1993). Changes in leaf ultra-structure and carbohydrates in *Arabidopsis thaliana* L. (Heyn) cv. Columbia during rapid cold acclimation. Protoplasma 172, 111–123.

Sakai, A. (1960). Survival of the twigs of woody plants at −196°C. Nature 185, 393–394.

Sakamoto, A., Valverde, R., Alia, Chen, T.H. and Murata, N. (2000). Transformation of *Arabidopsis* with the codA gene for choline oxidase enhances freezing tolerance of plants. Plant J. 22, 449–453.

Sangwan, V., Foulds, I., Singh, J. and Dhindsa, R.S. (2001). Cold-activation of *Brassica napus* BN115 promoter is mediated by structural changes in membranes and cyto-skeleton, and requires Ca^{2+} influx. Plant J. 27, 1–12.

Shaffer, M.A. and Fischer, R.L. (1988). Analysis of mRNAs that accumulate in response to low temperature identifies a thiol protease gene in tomato. Plant Physiol. 87, 431–436.

Sheen, J. (1996). Ca^{2+}-dependent protein kinases and stress signal transduction in plants. Science 274, 1900–1902.

Shinozaki, K. and Yamaguchi-Shinozaki, K. (2000). Molecular responses to dehydration and low temperature: differences and cross-talk between two stress signaling pathways. Curr. Opinion Plant Biol. 3, 217–223.

Steponkus, P.L. (1984). Role of the plasma membrane in freezing injury and cold acclimation. Annu. Rev. Plant Physiol. 35, 543–584.

Steponkus, P.L., Uemura, M., Balsamo, R.A. and Arvinte, T.L.D.V. (1988). Transformation of the cryobehavior of rye protoplasts by modification of the plasma membrane lipid composition. Proc. Natl. Acad. Sci. USA 85, 9026–9030.

Steponkus, P.L., Uemura, M., Joseph, R.A., Gilmour, S.J. and Thomashow, M.F. (1998). Mode of action of the COR15a gene on the freezing tolerance of *Arabidopsis thaliana*. Proc. Natl. Acad. Sci. USA 95, 14570–14575.

Stockinger, E.J., Gilmour, S.J. and Thomashow, M.F. (1997). *Arabidopsis thaliana* CBF1 encodes an AP2 domain-containing transcriptional activator that binds to the C-repeat/DRE, a cis-acting DNA regulatory element that stimulates transcription in response to low temperature and water deficit. Proc. Natl. Acad. Sci. USA 94, 1035–1040.

Suzuki, I., Kanesaki, Y., Mikami, K., Kanehisa, M. and Murata, N. (2001). Cold-regulated genes under control of the cold sensor Hik33 in *Synechocystis*. Mol. Microbiol. 40, 235–244.

Suzuki, I., Los, D.A., Kanesaki, Y., Mikami, K., and Murata, N. (2000). The pathway for perception and transduction of low-temperature signals in *Synechocystis*. EMBO J. 19, 1327–1334.

Tahtiharju, S. and Palva, E.T. (2001). Antisense inhibition of protein phosphatase 2C accelerates cold acclimation in *Arabidopsis thaliana*. Plant J. 26, 461–470.

Tahtiharju, S., Sangwan, V., Monroy, A.F., Dhindsa, R.S. and Borg, M. (1997). The induction of kin genes in cold-acclimating *Arabidopsis thaliana*. Evidence of a role for calcium. Planta 203, 442–447.

Tao, D.L., Oquist, G. and Wingsle, G. (1998). Active oxygen scavengers during cold acclimation of Scots pine seedlings in relation to freezing tolerance. Cryobiology 37, 38–45.

Thomashow, M.F. (1994) *Arabidopsis thaliana* as a model for studying mechanisms of plant cold tolerance. In: *Arabidopsis* (Meyerowitz, E.M. and Somerville, C.R., Eds), pp. 807–834. Cold Spring Harbor Laboratory Press, Cold Spring Harbor.

Thomashow, M.F. (1999). Plant cold acclimation: Freezing tolerance genes and regulatory mechanisms. Annu. Rev. Plant Physiol. Plant Mol. Biol. 50, 571–599.

Uemura, M., Joseph, R.A. and Steponkus, P.L. (1995). Cold acclimation of *Arabidopsis thaliana*. Effect on plasma membrane lipid composition and freeze-induced lesions. Plant Physiol. 109, 15–30.

Uemura, M. and Steponkus, P.L. (1994). A contrast of the plasma membrane lipid composition of oat and rye leaves in relation to freezing tolerance. Plant Physiol. 104, 479–496.

Ukaji, N., Kuwabara, C., Takezawa, D., Arakawa, K., Yoshida, S. and Fujikawa, S. (1999). Accumulation of small heat-shock protein homologs in the endoplasmic reticulum of cortical parenchyma cells in mulberry in association with seasonal cold acclimation. Plant Physiol. 120, 481–490.

Urao, T., Yakubov, B., Satoh, R., Yamaguchi-Shinozaki, K., Seki, M., Hirayama, T. and Shinozaki, K. (1999). A transmembrane hybrid-type histidine kinase in *Arabidopsis* functions as an osmosensor. Plant Cell 11, 1743–1754.

Urao, T., Yamaguchi-Shinozaki, K. and Shinozaki, K. (2000). Two-component systems in plant signal transduction. Trends Plant Sci. 5, 67–74.

Van Swaaij, A.C., Jacobsen, E., Kiel, J.A.K.W. and Feenstra, W.J. (1986). Selection, characterization and regeneration of hydroxyproline-resistant cell lines of *Solanum tuberosum*: tolerance to NaCl and freezing stress. Physiol. Plant. 68, 359–366.

Vigh, L., Los, D.A., Horvath, I. and Murata, N. (1993). The primary signal in the biological perception of temperature: Pd-catalyzed hydrogenation of membrane lipids stimulated the expression of the desA gene in *Synechocystis* PCC6803. Proc. Natl. Acad. Sci. USA 90, 9090–9094.

Wanner, L.A. and Junttila, O. (1999). Cold-induced freezing tolerance in *Arabidopsis*. Plant Physiol. 120, 391–400.

Warren, G., McKown, R., Marin, A.L. and Teutonico, R. (1996). Isolation of mutations affecting the development of freezing tolerance in *Arabidopsis thaliana* (L.) Heynh. Plant Physiol. 111, 1011–1019.

Webb, M.S., Uemura, M. and Steponkus, P.L. (1994). A comparison of freezing injury in oat and rye: two cereals at the extremes of freezing tolerance. Plant Physiol. 104, 467–478.

Wood, A.J., Saneoka, H., Rhodes, D., Joly, R.J. and Goldsbrough, P.B. (1996). Betaine aldehyde dehydrogenase in sorghum. Molecular cloning and expression of two related genes. Plant Physiol. 110, 1301–1308.

Xin, Z. and Browse, J. (1998). Eskimo1 mutants of *Arabidopsis* are constitutively freezing-tolerant. Proc. Natl. Acad. Sci. USA 95, 7799–7804.

Xin, Z. and Browse, J. (2000). Cold comfort farm: the acclimation of plants to freezing temperatures. Plant Cell Environ. 23, 893–902.

Xiong, L., Ishitani, M., Lee, H. and Zhu, J.K. (1999). HOS5—a negative regulator of osmotic stress-induced gene expression in *Arabidopsis thaliana*. Plant J. 19, 569–578.

Xiong, L., Ishitani, M., Lee, H. and Zhu, J.K. (2001a). The *Arabidopsis* los5/aba3 locus encodes a molybdenum cofactor sulfurase and modulates cold stress- and osmotic stress-responsive gene expression. Plant Cell 13, 2063–2083.

Xiong, L., Lee, B., Ishitani, M., Lee, H., Zhang, C. and Zhu, J.K. (2001b). FIERY1 encoding an inositol polyphosphate 1-phosphatase is a negative regulator of abscisic acid and stress signaling in *Arabidopsis*. Genes Dev 15, 1971–1984.

Yamaguchi-Shinozaki, K. and Shinozaki, K. (1994). A novel cis-acting element in an *Arabidopsis* gene is involved in responsiveness to drought, low-temperature, or high-salt stress. Plant Cell 6, 251–264.

Yang, W.J., Nadolska-Orczyk, A., Wood, K.V., Hahn, D.T., Rich, P.J., Wood, A.J., Saneoka, H., Premachandra, G.S., Bonham, C.C., Rhodes, J.C. et al. (1995). Near-isogenic lines of maize differing for glycinebetaine. Plant Physiol. 107, 621–630.

Yoshiba, Y., Kiyosue, T., Katagiri, T., Ueda, H., Mizoguchi, T., Yamaguchi-Shinozaki, K., Wada, K., Harada, Y. and Shinozaki, K. (1995). Correlation between the induction of a gene for delta 1-pyrroline-5-carboxylate synthetase and the accumulation of proline in *Arabidopsis thaliana* under osmotic stress. Plant J. 7, 751–760.

Zhu, B., Chen, T.H.H. and Li, P.H. (1996). Analysis of late-blight disease resistance and freezing tolerance in transgenic potato plants expressing sense and antisense genes for an osmotin-like protein. Planta 198, 70–77.

Sensing, Signaling and Cell Adaptation. Edited by K.B. Storey and J.M. Storey
© 2002 Elsevier Science B.V. All rights reserved.

CHAPTER 10

Sensing and Responses to Low Temperature in Cyanobacteria

Dmitry A. Los[1] and Norio Murata[2]*
[1]*Institute of Plant Physiology, Moscow, Russia*
[2]*National Institute for Basic Biology, Okazaki, Japan*

1. Introduction

Cyanobacteria are photosynthetic prokaryotes that are one of the most ancient extant forms of life on Earth (Schopf et al., 1965; Margulis, 1975). Cyanobacteria are widely distributed in nature and can be found in almost all environments from Antarctica where temperatures never exceed −20°C (Psenner and Sattler, 1998) to hot springs where temperatures can exceed 70°C (Ward et al., 1998). Cyanobacteria in the water pockets of Antarctic lake ice, where temperatures are always below 0°C, are metabolically active and retain the capacity for oxygenic photosynthesis (Paerl and Priscu, 1998). Photosynthesis also occurs in some species of *Synechococcus* for which the optimum growth temperature is 55–60°C (Meeks and Castenholtz, 1971). Thus, the genus Cyanobacterium includes psychrophilic, psychrotrophic, mesophilic and thermophilic species that differ from one another with respect to the optimal temperature for growth and the extent to which they can tolerate temperature stress.

Cyanobacteria are unicellular, filamentous, and colonial microorganisms with several features that make them particularly suitable for studies of stress responses at the molecular level. The general features of the plasma membranes and thylakoid membranes of cyanobacterial cells are similar to those of chloroplast membranes of higher plants in terms of lipid composition and assembly of

membranes and, indeed, cyanobacteria have provided a powerful model system for studies of the molecular mechanisms of stress responses and acclimation (Murata and Wada, 1995; Glatz et al., 1999).

Some cyanobacterial strains, such as *Synechocystis* sp. PCC 6803, *Synechococcus* sp. PCC 7942, and *Synechococcus* sp. PCC 7002, are naturally competent with respect to the incorporation of foreign DNA. Such DNA is taken up by cells and integrated at high frequency into the genome by homologous recombination (Williams, 1988; Haselkorn, 1991). For transformation of other strains, such as filamentous *Anabaena*, another method has been developed that is based on bacterial conjugation and the use of plasmids with a broad host range (Elhai and Wolk, 1988). Homologous recombination is particularly active in cyanobacteria (Williams and Szalay, 1983; Dolganov and Grossman, 1993) and, therefore, these cells have been widely used for the generation of mutants in which genes of interest have been disrupted (for review, see Vermaas, 1998).

The complete nucleotide sequences of the genomes of *Synechocystis* (Kaneko et al., 1996) and *Anabaena* sp. PCC 7120 (Kaneko et al., 2002), as well as those of *Synechococcus* WH8102, *Nostoc punctiforme* and *Prochlorococcus marinus* (Joint Genome Institute; http:///www.jgi.doe.gov), have been determined and the annotated data are available via the Internet. Moreover, since cyanobacterial genes can be randomly disrupted by random mutagenesis with transposons (Bhaya et al.,

*Corresponding author

1999) or with antibiotic-resistance cartridges (Suzuki et al., 2000a), their functions can be examined and determined under specific stress conditions. The availability of the complete sequence of the genome allows easy localization of the sites of mutations and identification of disrupted genes. Furthermore, DNA microarray analysis of gene expression now allows us to examine the expression of individual genes under stress conditions and the effects of specific mutations on the expression of such genes (Hihara et al., 2001; Suzuki et al., 2001; Kanesaki et al., 2002).

2. Cellular responses to low-temperature stress

The responses of cyanobacterial cells to low-temperature stress fall into two general categories, namely, the cold-induced desaturation of fatty acids in membrane lipids, which fluidizes membranes to compensate for decreases in membrane fluidity at low temperatures (Murata and Los, 1997), and the cold-induced synthesis of certain enzymes that are involved in transcription and translation, which compensates for decreases in the efficiency of transcription and translation at low temperatures (Sato, 1995). Both types of response act to protect cellular functions, in particular photosynthesis, against the adverse effects of cold.

The mesophilic cyanobacterium *Synechocystis* sp. PCC 6803 (hereafter *Synechocystis*) grows well at temperatures from 20 to 38°C. When the ambient temperature is changed from 35 to 20°C, the rate of cell division decreases but cells acclimate and continue to grow (Tasaka et al., 1996). However, at low temperatures such as 1–2°C, cell growth stops completely even though all bioenergetic reactions, for example, electron transport, generation of ATP, and the uptake of protons by isolated thylakoid membranes, remain active (Dilley et al., 2001). However, respiratory activity does decrease with decreases in temperature (Tasaka et al., 1996).

The thermophilic species *Synechococcus vulcanus* has optimal growth temperatures as high as 55°C. Proliferation of cells is inhibited but not arrested at 35°C (Kiseleva et al., 2000) and cells

can survive for several months even when stored at 4°C.

These profiles of temperature-dependent cell growth and cell function can be explained, at least in part, in terms of the extent of unsaturation of fatty acids in membrane lipids. When cyanobacterial cells are grown at temperatures below those that are normal for their growth, the unsaturation of fatty acids in their membrane lipids increases. Details of such changes in the unsaturation of fatty acids have been examined in cyanobacteria that include *Anabaena variabilis, Synechocystis, Synechococcus* (for review, see Los and Murata, 1998), and the thermophilic species *Synechococcus vulcanus* (Kiseleva et al., 1999).

In cyanobacterial cells, the glycerolipids that are generated initially contain saturated fatty acids extensively, and all desaturation reactions occur after these fatty acids have been incorporated into glycerolipids. Upon a downward shift in temperature of 10–15°C, the proliferation of cells and the synthesis of fatty acids *de novo* ceases for about 10 h. During this period, the fatty acids of membrane lipids are desaturated and the cells begin to grow again when the degree of unsaturation reaches a certain level. Analysis of gene transcripts revealed that the increase in desaturation that occurs after a downward shift in temperature is caused by stimulation of the expression of genes for the appropriate desaturases (Los et al., 1997; Sakamoto and Bryant, 1997; Kiseleva et al., 2000).

3. Cold-inducible genes and their regulation

There are several families of cold-inducible genes in cyanobacteria, as summarized in Table 10.1. The first cold-inducible genes to be characterized in cyanobacteria were the genes that encode fatty acid desaturases, namely, the enzymes that convert a C–C single bond (saturated bond) into a C=C double bond (unsaturated bond). The activities of fatty acid desaturases maintain membranes in the appropriate physical state for a given temperature (Vigh et al., 1993; Murata and Los, 1997). Subsequently, genes for RNA-binding proteins (Rbps)

Table 10.1. Genes that are induced by cold in cyanobacteria

Gene	Gene product	Cyanobacterium	Reference
Genes for desaturases			
desA	Δ12 desaturase	*Synechocystis* sp. PCC 6803	Wada et al. (1990)
		Synechocystis sp. PCC 6714	Sakamoto et al. (1994a)
		Synechococcus sp. PCC 7002	Sakamoto & Bryant (1997)
		Spirulina platensis	Murata et al. (1996)
desB	Δ3 desaturase	*Synechocystis* sp. PCC 6803	Sakamoto et al. (1994b)
		Synechococcus sp. PCC 7002	Sakamoto & Bryant (1997)
desC	Δ9 desaturase	*Synechocystis* sp. PCC 6803	Sakamoto et al. (1994c)
		Synechococcus sp. PCC 6301	Ishizaki-Nishizawa et al. (1996)
		Synechococcus sp. PCC 7002	Sakamoto & Bryant (1997)
		Synechococcus vulcanus	Kiseleva et al. (2000)
desD	Δ6 desaturase	*Synechocystis* sp. PCC 6803	Los et al. (1997)
		Spirulina platensis	Murata et al. (1996)
Genes for RNA-binding proteins			
rbpA1	RNA-binding protein (Rbp1)	*Anabaena variabilis* M3	Sato and Nakamura (1998)
		Synechocystis sp. PCC 6803	Suzuki et al. (2001)
rbpA2	RNA-binding protein (Rbp2)	*Anabaena variabilis* M3	Sato (1995)
rbpA3	RNA-binding protein (Rbp3)	*Anabaena variabilis* M3	Sato & Maruyama (1997)
rbpB	RNA-binding protein (RbpB)	*Anabaena variabilis* M3	Maruyama et al. (1999)
rbpC	RNA-binding protein (RbpC)	*Anabaena variabilis* M3	Sato (1995)
rbpE	RNA-binding protein (RbpE)	*Anabaena variabilis* M3	Maruyama et al. (1999)
rbpF	RNA-binding protein (RbpF)	*Anabaena variabilis* M3	Maruyama et al. (1999)
Genes for RNA helicases			
crhB	RNA helicase (CrhB)	*Anabaena* sp. PCC 7120	Chamot et al. (1999)
crhC	RNA helicase (CrhC)	*Anabaena* sp. PCC 7120	Yu & Owttrim (2000)
deaD	RNA helicase (DeaD)	*Synechocystis* sp. PCC 6803	Suzuki et al. (2001)
Genes for caseinolytic proteases			
clpB	Molecular chaperone (ClpB)	*Synechococcus* sp. PCC 7942	Porankiewicz & Clarke (1997)
clpP1	Protease (ClpP)	*Synechococcus* sp. PCC 7942	Porankiewicz et al. (1998)
clpX	Unknown	*Synechococcus* sp. PCC 7942	Porankiewicz et al. (1998)
Genes for ribosomal proteins			
rpsU	30S ribosomal subunit (S21)	*Anabaena variabilis* M3	Sato (1994)
rps12	30S ribosomal subunit (S12)	*Synechocystis* sp. PCC 6803	Suzuki et al. (2001)
rps13	30S ribosomal subunit (S13)	*Synechocystis* sp. PCC 6803	Suzuki et al. (2001)
rpl1	50S ribosomal subunit (L1)	*Synechocystis* sp. PCC 6803	Suzuki et al. (2001)
rpl3	50S ribosomal subunit (L3)	*Synechocystis* sp. PCC 6803	Suzuki et al. (2001)
rpl4	50S ribosomal subunit (L4)	*Synechocystis* sp. PCC 6803	Suzuki et al. (2001)
rpl11	50S ribosomal subunit (L11)	*Synechocystis* sp. PCC 6803	Suzuki et al. (2001)

continued

Table 10.1 (*continuation*)

Gene	Gene product	Cyanobacterium	Reference
rpl20	50S ribosomal subunit (L20)	*Synechocystis* sp. PCC 6803	Suzuki et al. (2001)
rpl21	50S ribosomal subunit (L21)	*Synechocystis* sp. PCC 6803	Suzuki et al. (2001)
rpl23	50S ribosomal subunit (L23)	*Synechocystis* sp. PCC 6803	Suzuki et al. (2001)
fus	Elongation factor EF-G	*Synechocystis* sp. PCC 6803	Suzuki et al. (2001)
Genes for other proteins			
rpoA	a subunit of RNA polymerase	*Synechocystis* sp. PCC 6803	Suzuki et al. (2001)
rpoD	σ^{70} factor of RNA polymerase	*Synechocystis* sp. PCC 6803	Suzuki et al. (2001)
cbiM	Cobalamin biosynthetic prot.	*Synechocystis* sp. PCC 6803	Suzuki et al. (2001)
cytM	Cytochrome c_M	*Synechocystis* sp. PCC 6803	Malakhov et al. (1999)
ndhC	NADH dehydrogenase subunit 3	*Synechocystis* sp. PCC 6803	Suzuki et al. (2001)
ndhD2	NADH dehydrogenase subunit 4	*Synechocystis* sp. PCC 6803	Suzuki et al. (2001)
ndhD6	NADH dehydrogenase subunit 6	*Synechocystis* sp. PCC 6803	Suzuki et al. (2001)
folE	GTP cyclohydrolase I	*Synechocystis* sp. PCC 6803	Suzuki et al. (2001)
ssl1633	CAB/ELIP/FLIP superfamily	*Synechocystis* sp. PCC 6803	Suzuki et al. (2001)
hliA	High-light inducible protein	*Synechocystis* sp. PCC 6803	Suzuki et al. (2001)
ssr2595	High-light inducible protein	*Synechocystis* sp. PCC 6803	Suzuki et al. (2001)
Genes for proteins of as yet unknown function			
slr1544		*Synechocystis* sp. PCC 6803	Suzuki et al. (2001)
slr0082		*Synechocystis* sp. PCC 6803	Suzuki et al. (2001)
sll0551		*Synechocystis* sp. PCC 6803	Suzuki et al. (2001)
sll0668		*Synechocystis* sp. PCC 6803	Suzuki et al. (2001)
slr1974		*Synechocystis* sp. PCC 6803	Suzuki et al. (2001)
slr0955		*Synechocystis* sp. PCC 6803	Suzuki et al. (2001)
		Synechocystis sp. PCC 6803	Suzuki et al. (2001)

The complete list of genes is available on the Internet at http://www.genome.ad.jp/kegg/expression.

were identified as cold-inducible genes (Sato, 1995). The Rbps appear to act similarly to the Csp RNA chaperones of *Escherichia coli* and *Bacillus subtilis*. Next, it was demonstrated that the expression of genes for RNA helicases (Chamont et al., 1999), which are involved in determining the appropriate secondary structures of mRNAs, are also induced by cold. In addition, the expression of several genes for ribosomal proteins is also induced by cold (Sato et al., 1997; Suzuki et al., 2001). Most recently, it was demonstrated that caseinolytic proteases (Clps) act as cold-shock chaperones and proteases (Celerin et al., 1998). One of these Clps seems to be essential for maintenance of the photochemical reaction centers of the photosystem II complex (PSII) during cold stress. Many other genes of known or as yet unknown functions were found to be inducible by cold when patterns of cold-induced gene expression were analyzed using DNA microarrays (Suzuki et al., 2001). The roles of these genes and their products in responses and acclimation to low temperature remain to be clarified.

3.1. Desaturases

3.1.1. Acyl–lipid desaturases of cyanobacteria
The acyl–lipid desaturases are one of three types of

fatty acid desaturase (Murata and Wada, 1995; Los and Murata, 1998). They catalyze the conversion to double (unsaturated) bonds of single bonds at specific positions of fatty acyl chains that have been esterified to membrane glycerolipids (Murata and Wada, 1995). The desaturation of fatty acids and the expression of genes for desaturases have been studied extensively in *Synechocystis*, which has four acyl–lipid desaturases (Fig. 10.1A). These enzymes catalyze desaturation at the Δ9, Δ12, Δ6 and ω3 positions, respectively, of fatty acids that are esterified at the *sn*-1 position of the glycerol moiety of glycerolipids (Murata et al., 1992;

Higashi and Murata, 1993). The Δ9 desaturase introduces the first unsaturated bond into stearic acid to produce oleic acid, which is further desaturated to linoleic acid by the Δ12 desaturase. The Δ6 and ω3 desaturases then introduce unsaturated bonds to generate tri- and tetra-unsaturated fatty acids (Fig. 10.1A).

3.1.2. *Expression of genes for fatty acid desaturases*

The *desA* gene for the Δ12 desaturase in *Synechocystis* was the first gene for an acyl–lipid desaturase to be cloned (Wada et al., 1990).

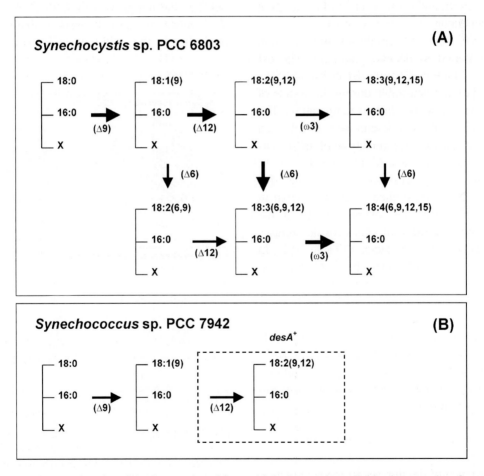

Fig. 10.1. Schematic representation of the desaturation of fatty acids in the membrane lipids of two species of cyanobacteria. First, saturated fatty acids are esterified to a glycerol backbone to generate glycerolipids. The saturated fatty acids are desaturated by the Δ9 desaturase to produce mono-unsaturated fatty acids. Polyunsaturated fatty acids are generated by reactions catalyzed by additional acyl–lipid desaturases (Murata and Wada, 1995). (A) *Synechocystis* sp. PCC 6803. Thick and thin arrows indicate major and minor pathways, respectively. (B) *Synechococcus* sp. PCC 7942. The reaction enclosed by a dotted line was introduced into cells by transformation with the *desA* gene for the Δ12 desaturase from *Synechocystis*. X represents a polar head group. Adapted from Murata et al. (1992) and Los and Murata (1999).

Expression of the *desA* gene is induced by cold and the level of its transcript increases 10-fold within 30 min of exposure of cells to a low temperature (e.g., 25°C), and the corresponding enzyme then accumulates in cells for several hours (Los et al., 1993; 1997). Expression of the *desD* gene for Δ6 desaturase is also induced by low temperatures, as demonstrated by Northern and Western blotting analysis (Los et al., 1997).

Among three cold-inducible genes for desaturases in *Synechocystis*, it is the *desB* gene for the ω3 desaturase that responds most dramatically to a decrease in temperature. Comparing the kinetics of accumulation of *desB* mRNA, the ω3 desaturase itself, and ω3-unsaturated fatty acids at a low temperature, we observed that the level of *desB* mRNA increased rapidly for about 10 min and then started to decline gradually. The ω3 desaturase, which was barely detectable at 35°C, gradually became detectable during the course of several hours after the shift in temperature. The accumulation of the ω3 desaturase was followed by the even slower accumulation of ω3-unsaturated fatty acids (Los and Murata, 1999).

Although the expression of the genes for desaturases has been studied in greatest detail in *Synechocystis*, there are several reports of the cold-induced expression of genes for desaturases in other species of cyanobacteria (Table 10.1). In *Synechococcus* sp. PCC 7002, the *desC, desA* and *desB* genes are cold-inducible (Sakamoto and Bryant, 1997). In *Synechococcus* sp. PCC 6301 (Ishizaki-Nishizawa et al., 1996) and *Synechococcus vulcanus* (Kiseleva et al., 2000), which normally produce only monounsaturated fatty acids, expression of the *desC* gene for the Δ9 desaturase is induced rapidly and transiently after a downward shift in temperature.

3.1.3. Biological functions of fatty acid desaturases

The importance of desaturases in the acclimation of cyanobacteria to low temperatures has been well documented; for reviews, see Murata and Wada (1995) and Los and Murata (1998, 1999). The roles of individual desaturases in such acclimation have been elucidated by genetic manipulation of the

genes for these desaturases in cyanobacteria. A series of mutants of *Synechocystis* that are defective, in a stepwise manner, in the desaturation of membrane lipids was generated by targeted mutagenesis of the genes for the individual desaturases (Tasaka et al., 1996). Targeted mutagenesis of the *desA* gene for the Δ12 desaturase and the *desD* gene for the Δ6 desaturase resulted in dramatic changes in fatty acid composition: there was a considerable increase in the level of mono-unsaturated oleic acid at the expense of polyunsaturated fatty acids, such as di-, tri-, and tetra-unsaturated fatty acids (Fig. 10.2A). At 35°C, *desA⁻/desD⁻* mutant cells proliferated at the same rate as wild-type cells but they did not grow well at 25°C. At 20°C, the *desA⁻/desD⁻* mutant cells were unable to proliferate, whereas the wild-type cells grew relatively

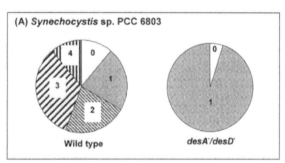

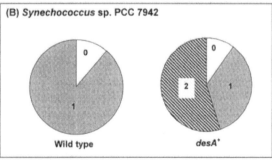

Fig. 10.2. Changes in the unsaturation of fatty acids in cyanobacterial cells by genetic manipulation of acyl–lipid desaturases. (A) The fatty acid composition of glycerolipids in wild-type and *desA⁻/desD⁻* cells of *Synechocystis* sp. PCC 6803 after growth at 25°C. (B) The fatty acid composition of glycerolipids in wild-type and *desA⁺* cells of *Synechococcus* sp. PCC 7942 after growth at 25°C. Numbers in pie charts represent numbers of double bonds in individual molecular species of lipids. Adapted from Tasaka et al. (1996) and Los and Murata (1998, 1999).

well (Tasaka et al., 1996). Furthermore, the *desA⁻/ desD⁻* mutant cells were unable to recover from photo-induced damage to the PSII complex at low temperatures. This failure was the result of the inability of these mutant cells to convert the precursor of the D1 protein to the mature form of the D1 protein which, together with the D2 protein, constitutes the heterodimeric reaction center of PSII (Kanervo et al., 1997). Thus, the ability of *Synechocystis* to tolerate cold stress is determined by the ability to synthesize polyunsaturated fatty acids and, as a result, by the ability to modulate the fluidity of the cell membranes (Szalontai et al., 2001).

These conclusions are supported by the results of another set of experiments that involved transformation of *Synechococcus* sp. PCC 7942, which normally contains only saturated and monounsaturated fatty acids, with the *desA* gene for the Δ12 desaturase from *Synechocystis* (Fig. 10.2B). The transformed *desA⁺* cells synthesized diunsaturated fatty acids at the expense of monounsaturated fatty acids and they were able to tolerate lower temperatures than wild-type cells (Wada et al., 1990). In addition, the *desA⁺* cells appeared to be more tolerant than wild-type cells to strong-light stress (Sippola et al., 1998).

3.2. RNA-binding proteins

RNA-binding proteins (Rbps) are involved in various aspects of the metabolism of RNA, such as splicing, modification, stability and translation (Kenan et al., 1991). The Rbps in chloroplasts of higher plants include two RNA-recognition motifs, an amino-terminal acidic domain, and a carboxy-terminal glycine-rich domain (Ye and Sugiura, 1992). Sato (1995) characterized the family of cold-inducible genes that encode Rbps in *A. variabilis*. The Rbps of the cyanobacterium can be divided into two subgroups: those with a glycine-rich carboxy-terminal domain and those without such a carboxy-terminal domain.

There are eight *rbp* genes in *A. variabilis* and the expression of seven of these genes is induced by low temperatures (Table 10.1) (Maruyama et al., 1999). All cold-regulated *rbp* genes encode Rbps with a glycine-rich carboxy-terminal domain. Transcripts of cold-inducible *rbp* genes are barely detectable at 38°C. However, they become easily detectable within 15–30 min after cells are transferred from 38°C to 22°C. The levels of corresponding proteins also increase dramatically after this change in temperature. However, while the levels of transcripts of cold-inducible *rbp* genes increase transiently, the levels of the encoded proteins increase gradually and remain at high plateau values for 24 h after the start of incubation at low temperatures (Sato, 1995).

Using a *lacZ* reporter gene fused to a number of modified promoter regions, Sato and Nakamura (1998) localized a putative cold-responsive *cis*-acting element in the 5′-untranslated region of the *rbpA1* gene. The 150-bp regulatory region of the gene, located between the site of initiation of transcription and a ribosome-binding site, was absolutely necessary for the cold-induced transcription of the *rbpA1* gene. Deletions within this region resulted in constitutive transcription from the promoter at both high and low temperatures. The existence of a putative repressor protein(s) was confirmed by gel mobility shift assays with the target DNA sequence as probe and extracts of proteins from cells that had been grown at high and low temperatures. The assay revealed the presence of a protein(s) that bound to the regulatory region of the *rbpA1* gene in extracts of cells grown at the high, but not at the low, temperature (Sato and Nakamura, 1998).

The discovery of cyanobacterial Rbps whose expression is regulated by cold revealed a new class of stress-inducible RNA-binding proteins. A survey of the sequences of the cyanobacterial genome indicated that cyanobacteria do not have the cold-shock proteins (Csps) that are characteristic of some eubacteria (Jones and Inoue, 1994). These Csps are synthesized rapidly in response to cold and are thought to play important roles in the regulation of transcription under low temperature stress in *E. coli* and *B. subtilis* (Jones and Inoue, 1994; Schnuchel et al., 1993). Some Csps, known as RNA chaperones, bind to β-sheet structures generated by single-stranded nucleic acids, and such structures are analogous to RNA-recognition motifs (Thieringer et al., 1998). The Rbps of cyanobacteria and the Csps of *E. coli* and *B. subtilis* are induced by cold and both classes of proteins bind RNA. Therefore, it is possible that the functions of Rbps are similar to those of Csps.

3.3. Cold-inducible RNA helicases

RNA helicases are responsible for modification of the secondary structure of mRNAs, a crucial aspect of the regulation of translation. In *E. coli,* RNA helicases are involved in the assembly of ribosomes, the turnover of RNA, and acclimation to low temperatures (Fuller-Pace, 1994). Two genes, *crhB* and *crhC,* that encode RNA helicases have been identified in *Anabaena* sp. PCC 7120 (Chamot et al., 1999). The *crhB* gene is expressed under a variety of stress conditions (e.g., cold stress, salt stress and nitrogen limitation) but the expression of the *crhC* gene is induced exclusively by cold stress. The *crhC* gene is specifically expressed after *Anabaena* cells are transferred from 30°C to 20°C (Chamot et al., 1999). In *Synechocystis,* strong induction by cold of expression of genes for helicases was also detected using DNA microarrays (Suzuki et al., 2001).

The CrhC protein is a member of the DEAD-box family of helicases (Yu and Owttrim, 2000). It has been suggested that CrhC participates in cold acclimation by destabilizing the secondary structures of mRNAs, thereby overcoming the cold-induced inhibition of the initiation of translation at low temperatures (Yu and Owttrim, 2000).

3.4. Ribosomal proteins

Cyanobacterial ribosomes resemble those in *E. coli* (Gray and Herson, 1976; Sato et al., 1997). The first report of the cold-induced expression of the *rpsU* gene for the ribosomal small-subunit protein S21, which is located just downstream of the *rbpA1* gene, came from studies of responses of *A. variabilis* to cold (Sato, 1995). A cold-induced increase in the level of the S21 protein was also observed in *Synechocystis.* However, in *Synechocystis,* the *rpsU* gene is not located close to the

rbpA gene but is found downstream of the rRNA operon (Sato et al., 1997).

Systematic analysis of expression of all the genes in the genome of *Synechocystis* using DNA microarrays (Suzuki et al., 2001) revealed sets of genes for ribosomal proteins that are induced by cold (Table 10.1). Such enhancement of expression suggests that an excess of ribosomal proteins is necessary for acclimation of the translational apparatus to cold, since translational activity always appears to decrease in prokaryotes upon exposure to low temperatures.

3.5. Caseinolytic proteases

Bacterial caseinolytic proteases (Clps) are novel molecular chaperones and this class of proteases includes constitutively expressed and stress-inducible proteins (Maurizi et al., 1990). The sequence of the *Synechocystis* genome (Kaneko et al., 1996) appears to include genes for ClpB, ClpC, ClpP, and ClpX, with ClpP being encoded by a multigene sub-family with as many as four homologs.

In *Synechococcus* sp. PCC 7942, ClpP1 accumulates rapidly under cold stress or strong ultraviolet (UV-B) light. A 15-fold increase in the level of ClpP1 was detected within 24 h of the start of cold stress (Porankiewicz et al., 1998). Growth of a *clpP1* null mutant, Δ*clpP1,* was severely inhibited at low temperatures (Porankiewicz and Clarke, 1997).

During acclimation to cold, wild-type *Synechococcus* cells replace the standard form of the D1 protein in the PSII reaction center (D1:1) with an alternative form, D1:2, within a few hours. Once cells have acclimated to the low temperature, the D1:2 form is replaced by the D1:1 form. In Δ*clpP1* cells, this latter event does not occur (Porankiewicz et al., 1998). The mechanisms responsible for the alteration between the D1:1 and D1:2 forms are poorly understood but it is clear that ClpP1 is indispensable for acclimation of this cyanobacterium to low temperatures.

ClpB (Hsp100) was first identified in *Synechococcus* sp. PCC 7942 as a heat-inducible molecular chaperone that is essential for tolerance to high temperatures (Eriksson and Clarke, 1996). However, the synthesis of ClpB is also strongly induced under cold stress (Porankiewicz and Clarke, 1997; Celerin et al., 1998). Targeted mutagenesis of the *clpB* gene resulted in accelerated inhibition of the activity of the PSII complex at low temperatures and reduced the ability of the mutant cells to acclimate to low temperatures in terms of both proliferation and survival. Porankiewicz and Clarke (1997) suggested that ClpB might renature and solubilize aggregated proteins under cold-stress conditions.

3.6. Cytochrome c_M

The level of expression of the *cytM* gene for cytochrome c_M in *Synechosystis* (Malakhov et al., 1994) is very low under normal growth conditions but it increases greatly at low temperatures and under high-intensity light. This cytochrome replaces plastocyanin and/or cytochrome c_6, which donate electrons to PSI under optimal growth conditions (Malakhov et al., 1999). The genes for plastocyanin (*petE*) and cytochrome c_6 (*petJ*) are strongly repressed at low temperatures. The replacement of the regular electron donors by cytochrome c_M might be advantageous under stress conditions, such as low temperature and high intensity light.

3.7. Genes that are down-regulated by low temperature

Some groups of genes that are repressed by cold are listed in Table 10.2. The genes that are most sensitive to cold stress are the genes that are involved in the assembly and function of the photosynthetic machinery. The expression of genes for subunits of PSI and PSII, and phycobilisomes, as well as some other light-regulated genes, is repressed soon after a decrease in temperature. However, such repression seems to be transient and normal levels of expression are restored within 60 min.

4. Membrane fluidity as a link between low temperatures and the induction of gene expression

There is considerable interest in the mechanisms whereby organisms, and in particular cyanobacterial cells, detect temperature and changes in temperature that lead to the cold-induced regulation of gene expression. The most detailed studies on this topic have involved the relationship between the expression of genes for desaturases and the dynamics of membrane structure in *Synechocystis*. Several lines of evidence suggest that membrane fluidity contributes to the perception of temperature and changes in temperature.

(1) In *Synechocystis* cells that have been acclimated to 36°C, induction of the expression of the *desA* gene starts when the temperature is lowered to 28°C, whereas in cells that have been acclimated at 32°C, induction of the *desA* gene starts at 26°C (Los et al., 1993). This observation indicates that cyanobacterial cells perceive a change in temperature and not the absolute temperature and, furthermore, that cells sense a change in temperature only when this changes exceeds approximately 5°C. In terms of physical responses, this biological perception of temperature is very sensitive since a change in temperature of 5°C represents a reduction of only 2% in molecular motion.

(2) A decrease in the physical motion of membrane lipids was induced artificially by the catalytic hydrogenation of fatty acids in the plasma membrane of *Synechocystis* (Vigh et al., 1993). This chemically induced decrease in the extent of unsaturation of fatty acids in membrane lipids and, therefore, in the fluidity of membrane lipids *in vivo* under isothermal conditions, minimized the contribution of additional effects that might result from a change in temperature. Hydrogenation of cells for 4 min converted 5% of the unsaturated fatty acids to saturated fatty acids in the glycerolipids of plasma membranes, but not of thylakoid membranes. Decreases in membrane fluidity, caused either by cold stress or by hydrogenation, resulted in the rapid induction of expression of the *desA* gene (Vigh et al., 1993). These findings suggested that the primary signal in the perception of

Table 10.2. Genes that are repressed at low temperatures in *Synechocystis* sp. PCC 6803*

	Gene	Gene product
Genes for components of photosystem I	*psaC*	Photosystem I subunit VII
	psaD	Photosystem I subunit II
	psaE	Photosystem I subunit IV
	psaF	Photosystem I subunit III
	psaJ	Photosystem I subunit IX
	psaL	Photosystem I subunit XI
	psaK	Photosystem I subunit X
Genes for components of photosystem II	*psbO*	Manganese-stabilizing polypeptide
	psbD	D2 protein
	psbC	CP43 protein
	psbU	12-kDa extrinsic protein
	psbJ	PsbJ protein
	psbE	Cytochrome b_{559}
Genes for components of phycobilisomes	*apcA*	Allophycocyanin a chain
	apcB	Allophycocyanin b chain
	apcC	Phycobilisome LC linker polypeptide
	apcD	Allophycocyanin B
	apcE	Phycobilisome LCM core-membrane linker polypeptide
	apcF	Phycobilisome core component
	cpcA	Phycocyanin a subunit
	cpcB	Phycocyanin b subunit
	cpcC	Allophycocyanin-associated linker polypeptide
	cpcE	Phycocyanin a phycocyanin lyase CpcE
	cpcF	Phycocyanin a phycocyanin lyase CpcF
	cpcG	Phycobilisome rod-core linker polypeptide
Genes for proteins involved in the biosynthesis of cofactors and chlorophyll	*chlP*	Chlorophyll synthase subunit
	sll1184	Heme oxygenase holoenzyme
	hemA	Transfer RNA-Gln reductase
	hemF	Coproporphyrinogen III oxidase
	slr0506	Protochlorophyllide oxidoreductase
	chlB	Protochlorophyllide reductase subunit
	slr0749	Light-independent protochlorophyllide reductase
Genes for other proteins	*lrtA*	Light-repressed protein
	sll1214	Phytochrome-regulated protein
	aqpZ	Water channel
	sll1920	Cation-transporting ATPase
	slr1856	Anti-sigma B factor antagonist
	isiB	Flavodoxin
	petE	Plastocyanin
	petF	Ferredoxin
	pilA1	Pilin
	slr0374	Cell-division-cycle protein

*From Suzuki et al. (2001). The complete list of genes is available on the Internet at http://www.genome.ad.jp/kegg/expression

temperature might be a change in the fluidity of plasma membrane.

5. Sensors and transducers of low-temperature signals

The extent of unsaturation of membrane lipids is the major factor that determines the fluidity of membrane lipids (Cossins, 1994). The discovery of feedback between membrane fluidity and the expression of genes for desaturases suggested the presence, in the cyanobacterial cytoplasmic membrane, of a temperature sensor that perceives a change in the physical motion of membrane lipids and transmits the signal to a mediator(s) that activates the expression of genes for desaturases (Murata and Wada, 1995; Murata and Los, 1997). The desaturases are then synthesized *de novo* and targeted to both the plasma membrane and thylakoid membranes (Mustardy et al., 1996), where they catalyze the desaturation of the fatty acids of the membrane lipids to compensate for the decrease in membrane fluidity that has been caused by exposure to a low temperature.

If the pathway for perception and transduction of low-temperature signals were similar to other bacterial stress-response pathways, which are generally two-component regulatory systems consisting of a histidine kinase for signal perception and a response regulator, it would be reasonable to postulate the existence of a cold-sensing histidine kinase system in cyanobacteria. Such a histidine kinase would presumably contain a primary perception module that is bound directly to the membrane and can sense changes in the membrane's physical state. Such a histidine kinase might be able to regulate all the gene expression responses to low-temperature shock.

To pursue this hypothesis about the possible nature of the temperature sensor in cyanobacterial cells, we systematically inactivated each of the putative histidine kinases (corresponding to 43 open-reading frames) in *Synechocystis* (Suzuki et al., 2000a). We identified two histidine kinases that affected the inducibility of the *desB* gene. A combination of targeted mutagenesis and random mutagenesis allowed us to identify the two

histidine kinases (membrane-bound Hik33 and soluble Hik19) and a putative response regulator Rer1 (Suzuki et al., 2000a,b). Hik33 has a highly conserved histidine kinase domain at its carboxyl-terminus, two membrane-spanning domains at its amino terminus, and a type-P linker and leucine zipper in the middle region (Fig. 10.3A). Type-P linkers have been found between the signal-input domain and the kinase domain of several membrane-bound histidine kinases from various organisms (Park et al., 1998). The rigidification of membrane lipids around Hik33 might change the structure of Hik33, perhaps by altering the conformation of the membrane-spanning domains of this enzyme. The conformational changes in the transmembrane domains and periplasmic loop might affect the structure of the type-P linker and thereby, influence the ability of the leucine zipper and/or PAS domain to generate an active form of Hik33 (Fig. 10.3B; Taylor and Zhulin, 1999; Suzuki et al., 2000a).

Recent results of experiments with DNA microarrays and cells with the mutant form of Hik33 suggest that not all cold-inducible genes are controlled by this histidine kinase (Suzuki et al., 2001). Thus, it is likely that some other cold sensor(s) exists that might control gene expression by some as yet unknown mechanism(s) (Browse and Xin, 2001; Suzuki et al., 2001).

6. Conclusions and perspectives

The acclimation of cyanobacteria to low temperatures involves changes in membrane fluidity and in the transcriptional and translational machinery that serve to protect cell functions, most notably photosynthesis, from damage and ensure its continued operation. Upon exposure of cells to low temperature, the fatty acids of membrane lipids are desaturated, the expression of genes for components of the transcriptional and translational machinery is enhanced, and genes for components of PSI and PSII are transiently repressed until the restructured membranes are ready to support the activities of the photosystem complexes at the new low ambient temperature.

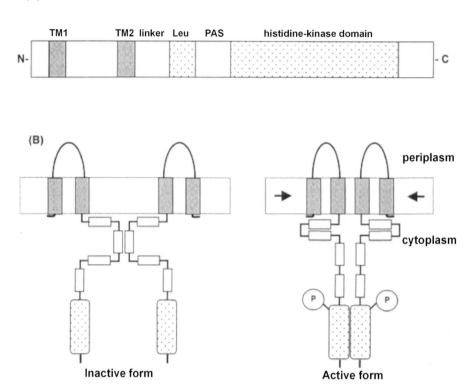

Fig. 10.3. Schematic representation of the putative effects of low temperature on the histidine kinase Hik33. (A) Domains of Hik33. TM1 and TM2, Transmembrane domains; Leu, leucine zipper; PAS, PAS domain, which contains amino acid motifs Per, Arnt, Sim, and phytochrome (Taylor and Zhulin, 1999). (B) Hypothetical model of the cold-induced activation of Hik33. At the optimal growth temperature, the homodimer composed of two molecules of Hik33 is in an inactive form. A decrease in temperature causes a decrease in membrane fluidity and lipid molecules move closer together (indicated by arrows). This movement pushes the linker domains of Hik33 closer together, with resultant autophosphorylation of the histidine kinase domains (P). As a consequence, Hik33 is activated and, possibly, phosphorylates the response regulator Rer1.

In cyanobacteria, there is very tight coordination between the physical state of the membranes and the expression of cold-inducible genes. A sophisticated system for desaturation of C–C bonds ensures the maintenance of appropriate membrane fluidity in spite of changes in ambient temperature and controls gene expression. This phenomenon allowed us to identify the cold-sensing histidine kinase Hik33. However, many cold-inducible genes are not controlled by Hik33 and further efforts should be focused on the identification of additional cold-sensing molecules and on characterization of the details of signal-transduction pathways from such low-temperature sensors to the expression of specific sets of individual genes.

References

Bhaya, D., Takahashi, A., Shahi, P. and Grossman, A.R. (2001). Novel motility mutants of *Synechocystis* strain PCC 6803 generated by *in vitro* transposon mutagenesis. J. Bacteriol. 183, 6140–6143.

Browse, J. and Xin, Z. (2001). Temperature sensing and cold acclimation. Curr. Opin. Plant Biol. 4, 241–246.

Celerin, M., Gilpin, A.A., Schisler, N.J., Ivanov, A.G., Miskiewicz, E., Krol, M. and Laudenbach, D.E. (1998). ClpB in a cyanobacterium: predicted structure, phylogenetic relationships, and regulation by light and temperature. J. Bacteriol. 180, 5173–5182.

Chamot, D., Magee, W.C., Yu, E. and Owttrim, G.W. (1999). A cold shock-induced cyanobacterial RNA helicase. J. Bacteriol. 181, 1728–1732.

Cossins, A.R. (1994). Homeoviscous adaptation of biological membranes and its functional significance. In: Temperature Adaptation of Biological Membranes (Cossins, A.R., Ed.), pp. 63–76. Portland Press, London.

Dilley, R.A., Nishiyama, Y., Gombos, Z. and Murata, N. (2001). Bioenergetic responses of *Synechocystis* 6803 fatty acid desaturase mutants at low temperatures. J. Bioenerg. Biomembr. 33, 135–141.

Dolganov, N. and Grossman, A.R. (1993). Insertional inactivation of genes to isolate mutants of *Synechococcus* sp. strain PCC 7942: isolation of filamentous strains. J. Bacteriol. 175, 7644–7651.

Elhai, J. and Wolk, C.P. (1988). Conjugal transfer of DNA into cyanobacteria. Methods Enzymol. 167, 747–765.

Eriksson, M.J. and Clarke, A.K. (1996). The heat shock protein ClpB mediates the development of thermotolerance in the cyanobacterium *Synechococcus* sp. strain PCC 7942. J. Bacteriol. 178, 4839–4846.

Fuller-Pace, F.V. (1994). RNA helicases: modulators of RNA structure. Trends Cell Biol. 4, 271–274.

Glatz, A., Vass, I., Los, D.A. and Vigh, L. (1999). The *Synechocystis* model of stress: from molecular chaperons to membranes. Plant Physiol. Biochem. 37, 1–12.

Gray, J.E. and Herson, D.S. (1976). Functional 70S hybrid ribosomes from blue-green algae and bacteria. Arch. Microbiol. 109, 95–99.

Haselkorn, R. (1991). Genetic systems in cyanobacteria. Methods Enzymol. 204, 418–430.

Higashi, S. and Murata, N. (1993). An in vivo study of substrate specificities of acyl–lipid desaturases and acyltransferases in lipid synthesis in *Synechocystis* PCC6803. Plant Physiol. 102, 1275–1278.

Hihara, Y., Kamei, A., Kanehisa, M., Kaplan, A. and Ikeuchi, M. (2001). DNA microarray analysis of cyanobacterial gene expression during acclimation to high light. Plant Cell 13, 793–806.

Ishizaki-Nishizawa, O., Fujii, T., Azuma, M., Sekiguchi, K., Murata, N., Ohtani, T. and Toguri, T. (1996). Low-temperature-enhanced resistance of higher plants is significantly enhanced by a nonspecific cyanobacterial desaturase. Nature Biotech. 14, 1003–1006.

Jones, P.G. and Inoue, M. (1994). The cold-shock response—a hot topic. Mol. Microbiol. 11, 811–818.

Kaneko, T., Sato, S., Kotani, H., Tanaka, A., Asamizu, E., Nakamura, Y., Miyajima, N., Hirosawa, M., Sugiura, M., Sasamoto, S., Kimura, T., Hosouchi, T., Matsuno, A., Muraki, A., Nakazaki, N., Naruo, K., Okumura, S., Shimpo, S., Takeuchi, C., Wada, T., Watanabe, A., Yamada, M., Yasuda, M. and Tabata, S. (1996). Sequence analysis of the genome of the unicellular cyanobacterium *Synechocystis* sp. strain PCC6803, II. Sequence determination of the entire genome and assignment of potential protein-coding regions. DNA Res. 3, 109–136.

Kaneko, T., Nakamura, Y., Wolk, C.P., Kuritz, N., Sasamoto, S., Watanabe, A., Iriguchi, M., Ishikawa, A., Kawashima, K., Kimura, T., Kishida, Y., Kohara, M.,

Matsumoto, M., Matsuno, A., Muraki, M., Nakazaki, N., Shimpo, S., Sugimoto, M., Takazawa, M., Yamada, M., Yasuda, M. and Tabata, S. (2001). Complete genomic sequence of the filamentous nitrogen-fixing cyanobacterium *Anabaena* sp. strain PCC 7120. DNA Res. 8, 205–213.

Kanesaki, Y., Suzuki, I., Allakhverdiev, S.I., Mikami, K. and Murata, N. (2002). Salt stress and hyperosmotic stress regulate the expression of different sets of genes in *Synechocystis* sp. PCC 6803. Biochem. Biophys. Res. Commun. 290, 339–348.

Kanervo, E., Tasaka, Y., Murata, N. and Aro, E.-M. (1997). Membrane lipid unsaturation-modulated processing of photosystem II reaction-center protein D1 at low temperature. Plant Physiol. 114, 841–849.

Kenan, D.J., Query, C.C. and Keene, J.D. (1991). RNA recognition: towards identifying determinants of specificity. Trends Biochem. Sci. 16, 214–220.

Kiseleva, L.L., Horvath, I., Vigh, L. and Los, D.A. (1999). Temperature-induced specific lipid desaturation in the thermophilic cyanobacterium *Synechococcus vulcanus*. FEMS Microbiol. Lett. 175, 179–183.

Kiseleva, L.L., Serebriiskaya, T.S., Horvath, I., Vigh, L., Lyukevich, A.A. and Los, D.A. (2000). Expression of the gene for the Δ9 acyl–lipid desaturase in the thermophilic cyanobacterium. J. Mol. Microbiol. Biotechnol. 2, 331–338.

Los, D.A. and Murata, N. (1998). Structure and expression of fatty acid desaturases. Biochim. Biophys. Acta 1394, 3–15.

Los, D.A. and Murata, N. (1999). Responses to cold shock in cyanobacteria. J. Mol. Microbiol. Biotechnol. 1, 221–230.

Los, D.A., Horvàth, I., Vigh, L. and Murata, N. (1993). The temperature-dependent expression of the desaturase gene *desA* in *Synechocystis* PCC6803. FEBS Lett. 318, 57–60.

Los, D.A., Ray, M. K. and Murata, N. (1997). Differences in the control of the temperature-dependent expression of four genes for desaturases in *Synechocystis* sp. PCC 6803. Mol. Microbiol. 25, 1167–1176.

Malakhov, M.P., Wada, H., Los, D.A., Semenenko, V.E. and Murata, N. (1994). A new type of cytochrome *c* from *Synechocystis* PCC6803. J. Plant Physiol. 144, 259–264.

Malakhov, M.P., Malakhova, O.A. and Murata, N. (1999). Balanced regulation of expression of the gene for cytochrome c_M and that of genes for plastocyanin and cytochrome c_6 in *Synechocystis*. FEBS Lett. 444, 281–284.

Margulis, L. (1975). Symbiotic theory of the origin of eukaryotic organelles; criteria for proof. Symp. Soc. Exp. Biol. 29, 21–38.

Maruyama, K., Sato, N. and Ohta, N. (1999). Conservation of structure and cold-regulation of RNA-binding proteins in cyanobacteria: probable convergent evolution with eukaryotic glycine-rich RNA-binding proteins. Nucleic Acids Res. 27, 2029–2036.

Maurizi, M.R., Clark, W.P., Kim, S.-H. and Gottesman, S. (1990). ClpP represents a unique family of serine proteases. J. Biol. Chem. 265, 12546–12552.

Meeks, J.C. and Castenholtz, R.W. (1971). Growth and photosynthesis in an extreme thermophile, *Synechococcus lividus* (Cyanophyta). Arch. Microbiol. 78, 25–41.

Murata, N. and Los, D.A. (1997). Membrane fluidity and temperature perception. Plant. Physiol. 115, 875–879.

Murata, N. and Wada, H. (1995). Acyl–lipid desaturases and their importance in the tolerance and acclimatization to cold of cyanobacteria. Biochem. J. 308, 1–8.

Murata, N., Wada, H. and Gombos, Z. (1992). Modes of fatty-acid desaturation in cyanobacteria. Plant Cell Physiol. 33, 933–941.

Murata, N., Deshnium, P. and Tasaka, Y. (1996). Biosynthesis of γ-linoleic acid in the cyanobacterium, *Spirulina platensis*. In: γ-Linolenic Acid (Huang, Y.-S. and Millis, D.E., Eds.), pp. 22–32. AOCS Press, Champaign, IL.

Mustardy, L., Los, D.A., Gombos, Z. and Murata, N. (1996). Immunocytochemical localization of acyl–lipid desaturases in cyanobacterial cells: Evidence that both thylakoid membranes and cytoplasmic membranes are sites of lipid desaturation. Proc. Natl. Acad. Sci. USA 93, 10524–10527.

Paerl, H.W. and Priscu, J.C. (1998). Microbial phototrophic, heterotrophic, and diazotrophic activities associated with aggregates in the permanent ice cover of Lake Bonney, Antarctica Microb. Ecol. 36, 221–230.

Park, H., Saha, S.K. and Inouye, M. (1998). Two-domain reconstitution of a functional protein histidine kinase. Proc. Natl. Acad. Sci. USA 95, 6728–6732.

Porankiewicz, J. and Clarke, A.K. (1997). Induction of a heat shock protein ClpB affects cold acclimation in the cyanobacterium *Synechococcus* sp. strain PCC 7942. J. Bacteriol. 179, 5111–5117.

Porankiewicz, J., Schelin, J. and Clarke, A.K. (1998). The ATP-dependent Clp protease is essential for acclimation to UV-B and low temperature in the cyanobacterium *Synechococcus*. Mol. Microbiol. 29, 275–283.

Psenner, R. and Sattler, B. (1998). Life at freezing point. Science 280, 2073–2074.

Sakamoto, T. and Bryant, D.A. (1997). Temperature-regulated mRNA accumulation and stabilization for fatty acid desaturase genes in the cyanobacterium *Synechococcus* sp. strain PCC 7002. Mol. Microbiol. 23, 1281–1292.

Sakamoto, T., Wada, H., Nishida, I., Ohmori, M. and Murata, N. (1994a). Identification of conserved domains in the Δ12 desaturase of cyanobacteria. Plant Mol. Biol. 24, 643–650.

Sakamoto, T., Los, D.A., Higashi, S., Wada, H., Nishida, I., Ohmori, M. and Murata, N. (1994b). Cloning of ω3

desaturase from cyanobacteria and its use in altering the degree of membrane-lipid unsaturation. Plant Mol. Biol. 26, 249–263.

Sakamoto, T., Wada, H., Nishida, I., Ohmori, M. and Murata, N. (1994c). Δ9 acyl–lipid desaturase of cyanobacteria: molecular cloning and substrate specificities in terms of fatty acids, *sn*-positions, and polar head groups. J. Biol. Chem. 269, 25576–25580.

Sato, N. (1994). A cold-regulated cyanobacterial gene cluster encodes RNA-binding protein and ribosomal protein S21. Plant Mol. Biol. 24, 819–823.

Sato, N. (1995). A family of cold-regulated RNA-binding protein genes in the cyanobacterium *Anabaena variabilis* M3. Nucleic Acids Res. 23, 2161–2167.

Sato, N. and Maruyama, K. (1997). Differential regulation by low temperature of the gene for an RNA-binding protein, *rbpA3*, in the cyanobacterium *Anabaena variabilis* strain M3. Plant Cell Physiol. 38: 81–86.

Sato, N. and Nakamura, A. (1998). Involvement of the 5′-untranslated region in cold-regulated expression of the *rbpA1* gene in the cyanobacterium *Anabaena variabilis* M3. Nucleic Acids Res. 26, 2192–2199.

Sato, N., Tachikawa, T., Wada, A. and Tanaka, A. (1997). Temperature-dependent regulation of the ribosomal small-subunit protein S21 in the cyanobacterium *Anabaena variabilis* M3. J. Bacteriol. 179, 7063–7071.

Schnuchel, A., Wiltscheck, R., Czisch, M., Herrler, M., Willimsky, G., Graumann, P., Marahiel, M.A. and Holak, T.A. (1993). Structure in solution of the major cold-shock protein from *Bacillus subtilis*. Nature 364, 169–171.

Schopf, J.W., Barghoorn, E.S., Maser, M.D. and Gordon, R.O. (1965). Electron microscopy of fossil bacteria two billion years old. Science 149, 1365–1367.

Sippola, K., Kanervo, E., Murata, N. and Aro, E.-M. (1998). A genetically engineered increase in fatty acid unsaturation in *Synechococcus* sp. PCC 7942 allows exchange of D1 protein forms and sustenance of photosystem II activity at low temperature. Eur. J. Biochem. 251, 641–648.

Suzuki, I., Los, D.A., Kanesaki, Y., Mikami, K. and Murata N. (2000a). The pathway for perception and transduction of low-temperature signals in *Synechocystis*. EMBO J. 19, 1327–1334.

Suzuki, I., Los, D.A. and Murata N. (2000b). Perception and transduction of low-temperature signals to induce desaturation of fatty acids. Biochem. Soc. Trans. 28, 628–630.

Suzuki, I., Kanesaki, Y., Mikami, K., Kanehisa, M. and Murata, N. (2001). Cold-regulated genes under control of the cold sensor Hik33 in *Synechocystis*. Mol. Microbiol. 40, 235–244.

Szalontai, B., Nishiyama, Y., Gombos, Z. and Murata, N. (2000). Membrane dynamics as seen by Fourier transform infrared spectroscopy in a cyanobacterium,

Synechocystis PCC 6803. The effects of lipid unsaturation and the protein-to-lipid ratio. Biochim. Biophys. Acta 1509, 409–419.

Tasaka, Y., Gombos, Z., Nishiyama, Y., Mohanty, P., Ohba, T., Ohki, K. and Murata, N. (1996). Targeted mutagenesis of acyl–lipid desaturases in *Synechocystis*: Evidence for the important roles of polyunsaturated membrane lipids in growth, respiration and photosynthesis. EMBO J. 15, 6416–6425.

Taylor, B.L. and Zhulin, I.B. (1999). PAS domains: internal sensors of oxygen, redox potential, and light. Microbiol. Mol. Biol. Rev. 63, 479–506.

Thieringer, H.A., Jones, P.G. and Inoue, M. (1998). Cold shock and adaptation. Bioessays 20, 49–57.

Vermaas, W.F. (1998). Gene modifications and mutation mapping to study the function of photosystem II. Methods Enzymol. 297, 293–310.

Vigh, L., Los, D.A., Horvath, I. and Murata, N. (1993). The primary signal in the biological perception of temperature: Pd-catalyzed hydrogenation of membrane lipids stimulated the expression of the *desA* gene in *Synechocystis* PCC6803. Proc. Natl. Acad. Sci. USA 90, 9090–9094.

Wada, H., Gombos, Z. and Murata, N. (1990). Enhancement of chilling tolerance of a cyanobacterium by genetic manipulation of fatty acid desaturation. Nature 347, 200–203.

Ward, D.M., Ferris, M.J., Nold, S.C. and Bateson, M.M. (1998). A natural view of microbial biodiversity within hot spring cyanobacterial mat communities. Microbiol. Mol. Biol. Rev. 62, 1353–1370.

Williams, J.G.K. (1988). Construction of specific mutations in photosystem II photosynthetic reaction center by genetic engineering methods in *Synechocystis* PCC6803. Methods Enzymol. 167, 766–778.

Williams, J.G. and Szalay, A.A. (1983). Stable integration of foreign DNA into the chromosome of the cyanobacterium *Synechococcus* R2. Gene 24, 37–51.

Ye, L. and Sugiura, M. (1992). Domain required for nucleic acid-binding activities in chloroplast ribonucleoproteins. Nucleic Acids Res. 20, 6275–6279.

Yu, E. and Owttrim, G.W. (2000). Characterization of the cold stress-induced cyanobacterial DEAD-box protein CrhC as an RNA helicase. Nucleic Acids Res. 28, 3926–3934.

Sensing, Signaling and Cell Adaptation. Edited by K.B. Storey and J.M. Storey
© *2002 Elsevier Science B.V. All rights reserved.*

CHAPTER 11

Dehydrins

Jan Svensson[1,3], Abdelbagi M. Ismail[2], E. Tapio Palva[1] and Timothy J. Close[3]

[1]*Department of Biosciences, Division of Genetics, University of Helsinki, Helsinki, Finland*
[2]*CSWS Division, International Rice Research Institute, Metro Manila, Philippines*
[3]*Department of Botany & Plant Sciences, University of California, Riverside, CA, USA*

1. Introduction

Adaptation of plants to environmental perturbation is crucial to growth and survival. In crop plants, environmental adaptation can help reduce yield losses. Average yields of field crops are depressed nearly 70% by abiotic stresses imposed by unfavorable environments or suboptimal management practices. Annual economic losses caused by freezing alone are estimated at US $14 billion worldwide (Boyer, 1982; Steponkus et al., 1993). Plants cope with environmental stresses through a number of biochemical and physiological modifications reflected at the gene expression level. Recently, considerable attention has been devoted to plant responses to abiotic stresses, particularly drought, salinity and low temperature, in an effort to untangle fundamental mechanisms. One commonality among these stresses is cellular dehydration. This commonality is evident in the overlap of the molecular responses to these stresses in many plant species. Overlaps include: (1) genes induced in common by more than one stress, (2) cross-adaptation to multiple stresses attained by acclimation to any one stress, and (3) a role of the phytohormone abscisic acid (ABA) in adaptation to each of these stresses (Guy, 1990; Arora et al., 1998).

Stress-induced genes encode enzymes required for biosynthesis of osmoprotectants, late embryogenesis abundant (LEA) proteins, antifreeze proteins, chaperones, detoxification enzymes, transcription factors, kinases, enzymes involved in phosphoinositide metabolism, and other proteins with various known and unknown functions (Bray, 1993; 1997; Ingram and Bartels, 1996; Thomashow, 1998). One group of proteins commonly found in response to any osmotic or low temperature stress is the subset of LEA proteins known also as dehydrins (DHNs).

2. Dehydrins

DHNs are a family of proteins thought to play a protective role in plants during cellular dehydration (Close, 1996, 1997; Campbell and Close, 1997). In addition to their production during the later stages of embryogenesis and in mature seeds, DHNs accumulate due to salinity, water deficit, low temperature and in response to ABA treatment. DHNs can also be induced in response to changes in photoperiod. The term "dehydrin" was first introduced in 1989 and was initially intended to mean "dehydration-induced proteins" (Close et al., 1989). But, since then this term has become widely used to include all proteins with specific sequence homology rather than expression characteristics. The first two *Dhn* genes cloned and characterized were *RAB21* (responsive to ABA) from rice and *D-11* from cotton (Baker et al., 1988; Mundy and Chua, 1988). Dure et al. (1989) classified LEA proteins into three groups with the DHNs classified as LEA group 2, often referred to as LEA-D11 or LEA (II) proteins. Currently many different names are used for this family of proteins,

(a)

D_{71}	E_{52}	Y_{59}	G_{73}	N_{70}	P_{69}
E_2	Q_{15}	H_5	R_1	Q_2	V_3
G_1	A_4	F_4		D_1	A_1
	V_2	R_3		E_1	H_1
	T_1	K_2			
		Q_1			

(b)

L_{54}	H_{39}	R_{57}	S_{51}	G_{37}	S_{56}	S_{55}	$SSSSS_{31}$	E_{42}	D_{37}	D_{38}
H_2	R_{12}		T_5	S_9	K_1	G_1	$SSSSSS_{16}$	D_{13}	E_{17}	E_{19}
I_1	Q_5		R_1	D_5		I_1	SSS_3	A_1	S_2	
	A_1			N_4			$SSSSSSS_3$	L_1	T_1	
				H_2			$SSSSSSSS_2$			
							$SSSS_1$			
							SS_1			

(c)

E_{140}	K_{216}	K_{197}	G_{215}	I_{80}	M_{99}	E_{116}	K_{205}	I_{213}	K_{227}	E_{168}	K_{235}	L_{196}	P_{223}	G_{235}
K_{35}	N_{11}	E_{18}	S_{22}	L_{58}	K_{44}	D_{106}	N_{26}	V_{22}	M_{11}	D_{62}	Q_8	I_{39}	H_{12}	V_3
R_{31}	E_6	G_{10}	K_2	V_{52}	L_{41}	G_{13}	Q_7	L_9	Q_3	Q_{10}	M_1	V_4	S_8	H_3
H_{15}	G_6	D_5	E_2	M_{24}	V_{37}	Q_3	R_4	M_1	G_2	G_2	T_1	M_3	G_3	S_2
D_7	R_3	R_5	I_1	F_{22}	I_{10}	A_2	S_3	E_1	N_2	K_1	S_1	F_3	V_1	L_2
G_4	D_2	A_3	D_1	A_5	T_8	T_3	M_1	F_1	E_1	A_1	R_1	A_1		A_1
Q_7	Q_2	V_3	P_1	T_2	F_2	K_1	G_1		G_1	N_1		T_1		R_1
N_4	M_1	T_2	N_1	D_1	G_2	Y_1				M_1				
L_1		Q_1	C_1	S_2	L_1					K_1				
A_1		S_1	R_1	A_1	G_1									
V_1		M_1	V_1	D_1										
P_1		P_1												

Fig. 11.1. The consensus dehydrin Y (a), S (b) and K (c) segments from 92 dehydrins present in the Swiss Protein database as of December, 2001. Letters are single letter amino acid abbreviations and subscripts indicate the number of occurrences of each residue at each position.

including RAB, LEA D-11, LEA (II) and DHNs, among others.

DHNs share no sequence similarity with any enzyme or protein of known function, have a wide size range (82–648 amino acids), can accumulate to high levels (1–5% of soluble protein) (Close, 1996) and appear to be confined to photosynthetic organisms. During the last decade DHNs have been studied in numerous plant species in an attempt to reveal their presumed biological functions during cellular dehydration. Genes encoding DHNs have been cloned from angiosperms and gymnosperms, as well as mosses (Velten and Oliver, 2001) and lycopods (GenBank U96715). There is also immunological evidence of DHNs from ferns and liverworts (Table 2 of Campbell and Close, 1997).

2.1. Conserved sequence motifs

DHNs are unified by the presence of a highly conserved lysine-rich domain called the K-segment, together with two relatively less common domains called the Y- and S-segments (Close, 1996; 1997). In the N-terminal part of the protein, the Y-segment (DEYGNP) can be found in one to three copies. Analysis of 92 DHNs in the Swiss Protein database found 76 conserved Y-segments (Fig. 11.1a). Positions 1, 4, 5, and 6 are remarkably conserved and the amino acid residues immediately preceding and following the Y-segment are somewhat conserved, with the predominant amino acid residues V/T and V/I, respectively.

The S-segment has the consensus sequence, $LHRSGS_{4-10}(E/D)_3$ (Fig. 11.1b), which in most cases precedes the first K-segment. Out of 57 S-segments analyzed, 43 have a glycine after the three acidic amino acids, whereas the remaining 14 generally have a longer stretch of acidic amino acids (up to seven) followed by V/I/G. The S-segment is very conserved in all positions apart from position five (Fig. 11.1b). The S-segment has been shown to be phosphorylated (see Section 2.3). Seven of the 92 DHNs analyzed contain a stretch of four to five serine residues in the C-terminus followed by the sequence DSD. Apart from these serine residues, C-terminal regions that contain a tract of serines show no other homology to the S-segment presented in Fig. 11.1b.

The K-segment (EKKGIME/DKIKEKLPG) is typically repeated one to eleven times and is the only conserved motif present in all DHNs discovered thus far. Analysis of the 92 DHNs available in the Swiss Protein database disclosed 247 K-segments. The K-segment has a high degree of sequence conservation and most of the alternate amino acids at each position maintain hydrophobicity or charge (Fig. 11.1c). The K-segment has been proposed to form an amphipathic α-helix (Dure, 1993). Furthermore, by reducing the 15 amino acid consensus shown above to 10 (IMDKIKEKLP) or 12 (GIMDKIKEKLPG) residues, the K-segment was proposed to form a class A amphipathic α-helix (Close, 1996). Class A amphipathic helices have well demarcated polar and non-polar faces with negatively charged residues opposite the hydrophobic face and positively charged residues at the polar/non-polar interface (Segrest et al., 1990).

The amino acids separating the conserved segments are termed ϕ-segments. These segments are less preserved and show considerable variation between different DHNs, and sometimes occur as tandem repeats. The ϕ-segments are generally rich in glycine (up to 30%), threonine and charged amino acids.

Based on the YSK shorthand (Close, 1996), five distinct structural types of DHNs can be recognized among higher plant sequences. These comprise Y_nSK_n (35), SK_n (21), K_n (17), Y_nK_n (10), and K_nS (7) types, with the numbers indicating the occurrence of each structural type within the DHNs found in the Swiss Protein database. A search of the *Arabidopsis* proteome revealed ten DHNs, including at least one DHN from each of the five structural types, SK_n (3), Y_nSK_n (3), K_n (2), Y_nK_n (1) and K_nS (1). If the YSK structural types bear a functional meaning, then perhaps it is valid to expect that all plants contain at least one DHN of each type. In addition to these ten proteins that clearly fall within the definition of DHNs, we found six other *Arabidopsis* proteins, each of about 250 amino acids length, that contain a reasonable K-segment-like sequence. The consensus for these proteins is ELMEKISEKIH, which has high homology to position 5–13 of the DHN K-segment (IMD/EKIKEKL). These proteins have the K-segment close to the N-terminal followed almost immediately by a serine stretch resembling the stretch found in the K_nS type of DHNs. The C-terminal 200 amino acids of these proteins show a high degree of homology to reticulons. Reticulons are predominantly associated with the endoplasmic reticulum and were originally described as a group of three alternative transcripts from a single gene expressed in human neuroendocrine cells (Roebroek et al., 1993). It is tempting to speculate that DHNs and reticulons are functionally related, based on this sequence motif similarity.

Two of the DHNs that have been analyzed are from mosses and are of a type possibly not found in higher plants (X_nK). These have one K-segment located in the C-terminus, while most of the remainder of each protein is composed of a series of repeated sequence elements (designated "X"). The DHN cloned from *Tortula ruralis* termed "rehydrin" is composed of 15 repeated sequence elements, each containing a 15 amino acid motif named "GPN" after the highly conserved final three amino acids within the motif (Velten and Oliver, 2001). A section of the GPN segment shows some degree of similarity to the K-segment consensus, and is also predicted to form an amphipathic α-helix (Velten and Oliver, 2001), so these proteins could possibly also be described (liberally) as K_n types. However, the motifs common to higher plant DHNs seem to be combined in a somewhat different manner in moss DHNs. Another moss DHN, *Physcomitrella patens* PpDHNA, contains eleven repetitive sequence elements, each being 35 amino acids, and each repeat contains the GPN motif. But, the N-terminal part of the repeat of PpDHNA contains a Y-segment-like sequence (DNYGNR/P). One is a perfect Y-segment "DEYGNP" and an additional four have only one substitution relative to the higher plant consensus Y-segment (DN/SEYGNP)

(Svensson, 2001). All substitutions occur at the second position, which is the least conserved position of the Y-segment. In the EMBL database we also found a partial *P. patens Dhn* sequence (BI74057) encoding two K-segments. Taken together, it appears that mosses contain multiple types of DHNs, as do higher plants.

2.2. Biochemical properties

A notable characteristic of DHNs is that they remain in solution upon boiling. A biased amino acid composition with a high percentage of glycine, charged and polar residues renders them highly hydrophilic, which may be part of the basis of high temperature solubility. In addition, there are few or no cysteine or tryptophan residues in DHNs, which further suggests that these proteins do not tend to fold to form a hydrophobic core. From the first sequences obtained, it appeared as if DHNs were devoid of cysteine and tryptophan (Close et al., 1989; Close, 1996, 1997). However, with a larger number of examples (92 DHNs from the Swiss Protein database), it now appears that about 25 % of the deduced polypeptide sequences harbor one or more cysteine residues and about 8.5% contain one or two tryptophan residues. Assuming that these reflect actual amino acid sequences rather than DNA sequencing errors, the knowledge that DHNs can tolerated cysteine or tryptophan residues at certain positions can be put to use experimentally. Cysteine residues provide opportunities to couple tags of various sorts using the sulfhydryl group. The intrinsic fluorescence of tryptophan can be used to study the conformation of DHNs in different physicochemical environments.

DHNs have been purified from plants and by expression of recombinant DHNs in *Escherichia coli* (summarized in Svensson, 2001). Purification of *Arabidopsis* DHNs with immobilized metal affinity chromatography revealed strong binding to Cu-charged columns. Since DHNs do not possess any known metal binding motif, the strong binding observed is probably due to coordination complexes formed by multiple histidine residues at various surface-accessible positions within the DHN protein (Svensson et al., 2000).

Structural analyses of four DHNs have indicated that they are intrinsically unstructured proteins. Circular dichroism analysis has shown that maize DHN1 from inbred G50 (a.k.a. RAB17, YSK_2), cowpea DHN1 (Y_2K) and *Citrus unshiu* DHN CuCor19 (K_3S) are essentially unstructured in aqueous solution (Ceccardi et al., 1994; Ismail et al., 1999a; Hara et al., 2001). Additionally, ^{1}H-NMR spectra demonstrated that DSP16 (YSK_2) from *Craterostigma plantagineum* is essentially unstructured in aqueous solution (Lisse et al., 1996). The lack of a defined structure in aqueous solution may be misleading, however, since DHNs *in vivo* may fold into a more ordered structure by interacting with any number of other molecules or intracellular structures such as membranes or nucleoprotein complexes. Intrinsic unfoldedness of DHNs may be a reflection of thermodynamic controls on DHN folding *in vivo*, perhaps influenced by such factors as the concentration of solutes, phosphorylation of the S-segment, or modulations of binding partners such as alterations in the phospholipid content of membranes (see Wright and Dyson, 1999 for a review of intrinsically unstructured proteins). In the presence of SDS, α-helical structure has been induced in cowpea DHN1 and *C. unshiu* DHN CuCor19 (Ismail et al. 1999a; Hara et al., 2001). This might be due to the adoption of structure by the K-segment when bound to the surface of SDS-micelles, which would be consistent with DHN-membrane interactions *in vivo*. Choi et al. (1999) compared allelic differences in barley DHNs and noted that most differences between allelic forms of DHNs are duplications or deletions of the ϕ-sequences separating the conserved repeats rather than amino acid substitutions. Similar types of differences were noted between alleles of cowpea *Dhn1* (Ismail et al., 1999b). Choi et al. (1999) suggested that the Y-, S-, K-, and ϕ-segments represent domains with considerable constraint and therefore DHNs in their native state do have a specific structure.

An apparently oligomeric behavior in gel filtration was observed for maize DHN1 (17.7 kDa, YSK_2), which eluted with an apparent molecular weight (MW) of 42 kDa (Ceccardi et al., 1994), and spinach COR85 (a.k.a. CAP85, K_{11}), which eluted with an apparent MW of 350 kDa (Kazuoka and Oeda, 1994). In addition, four *Arabidopsis* DHNs (LTI29, SK$_3$; COR47, SK$_3$; RAB18, Y$_2$SK$_2$; and LTI30, K$_6$) were analyzed by gel filtration and eluted at a much higher apparent MW values than predicted from their monomeric MWs calculated from their amino acid sequences. However, since gel filtration is influenced by shape in addition to mass, these gel filtration data do not provide rigorous proof of oligomerization. In fact, the *Arabidopsis* DHNs had a slightly lower elution volume (v_e) under denaturing conditions compared to native conditions, further indicating that the DHNs tested are not oligomeric (Svensson et al., 2000). The anomalous behavior of DHNs in the aqueous conditions used for gel filtration seems consistent with the apparent lack of intrinsic structure determined by CD and NMR methods, as described above. The implication is that DHNs in aqueous solution may not be compact, folded proteins or oligomers but rather unstructured monomers. Consistent with this alternative interpretation, DSP16 (YSK$_2$) was found to behave as a monomer in dynamic light scattering and sedimentation analyses (Lisse et al., 1996); these measurements are less influenced by the shape of the protein than is gel filtration.

2.3. Post-translational modifications

Phosphorylation of a DHN was first reported for maize RAB17 (YSK$_2$) (Vilardell et al., 1990). Plana et al. (1991) later demonstrated by *in vitro* and *in vivo* phosphorylation of RAB17 and protein fragmentation that a phosphorylated polypeptide fragment contained the S-segment. The three acidic amino acids following the serine residues were identified as a putative casein kinase 2 substrate sequence, further supporting these results; mutations in this sequence resulted in no or extremely low incorporation of phosphorus (Jensen et al., 1998). Two other DHNs have also been shown to be phosphorylated, tomato TAS14 (YSK$_2$) (Godoy et al., 1994) and *C. plantagineum* DSP16 (YSK$_2$; Lisse et al., 1996). All DHNs shown to be phosphorylated contain the consensus S-segment (Fig. 11.1b). No data are available on whether the K$_n$S type of DHN is phosphorylated.

DHN-like proteins from blueberry and pistachio have recently been found to be glycosylated, a post-translational modification not previously reported (Golan-Goldhirsh, 1998; Levi et al., 1999). The cDNAs for glycosylated blueberry and pistachio encode K$_5$ DHNs.

3. Dehydrin gene expression

How are abiotic stress responsive genes activated? Nordin et al. (1993) originally suggested the identity of a low temperature responsive promoter element (LTRE). The presence of this element was subsequently demonstrated by deletion analysis and it was shown that the 9-bp element, TACCGACAT, confers responsiveness to low temperature, drought and high salinity, but not to ABA (Yamaguchi-Shinozaki and Shinozaki, 1994). This low-temperature and dehydration-responsive element (LTRE/DRE) occurs in several *Dhn* promoters as well as in promoters of other cold and drought responsive genes and has also been referred to as the C-repeat (CRT) (Baker et al., 1994). A small family of *Arabidopsis* proteins called CBF1, CBF2 and CBF3 (CRT binding factor) or DREB1B, DREB1C and DREB1A (DRE-binding protein) activate the low temperature induced expression of genes carrying the DRE/CRT/LTRE element (for review see Shinozaki and Yamaguchi-Shinozaki, 2000). The DRE/CRT/LTRE element is also recognized by a family of related transcription factors (DREB2A and DREB2B) specific to drought (Liu et al., 1998).

Most of the low temperature and drought responsive *Dhn* genes are also induced by exogenous ABA. Consequently, their promoters would be expected to contain *cis*-elements mediating this response. Indeed, sequences closely resembling ABA response elements (ABREs) exist in these promoters, with the consensus motif of C/TACGTGGC. The ABREs have been shown to confer ABA-regulated expression of many genes (Guiltinan et al., 1990; Leung and Giraudat 1998). Proteins that can specifically bind to ABREs contain the basic domain/leucine zipper (bZIP) motif found in many transcription factors (e.g., Guiltinan et al., 1990). Two bZIP proteins that bind specifically to the ABRE elements were recently characterized (Uno et al., 2000). These genes, encoding AREB1 and AREB2 (ABA-responsive element binding protein), are drought, salt and ABA responsive. In another study a small family of ABRE binding factors (ABFs) was characterized, also from *Arabidopsis* (Choi et al., 2000). *ABF*s respond differently to various environmental stresses including low temperature, suggesting that they act in different stress-responsive pathways.

During seed development, ABA plays an important role in embryo maturation, acquisition of desiccation tolerance and induction and maintenance of dormancy. The ABRE elements are also involved in developmental control of *Dhn* expression in seeds. This expression appears to be ABA-mediated involving the ABRE promoter elements. Different ABRE element combinations are active in vegetative tissues and in seeds, as shown by analysis of the maize *RAB17* promoter (Busk et al., 1997). The transcriptional activator ABI3 in *Arabidopsis* (related to VP1 in maize) is the key regulator of this response in seeds (Parcy et al., 1994) and *abi3* mutants remain non-dormant, do not acquire desiccation tolerance and fail to accumulate several seed maturation related transcripts. Expression of the normally seed specific *ABI3* gene in transgenic *Arabidopsis* results in ABA-induced expression of seed specific genes in vegetative tissues and influences stress responses in these tissues (Parcy and Giraudat, 1997).

In summary, *Dhn* genes are controlled by a number of distinct but interacting signal pathways that may also exhibit tissue specificity.

4. Association with abiotic stresses and anticipated functions

4.1. *Low temperature*

Survival of temperate plants during cold winters is dependent on their ability to acclimate at low non-freezing temperatures, which will then prepare them to withstand freezing temperatures as low as −5 to −30°C (Thomashow, 1998). Numerous physiological changes are induced during

acclimation including increased levels of sugars, soluble proteins, proline, organic acids, and various proteins including DHNs, as well as alterations in membrane lipid composition (Hughes and Dunn, 1990).

DHNs have been extensively studied in barley, where a dispersed family of 12 *Dhn* genes were characterized (Choi et al., 1999; Choi and Close, 2000; Zhu et al., 2000). Most of the barley *Dhn* genes are upregulated by dehydration and ABA, while others are cold-induced or embryo-specific. One common association is that all of the *Dhn* genes that encode an YSK$_2$-type DHN are alkaline and upregulated by dehydration and ABA, but not by low temperature. A converse association is that barley DHNs upregulated by low temperature are neutral or acidic. Barley *Dhn5* and *Dhn8* were induced by cool temperatures both under controlled conditions and in the field, and the extent of expression of these genes was strongly dependent on the extent of cold encountered in the field. Expression was higher under cold, dry conditions and lower under warm, wet conditions. However, the induction of *Dhn5* steady state levels was faster, occurring within 12 h of cold treatment, compared to *Dhn8*, which required about 4 d to reach steady state, indicating an influence of multiple control pathways. In another study, the DHN5 protein (P-80) accumulated specifically in response to low temperature during cold acclimation in the barley cultivar "Aramir" (Bravo, et al., 1999). The level of P-80 did not increase with drought, ABA or high temperature and a period of 48 h was required before its induction to detectable levels under low temperature. The level of P-80 decreased when acclimated plants were returned to normal conditions. Involvement of this DHN in preventing injury during freeze-induced dehydration was envisaged from its specific localization to tissues that are most vulnerable to ice nucleation during freezing, such as vascular and epidermal tissues. Most of the remaining barley DHNs were induced by water stress, and most of these (*Dhn*1, 2, 3, 4, 7 and 9) were also induced by freeze-thaw treatments (Zhu et al., 2000). *Dhn*12 is unique in having a different expression pattern compared to the other 11 *Dhn* genes, being apparently embryo specific instead of stress responsive (Choi and Close, 2000).

Among the 12 *Dhn* genes in barley, a cluster on chromosome 5H is of particular significance. Three *Dhn* genes (*Dhn1*, *Dhn2* and *Dhn9*) were mapped within the 99% confidence interval of a major quantitative trait locus (QTL) for winter hardiness (Pan et al., 1994; Choi et al., 1999). Expression of these genes was higher under cooler acclimation temperatures and in a freezing tolerant cultivar, suggesting that the cold-induced *Dhn* genes may be involved in adaptation to cellular dehydration incurred during freezing. However, none of these *Dhn* genes seems to be very tightly associated with the *Vrn-1H* locus, which is the strongest component of what appears to be a complex QTL on chromosome 5H involved in winter hardiness (Zhu et al., 2000).

Using antibodies specific for wheat WCS120 (K_6) and WCOR410 (SK_3), which are responsive to low temperature, Fowler et al. (2001) found that homologs of these proteins in barley (DHN5 and DHN8, respectively) have similar accumulation kinetics. Accumulation of the barley homologs were much higher in plants that were exposed to low temperature under short day photoperiod than under long day photoperiod. Further, the level of differential expression paralleled the level of low temperature tolerance. Taken together, these results make a case that developmental pathway regulators also are important in the control of freezing tolerance mechanisms (Fowler et al., 2001).

The *Dhn* gene family in wheat is quite similar to that found in barley and includes members that are coordinately regulated by low temperature. This gene family encodes a group of highly abundant proteins with apparent MWs ranging from 12 to 200 kDa (Sarhan et al., 1997). Accumulation of wheat WCS120 and WCOR410 proteins is correlated with the capacity to develop freezing tolerance (Danyluk et al., 1994, 1998). The specific accumulation of WCS120 and closely related DHNs in the vascular transition zone of the meristematic crown tissues of acclimated plants suggested a role for these proteins in protecting sensitive tissues during freezing (Houde et al., 1995; Sarhan et al., 1997). Houde et al. (1995)

argued that these proteins probably act by surrounding vital cellular proteins and protecting them during dehydration or freezing. The specific localization of WCOR410 peripheral to the plasma membrane led the authors to propose that these proteins are involved in cryoprotection of the plasma membrane against freezing or dehydration by preventing destabilization of the plasma membrane (Danyluk et al., 1998). The exact mechanism of how WCOR410 proteins protect the plasma membrane is not known and three possible models were suggested: (1) replacement of water, thereby solvating membranes, (2) prevention of interactions between membrane bilayers, thereby reducing fusion of membranes and lamellar to hexagonal II phase transitions, or (3) the formation of salt bridges, thereby preventing the damaging effect of increased ionic concentrations (Danyluk et al., 1998).

A comparison of cowpea (*Vigna unguiculata* L.) near isogenic lines with allelic variation in the gene encoding DHN1 (35 kDa, Y_2K) demonstrated that the presence of this protein in mature seeds is associated with chilling tolerance during seedling emergence (Ismail et al, 1997; 1999b). The increased chilling tolerance observed was not correlated with reduced electrolyte leakage, signifying that the effect of cowpea DHN1 might not be protection of the plasma membrane. The accumulation of this DHN is coordinated with the onset of the dehydration phase of embryo development. The ability of this protein to adopt α-helical structure in the presence of SDS supports the hypothesis that DHNs may bind to membranes (Ismail et al., 1999a). It was suggested that DHN1 interacts with membranes in the interior of the cells and reduces dehydration-induced damage. Furthermore, it was suggested that this DHN may act in a manner similar to α-synuclein (Ismail et al., 1999b), which is a protein found in presynaptic nerve terminals that is unstructured in aqueous solution but gains α-helicity upon binding to lipid bilayers (Davidson et al., 1998).

Five *Arabidopsis* DHNs: ERD14 (SK_2), LTI30 (K_6), LTI29 (SK_3), COR47 (SK_3), and RAB18 (Y_2SK_2) were studied in unstressed and low-temperature treated *Arabidopsis* plants (Nylander et al., 2001). LTI30 protein was induced only by low temperature treatment, whereas LT129 and COR47 proteins were cold-induced, but also present in low concentration in unstressed plants. ERD14 protein was constitutive, but rose to elevated levels upon cold treatment. RAB18 protein did not accumulate upon low temperature treatment (Nylander et al., 2001). The corresponding transcript accumulation correlated well with protein levels, with the exception of *LTI30*. Although LTI30 protein was not detected in unstressed plants, *LTI30* transcripts were present in the same material. Therefore, *LTI30* expression seems to be both transcriptionally and post-transcriptionally regulated.

DHNs have also been studied in several fruit trees and other woody perennials. In trees of temperate origin the amount of DHNs fluctuates seasonally, being high in winter and low during the active growth period (Wisniewski et al., 1996; Sauter et al., 1999). Two photoperiod-induced DHN-like proteins with molecular masses of 34 and 36 kDa were identified in birch (*Betula pubescens*) and their accumulation correlated with cellular dehydration under short days. This led to the hypothesis that DHNs play a role in freeze tolerance of this woody species (Welling et al., 1997). Rinne et al. (1999) identified two additional DHNs that accumulate during cold hardening in nuclei, storage protein bodies and starch-rich amyloplasts of the same species. The authors hypothesized that DHNs might create local pools of water in dehydrated cells to maintain activity of enzymes required for degradation of starch and protein reserves that are gradually consumed during winter. To substantiate their hypothesis they further reported that under conditions of dehydration, the activity of α-amylase could be improved by a partially purified DHN.

A 60 kDa DHN was detected in Scots pine (*Pinus sylvestris*) seedlings during deacclimation, and the quantity decreased with resumption of growth. The concentration of this DHN also decreased during de-hardening and rapidly decreased when seedlings were fertilized with nitrogen (Kontunen-Soppela et al., 2000). In the evergreen *Rhododendron* species, a cold-induced

25 kDa DHN was found to co-segregate with freezing tolerance of cold-acclimated plants that account for up to 78% of the variation in leaf freezing tolerance among parents, F_1 and F_2 progenies (Lim et al., 1999). The authors argued that this DHN could be used as a genetic marker for cold hardening due to the parallel changes observed in protein levels and leaf freezing tolerance.

In *C. unshiu*, a cold-responsive DHN (CuCOR19, K_3S) was vigorously induced by cold stress (Hara et al., 2001). The authors hypothesized that the random coil structure adopted by this DHN in solution may have a role in cryoprotection of enzymes by forming cohesive layers with the surfaces of oligomeric enzymes, thereby preventing their dissociation at low temperature and preserving their activity. As suggested by others, another idea was that random coils could also bind water molecules and prevent excessive dehydration of cells (Ingram and Bartels, 1996).

Three DHN-like proteins of 65, 60, and 14 kDa apparent MW accumulated during cold acclimation in blueberry (Arora et al., 1997; Levi et al., 1999). The levels of these DHNs increased considerably during low temperature acclimation and decreased during deacclimation or resumption of growth. Cold hardiness levels of dormant buds were positively correlated with the relative levels of these DHNs. Using two blueberry cultivars and a wild relative, Panta et al. (2001) found that blueberry DHNs are responsive to photoperiod and can be induced in plant organs other than floral buds, but with greater accumulation in stems and roots than in leaves under low temperature.

When taken together, the above studies provide strong circumstantial evidence of an involvement of DHNs in the acquisition of freezing tolerance in temperate plant species. Most of these DHNs accumulate slowly during an acclimation period when plants are exposed to low, non-freezing temperatures, and their levels decrease with deacclimation or resumption of growth. These proteins possibly wield their protective role through maintaining the integrity of macromolecules and cellular membranes as well as preserving activity of crucial enzymes involved in energy metabolism, mobilization of stored reserves needed for maintenance during winter, or in cellular protection and repair mechanisms.

4.2. Water deficit

"Drought-induced" DHNs have been identified in many plant species, among which we summarize just a few examples to illustrate the commonly observed trends and some exceptions. In general, DHN accumulation is highly associated with water deficit and increased ABA levels.

In *Arabidopsis*, RAB18 accumulated to high levels under water deficit (Mäntylä et al., 1995) and also in response to ABA (Nylander et al., 2001). In sorghum, a 21 kDa DHN accumulated in response to water deficit in both drought tolerant and intolerant genotypes. The isolated cDNA of this protein encodes a hydrophilic alkaline K_2 DHN of approximately 16.3 kDa. The protein accumulated in leaves and roots, with greater transcript expression in mature plants than in seedlings (Wood and Goldsbrough, 1997). The effect of day/night cycle on the expression of two stress-induced *Dhn* genes *HaDhn1* (Y_3SK_2) and *HaDhn2* (SK_2) was investigated in leaves of sunflower subjected to water deficit (Cellier et al., 2000). *HaDhn1* transcript levels oscillate diurnally, with the peak of mRNA accumulation at midday. However, the accumulation of *HaDhn2* transcripts did not fluctuate diurnally, instead increasing with water deficit intensity. This suggests that these two DHNs have different roles during drought stress. In tall fescue grass a set of DHNs (23–60 kDa) were progressively induced by water deficit. Among them, a 35 kDa DHN was induced by stress, and this occurred with or without ABA treatment (Jiang and Huang, 2002).

Accumulation of *Dhn* transcripts in leaf bases of cocksfoot, a perennial forage grass, occurred only under severe water deficit conditions, increased progressively with deficit intensity, and disappeared within 8 days of rewatering. The level of ABA in leaf bases followed a similar pattern, suggesting that this induction might be responsive to ABA. However, since this stress response was more pronounced in a drought susceptible than a resistant cultivar, the authors speculated that

survival of the leaf bases and meristematic tissues in this species is associated with low levels of *Dhn* transcripts and ABA (Volaire et al., 1998).

Woody plants also retain the capacity to accumulate DHNs under water deficit. In aspen (*Populus tremula*), a 43 kDa DHN was constitutively present in roots but only induced in response to water deficit in shoots. Two other DHNs with apparent molecular weights of 31 and 33 kDa were differentially expressed in response to moisture deficit and ABA treatment, respectively, with the former specific only to shoot and the latter expressed both in shoots as well as roots (Pelah et al., 1997). In blueberry, drought responsive DHNs accumulated to a greater extent in stems than in roots and leaves. However, their accumulation occurred prior to significant changes in relative water content (Panta et al., 2001).

4.3. Salinity

Salinity can cause damage to plants through cellular dehydration by osmosis, accumulation of ions to toxic levels and by causing nutritional imbalances. Cellular dehydration may arise due to low osmotic potential imposed externally, when ion uptake is inadequate to balance reduction in osmotic potential in the soil, or from high salts in the intercellular spaces of leaves causing dehydration of the symplast. Salinity responsive LEA genes probably are induced due to cellular dehydration caused by high osmotic stress, explaining why "salinity responsive" DHNs have been observed in a number of studies (Godoy et al., 1994; Danyluk et al., 1994). Genetic variability in the extent to which salt-responsive genes are expressed has also been noted. For example, Moons et al. (1995) found that levels of DHNs are significantly higher in roots of salt tolerant rice genotypes than in intolerant cultivars. In a recent study with finger millet and rice, a positive correlation between levels of DHNs and stress tolerance was observed (Jayaprakash et al., 1998). These DHNs were also responsive to partial dehydration and ABA, and they accumulated to higher levels in tolerant cultivars than in the intolerant cultivars of both species, suggesting an association between stress tolerance and the quantitative expression of these proteins.

In *Arabidopsis*, only the constitutive but stress responsive DHN ERD14 accumulated to large amounts upon salt stress, with the highest amount in stems and leaves. LTI30 accumulated to low levels in roots but not in stems, leaves and flowers of salt treated plants. In plants grown under saline conditions, low levels of LTI29 were found in roots, stems, leaves and flowers whereas COR47 was detected only in roots and stems. RAB18 accumulated in salt stressed flowers but not in roots, stems or leaves (Nylander et al. 2001).

In general, the induction of DHNs by the osmotic component of salinity is commonplace, and as with drought and freezing tolerance a number of studies have associated higher DHN induction levels with more tolerant genotypes.

5. Seed dehydrins

Acquisition and maintenance of desiccation tolerance during embryo maturation in orthodox (desiccation tolerant) seeds entail the activation of a suite of mechanisms. These include antioxidant systems, oligosaccharides and amphipathic molecules, and LEA proteins (Oliver and Bewley, 1997; Pammenter and Berjak, 1999). DHNs are probably the most commonly observed LEA proteins during seed maturation in orthodox seeds (Ingram and Bartels, 1996; Close, 1997).

Cowpea DHN1 accumulated to a detectable level in seeds when seed moisture content decreased to about 60%, and its level increased progressively with seed maturation drying (Ismail et al., 1999a). Accumulation of *Dhn1* transcripts began earlier, approximately 10 days before DHN1 protein reached detectable level (Ismail, Hall and Close, unpublished). In wheat, a 25 kDa DHN was strongly induced in developing grains at the time when desiccation tolerance developed, and this protein could also be induced at earlier stages by detachment of the grains (Black et al., 1999). In this study, accumulation of oligosaccharides occurred later and was not correlated with the development of desiccation tolerance. From this temporal separation, the authors argued that

induction of desiccation tolerance does not require interaction of DHNs with polysaccharides.

Greggains et al. (2000) attributed the greater desiccation tolerance of orthodox seeds in some species of the genus *Acer* to the accumulation of greater amounts of DHNs and soluble sugars in the embryo during maturation drying than *Acer* species with recalcitrant (desiccation sensitive) seeds. DHNs are also detected in recalcitrant plant species and some evidence suggests that DHN-like proteins are more abundant in seeds from plants of temperate origin than in seeds of tropical wetland species (Farrant et al., 1996). However, Han et al. (1997) reported variation within tropical recalcitrant species in the ability to accumulate DHNs during seed development and in response to various abiotic stresses or ABA treatment. Their studies support the contention that desiccation sensitivity of recalcitrant seeds is due, in part, to inability to accumulate sufficient DHNs and/or other LEA proteins. It is not certain from these studies to what extent DHN expression in recalcitrant seeds is related to their ability to tolerate dehydration.

In *Arabidopsis*, RAB18 accumulated in drying seeds and its levels remained high even after two days of imbibition (Nylander et al. 2001). *Arabidopsis* XERO1 (YSK$_2$) expression was observed in seeds and young seedling up to 3–4 days post-germination, but not during later stages of seedling development or in mature plants (Rouse, et al., 1996). Yet, in another study XERO1 expression was constitutive in 2 week old seedlings (Welin et al., 1994). Another example of DHNs in seeds of *Arabidopsis* is the PAP310 (X91920) cDNA, encoding a Y$_3$SK$_2$ DHN, that was isolated from a dry seed library. So far, it seems that only YSK types of DHNs are present in *Arabidopsis* seeds.

6. Localization of dehydrins

6.1. *Tissue distribution*

The tissue distribution of DHNs has been analyzed in several plant species by immuno-histochemical methods. DHN5 in cold acclimated barley was

Table 11.1. Summary of immuno-histochemical localization studies of the LTI29, ERD14, RAB18 and LTI30 dehydrins in *Arabidopsis*. Adapted from Nylander et al., 2001.

DHN	Treatment	Root	Stem	Leaf	Flower
LTI29	non-stressed	VT, RT	VT	nd	nd
ERD14	non-stressed	VT, RT	VT	VT	VT
RAB18	non-stressed	nd	VT, ST[1]	ST[1]	ST[1]
LTI30	non-stressed	nd	nd	nd	nd
LTI29	low temperature	VT, GS, RT	VT, GS	VT, M	VT, GS
ERD14	low temperature	VT, GS, RT	VT, GS	VT, M	VT
RAB18	ABA-treatment	VT, GS, RT	VT, GS, ST[2]	VT, M, ST[2]	VT, GS, T[2]
LTI30	low temperature	VT, GS	VT	VT	VT, PS

[1]Staining detected in the nuclei.
[2]Staining detected both in cytoplasm and nuclei.
ABA: abscisic acid, DHN: dehydrin, GS: general staining, M: mesophyll cells, nd: not detected, PS: pollen sacks, RT: root tip, ST: stomatal guard cells, VT: vascular tissue.

localized in vascular tissue and epidermis of shoots and in non-acclimated leaves in the vascular bundle (Bravo et al., 1999), as mentioned earlier in this review. Schneider et al. (1993) analyzed the localization of DSP16 and DSP16-like DHNs in drought-treated *C. plantagineum.* Staining was detected in all types of cells, but preferentially in cytoplasm-rich cells like phloem sieve tube elements in leaves and embryonic cells in the seed. ECP40 (Y_3SK_2), a carrot DHN, was localized in the endosperm and zygotic embryos in mature seeds (Kiyosue et al., 1993). Proteins of the WCS120 DHN (K_6) family were localized mainly in the vascular bundle and bordering parenchyma cells in cold acclimated wheat crown tissues (Houde et al., 1995). Wheat DHN WCOR410 (SK_3) was also found to accumulate preferentially in the vascular transition area (Danyluk et al., 1998). Peach PCA60 (Y_2K_9) appeared to be generally distributed in cells of all tissues of shoots collected in winter, including epidermal, cortical, phloem and xylem tissues (Wisniewski et al., 1999). Immunolocalization in salt-stressed tomato plants showed strong accumulation of TAS14 in developing adventitious root primordia, vascular tissue of the shoot and differentiated cortical cells of the stems and leaves (Godoy et al., 1994). The maize DHN RAB17 was localized in all cell types of mature embryos and in aleurone layers (Asghar et al., 1994; Goday et al., 1994).

In *Arabidopsis* LTI29 and ERD14 accumulated in the root tip and vascular tissues, whereas RAB18 was localized to stomatal guard cells in unstressed plants (Table 11.1). Localization of these DHNs in stressed plants was not restricted to certain tissues or cell types; LTI29, ERD14 and RAB18 were present in most cells although cells within and surrounding the vascular tissue showed more intense staining (Table 11.1). In contrast, LTI30 was not detected in plants grown under control conditions and was mainly localized to vascular tissues and anthers of cold treated plants (Table 11.1) (Nylander et al., 2001).

In general, localization studies, particularly with *Arabidopsis* (Table 11.1), have revealed tissue and cell type specific accumulation of DHNs in unstressed plants. This suggests that DHNs under non-stressed conditions carry out some basic protective function in certain cell types (e.g., vascular tissue), and upon stress conditions similar functions are extended to most cells and tissues. Overall, DHNs consistently have been observed in vascular tissues and surrounding cells, which are generally more vulnerable to ice nucleation and damage at low temperature, as well as other cells that often enter and exit a state of dehydration without losing viability (e.g. aleurone). The variation in spatial distribution and stress specificity suggests functional specialization of different DHNs or different sub-families.

6.2. *Subcellular localization*

Several studies have demonstrated that DHNs are present in the cytoplasm and nucleus. Examples of such studies are from maize embryos (Asghar et al., 1994; Goday et al., 1994; Egerton-Warburton et al., 1997), developing tomato root primordia (Godoy et al., 1994), pea root tip meristems subjected to slow dehydration (Bracale et al., 1997), wheat crown tissues (Houde et al., 1995) and peach shoots (Wisniewski et al., 1999). The nuclear localization of maize RAB17 seems to be enhanced by phosphorylation of the S-segment (Jensen et al., 1998). Most DHNs contain an S-segment and NLS sequences similar to those found in RAB17. Yet, some studies have found such DHNs only in the cytoplasm, including phosphorylated DSP16 in desiccated leaves of the resurrection plant *C. plantagineum* (Schneider et al., 1993) and rice RAB21 (YSK$_2$, Mundy and Chua, 1988). Also, the wheat DHN WCOR410 was found to accumulate in the vicinity of the plasma membrane of cells in the vascular transition area (Danyluk et al., 1998). From the aggregate of these studies it can be inferred that phosphorylation of serines in the S-segment contributes to nuclear localization, but other factors control the subcellular location of DHNs as well.

A few studies also seem to provide examples of the presence of DHNs in or in association with organelles or cytoplasmic endomembrane structures. The peach DHN PCA60 was reported to be associated with chloroplasts in addition to the cytoplasm and nucleus (Wisniewski et al., 1999), and fractionation studies of winter wheat, winter rye and maize were interpreted as showing the presence of two DHN-like proteins associated with mitochondria (Borovskii et al., 2000). Another fractionation study showed the presence of spinach CAP85 predominately in the soluble fraction of the cytoplasm, but also associated with the endoplasmic reticulum (Neven et al., 1993). In nonacclimated birch a 16 kDa constitutive DHN was localized in the cytoplasm, whereas after cold acclimation two additional DHNs were detected and the three DHNs were localized in the nuclei, storage protein bodies and starch-rich amyloplasts

(Rinne et al., 1999). In maize, one study has demonstrated DHNs surrounding the periphery of protein and lipid bodies in addition to more general dispersion throughout the cytoplasm and nucleus (Egerton-Warburton et al., 1997). In summary, DHNs appear to be found within or surrounding numerous subcellular organelles endomembranes. Again, there may be specificity of intracellular location for each sub-type of DHN.

7. Functional studies of dehydrins

The contribution of DHNs to abiotic stress tolerance has also been investigated by exploring natural genetic variation in *Dhn* genes, and by using transgenic or in vitro cryoprotective approaches.

The study with cowpea DHN1 discussed in Section 5.1 provided evidence that natural allelic variation in a single *Dhn* gene can account for one aspect of chilling tolerance during seedling emergence (Ismail et al., 1999b). Numerous other genetic studies have been conducted using transgenic approaches that overproduce single DHNs in plants or yeast and assess effects on stress tolerance. Tobacco plants that overproduce spinach CAP85 were evaluated for relative freezing tolerance by measurement of electrolyte leakage on detached leaves subjected to controlled freezing. The authors concluded that CAP85 had no profound influence on stress tolerance (Kaye et al., 1998). Similar results were obtained when overproducing RAB18 or LTI29 in *Arabidopsis* and tobacco (Lång, 1993; Welin, 1994). Drought tolerance of tobacco plants that overproduced *C. plantagineum* DSP16 was evaluated by mild drought stress on detached leaves followed by determination of electrolyte leakage. No increase in drought tolerance was detected (Iturriaga et al., 1992). The authors suggested that this could be due to unsuitability of the ion leakage test or a requirement for simultaneous expression of several drought-related proteins. Likewise, no improvement in drought tolerance was detected by overproducing LTI29 in *Arabidopsis* using a similar test system (Lång, 1993; Welin, 1994). Drought tolerance of tobacco and *Arabidopsis* plants that overproduced RAB18 was estimated visually on detached leaves and no difference was detected compared to wild type plants (Lång, 1993). In contrast, constitutive expression of barley DHN1 (*ABA2*) in *Arabidopsis* enhanced the germination rate under salt and osmotic stress, though it was without a significant effect on germination at low temperature or freezing damage of adult plants (Calestani et al., 1998). Expression of the tomato DHN *LE4* (a.k.a. TAS14) (YSK$_2$) in yeast conferred tolerance to high concentrations of KCl, but not to NaCl or sorbitol. In addition, the yeast strain producing LE4 had increased freezing tolerance (Zhang et al., 2000).

Taken together, overproduction of DHNs in transgenic plants has so far not improved freezing or drought tolerance, the only exception being an improvement in germination under salt and osmotic stress, as noted above. In contrast, over-expression of *CBF1/DREB1B* or *CBF3/DREB1A* led to the constitutive expression of DHN (COR47) and several other normally stress-induced genes that carry promoters containing the DRE/CRT/LTRE element, and the result was improved freezing, drought and salt tolerance of non-acclimated plants under laboratory conditions (Jaglo-Ottosen et al., 1998; Kasuga et al., 1999). Whether or not DHN gene expression is a major factor in this general increase in stress tolerance has not been studied.

Biochemical analyses of DHNs have shown that spinach COR85, maize DHN1, wheat WSC120, peach PCA60 and *Citrus unshiu* CuCor19 have cryoprotective activity *in vitro* (Kazuoka and Oeda, 1994; Houde et al., 1995; Close, 1996; Wisniewski et al., 1999; Hara et al., 2001). In addition to cryoprotection, PCA60 was demonstrated to possess antifreeze activity by modifying the normal growth of ice and exhibiting thermal hysteresis (Wisniewski et al., 1999).

Proper functioning of DHNs during cellular dehydration might involve complex processes involving the concerted action of a group of DHNs or interaction with other protective molecules such as other LEA proteins or compatible solutes (Hoekstra et al., 2001). Nevertheless, this complexity would neither preclude the possibility that specific alleles of specific DHN genes can be the basis of naturally occurring trait variation, nor mandate that allelic variation in master regulatory genes control any specific stress tolerance trait. At the present time, where mechanisms are only beginning to be understood, there is much to be gained from striving to understand the mechanistic bases of genetic variation in stress tolerance that nature has provided.

8. Conclusions

Strong circumstantial evidence, in the form of gene expression patterns and immunolocalization, has accrued that support a role of DHNs in abiotic stress adaptation in plants. Several *in vitro* studies have led to a consistent observation of cryoprotective properties of DHNs. Most of the evidence to date points to a role of DHNs in stabilizing membranes and macromolecules, preventing structural damage during cellular dehydration, and maintaining the activity of essential enzymes. The biased amino acid composition of DHNs towards extreme hydrophilicity and their abundance argue in favour of a mass action rather than direct enzymatic role. This notion is supported by the absence of similarity to any known enzymes. In addition, genetic evidence of a role of *Dhn* genes in natural variation in stress tolerance and in transgenic plants has started to accumulate, though none of the genetic evidence yet is entirely compelling.

Despite considerable research and steady progress to discern the functional significance of DHNs, knowledge of their fundamental biochemical properties and the phenological consequences of their expression is still incomplete. Given their practically universal association with abiotic stress adaptation, DHNs certainly merit further investigation.

References

Arora, R., Rowland, L.J. and Panta, G.R. (1997). Chill-responsive dehydrins in blueberry: are they associated with cold hardiness or dormancy transitions? Physiol. Plant. 101, 8–16.

Arora, R., Pitchay, D.S. and Bearce, B.C. (1998). Water-stress-induced heat tolerance in geranium leaf tissues: a possible linkage through stress proteins? Physiol. Plant. 103, 24–34.

Asghar, R., Fenton, R.D., DeMason, D.A. and Close, T.J. (1994). Nuclear and cytoplasmic localization of maize embryo and aleurone dehydrin. Protoplasma 177, 87–94.

Baker, J., Steele, C. and Dure, L.I. (1988). Sequence and characterization of 6 *Lea* proteins and their genes from cotton. Plant Mol. Biol. 11, 277–291.

Baker, S.S., Wilhelm, K.S. and Thomashow, M.F. (1994). The 5′-region of *Arabidopsis thaliana cor15a* has *cis*-acting elements that confer cold-, drought- and ABA-regulated gene expression. Plant Mol. Biol. 24, 701–713.

Black, M., Corbineau, F., Gee, H. and Côme, D. (1999). Water content, raffinose, and dehydrins in the induction of desiccation tolerance in immature wheat embryos. Plant Physiol. 120, 463–471.

Borovskii, G.B., Stupnikova, I.V., Antipina, A.I., Downs, C.A. and Voinikov, V.K. (2000). Accumulation of dehydrin-like protein in the mitochondria of cold treated plants. J. Plant Physiol. 156, 797–800.

Boyer, J.S. (1982). Plant productivity and environment. Science 218, 443–448.

Bracale, M., Levi, M., Savini, C., Dicorato, W. and Galli, M.G. (1997). Water deficit in pea root tips: effects on the cell cycle and on the production of dehydrin-like proteins. Ann. Bot. 79, 593–600.

Bravo, L.A., Close, T.J., Corcuera, L.J. and Guy, C.L. (1999). Characterization of an 80 kDa dehydrin-like protein in barley responsive to cold acclimation. Physiol. Plant. 106, 177–183.

Bray, E.A. (1993). Molecular responses to water deficit. Plant Physiol. 103, 1035–1040.

Bray, E.A. (1997). Plant responses to water deficit. Trends Plant Sci. 2, 48–54.

Busk, P.K., Jensen, A. B. and Pages, M. (1997). Regulatory elements in the promoter of the abscisic acid responsive gene *rab17* from maize. Plant J. 11, 1285–1295.

Calestani, C., Bray, A.E., Close, T.J. and Marmiroli, M. (1998). Constitutive expression of a dehydrin in *Arabidopsis thaliana* enhances tolerance to salt stress. 9th International Conference on *Arabidopsis* Research, Univ. Wisconsin, Madison, USA.

Campbell, S.A. and Close, T.J. (1997). Dehydrins: genes, proteins and associations with phenotypic traits. New Phytol. 137, 61–74.

Ceccardi, T.L., Meyer, N.C. and Close, T.J. (1994). Purification of a maize dehydrin. Prot. Expr. Purif. 5, 266–269.

Cellier, F., Conéjéro, G. and Casse, F. (2000). Dehydrin transcript fluctuations during day/night cycle in drought-stressed sunflower. J. Exp. Bot. 51, 299–304.

Choi, D.-W., and Close, T.J. (2000). A newly identified barley gene, *Dhn12*, encoding a YSK_2 dehydrin, is located on chromosome 6H and has embryo-specific expression. Theor. Appl. Genet. 100, 1274–1278.

Choi, D.-W., Zhu, B. and Close, T.J. (1999). The barley (*Hordeum vulgare* L.) dehydrin multigene family: sequences, allelic types, chromosome assignments, and expression characteristics of 11 *Dhn* genes of cv. Dicktoo. Theor. Appl. Genet. 98, 1234–1247.

Choi, H., Hong, J., Ha, J., Kang, J. and Kim, S.Y. (2000). ABFs, a family of ABA-responsive element binding factors. J. Biol. Chem. 275, 1723–1730.

Close, T.J. (1996). Dehydrins. Emergence of a biochemical role of a family of plant dehydration proteins. Physiol. Plant. 97, 795–803.

Close, T.J. (1997). Dehydrins: a commonality in the response of plants to dehydration and low temperature. Physiol. Plant. 100, 291–296.

Close, T.J., Kortt, A.A. and Chandler, P.M. (1989). A cDNA-based comparison of dehydration-induced proteins (dehydrins) in barley and corn. Plant Mol. Boil. 13, 95–108.

Danyluk, J., Houde, M., Rassart, E. and Sarhan, F. (1994). Differential expression of a gene encoding an acidic DHN in chilling sensitive and freezing tolerant Gramineae species. FEBS Lett. 244, 20–24.

Danyluk, J., Perron, A., Houde, M., Limin, A. Fowler, B., Benhamou, N. and Sarhan, F. (1998). Accumulation of an acidic dehydrin in the vicinity of the plasma membrane during cold acclimation of wheat. Plant Cell 10, 623–638.

Davidson, W.S., Jonas, A., Clayton, D.F. and George, J.M. (1998). Stabilization of α-synuclein secondary structure upon binding to synthetic membranes. J. Biol. Chem. 273, 9443–9449.

Dure, L.I., (1993). Structural motifs in LEA proteins of higher plants. In: Response of Plants to Cellular Dehydration during Environmental Stress (Close, T.J. and Bray, E.A., Eds.), pp. 91–103. American Society of Plant Physiologists, Rockville, MD.

Dure, L.I., Crouch, M., Harada, J., Ho, T.-H.D., Mundy, J., Quatrano, R., Thomas, T. and Sung, Z.R. (1989). Common amino acid sequence domains among LEA proteins of higher plants. Plant Mol. Biol. 12, 475–486.

Egerton-Warburton, L.M., Balsamo, R.A. and Close, T.J. (1997). Temporal accumulation and ultrastructural localization of dehydrins in *Zea mays*. Physiol Plant. 101, 545–555.

Farrant, J.M., Pammenter, N.W., Berjak, P., Farnsworth, E.J. and Vertucci, C.W. (1996). Presence of dehydrin-like proteins and levels of abscisic acid in recalcitrant (desiccation sensitive) seeds may be related to habitat. Seed Sci. Res. 6, 175–182.

Fowler, D.B., Breton, G., Limin, A.E., Mahfoozi, S. and Sarhan, F. (2001). Photoperiod and temperature interactions regulate low-temperature-induced gene expression in barley. Plant Physiol. 127, 1676–1681.

Goday, A., Jensen, A.B., Culianezmacia, F.A., Alba, M.M., Figueras, M., Serratosa, J., Torrent, M. and Pages, M. (1994). The maize abscisic acid-responsive protein Rab17 is located in the nucleus and interacts with nuclear-localization signals. Plant Cell 6, 351–360.

Godoy, J.A., Lunar, R., Torresschumann, S., Moreno, J., Rodrigo, R.M. and Pintortoro, J.A. (1994). Expression, tissue distribution and subcellular-localization of dehydrin Tas14 in salt-stressed tomato plants. Plant Mol. Biol. 26, 1921–1934.

Golan-Goldhirsh, A. (1998). Developmental proteins of *Pistacia vera* L. bark and bud and their biotechnological properties: a review. J. Food Biochem. 22, 375–382.

Greggains, V., Finch-Savage, W.E., Quick, W.P. and Atherton, N.M. (2000). Putative desiccation tolerance mechanisms in orthodox and recalcitrant seeds of the genus *Acer*. Seed Sci. Res. 10, 317–327.

Guiltinan, M.J., Marcotte, W.R. and Quatrano, R.S. (1990). A plant leucine zipper protein that recognizes an abscisic acid response element. Science 250, 267–271.

Guy, C.L. (1990). Cold acclimation and freezing stress tolerance: Role of protein metabolism. Ann. Rev. Plant Physiol. Plant Mol. Biol. 41, 187–223.

Han, B., Berjak, P., Pammenter, N., Farrant, J. and Kermode, A.R. (1997). The recalcitrant plant species, *Castanospermum australe* and *Trichilia dregeana*, Differ in their ability to produce dehydrin-related polypeptides during seed maturation and in response to ABA or water-deficit-related stresses. J. Exp. Bot. 48, 1717–1726.

Hara, M., Terashima, S. and Kuboi, T. (2001). Characterization and cryoprotective activity of cold-responsive dehydrin from *Citrus unshiu*. J. Plant Physiol. 158, 1333–1339.

Hoekstra, F.A., Golovina, E.A. and Buitink J. (2001). Mechanisms of plant desiccation tolerance. Trends Plant Sci. 6, 431–438.

Houde, M., Daniel, C., Lachapelle, M., Allard, F., Laliberté, S. and Sarhan, F. (1995). Immunolocalization of freezing-tolerance-associated proteins in the cytoplasm and nucleoplasm of wheat crown tissue. Plant J. 8, 583–593.

Hughes, M.A. and Dunn, M.A. (1990). The effect of temperature on plant growth and development. Biotechnol. Genet. Eng. Rev. 8, 161–188.

Ingram, J. and Bartels, D. (1996). The molecular basis of dehydration tolerance in plants. Ann. Rev. Plant Physiol. Plant Mol. Biol. 47, 377–403.

Ismail, A.M., Hall, A.E. and Close, T.J. (1997). Chilling tolerance during emergence of cowpea associated with a dehydrin and slow electrolyte leakage. Crop Sci. 37, 1270–1277.

Ismail, A.M., Hall, A.E. and Close, T.J. (1999a). Purification and partial characterization of a dehydrin involved

in chilling tolerance during seedling emergence of cowpea. Plant Physiol. 237–244.

Ismail, A.M., Hall, A.E. and Close, T.J. (1999b). Allelic variation of a dehydrin gene cosegregates with chilling tolerance during seedling emergence. Proc. Nat. Acad. Sci. USA. 96, 13566–13570.

Iturriaga, G., Schneider, K., Salamini, F. and Bartels, D. (1992). Expression of desiccation related proteins from the resurrection plant *Craterostigma plantagineum* in transgenic tobacco. Plant Mol. Biol. 20, 555–558.

Jaglo-Ottosen, K.R., Gilmour, S., Zarka, D.G., Schabenberger, O. and Thomashow, M.F. (1998). *Arabidopsis CBF1* overexpression induces *COR* genes and enhances freezing tolerance. Science 280, 104–106.

Jayaprakash, T.L., Ramamohan, G., Krishnaprasad, B.T., Ganeshkumar, Prasad, T.G., Mathew, M.K. and Udayakumar, M. (1998). Genotypic variability in differential expression of lea2 and lea3 genes and proteins in response to salinity stress in fingermillet (*Eleusine coracana* Gaertn) and rice (*Oryza sativa* L.) seedlings. Ann. Bot. 82, 513–522.

Jensen, A.B., Goday, A., Figueras, M., Jessop, A.C. and Pages, M. (1998). Phosphorylation mediates the nuclear targeting of the maize Rab17 protein. Plant J. 13, 691–697.

Jiang, Y. and Huang, B. (2002). Protein alterations in tall fescue in response to drought stress and abscisic acid. Crop Sci. 42, 202–207.

Kasuga, M., Liu, Q., Miura, S., Yamaguchi-Shinozaki, K. and Shinozaki, K. (1999). Improving plant drought, salt, and freezing tolerance by gene transfer of a single stress-inducible transcription factor. Nature Biotechnol. 19, 287–291.

Kaye, C., Neven, L., Hafig, A., Li, Q.B., Haskell, D. and Guy, C. (1998). Characterization of a gene for spinach CAP160 and expression of two spinach cold-acclimation proteins in tobacco. Plant Physiol. 116, 1367–1377.

Kazuoka, T. and Oeda, K. (1994). Purification and characterization of Cor85-oligomeric complex from cold-acclimated spinach. Plant Cell Physiol. 35, 601–611.

Kiyosue, T., Yamaguchi-Shinozaki, K., Shinozaki, K., Kamada, H. and Harada, H. (1993). cDNA cloning of Ecp40, an embryonic-cell protein in carrot, and its expression during somatic and zygotic embryogenesis. Plant Mol. Biol. 21, 1053–1068.

Kontunen-Soppela, S., Taulavuori, K., Taulavouri E., Lähdesmäki P. and Laine K. (2000). Soluble proteins and dehydrins in nitrogen-fertilized Scots pine seedlings during deacclimation and the onset of growth. Physiol. Plant. 109, 404–409.

Lång, V. (1993). The role of ABA and ABA-induced gene expression in cold acclimation of *Arabidopsis thaliana*. Molecular Genetics. PhD Thesis, Swedish University of Agricultural Sciences, Uppsala, Sweden.

Leung, J. and Giraudat, J. (1998). Abscisic acid signal transduction. Annu. Rev. Plant. Physiol. Plant Mol. Biol. 49, 199–222.

Levi, A., Panta, G.R., Parmentier, C.M., Muthalif, M.M., Arora, R., Shanker, S. and Rowland, L.J. (1999). Complementary DNA cloning, sequencing and expression of an unusual dehydrin from blueberry floral buds. Physiol. Plant. 107, 98–109.

Lim, C.C., Krebs, S.L. and Arora, R. (1999). A 25 kDa dehydrin associated with genotype-and age-dependent leaf freezing-tolerance in Rhododendron: a genetic marker for cold hardiness? Theor. Appl. Genet. 99, 912–920.

Lisse, T., Bartels, D., Kalbitzer, H.R. and Jaenicke, R. (1996). The recombinant dehydrin-like desiccation stress protein from the resurrection plant Craterstigma plantagineum displays no defined three-dimensional structure in its native state. Biol. Chem. 377, 555–561.

Liu, Q., Kasuga, M., Sakuma, Y., Abe, H., Miura, S., Yamaguchi-Shinozaki, K. and Shinozaki, K. (1998). Two transcription factors, DREB1 and DREB2, with an EREBP/AP2 DNA binding domain separate two cellular signal transduction pathways in drought- and low-temperature-responsive gene expression, respectively, in Arabidopsis. Plant Cell 10, 1391–1406.

Mäntylä, E., Lång, V. and Palva, E.T. (1995). Role of abscisic acid in drought-induced freezing tolerance, cold acclimation, and accumulation of Lti78 and Rab18 proteins in Arabidopsis thaliana. Plant Physiol.107, 141–148.

Moons, A., Bauw, G., Dekeyser, R., Von Montagu, M. and Van Der Straeten, D. (1995). Novel, ABA responsive proteins in vegetative rice tissue. Current Top. Plant Physiol. 10, 288–289.

Mundy, J. and Chua, N.H. (1988). Abscisic acid and water-stress induce the expression of a novel rice gene. EMBO J. 7, 2279–2286.

Neven, L.G., Haskell, D.W., Hofig, A., Li, Q.B. and Guy, C.L. (1993). Characterization of a spinach gene responsive to low-temperature and water-stress. Plant Mol. Biol. 21, 291–305.

Nordin, K., Vahala, T. and Palva, E.T. (1993). Differential expression of two related, low-temperature-induced genes in Arabidopsis thaliana (L.) Heynh. Plant Mol. Biol. 21, 641–653.

Nylander, M., Svensson, J., Palva, E.T. and Welin, B.V. (2001). Stress-induced accumulation and tissue-specific localization of dehydrins in Arabidopsis thaliana. Plant Mol. Biol. 45, 263–279.

Oliver, M.J. and Bewley, J.D. (1997). Desiccation-tolerance of plant tissues: A mechanistic overview. Hort. Rev. 18, 171–213.

Pammenter, N.W. and Berjak, P. (1999). A review of recalcitrant seed physiology in relation to desiccation-tolerance mechanisms. Seed Sci. Res. 9, 13–37.

Pan, A., Hayes, P.M., Chen, F., Chen, T.H.H., Blake, T., Wright, S., Karsai, I. and Bedo, Z. (1994). Genetic analysis of the components of winterhardiness in barley (Hordeum vulgare L.). Theor. Appl. Genet. 89, 900–910.

Panta, G.R., Rieger, M.W. and Rowland, L.J. (2001). Effect of cold and drought stress on blueberry dehydrin accumulation. J. Hort. Sci. Biotechnol. 76, 549–556.

Parcy, F., Valon, C., Raynal, M., Gaubier-Comella, P., Delseny, M. and Giraudat, J. (1994). Regulation of gene expression program during Arabidopsis seed development: roles of the ABI3 locus and of endogenous abscisic acid. Plant Cell 6, 1567–1582.

Parcy, F. and Giraudat, J. (1997). Interactions between the ABI1 and the ectopically expressed ABI3 genes in controlling abscisic acid responses in Arabidopsis vegetative tissues. Plant J. 11, 693–702.

Pelah, D., Shoseyov, O., Altman, A. and Bartels, D. (1997). Water-stress response in Aspen (Populus tremula): Differential accumulation of dehydrin, sucrose synthase, GAPDH homologues, and soluble sugars. J. Plant Physiol. 151, 96–100.

Plana, M., Itarte, E., Eritja, R., Goday, A., Pages, M. and Martinez, M.C. (1991). Phosphorylation of maize Rab-17 protein by casein kinase-2. J. Biol. Chem. 266, 22510–22514.

Rinne, P.L.H., Kaikuranta, P.L.M., van der Plas, L.H.W. and van der Schoot, C. (1999). Dehydrins in cold-acclimated apices of birch (Betula pubscens Ehrh.): production, localization and potential role in rescuing enzyme function during dehydration. Planta 209, 377–388.

Roebroek, A.J.M., Van den Velde, H.J.K., Van Bokhoven, A., Broers, J.L.V., Ramamakers, F.C.S. and Van de Ven, W.J.M. (1993). Cloning and expression of alternative transcripts of a novel neuroendocrine-specific gene and identification of its 135 kDa translational product. J. Biol. Chem. 268, 13439–13447.

Rouse, D.T., Marotta, R. and Parish, R.W. (1996). Promoter and expression studies on an Arabidopsis thaliana dehydrin gene. FEBS Lett. 381, 252–256.

Sarhan, F., Ouellet, F. and Vazquez-Tello, A. (1997). The wheat wcs120 gene family. A useful model to understand the molecular genetics of freezing tolerance in cereals. Physiol. Plant. 101, 439–445.

Sauter, J.J., Westphal, S. and Wisniewski, M. (1999). Immunological identification of dehydrin-related proteins in the wood of five species of Populus and in Salix caprea L. J. Plant Physiol. 154, 781–788.

Schneider, K., Wells, B., Schmelzer, E., Salamini, F. and Bartels, D. (1993). Desiccation leads to the rapid accumulation of both cytosolic and chloroplastic proteins in the resurrection plant Craterostigma plantagineum Hochst. Planta 189, 120–131.

Segrest, J.P., Deloof, H., Dohlman, J.G., Brouilette, C.G.

and Anantharamaiah, G.M. (1990). Amphipathic helix motif: classes and properties. Protein Struct. Funct. Genet. 8, 103–117.

Shinozaki, K. and Yamaguchi-Shinozaki, K. (2000). Molecular responses to dehydration and low temperature: differences and cross-talk between two stress signaling pathways. Curr. Opin. Plant Biol. 3, 217–223.

Steponkus, P.L., Uemura, M. and Webb, M.S. (1993). A contrast of the cryostability of the plasma membrane of winter and spring oat: two species that widely differ in their freezing tolerance and plasma membrane lipid composition. Advances in Low Temperature Biology. (Steponkus, P.L., Ed.) pp. 211–213. London, JAI Press.

Svensson, J. (2001). Functional studies of the role of plant DHNs in tolerance to salinity, desiccation and low temperature. Ph.D. thesis, Swedish University of Agricultural Sciences, Uppsala, Sweden.

Svensson, J., Palva, E.T. and Welin, B. (2000). Purification of recombinant *Arabidopsis thaliana* dehydrins by metal ion affinity chromatography. Protein Expr. Purif. 20, 169–178.

Thomashow, M. F. (1998). Role of cold-responsive genes in plant freezing tolerance. Plant Physiol. 118, 1–7.

Uno, Y., Furihata, T., Abe, H., Yoshida, R., Shinozaki, K. and Yamaguchi-Shinozaki, K. (2000). Arabidopsis basic leucine zipper transcription factors involved in an abscisic acid-dependent signal transduction pathway under drought and high-salinity conditions. Proc. Natl. Acad. Sci. USA 97, 11632–11637.

Velten, J. and Oliver, M.J. (2001). Tr88, a rehydrin with a dehydrin twist. Plant Mol. Biol. 48, 713–722.

Vilardell, J., Goday, A., Freire, M.A., Torrent, M., Martinez, M.C., Torne, J.M. and Pages, M. (1990). Gene sequence, developmental expression, and protein phosphorylation of RAB17 in maize. Plant Mol. Biol. 14, 423–432.

Volaire, F., Thomas, H., Bertagne, N., Bourgeois, E., Gautier, M.F. and Lelièvre, F. (1998). Survival and recovery of perennial forage greases under prolonged Mediterranean drought II: water status, solute accumulation, abscisic acid concentration and accumulation of dehydrin transcripts in bases of immature leaves. New Phytol. 140, 451–460.

Welin, B. (1994). Molecular analysis of cold acclimation in *Arabidopsis thaliana* and engineering of biosynthetic pathways leading to enhanced osmotolerance. PhD Thesis. Swedish University of Agricultural Sciences, Uppsala, Sweden.

Welin, B.V., Olsson, Å., Nylander, M. and Palva, E.T. (1994). Characterization and differential expression of *dhn/lea/rab*-like genes during cold acclimation and drought stress in *Arabidopsis thaliana*. Plant Mol. Biol. 26, 131–144.

Welling, A., Kaikuranta P. and Rinne, P. (1997). Photoperiodic induction of dormancy and freezing tolerance in *Betula pubescens*. Involvement of ABA and dehydrins. Physiol. Plant. 100, 119–125.

Wisniewski, M., Close, T.J., Artlip, T. and Arora, R. (1996). Seasonal patterns of dehydrins and 70-kDa heat-shock proteins in bark tissues of eight species of woody plants. Physiol. Plant. 96, 496–505.

Wisniewski, M., Webb, R., Balsamo, R., Close, T.J., Yu, X.M. and Griffith, M. (1999). Purification, immunolocalization, cryoprotective, and antifreeze activity of PCA60: A dehydrin from peach (*Prunus persica*). Physiol. Plant. 105, 600–608.

Wood, A.J. and Goldsbrough, P.B. (1997). Characterization and expression of dehydrins in water-stressed *Sorghum bicolor*. Physiol. Plant. 99, 144–152.

Wright, P.E. and Dyson, H.J. (1999). Intrinsically unstructured proteins: Re-assessing the protein structure–function paradigm. J. Mol. Biol. 293, 321–331.

Yamaguchi-Shinozaki, K. and Shinozaki, K. (1994). A novel *cis*-acting element in an *Arabidopsis* gene is involved in responsiveness to drought, low-temperature, or high-salt stress. Plant Cell 6, 251–264.

Zhang, L., Ohta, A., Takagi, M. and Imai, R. (2000). Expression of plant group 2 and group 3 lea genes in *Saccharomyces cerevisiae* revealed functional divergence among *LEA* proteins. J. Biochem. (Tokyo) 127, 611–616.

Zhu, B., Choi, D.-W., Fenton, R. and Close, T.J. (2000). Expression of the barley dehydrin multigene family and the development of freezing tolerance. Mol. Gen. Genet. 264, 145–153.

Sensing, Signaling and Cell Adaptation. Edited by K.B. Storey and J.M. Storey
© 2002 Elsevier Science B.V. All rights reserved.

CHAPTER 12

Dual Role of Membranes in Heat Stress: As Thermosensors They Modulate the Expression of Stress Genes and, by Interacting with Stress Proteins, Re-organize Their Own Lipid Order and Functionality

László Vígh[1] and Bruno Maresca[2]

[1]*Hungarian Academy of Sciences, Biological Research Centre, Institute of Biochemistry, Szeged, Hungary*

[2]*University of Salerno, Faculty of Pharmacy, Department of Pharmaceutical Sciences, Fisciano-Salerno, Italy*

1. Introduction

In this chapter the terms "membrane fluidity", "membrane order" and "non-lamellar phases", are defined as follows.

Membrane fluidity: A widely used but subjective term that describes the relative diffusional motion of molecules within membranes. Fluidity is used rather than viscosity, because membranes are planar, asymmetric structures, and their properties are not comparable to bulk phases. The term fluidity is meant to convey the impression of lateral diffusion, molecular wobbling and chain flexing, that are found in functional membranes where the lipids are in the liquid-crystalline lamellar phase.

Membrane order: The motional movement of molecules or molecular domains within the membrane. Membrane order can be quantified by estimating the motion of paramagnetic probes and calculating an order parameter from the ESR or NMR spectrum.

Non-lamellar phases: Non-bilayer arrangements of lipids in aqueous media. These can be hexagonal (H_I) or inverted hexagonal (H_{II}) arrangements; H_I phase is seldom found in membranes.

Thus far, the molecular effects of high temperature on most living systems have received surprisingly biased attention. For example, the consequences of heat stress on cell adaptation, cell malfunction and lethality have been investigated predominantly from the perspective of the molecular properties of heat-shock proteins (HSPs) and the regulation of their coding sequences. Thus, there is a wealth of information on the regulation of heat-shock (HS) genes, including their regulatory transcriptional factors, and on the role of their protein products. Appealing, but nonetheless speculative, theories on their possible functions as chaperones *in vivo* have been proposed, but the identification of the primary sensor and the sequences of heat-events leading either to cell death or to repair/recovery have been essentially neglected. Several laboratories have focused their attention on the crucial role of membranes as primary targets of heat stress and have attempted to understand how proper lipid/protein interaction within the membrane determines the regulation of HS genes. Such molecular interactions have been shown to be critically involved in the conversion of physical and chemical signals from the environment into sequential processes culminating, in a specific manner, in the transcriptional activation of stress regulated genes. In turn, the interactions between HSPs and specific regions (domains) of membranes remodel the status of membrane physical order. We propose that the specificity of gene expression is obtained by the uneven distribution of these membrane domains that precisely sense biological and physical environmental regulating signals and different forms of stresses. These studies have strongly modified our vision of the functions of biological membranes. With a critical re-evaluation of the current literature data and with our recent findings, we propose that the composition, organization and physical state of membranes play central and determining roles in the cellular responses during acute heat stress and pathological states.

The model that we propose here is based on three main substantiated assumptions:

1. The lipid composition and the physical state of membranes are decisive factors in the processes of perception and transduction of heat stress into a proper biological signal that triggers the transcriptional activation of HS genes.

2. The pre-existing composition and physical state of membranes determine the extent of HS-induced damage to membranes and has profound consequences on cell viability under stress condition.

3. Besides limited changes in lipid synthesis and/or modification, specific members of HSPs are actively involved in the restoration/repair of the damaged membranes in heat-stressed cells by precise association at (or within) the membranes. Temporary association of certain HSPs with membranes re-establishes the lateral packing order (fluidity), bilayer stability and membrane permeability and restores membrane functionality during heat stress. Thus, HSPs participate directly in the biophysical recovery of membranes to regain the pre-stress state, cause inactivation of the membrane-perturbating signal, and thereby switch off the synthesis of HS genes in a highly responsive feedback loop.

2. Membranes as "cellular thermometers"

The large number of stress conditions that induce a HS response (HSR) led to the hypothesis that multiple cellular targets act as sensors to generate the inducing signal. One of the models proposes that the accumulation of denatured proteins under heat stress triggers the activation of the stress-response (Parsell and Lindquist, 1993; Roussou et al., 2000). However, this model does not take into proper consideration the fact that warm-acclimated organisms that constitute more than 95% of all living species on Earth, do not induce HSPs when their physiological temperature increases during seasonal changes. Furthermore, it is well known that HSPs are present in abnormal levels in a variety of human degenerative diseases. In these pathological disorders, while HSP response is either higher or lower than normal, there is no evidence for modification of the kinetics or of the accumulation of denatured proteins that could justify the changes observed in HS gene transcription. Thus, whereas protein denaturation represents a mechanism for the recruitment of HSPs, it does not explain several physiological and pathological conditions in which the HSR is altered.

We have been working on a more complex model that could explain not only the above temperature paradox, but also the changes in the pattern of gene expression in disease states. Our alternative, but not necessarily exclusive, view is that the temperature-sensing mechanism is intimately associated with the composition and physical state of membranes. It is well known that, on a seasonal scale, membranes are the main targets of temperature adaptation (Hazel, 1995). We propose that during abrupt temperature fluctuations membranes represent the most thermally sensitive macromolecular structures. Our theory is supported by a number of independent investigations (reviewed recently by Vígh et al., 1998) and is historically stimulated by attempts to identify components of the pathway for the perception and transduction of low temperature signals in cyanobacteria and heat shock in yeast. We have demonstrated that the decrease in the degree of unsaturation of fatty acids in the plasma membrane of the cyanobacterium *Synechocystis* PCC 6803 (hereafter *Synechocystis*) attained by catalytic lipid hydrogenation under isothermal conditions *in vivo* (Vígh and Joó, 1983; Joó et al., 1991) enhances the expression of a gene for an acyl–lipid desaturase, which is normally inducible by a low-temperature shift (Vígh et al., 1993). Further, we have shown that genetic modification of lipid unsaturation and membrane fluid state in the yeast *Saccharomyces cerevisiae* resets the optimal temperature of heat shock response (Carratù et al., 1996). A common denominator for the effect of cold and heat stress, i.e. a change in membrane fluidity, clearly appears to play the primary role in the perception of temperature change causing transcriptional induction of desaturases and a reset of heat shock gene transcription. Murata and his co-workers provided novel data showing that cold shock induces an abrupt change in membrane fluidity in *Synechocystis* and simultaneously a signal is transduced from the membrane to the chromosomes. These authors identified by selective gene knockouts two membrane histidine kinases together with a response regulator as key components of the signal cascade in cold shock conditions (Suzuki et al., 2000).

Recently, it has been shown that the physical state of the membrane of *Synechocystis* (the thylakoid) (Horváth et al., 1998) and those of *Salmonella* and *Mycobacterium marinum* affects the temperature-induced expression of HS genes. The physical order of thylakoids—or cytoplasmic membrane—was reduced in response to either a downshift of the growth temperature or administration of benzyl alcohol which was paralleled by an enhanced thermosensitivity of the photosynthetic and cytoplasmic membranes in both model systems. We proposed that there is a close correlation between the physical state of the thylakoid membrane and the threshold temperatures required for maximal activation of HS-inducible genes, i.e., dnaK, groESL, cpn60 and *hsp17* (Horváth et al., 1998; Glatz et al., 1999). Therefore, under physiological temperatures, membrane lipid fluidity, regulated by the environmental temperature [or in human diseases by changes in unsaturated to saturated fatty acids (UFA/SFA) or protein–lipid ratio], determines the temperature at which HS genes are transcribed.

Unbalanced membrane phospolipid composition was shown to affect the expression of several regulatory genes in *E. coli* (Inoue et al., 1997).

Recent evidence suggests that overproduction of a *Synechocystis* Δ^{12}-desaturase, inactive enzymatically under our experimental conditions, inserts in the lipid bilayer and is able to cause a resetting of the HSR in *Salmonella*. A substantially higher membrane protein content, ie. an unbalanced protein/phospholipid ratio, is found in the membranes of transformed cells. As evidence that the desaturase-transformed cells are unable to properly accommodate the extra membrane protein, they display a greatly elevated permeability in their outer membrane even under non-stressed conditions. Similar to the above finding, overproduction of the membrane-bound sn-glycerol-acyltransferse in *E. coli* did trigger the HSR (Wilkinson and Bell, 1988).

Genetic manipulation of the ratio of UFA/SFA, obtained by over-expression of a desaturase gene, had a significant effect in *Saccharomyces cerevisiae* on the expression of the Hsp70 and Hsp82 genes (Carratù et al., 1996). If, as postulated, the alteration of the membrane physical state is a sort of "cellular thermometer" by which yeast (and other cells) sense a change in temperature, then there must be a molecular link between the sensory membrane, the downstream signalling pathways and specific DNA sequences. Yeast cells have two independent and differentially regulated stress response pathways, the HSR and the general stress response (GSR) (Chatterjee et al., 2000). To be effective, the HSR system requires the activation of heat shock elements (HSEs) located within the HS gene promoters by the heat shock transcription factors (HSTFs). Genes in the GSR system contain a different GSR promoter element (STRE) that has the ability to bind transcription factors MSN2 and MSN4 (Chatterjee et al., 2000). By using MSN2-GFP (green fluorescent protein) fusions, a variety of stresses including a temperature up-shift or the presence of agents that have the ability to perturb membranes, have been shown to trigger the translocation of this fusion protein from the cytoplasm to the nucleus (Moskvina et al., 1999). Since the activation of STRE signalling is independent of a decrease in the internal pH or a change in cAMP level, it was concluded that events occurring at the level of the plasma membrane (increased permeability, decreased membrane potential, elevated membrane fluidity, etc.) are sufficient to evoke a stress response and may be important components of a primary stress-sensing mechanism in yeast cells. In agreement with the above findings, Curran and co-workers demonstrated in a series of studies that the heat sensitivity of both the HSR (Chatterjee et al., 1997) and GSR (Chatterjee et al., 2000) pathways depends critically on the fatty acid composition of membrane lipids present in the yeast cells. Furthermore, they excluded the direct thermal denaturation of cellular proteins as the trigger for the activation of these major stress-signalling pathways in yeast, since different fatty acids added to the growth media gave different thresholds of HSR. Rather, they suggested that heat stress is detected by membrane-linked thermostat(s) whose activation is a consequence not only of the elevated temperature but also of the specific composition and physical state of the membrane lipids. Further, in agreement with these data, we have shown that a chemically synthesized pharmaceutical agent induces HS gene expression in mammalian cells and diabetic patients, most likely by interfering with membrane lipids (Vígh et al., 1997; Török et al., in preparation).

3. Pre-existing membrane lipid composition and physical state determines the heat induced membrane damage and has a primary and major effect on cell viability

As evidenced by the leakage of intracellular substances, by the loss of membrane components, by the sensitization to a variety of inhibitors, or by the adsorption of hydrophobic dyes, membranes are among the critical targets of heat stress in bacterial cells (Tsuchido et al., 1989) and likely in all eukaryotic cells as well. The outer membrane of Gram-negative bacteria constitutes the permeability barrier against the entry of hydrophobic compounds into cells. Heat treatment of *E. coli* has been shown to cause blebbing and vesiculation of the outer membrane with a simultaneous increase in cell surface hydrophobicity and damage to the permeability barrier (Myajake et al., 1993). During

post-heating recovery, the permeability barrier function is restored, however, the surface structure of restored cells is not identical to that existing before heat treatment (Tsuchido et al., 1989). In full support of the hypothesis that the fluidity of cell membranes plays a critical role in heat-killing, a positive correlation between the membrane microviscosity and the critical temperature required to kill *E. coli* was demonstrated almost 20 years ago through the use of UFA-requiring auxotrophic K1060 cells (Dennis and Yatvin, 1981). It is well known that *E. coli* membranes exhibit a variable composition of acyl chains in their phospholipids, depending on the growth temperature. As a general rule, the higher the growth temperature, the lower is the ratio of UFA/SFA (Marr and Ingraham, 1962). This ability to synthesize phospholipids that allows membranes to have almost identical fluidity at any given growth temperature has been termed homeoviscous adaptation (Sinensky, 1974). A large body of evidence suggests that such an adjustment of membrane fluidity at variable environmental temperatures is a general phenomenon in poikilothermic organisms and is needed to optimize the barrier functions, the permeability properties, the activities of membrane-bound enzymes and the signalling mechanism of the HSR, etc. (see above). On the other hand, whereas the fluidity is maintained at a certain level to preserve membrane functions, a sudden change in temperature causes an almost instantaneous, non-equilibrium state in the lipid order within the membrane. Before further discussion, however, we must introduce another essential attribute of biological membranes that is linked to the polymorphic characteristics of membrane lipids.

Virtually all membranes contain considerable amounts of lipids that do not spontaneously form bilayers in an aqueous buffer but, rather, have a strong preference to form nonlamellar structures, most commonly the inverted hexagonal phase, H_{II} (Cullis and Kruijff, 1979). The permanent presence of extensive non-bilayer structures could potentially interfere with the membrane barrier function and therefore be highly undesirable. The massive formation of non-bilayer lipid phase has been demonstrated in heat-stressed pea thylakoids

(Quinn et al., 1989). On the other hand, the formation of transient and local non-lamellar structures paradoxically seems to be vitally important to several functional processes, such as membrane fusion, cell division, the specific activation of several membrane enzymes, and the *trans*-bilayer movement of lipids and proteins (Rietveld et al., 1993). With the aid of recently developed anti-phospholipid monoclonal antibodies against non-bilayer phospholipid arrangements, direct evidence for the presence of non-bilayer structures in membranes of cultured mammalian cells has been obtained *in vivo* (Aguilar et al., 1999).

In a bilayer matrix consisting of a mixture of lipids of bilayer and non-bilayer types, there is a certain tendency to adopt a non-lamellar configuration, leading to packing constraints and/or optimal matching of the protein–lipid interface in the hydrophobic region (van der Does et al., 2000). If these physical forces are really important features of the membrane functions, as suggested by de Kruijff (1997) and others, then a strict regulation of membrane composition is required for the maintenance of a proper balance between the two kinds of lipids. There is evidence for such a polymorphic regulation of lipid composition in several microorganisms, such as *Acholeplasma laidlawii* and *Clostridium butyricum*, in which the ratio of lipids with a bilayer versus non-bilayer propensity was found to respond to changes in environmental conditions, including temperature (see references in de Kruijff, 1997). In *E. coli,* the non-bilayer structures appear to be essential for cell viability (Rietveld et al., 1993) and it was suggested that non-bilayer lipids are necessary for efficient protein transport across the plasma membrane (Rietveld et al., 1995). It was shown using NMR techniques (Killian et al., 1992) that membranes of *E. coli* K1059 cells grown in the presence of ^{2}H-labeled oleic acid at 37°C display a gel-to-liquid-crystalline phase transition in the range of 4–20°C, whereas a bilayer-to-non-bilayer transition occurs at about 42°C. Clearly, the maintenance of conditions in which the membrane is always close to the bilayer–non-bilayer transition is critical for the optimal functioning of *E. coli,* and most probably all living cells.

Besides temperature, the non-bilayer propensity of lipids is determined mostly by their head groups, although their acyl chain composition might also play a "fine-tuning" role (Quinn et al., 1989). In *E. coli,* the predominant lipid species, phosphatidylethanolamine (PE) has a non-bilayer propensity, while in photosynthetic organisms such as *Synechocystis* the major non-bilayer-forming lipid is monogalactosyl-diacylglycerol (MGDG), that accounts for about half of the total thylakoid lipids (Glatz et al., 1999). Saturation of thylakoid lipids obtained by catalytic hydrogenation increases the thermal stability of membranes (Thomas et al., 1986; Vígh et al., 1989), as indicated by an elevation of the temperature at which lipids (mostly MGDG) separate into the membrane-disrupting non-bilayer structures (Horváth et al., 1986). A strong support for this hypothesis comes from studies with transgenic tobacco plants with partially silenced desaturases. These plants have an improved rate of photosynthesis at 40°C and a markedly superior growth at 36°C (Murakami et al., 2000).

As pointed out earlier, it was shown in a variety of biological systems, that both the fluidity and the polymorphic phase behavior of the lipids are under close homeostatic regulation. Thus, whereas both the heat-induced membrane "hyperfluidity" (Yatvin and Cramp, 1993) and the formation of non-bilayer lipid structures are likely necessary to warn the cells to be in "alarm" state and to produce an HS signal, their lasting presence within membranes exerts harmful effects on membrane integrity and violates the rule of homeostatic control (Vígh et al., 1998; Quinn et al., 1989). The critical question is how cells reorganize the membranes right after an abrupt and temporary change of temperature (or exposure to membrane perturbing agents) so that the physical properties (acyl chain order, permeability, the proper balance between bilayer and non-bilayer lipids) compensate for the new temperature conditions. Obviously, during a rapid heat stress, the time span available is not sufficient for either a sizable remodelling of the lipid head-groups or a significant elevation of the level of more saturated lipid species. Notably, cells contain only "desaturases" that act in response to cold, but not "saturases" which might predictably be favored by heat stress. Rather, to stabilize the membrane during heat stress, a specific subset of HSPs associates transiently with the membranes. This macromolecular association represents a powerful tool with which cells can cope with rapidly fluctuating temperature conditions (or particular forms of stress) to achieve a temporary restructuring of the membrane physical state with the consequent preservation of membrane activities. However, this model does not exclude a limited heat-induced modulation of the lipid membrane, resulting from a "retailoring" of the lipid fatty acid molecular species by a de-acylation/re-acylation cycle, double bond re-isomerization or any other relatively rapid, energy-inexpensive means (Quinn et al., 1989).

4. HSPs are actively involved in the repair of the damaged membranes in heat-stressed cells by association with the membranes

Direct experimental evidence of membrane fluidity adjustment at high temperature was presented by Mejia et al. (1995) who carried out fluidity measurements of *E. coli* membranes obtained from cells grown at 30°C or 45°C or exposed to a temperature upshift from 30 to 45°C for various periods of time. These authors demonstrated that during heat shock there is a dramatic and immediate increase in membrane fluidity followed by a gradual decrease over time to reach after ~30 min a fluidity value corresponding to that of membranes from steady-state cultures grown at 45°C. In a separate study, these authors also measured the corresponding changes in the fatty acid profile of *E. coli* over the same length of HSR (Mejia et al., 1999). They concluded that the ratio of UFA/SFA decreases stepwise during HS and that 30 min after the temperature jump, the reduction is equivalent to 57% of the difference between the ratios found in steady-state cultures at 30° versus 45°C. These data demonstrate that about half of the value of membrane fluidity relaxation observed during HSR cannot be ascribed to lipid changes.

A correlation between heat-induced changes in membrane fluidity and lethality in mammalian cells has been reported by Dynlacht and Fox (1992). With a variety of cell lines with differing sensitivities to hyperthermia, these authors measured the initial and the post-heating membrane microviscosity. The most resistant cell lines were found to display the smallest change in fluidity during the heat treatment, and the largest increase in fluidity was observed in the most sensitive cell lines. Clearly, the ability of the cells to repair or replace membrane components is reflected by the fluidity profiles and explains the varying killing effects of heat. We have treated mammalian cells (K562) with benzyl alcohol and heptanol at concentrations that at normal growth temperature induce HSR, including the synthesis of the major stress proteins such as HSP70. The critical concentrations of each of the two fluidizers were selected so that their addition to cells caused an identical increase in the level of plasma membrane fluidity. Formation of isofluid states resulted in almost identical downshifts in the temperature thresholds of the HSR of the treated cells. As in the case of heat stress, the initial fluidity up-shift induced by the membrane perturbants was followed by a relaxation period. It is noted that the above treatments were not accompanied by changes in the lipid class and/or fatty acid composition. A similar rapid decrease in membrane fluidity was documented by Fujimoto et al. (1999) in a study aimed at assessing the relationship between membrane fluidity and apoptosis with the human leukemia cell line U937. By monitoring the enzymatic activity of luciferase expressed in murine cells recently we have demonstrated that benzyl alcohol, heptanol and the pharmacological agent bimoclomol do not cause protein denaturation at concentrations that do induce the HSR.

It has been reported that temperature-induced membrane disorganization is paralleled by heat stress induced translocation of nascent proteins to the outer membrane of *E. coli* (Yatvin, 1987). The translocation occurs within seconds in the absence of RNA synthesis and requires the presence of non-bilayer lipid phase. This phenomenon was interpreted as an adaptive response to an altered environment enabling the cell to respond to stress by stabilizing its outer membrane. In a separate study, the 28-kDa phage-shock protein A (PspA) a peripheral *E. coli* inner membrane protein was shown to be induced when cells were grown in the presence of n-hexane or cyclooctane (Kobayashi et al., 1998). The strong induction of PspA is seen under various stress conditions, such as HS, hyperosmotic shock, the inhibition of fatty acid synthesis by diazoborine, or upon the addition of other fluidizing agents, e.g. ethanol (see references in Kobayashi et al., 1998). It was suggested that the intercalation of hexanol in the membrane weakens the lipid–lipid interactions and disturbs the membrane structure and function, such as the maintenance of the proton-motive force, while the appearance of membrane-associated stress-response protein counterbalances these effects. The over-expression of PspA improves the survival of *E. coli* cells abruptly exposed to n-hexane.

A most significant earlier finding, relevant to the role of HSP in membrane protection, was that a subset of the 15-kDa HSPs of *E. coli* recovering from sublethal heat stress was found to interact with membranes (Myajake et al., 1993). This subset of proteins were sigma-32-dependent and two of them, designated C14.7 and G13.5 (at that time HSPs with unknown functions) were later identified as IbpA and IbpB (Kitagawa et al., 2000). It is noted that IbpA and IbpB are related to *Synechocystis* HSP17 which interacts with membranes both *in vivo* and *in vitro* right after HS.

Group I chaperonins, comprising the highly conserved GroEL from *E. coli*, are generally regarded as soluble proteins that function in the cytoplasm of prokaryotes and in the matrix compartment of mitochondria and chloroplasts. In contrast with this classical dogma, however, a number of reports suggest the existence of an additional, membrane-associated pool of GroEL homologs. An example is the study by Newman and Crooke (2000) who showed by means of immunogold cryothin-section EM and immunofluorescence that 16% of the labelled GroEL proteins of the cell were located in the membrane fraction. Moreover, the relative density of the gold particles bound to GroEL was significantly higher in the membrane

region than in the cytosol (2.98 versus 1.89). Similar to the immunogold EM results, the fluorescent patterns of cytosol versus membrane locations obtained by probing with anti-GroEL as the primary antibody again indicated strong membrane labeling. Clearly, a subset of *E. coli* GroEL is *ab ovo* associated with the membrane even under non-stressed conditions.

In *Mycobacterium leprae* and *Coxiella burnetii* a fraction of the GroEL chaperonins sediments out with the insoluble pellet following cell lysis (Gillis et al., 1985; Vodkin and Williams, 1988). The early finding that GroEL is localized in the cytoplasmic membrane in the photosynthetic prokaryote *Chromatium vinosum* led to speculation that membrane-associated chaperonins assist in the post-translational assembly of oligomeric proteins in the membrane (McFadden et al., 1989). Binding of a chloroplast 60 kDa HSP to the thylakoid membrane was also suggested in *Vigna sinensis* (Krishnashamy et al., 1988). Localization of chaperonins in the thylakoid region was demonstrated in the nitrogen-fixing cyanobacterium *Anabaena PCC7120* (Jäger and Bergman, 1990). Temperature was shown to control the subcellular distribution of HSP60 species in *Borrelia burgdorferi* (Scorpio et al., 1994), in which HSP60 and HSP70 are primarily involved in the processing of flagellin. Although in mammalian cells the majority of mtHSP60 is localized in the matrix compartment of the mitochondria, highly specific immunolabelling also locates mtHSP60 in the mitochondrial outer membrane, plasma membrane, endoplasmic reticulum and peroxisomes (Soltys and Gupta, 1996). In chloroplasts, stromal Cpn60 transiently associates with an integral membrane protein, the import intermediate associated protein (IAP100) (Kessler and Blobel, 1996). Cell surface translocation of GroEL was also suggested to be involved in the processing of proteins for antigen presentation (Blander and Horwitz, 1993). The heat-induced extracytoplasmic location of HSP60 chaperonins in *Legionella pneumophila* led Garduno et al. (1998) to suggest a potential role of chaperonins in pathogenesis.

The rapid, nonlethal heat exposure of *Synechocystis* cells induces an enhancement of photosystem II (PSII) thermotolerance in parallel with HSP synthesis and an increased molecular order (i.e., reduced fluidity) of the thylakoids. Analysis of *Synechocystis* HSPs revealed that the two GroEL homologs (Lehel et al., 1992, 1993; Kovács et al., 2001) are distributed both in the cytosol and in the highly purified thylakoid fractions (Kovács et al., 1994). The reorganization of membrane microdomains in parallel with the association of GroEL-type chaperonins was further supported on thylakoid membranes by using the fluorescence lifetime distribution technique (Török et al., unpublished). The thermoprotection induced by heat adaptation together with characteristic changes in the membrane physical state, seems to operate more effectively in the light than in the dark (Glatz et al., 1999). This finding is in agreement with the previously reported light-dependent transcriptional control of GroEL (Glatz et al., 1997). Since the cpn60 deletion mutant of *Synechocystis* cells fully preserves its ability to stabilize thylakoid membranes to sublethal heat, we concluded that the heat-induced membrane association of GroEL but not of Cpn60 is a necessary and sufficient condition for thermoadaptation. Alternatively, the membrane-stabilizing effects of the two GroEL analogs, independently of their apparently differing chaperone properties (Kovács et al., 2001) are interchangeable. Evidence relating to this alternative can be obtained via the isolation of groEL-deleted cells with intact GroES.

Encouraged by the above findings we initiated a systematic study to understand the mechanism of association of HSPs with membranes. (Török et al., 1997). We first demonstrated that *E. coli* GroEL chaperonin associates with lipid membranes. The binding is apparently ruled by the composition and physical state of the host bilayer. The limited proteolysis of GroEL oligomers by proteinase K leaves the chaperonin oligomer intact, but prevents its association with lipid membranes. Both of the physiologically relevant chaperonin hetero-oligomers $GroEL_{14}GroES_7$ and $GroEL_{14}(GroES_7)_2$ (Török et al., 1996) are located predominantly on the surface of the lipid bilayer. GroEL binding increases the membrane physical order, especially in the polar head-group region of

the lipids, as probed with different fluorophores. The GroEL-lipid interaction occurs almost exclusively in the liquid-crystalline ("fluid") state of the host model membrane and not in the gel state. Protein folding and ATPase activities are greatly modulated during lipid association apparently as a function of the composition of the interacting lipids. We concluded that GroEL chaperonins have dual functions to (a) assist the folding of both soluble and membrane-associated proteins and (b) rigidify and therefore stabilize lipid membranes during heat stress (Török et al., 1997).

We demonstrated that most of the newly synthesized small HSPs (sHSPs) of *Synechocystis* are associated with the thylakoid membranes (Horváth et al., 1998) and that sHSPs of this organism (HSP17) also influence membrane fluidity, predominantly in the deep hydrophobic region (Horváth et al., 1998; Török et al., 2001; Tsvetkova et al., submitted). We have shown that the transcription of *Synechocystis* hsp17 is strongly regulated by subtle changes in the physical order of the membranes (Horváth et al., 1998). Direct evidence of the physiological relevance of hsp17 thylakoid association was recently presented by Lee et al. (2000) who reported that inactivation of hsp17 results in a greatly reduced activity of photosynthetic oxygen evolution in heat-stressed *Synechocystis*. In line of the above findings, the expression of the sHSP Lo18 from the lactic bacterium *Oenococcus oeni* was shown to be induced by administration of benzyl alcohol. Moreover, it was demonstrated that a subset of Lo18 is localized in the membrane fraction and the actual level of its membrane association depends on the temperature upshift (Delmas et al., 2001).

Other observations support membrane localization of the sHSPs. In the chloroplasts of heat-stressed plants, for example, the sHSPs are associated with the thylakoids (Adamska and Kloppstech, 1991). Experiments with the green alga *Clamydomonas* have revealed that elevated levels of sHSPs increase the resistance of thylakoids to light and heat damage (Eisenberg-Domovich et al., 1994). In particular, sHSPs appear to protect the PSII electron transport system (Heckathorn et al., 1998) which is thought to be the most thermolabile element of the thylakoid.

Several members of the sHSP family, present in the mammalian eye lens, plants and bacteria, also associate with membranes, particularly under stressful conditions. It has been suggested that an increased membrane binding of α-crystallin is an integral step in the pathogenesis of many forms of cataracts (Cobb and Petrash, 2002). HSPB2, a new member of the sHSP family, expressed in heart and skeletal muscle, was shown to associate with the outer membrane of the mitochondria (Nakagawa et al., 2001). Therefore, we hypothesize that a subset of sHSPs may play a role in the control of thermal stress by acting as lipid-interacting membrane-stabilizing elements. Our studies with HSP17 in *Synechocystis* support this hypothesis (Török et al., 2001; Tsvetkova et al., in press).

We overexpressed the *Synechocystis* HSP17 gene in *E. coli* and purified its protein product to homogeneity. The existence of surface-exposed hydrophobic patches and the further temperature-driven exposure of buried hydrophobic regions were revealed by bis-ANS binding, in accordance with Das and Surewicz (1995). HSP17 interacts both with large unilamellar vesicles (LUVs) and with monolayers of various synthetic and *Synechocystis* lipids. The lipid interaction of HSP17 is specific with some preference for the fluid lipid matrix with acidic domains. Similar to our previous findings with GroEL (Török et al., 1997), LUV-HSP17 interaction generally (but not with negatively charged saturated lipid molecular species) causes an elevated membrane order as measured by DPH anisotropy. DSC studies revealed that even at an extremely low protein:lipid molar ratio (1:2000) HSP17 stabilizes the lamellar liquid-crystalline phase at the expense of the non-lamellar lipid phase, H_{II}, that is known to disrupt membranes under severe heat stress (see above) (Tsvetkova et al., in press). In agreement with fluorescence anisotropy studies, infrared spectroscopy (IR) revealed a clear upward shift in the gel-to-liquid-crystalline transition of DMPC-HSP17 membranes. IR studies with lipid/HSP17 systems, conducted according to Tsvetkova et al. (1998), indicated considerable changes in the regions of the lipid phosphate bands (polar head-group region), the C=O band (interfacial region) and the CH_2 band (hydrophobic core) reflecting the high complexity of the lipid–sHSP association.

HSP17 forms a stable complex with denatured substrates (Lee et al., 1997), like malate dehydrogenase (MDH) and, as shown for IbpB (Veinger et al., 1998), transfers the denatured proteins to DnaK/DnaJ/GrpE and GroEL/ES chaperones for subsequent active chaperone-mediated refolding. Depending on their composition, however, LUVs alter the kinetics of the HSP17-KJE-LS-mediated reactivation of denatured MDH. LUVs made of particular *Synechocystis* lipids strongly inhibited the refolding process. HSPs may represent a new class of amphitrophic proteins (Johnson and Cornell, 1999; Török et al., 2001). Lipid specificity implies that the membrane binding of HSPs through a specific HSP/lipid interaction confines the location of the HSPs to one or more *membrane lipid domains*. Vice versa, selective lipid interactions might also modify the oligomeric organization and other, yet unrevealed features of the HSPs.

To confirm that heat-induced membrane dynamical changes are causally related to the membrane lipid–HSP17 interaction, we analyzed *Synechocystis* hsp17⁻ mutant and hsp17⁺ revertants. The more heat-sensitive mutant displays a significantly reduced stability of thylakoid electron transfer, in parallel with a strong increase in membrane fluidity as measured by DPH anisotropy (Török et al., 2001). The expression of cyanobacterial HSP17 also results in a highly elevated thermoprotection of *E. coli*. Studies are in progress to assess the abundance of heterologous sHSP in *E. coli* membrane fractions and to relate this to the lipid composition, fluidity, phase state, and permeability properties of membrane during various stress conditions.

In agreement with our findings concerning the "lipid sensitivity" of the presumed refoldase activities of sHSPs together with other chaperones, it should be noted that there are several other observations that do not favour the classical model suggested either for the concerted operation of the DnaK/DnaJ/GrpE chaperone machine or for their cooperation with sHSPs. For instance, as shown by Delaney (1990), the grpE⁻ *E. coli* mutant is substantially more (and not less!) resistant to heat treatment than the wild type. Thermoadaptation is also more effective in the mutant compared to the revertant cells. We have recently isolated a partial dnaK2 deletion mutant of *Synechocystis*. In this cyanobacterium, dnaK2 is the only "active" of the three dnaK homolog genes and a significant portion of DnaK2 protein is associated with thylakoid during heat stress. One of the three DnaK proteins of *Synechococcus* sp. PCC7942 was also found to be located on the surface of the thylakoid membrane on the cytosol side and the peptide-binding domain of the protein was shown to be required for the membrane association (Nimura et al., 1996). Surprisingly, a partial deletion of *Synechocystis* dnaK2 causes an altered fatty acid composition and membrane physical state parallel with a highly elevated sensitivity of various membrane functions under stress conditions such as heat or ultraviolet B radiation. It is noted, that in H9c2 heart myoblasts, upon exposure to hydrogen peroxide, hsc-70 overexpressing cells exhibited a lower lipid peroxidation than their sham-transfected controls (Su et al., 1999). It was concluded, that membranes represent a selective target for HSC70-mediated protection during oxidative challenge and HSC70 is an effective protein to curtail membrane lipid peroxidation. Supporting their affinity for lipids, both HSC70 and HSP70 were co-purified with long chain saturated free fatty acids during their isolation (Guidon and Hightower, 1986). It was recently reported that HSC70 incorporates into phospholipid membranes forming a nucleotide-regulated ion conducting channel (Arispe and De Maio, 2000).

Finally, the chaperone and cellular thermo-protective properties of two closely related mammalian sHSPs have been compared by van der Klundert et al. (1999). In a clonal survival assay, Chinese hamster ovary cells stably overexpressing HSP20 survived equally well as αB-crystallin expressing cells, after a heat shock. In a transient assay, however, overexpression of HSP20 did not result in an enhanced recovery of co-expressed firefly luciferase after HS, in contrast to αB-crystallin. This might indicate that these highly homologous stress proteins are involved in at least

partially distinct protective activities in cultured cells. As described above, the addition of purified sHSPs to isolated thylakoids protects, without any cofactors, the thermolabile PSII (Heckatorn et al., 1998). Administration of sHSPs alone to sub-mitochondrial membrane vesicles, also revealed that sHSP function is totally responsible for the heat "acclimation" of complex I electron transport in pre-heat-stressed plants (Downs and Heckatorn, 1998).

In conclusion, combining literature data with our experimental studies on the *Synechocystis* GroEL-type chaperonins and the sHSP HSP17, we suggest that:

1. the thermally controlled membrane association of HSPs is, at least partially, lipid-mediated; and

2. some HSPs may function as membrane-stabilizing proteins rather than active chaperones by directly regulating key membrane features such as fluidity, permeability and the propensity to form a non-bilayer phase (Fig. 12.1).

Phospholipase A_2 (PLA_2) was examined for its role in the first phase of the transcriptional response to cellular stress in mammalian cells (Jurivich et al., 1996). These authors showed that when HeLa S3 and Jurkat cells were exposed to exogenous PLA_2, HSF1-DNA binding was induced. In addition, exposure to PLA_2 altered the thermal threshold for HSF1 activation and it was postulated that either conformational changes or other modifications of HSF1 were induced when cells were treated by PLA_2. Surprisingly, the monocyte-like cell line, U-937, was insensitive to the action of exogenous PLA_2. Neither HSF1-DNA binding or lowering of the temperature threshold for HSF1 activation was observed in PLA_2-treated U-937 cells. These data led these authors to suggest that PLA_2, the key enzyme of the inflammatory response, partially affects transcriptional switches mediating thermal stress in some cell types but not others. However, these data may be interpreted in a different way by assuming that exogenous PLA_2, by reshaping membrane lipids, affects the "membrane sensor" and thereby induces the transcriptional activation of HS genes. Furthermore, the finding that PLA_2

The dual role of membranes during heat stress

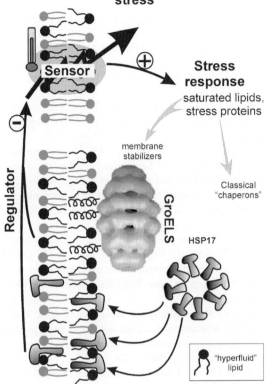

Fig. 12.1. Proposed model of cross talk between membrane physical state and HS gene expression. Temperature stress modifies membrane physical state that activates a membrane signal culminating in HS gene transcription. In turn, specific subsets of HSPs localize with specific membrane lipids, temporarily reestablishing proper membrane lipid order and phase state that then turns off HS gene transcription (see text for details).

treatment influences some but not other cells may be explained by the presence of different membrane domains that have different sensitivity to PLA$_2$ treatment.

The HSR was also examined in freshly isolated leukocytes from the pronephros of rainbow trout (*Oncorhynchus mykiss*) (Samples et al., 1999). In this study, it was shown that leukocytes isolated from rainbow trout acclimated to 5 or 19°C expressed elevated levels of HSP70 mRNA when heat shocked at 5°C above their respective acclimation temperature. Supplementation with exogenous docosahexaenoic acid or arachidonic acid

followed by heat shock enhanced the levels of HSP70 mRNA. These authors suggested that thermal stress initiates changes in membrane physical state, which when acted upon by phospholipases, may release lipid mediators that could serve as triggering signals during the heat shock response.

Attenuation of cardiac heat-shock protein expression has been shown in some pathological conditions, such as cardiac hypertrophy and aging. Recently we have shown that the expression of HSP70 in response to heat and ischemic stress is also diminished in hearts of rats fed by cholesterol-enriched diet (Csont el al., 2002). Since hyperlipidemia is known to alter lipid composition (Hexeberg et al., 1993) and thereby physical state of myocardial membranes, it likely also limits the capacity of cells to accumulate HSPs under various stress conditions. Notably, specific alterations in myocardial lipid molecular species after induction of the diabetic state have been shown (Han et al., 2000).

Finally, Khar et al. (2001) reported that curcumin, a well-known dietary pigment derived from *Curcuma longa* that has potent anti-inflammatory, antioxidant, and anti-carcinogenic effects, induces HSR in several mammalian cells. However, whereas curcumin showed the HS inducing response in transformed cell lines such as leukemia, breast, colon, hepatocellular, and ovarian carcinomas, it had no effect on nontransformed cell lines. These data indirectly corroborate our hypothesis that protein denaturation is not the main mechanism by which HSPs are transcriptionally regulated since no changes in the pattern of protein denaturation have been reported between normal and transformed cell lines. Thus, the dogma of the HSR based solely on denatured proteins is no longer realistic and independent evidence is accumulating to point to a more complex system in which more than one signal is involved.

We assume that, irrespective of their chaperone activity, the ability of HSPs to alter membrane organization (microdomain, asymmetry, etc.) and physical parameters is a rapid, reversible and powerful tool of cellular adaptation. It may antagonize the heat-induced lipid disorganization of the membrane and thus might serve to preserve, at least temporarily, membrane structure and function during heat stress (Vígh et al., 1998).

Furthermore, the association of HSPs with membranes likely causes inactivation of the membrane-perturbation signal induced by heat, thereby turning off the HS genes in a feedback loop. On the basis of the model proposed above, the modulation of membrane physical order may repress transcription of HS genes in the heat-modified state, explaining the known temporality of induction of the stress response. Therefore, such proposed "cross-talk" between the membrane-located sensor and the HSR suggests the existence of an as yet unknown feedback mechanism of HS gene regulation. We believe that this proposed mechanism of transcriptional HS gene activation and attenuation mediated by membrane physical state may also be operative for other genes regulated by signalling cascades. Thus, linking together membrane physical state and regulation of heat shock gene expression may represent a "unifying theory" in which membrane domains are key elements in a new important modality of gene expression. We are aware that this model, if it gains further experimental evidence, may suggest the existence a new mode of controlling the expression of genes regulated by signaling cascades in normal and pathological human conditions. In addition, the close interactions of two separate macromolecular compartments, membranes and gene activity, mediated by specific members of HSPs, make these proteins excellent potential targets for a new drug family.

Acknowledgements

The work was supported by grants from the Hungarian National Scientific Research Foundation (OTKA) to I.H. (T29883) and L.V. (T038334) and from NKFP to L.V. (OM 00570/2001) and by a Contract from the Italian Ministero dell' Universitá (MIUR), 2002, Potenziamento dell'Esistente Rete di Ricerca Scientifica sulle Proteine da Stress.

References

Adamska, I. and Kloppstech, K. (1991) Evidence for the localization of the nuclear-coded 22-kDa heat-shock protein in a subfraction of thylakoid membranes. Eur. J. Biochem. 198, 375–381.

Aguilar, L., Ortega-Pierres, G., Campos, B., Fonseca, R., Ibanez, M., Wong, C., Farfan, N., Naciff, J.M., Kaetzel, M.A., Dedman, J.R. and Baeza, I. (1999). Phospholipid membranes form specific nonbilayer molecular arrangements that are antigenic. J. Biol. Chem. 274, 25193–25196.

Arispe, N. and De Maio, A. (2000). ATP and ADP modulate a cation channel formed by Hsc70 in acid phospholipid membrane. J. Biol. Chem. 275, 30839–30843.

Blander, S.J. and Horwitz, M.A. (1993). Major cytoplasmic membrane protein of Legionella pneumophila, a genus common antigen and member of the hsp 60 family of heat shock proteins, induces protective immunity in a guinea pig model of Legionnaires' disease. J. Clin. Invest. 91, 717–721.

Carratù, L., Franceschelli, S., Pardini, C.L., Kobayashi, G.S., Horváth, I., Vígh, L. and Maresca, B. (1996). Membrane lipid perturbation sets the temperature of heat shock response in yeast. Proc. Natl. Acad. Sci. USA 93, 3870–3875.

Chatterjee, M.T., Khalawan, S.A. and Curran, B.P.G. (1997). Alterations in cellular lipids may be responsible for the transient nature of the yeast heat shock response. Microbiology 143, 3063–3068.

Chatterjee, M.T., Khalawan, S.A. and Curran, B.P.G. (2000). Cellular lipid composition influences stress activation of the yeast general stress response element (STRE). Microbiology 146, 877–884.

Cobb, B.A. and Petrash, J.M. (2002). α-Crystallin chaperone-like activity and membrane binding is age-related cataracts. Biochemistry 41, 483–490.

Cullis, P.R. and de Kruijff, B. (1979). Lipid polymorphism and the functional roles of lipids in biological membranes. Biochim. Biophys. Acta 559, 399–420.

Csont, T., Balogh, G., Csonka, Cs., Horváth, I., Vígh, L. and Ferdinandy, P. (2002). Hyperlipidemia induced by high cholesterol diet inhibits heat-shock response after ischemic and heat stress in rat hearts. Biochem. Biophys. Res. Comm. 290, 1535–1538.

Das, K.P. and Surewicz, W.K. (1995). Temperature-induced exposure of hydrophobic surfaces and its effect on the chaperone activity of alpha-crystallin. FEBS Lett. 369, 321–325.

Delmas, F., Pierre, F., Coucheney, F., Divies, C. and Guzzo, J. (2001). Biochemical and physiological studies of the small heat shock protein L018 from lactic acid bacterium Oenococcus oeni. J. Mol. Microbiol. Biotechnol. 3, 601–610.

De Kruijff, B. (1997). Biomembranes. Lipids beyond the bilayer. Nature 386, 129–130.

Delaney, J.M. (1990). A grpE mutant of Escherichia coli is more resistant to heat than the wild-type. J. General Microbiol. 136, 797–801.

Dennis, W.H. and Yatvin, M.B. (1981). Correlation of hyperthermic sensitivity and membrane microviscosity in *E. coli* K1060. Int. J. Radiat. Biol. 39, 265–271.

Downs, C.A. and Heckatorn, S.A. (1998). The mitochondrial small heat-shock protein protects NADH: ubiquinone oxidoreductase of the electron transport chain during heat stress in plants. FEBS Lett. 430, 246–250.

Dynlacht, J.R. and Fox, M.H. (1992). The effect of 45°C hyperthermia on the membrane fluidity of cells of several lines. Radiation Res. 130, 55–60.

Eisenberg-Domovich, Y., Kloppsteck, K. and Ohad, I. (1994). Reversible membrane association of heat-shock protein 22 in *Chlamydomonas reinhardtii* during heat shock and recovery. Eur. J. Biochem. 222, 1041–1046.

Fujimoto, K., Uwasaki, C., Kawaguchi, H., Yasugi, E. and Oshima, M. (1999). Cell membrane dynamics and the induction of apoptosis by lipid compounds. FEBS Lett. 446, 113–116.

Garduno, R.A., Faulkner, G., Trevors, M.A., Vats, N. and Hoffman, P.S. (1998). Immunolocalization of Hsp60 in *Legionella pneumophila*. J. Bacteriol. 180, 505–513.

Gillis, T.P., Miller, R.A., Young, D.B., Khanolkar, S.R. and Buchanan, T.M. (1985). Immunochemical characterization of a protein associated with *Mycobacterium leprae* cell wall. Infect. Immunol. 49, 371–377.

Glatz, A., Horváth, I., Varvasovszki, V., Kovács, E., Török, Zs. and Vígh, L. (1997). Chaperonin genes of the *Synechocystis* PCC 6803 are differentially regulated under light-dark transition during heat stress. Biochem. Biophys. Res. Commun. 239, 291–297.

Glatz, A., Vass, I., Los, D.A. and Vígh, L. (1999). The *Synechocystis* model of stress: from molecular chaperones to membranes. Plant Phys. Biochem. 37, 1–12.

Guidon, P.T. and Hightower, L.E. (1986) Purification and initial characterization of the 71-kilodalton rat heat-shock protein and its cognate as fatty acid binding protein. Biochemistry 25, 3231–3239.

Han, X., Abendschein, D.R., Kelley, J.G. and Gross, R.W. (2000). Diabetes-induced changes in specific lipid molecular species in rat myocardium. Biochem. J. 352, 79–89.

Hazel, J.R. (1995). Thermal adaptation in biological membranes: is homeoviscous adaptation the explanation? Ann. Rev. Physiol. 57, 19–42.

Heckathorn, S.A., Downs, C.A., Sharkey, T.D. and Coleman, J.S. (1998). The small, methionine-rich chloroplast heat-shock protein protects photosystem II electron transport during heat stress. Plant Physiol. 116, 439–444.

Hexeberg, S., Willumsen, N., Rotevatn, S., Hexeberg, E. and Berge, R.K. (1993). Cholesterol induced lipid accumulation in myocardial cells of rats. Cardiovasc. Res. 27, 442–446.

Horváth, I., Glatz, A., Varvasovszki, V., Török, Zs., Páli, T., Balogh, G., Kovács, E., Nádasdi, L., Benko, S., Joó, F.

and Vígh, L. (1998). Membrane physical state controls the signaling mechanism of the heat shock response in *Synechocystis* PCC 6803: Identification of *hsp17* as a "fluidity gene". Proc. Natl. Acad. Sci. USA, 95, 3513–3518.

Horváth, I., Mansourian, A.R., Vígh, L., Thomas, P.G., Joó, F. and Quinn, P.J. (1986). Homogeneous catalytic hydrogenation of the polar lipids of pea chloroplasts *in situ* and the effects on lipid polymorphism. Chem. Phys. Lipids 39, 251–264.

Inoue, K., Matsuzaki, H., Matsumoto, K. and Shibuya, I. (1997). Unbalanced membrane phospholipid compositions affect transcriptional expression of certain regulatory genes in *Escherichia coli*. J. Bacteriol. 179, 2872–2878.

Jäger, K.M. and Bergman, B. (1990). Localization of a multifunctional chaperonin (GroEL protein) in nitrogen-fixing *Anabaena* PCC 7120. Planta 183, 120–125.

Johnson, J.E. and Cornell, R.B. (1999). Amphitropic proteins: regulation by reversible membrane interactions (review). Mol. Membrane Biol. 16, 217–235.

Joó, F., Balogh, N., Horváth, L.I., Filep, Gy., Horváth, I. and Vígh, L. (1991). Complex hydrogenation/oxidation reactions of the water-soluble hydrogenation catalyst palladium di (sodium alizarinmonosulfonate) and details of homogeneous hydrogenation of lipids in isolated biomembranes and living cells. Anal. Biochem. 194, 34–40.

Jurivich, D.A., Pangas, S., Qiu, L. and Welk, J.F. (1996). Phospholipase A2 triggers the first phase of the thermal stress response and exhibits cell-type specificity. J. Immunol. 157, 1669–1677.

Khar, A., Ali, A.M., Pardhasaradhi, B.V., Varalakshmi, C.H., Anjum, R. and Kumar, A.L. (2001). Induction of stress response renders human tumor cell lines resistant to curcumin-mediated apoptosis: role of reactive oxygen intermediates. Cell Stress Chaperones 6, 368–376.

Kessler, F. and Blobel, G. (1996). Interaction of the protein import and folding machineries in the chloroplast. Proc. Natl. Acad. Sci. USA 93, 7684–7689.

Killian, A., Fabrie, C.H.J.P., Baart, W., Morein, S. and de Kruijff, B. (1992). Effects of temperature variation and phenethyl alcohol addition on acyl chain order and lipid organization in *Escherichia coli* derived membrane systems. A ^{2}H- and ^{31}P-NMR study. Biochim. Biophys. Acta 1105, 253–262.

Kitagawa, M., Matsumara, Y. and Tsuchido, T. (2000). Small heat shock proteins, IbpA and IbpB, are involved in resistances to heat and superoxide stresses in *Escherichia coli*. FEMS Microbiol Lett. 184, 165–171.

Klundert, F.A.J.M., van den Ijssel, P.R.L.A, Stege, G.J.J. and de Jong, W.W. (1999). Rat Hsp20 confers thermoresistance in a clonal survival assay, but fails to protect coexpressed luciferase in Chinese hamster ovary cells. Biochem. Biophys. Res. Comm. 254, 164–168.

Kobayashi, H., Yamamoto, M. and Aono, R. (1998). Appearance of a stress-response protein, phage-shock protein A, in *Escherichia coli* exposed to hydrophobic organic solvents. Microbiology 144, 353–359.

Kovács, E., van der Vies, S.M., Glatz, A., Török, Z., Varvasovszki, V., Horváth, I. and Vígh, L. (2001). The chaperonins of *Synechocystis* PCC 6803 differ in heat inducibility and chaperone activity. Biochem. Biophys. Res. Comm. 289, 908–915.

Kovács, E., Török, Z., Horváth, I. and Vígh, L. (1994). Heat stress induces association of the GroEL-analog chaperonin with thylakoid membranes in cyanobacterium, *Synechocystis* PCC 6803. Plant Physiol. Biochem. 32, 285–293.

Krishnasamy, S., Mannar Mannan, R., Krishnan, M. and Gnanam, A. (1988). Heat shock response of the chloroplast genome in *Vigna sinensis*. J. Biol. Chem. 263, 5104–5109.

Lee, G.J., Roseman, A.M., Saibil, H.R. and Vierling, E. (1997). A small heat shock protein stably binds heat-denatured model substrates and can maintain a substrate in a folding-competent state. EMBO J. 16, 659–671.

Lee, S., Owen, H.A., Prochaska, D.J. and Barnum, S. (2000). HSP16.6 is involved in the development of thermotolerance and thylakoid stability in the unicellular cyanobacterium, *Synechocystis* sp. PCC 6803. Curr. Microbiol. 40, 283–287.

Lehel, C., Los, D., Wada, H., Györgyei, J., Horváth, I., Kovács, E., Murata, N. and, Vígh, L. (1993). A second GroEL-like gene, organized in a *groESL* operon is present in the genome of *Synechocystis* sp. PCC 6803. J. Biol. Chem. 268, 1799–1804.

Lehel, C., Wada, H., Kovács, E., Török, Z., Gombos, Z., Horváth, I., Murata, N. and Vígh, L. (1992). Heat shock protein synthesis of the cyanobacterium *Synechocystis* PCC 6803: purification of the GroEL-related chaperonin. Plant Mol. Biol. 18, 327–336.

Marr, A.G. and Ingram, J.L. (1962). Effect of temperature on the composition of fatty acids in *Escherichia coli*. J. Bacteriol. 84, 1260–1267.

McFadden, B.A., Torres-Ruiz, J.A. and Franceschi, V.R. (1989). Localization of ribulose-bisphosphate carboxylase-oxygenase and its putative binding protein in the cell envelope of *Chromatium virosum*. Planta 178, 297–302.

Mejia, R., Gomez-Eichelman, M. and Fernandez, M.S. (1995). Membrane fluidity of *Escherichia coli* during heat-shock. Biochim. Biophys. Acta 1239, 195–200.

Mejia, R., Gomez-Eichelman, M. and Fernandez, M.S. (1999). Fatty acid profile of *Escherichia coli* during the heat-shock response. Biochem. Mol. Biol. Int. 47, 835–844.

Michels, A.A., Nguyen, V.T., Konings, A.W., Kampinga, H.H. and Bensaude, O. (1995) Thermostability of a nuclear-targeted luciferase expressed in mammalian cells. Destabilizing influence of the intranuclear micro-

environment. Eur. J. Biochem. 234, 382–389

Moskvina, E., Imre, E-M. and Ruis, H. (1999). Stress factors acting at the level of the plasma membrane induce transcription via the stress response element (STRE) of the yeast *Saccharomyces cerevisiae*. Mol. Microbiol. 32, 1263–1272.

Murakami, Y., Tsuyama, M., Kobayashi, Y., Kodama, H. and Iba, K. (2000). Trienoic fatty acids and plant tolerance of high temperature. Science 287, 476–479.

Myajake, T., Araki, S. and Tsuchido, T. (1993). Synthesis and sedimentation of a subset of 15 kDa heat shock proteins in *E. coli* cells recovering from sublethal heat stress. Biosci. Biotech. Biochem. 57, 578–583.

Nakagawa, M., Tsujimoto, N., Nakagawa, H., Iwaki, T., Fukumaki, Y. and Iwaki, A. (2001). Association of HSPB2, a member of the small heat shock protein family, with mitochondria. Exp. Cell Res. 271, 161–168.

Newman, G. and Crooke, E. (2000). DnaA, the initiator of *Escherichia coli* chromosomal replication, is located at the cell membrane. J. Bacteriol. 182, 2604–2610.

Nimura, K., Yoshikava, H. and Takahashi, H. (1996). DnaK3, one of the three proteins of cyanobacterium *Synechococcus* sp. PCC7942, is quantitatively detected in the thylakoid membrane. Biochem. Biophys. Res. Comm. 229, 334–340.

Parsell, D.A. and Lindquist, S. (1993). The function of heat-shock proteins in stress tolerance: degradation and reactivation of damaged proteins. Ann. Rev. Genet. 27, 437–496.

Quinn, P.J., Joó, F. and Vígh, L. (1989). The role of unsaturated lipids in membrane structure and stability. Prog. Biophys. Mol. Biol. 53, 71–103.

Rietveld, A.G., Killian, J.A., Dowhan, W. and de Kruijff, B. (1993). Polymorphic regulation of membrane phospholipid composition in *Escherichia coli*. J. Biol. Chem. 268, 12427–12433.

Roussou, I., Nguyen, T., Pagoulatos, G.N. and Bensaude, O. (2000) Enhanced protein denaturation in indomethacin-treated cells. Cell Stress Chaperones. 5, 8–13.

Samples, B.L., Pool, G.L. and Lumb, R.H. (1999). Polyunsaturated fatty acids enhance the heat induced stress response in rainbow trout (*Oncorhynchus mykiss*) leukocytes. Comp. Biochem. Physiol. B 123, 389–397.

Scorpio, A., Johnson, P., Laquerre, A. and Nelson, D.R. (1994). Subcellular localization and chaperone activities of *Borrelia burgdorferi* Hsp60 and Hsp70. J. Bacteriol. 176, 6449–6456.

Sinensky, M. (1974). Homeoviscous adaptation—a homeostatic process that regulates the viscosity of membrane lipids in *Escherichia coli*. Proc. Natl. Acad. Sci. USA 71, 522–525.

Soltys, B.J. and Gupta, R.S. (1996). Immunoelectron microscopic localization of the 60-kDa heat shock chaperonin protein (Hsp60) in mammalian cells. Exp. Cell Res. 222, 16–27.

Su, C-Y., Chong, K-Y., Edelstein, K., Lille, S., Khardori, R. and Lai, C-C. (1999) Constitutive hsp70 attenuates hydrogen peroxide-induced membrane lipid peroxidation. Biochem. Biophys. Res. Comm. 265, 279–284.

Suzuki, I., Los, D.A., Kanesaki, Y., Mikami, K. and Murata, N. (2000). The pathway for perception and transduction of low-temperature signals in *Synechocystis*. EMBO J. 19, 1327–1334.

Thomas, P.G., Dominy, P.J., Vígh, L. Mansourian, A.R. Quinn, P.J. and Williams, W.P. (1986). Increased thermal stability of pigment–protein complexes of pea thylakoids following catalytic hydrogenation of membrane lipids. Biochim. Biophys. Acta 849, 131–140.

Török, Z., Goloubinoff, P., Horváth, I., Tsvetkova, N.M., Glatz, A., Balogh, G., Varvasovszki, V., Los, D.A., Vierling, E., Crowe, J.H. and Vígh, L. (2001). *Synechocystis* HSP17 is an amphitropic protein that stabilizes heat-stressed membranes and binds denatured proteins for subsequent chaperone-mediated refolding. Proc. Natl. Acad. Sci. USA 98, 3098–3103.

Török, Z., Horváth, I., Goloubinoff, P., Kovács, E., Glatz, A., Balogh, G. and Vígh, L. (1997). Evidence for a lipochaperonin: association of active protein-folding GroESL oligomers with lipids can stabilize membranes under heat shock conditions. Proc. Natl. Acad. Sci. USA 94, 2192–2197.

Török, Zs., Vígh, L. and Golubinoff, P. (1996). Fluorescence detection of symmetric GroEL$_{14}$(GroES$_7$)$_2$ hetero-oligomers involved in protein release during the chaperonin cycle. J. Biol. Chem. 271, 16180–16186.

Tsvetkova., N.M., Horvath, I., Török, Z., Balogi, Z., Shigapova, N., Wolkers, W.F., Crowe, L.M., Tablin, F., Vierling, E., Crowe, J.H. and Vigh, L. (2002). Small heat-shock proteins regulate membrane lipid polymorphism. Proc. Natl. Acad. Sci. USA. In press.

Tsuchido, T., Aoki, I. and Takano, M. (1989). Interaction of the fluorescent dye 1-N-phenylnaphthylamine with *Escherichia coli* cells during heat stress and recovery from heat stress. J. Gen. Microbiol. 135, 1941–1947.

Tsvetkova, N.M., Phillips, B.L., Crowe, L.M., Crowe, J.H. and Risbud, S.H. (1998). Effect of sugars on headgroup mobility in freeze-dried dipalmitoyl-phosphatidylcholine bilayers: solid-state ^{31}P NMR and FTIR studies. Biophys. J. 75, 2947–2955.

van der Does, C., Swaving, J., van Klompenburg, W. and Driessen, A.J.M. (2000). Non-bilayer lipids stimulate the activity of the reconstituted bacterial protein translocase. J. Biol. Chem. 275, 2472–2478.

Veinger, L., Diamant, S., Buchner, J. and Goloubinoff, P. (1998). The small heat-shock protein IbpB from *Escherichia coli* stabilizes stress-denatured proteins for subsequent refolding by a multichaperone network. J. Biol. Chem. 273, 11032–11037.

Vígh, L., Gombos, Z., Horváth, I. and Joó, F. (1989). Saturation of membrane lipids by hydrogenation induces thermal stability in chloroplast inhibiting the heat-dependent stimulation of Photosystem I-mediated electron transport. Biochim. Biophys. Acta 979, 361–364.

Vígh, L. and Joó, F. (1983). Modulation of membrane fluidity by catalytic hydrogenation affects the chilling susceptibility of the blue-green alga, *Anacystis nidulans*. FEBS Lett. 162, 423–427.

Vígh, L., Literáti, N.P, Horváth, I., Török, Zs., Balogh, G., Glatz, A., Kovács, E., Boros, I., Ferdinandy, P., Farkas, B., Jaszlits, L., Jednákovits, A., Korányi, L. and Maresca, B. (1997). Bimoclomol: a novel, non-toxic, hydroxylamine derivative with stress protein inducing activity and wide cytoprotective effects. Nature Medicine 3, 1150–1154.

Vígh, L., Los, D.A., Horváth, I. and Murata, N. (1993). The primary signal in the biological perception of temperature: Pd-catalyzed hydrogenation of membrane lipids stimulated the expression of the desA gene in *Synechocystis* PCC 6803. Proc. Natl. Acad. Sci. USA, 90, 9090–9094.

Vígh, L., Maresca, B. and Harwood, J.L. (1998). Does the membrane physical state control the expression of heat shock and other genes? Trends. Biochem. Sci. 23, 369–373.

Vodkin, M.H. and Williams, J.C. (1988). A heat shock operon in *Coxiella burnetti* produces a mycobacteria and *Escherichia coli*. J. Bacteriol. 170, 1277–1234.

Wilkinson, W.O. and Bell, R.M. (1988). sn-Glycerol-3-phosphate acyl-transferase tubule formation is dependent upon heat shock proteins (htpR)*. J. Biol. Chem. 263, 14505–14510.

Yatvin, M.B. (1987). Influence of membrane–lipid composition on translocation of nascent proteins in heated *Escherichia coli*. Biochim. Biophys. Acta 901, 147–156.

Yatvin, M.B. and Cramp, W.A. (1993). Role of cellular membranes in hyperthermia: some observations and theories reviewed. Int. J. Hyperthermia 9, 165–185.

Sensing, Signaling and Cell Adaptation. Edited by K.B. Storey and J.M. Storey

CHAPTER 13

Cellular Adaptation to Amino Acid Availability: Mechanisms Involved in the Regulation of Gene Expression and Protein Metabolism

Sylvie Mordier, Alain Bruhat, Julien Averous and Pierre Fafournoux*
UNMP, INRA de Theix, Saint Genes Champanelle, France

1. Introduction

Mammals have the ability to adapt their own metabolic demand to survive in a variable and sometimes hostile environment. External stimuli to which they must be able to respond include thermal variations, rhythmic changes imposed by alteration of day and night, and the necessity to adjust to the intermittent intake of food. In addition, the animal itself provides internal metabolic variations, such as menstrual cycle or pregnancy in females and growth of tissues in young animals. All these external and internal factors demand metabolic responses, and associated regulatory mechanisms.

Regulation of metabolism is achieved by mechanisms operating at the cellular level (e.g., adaptive changes in enzyme activity) and also by coordinated actions between cells and tissues (e.g., endocrine signals in response to stimuli with consequent changes in the metabolism of target organs). These mechanisms involve the conditional regulation of specific genes in the presence or absence of appropriate nutrients. In multicellular organisms, the control of gene expression involves complex interactions of hormonal, neuronal and nutritional factors. Although not as widely appreciated, nutritional signals play an important role in controlling gene expression in mammals. It has

been shown that major (carbohydrates, fatty acids, sterols) and minor (minerals, vitamins) dietary constituents participate in the regulation of gene expression (Towle, 1995; Pégorier, 1998; Foufelle et al., 1998; Vaulont et al., 2000; Duplus et al., 2000; Grimaldi, 2001). However, the mechanisms involved in the amino acid control of gene expression have just begun to be understood in mammalians cells (Kilberg et al., 1994; Fafournoux et al., 2000; Bruhat and Fafournoux, 2001). This review summarizes recent work on the effect of amino acid availability in the regulation of biological functions. On the basis of the physiological concepts of amino acids homeostasis, we will discuss specific examples of the role of amino acids in the regulation of physiological functions, particularly focusing on the mechanisms involved in the amino acid regulation of gene expression and protein turnover.

2. Regulation of amino acid metabolism and homeostasis in the whole animal

Mammals are composed of a series of organs and tissues with different functions and different metabolic demands. As a consequence, the regulation of protein and amino acid metabolism in the whole animal is made up of the sum of the regulatory responses in all individual parts of the body and is achieved through a series of reactions that are both integrated and cooperative.

---

*Corresponding author

First, the reactions are integrated because there is a continuous exchange of amino acids between tissues, and metabolic regulation responds to this flow of compounds arriving at each cell. The impact of free amino acids varies from tissue to tissue. This is due to (i) the ability of the cells to respond to amino acid availability that can differ from one tissue to another, and (ii) the intracellular amino acid concentration which also depends on the degree to which protein turn-over contributes to the internal free amino acid pool.

Second, the metabolism of amino acids is cooperative because enzymes regulating the reactions of amino acid metabolism are distributed differently in various tissues. For example, arginine is mainly synthesised by liver and kidney but can be degraded by nitric oxide synthase in most of the tissues. Arginine is also catabolized by arginase which is primarily found in the liver (Cynober et al., 1995).

In contrast to other macronutrients (lipids or sugars) amino acids exhibit two important characteristics. First, in healthy adult humans, nine amino acids (valine, isoleucine, leucine, lysine, methionine, phenylalanine, threonine, histidine and tryptophan) are indispensable (or essential). In addition, under a particular set of conditions certain dispensable (non-essential) amino acids may become indispensable. These amino acids are called "conditionally indispensable". For example, enough arginine is synthesized by the urea cycle to meet the needs of an adult but not those of a growing child. Secondly, there are no large stores of amino acids. Consequently, when necessary, an organism has to hydrolyse muscle protein to produce free amino acids. This loss of protein will be at the expense of essential elements. Therefore, complex mechanisms that take into account these amino acid characteristics are needed for maintaining the free amino acid pools.

2.1. Free amino acids pools

The size of the pool of each amino acid is the result of a balance between input and removal (Fig. 13.1). The metabolic outlets for amino acids are protein synthesis and amino acid degradation

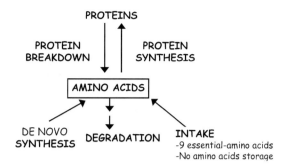

Fig. 13.1. Biochemical systems involved in the homeostasis of proteins and amino acids.

whereas the inputs are *de novo* synthesis (for non-essential amino acids), protein breakdown and dietary supply. Changes in the rates of these systems lead to an adjustment in nitrogen balance. For example, the level of amino acids in the plasma has been reported to rise after a protein-containing meal is given to an animal or a human subject. The concentration of leucine and certain other amino acids approximately doubles in peripheric blood after a protein rich meal (Aoki et al., 1976) and reaches much higher concentrations within the portal vein (Everson et al., 1989; Fafournoux et al., 1990). Although most of the essential amino acids are degraded in the liver, the branched chain amino acids and methionine are poorly used by hepatocytes. Consequently, after a protein-rich meal, these amino acids pass through the liver into the general circulation and cause a much greater increment in systemic blood levels than in the case of other essential amino acids, which are removed efficiently by the liver (Munro, 1970; Young et al., 1994). Data suggest that the effect of a protein-rich meal on protein turn-over is due to postprandial increases in the concentrations of circulating amino acids (Yoshizawa et al., 1995; Svanberg et al., 1997).

Another example of an adjustment of the nitrogen balance is the adaptation to an amino acid deficient diet. A dramatic diminution of the plasma concentrations of certain essential amino acids has been shown to occur following a dietary imbalance, a deficiency of any one of the essential amino acids or a deficient intake of protein (Grimble and Whitehead, 1970; Peng and Harper, 1970; Ozalp et al., 1972; Baertl et al., 1974). Long term feeding with such diets can lead to a negative nitrogen balance and clinical symptoms.

In addition to nutritional factors, various forms of stress (trauma, thermal burn, sepsis, fevers, etc.) can affect nitrogen metabolism, amino-acidemia and lead to a state of negative nitrogen balance and a significant loss of lean body mass. Cachexia and wasting syndromes are also observed during several chronic illnesses (e.g., chronic renal, cardiac, hepatic, and pulmonary diseases), acquired immunodeficiency syndrome (AIDS), and cancer (for references, see Jeejeebhoy, 1981; Jeevanandam et al., 1984; Wolfe et al., 1989; Ziegler et al., 1994; Attaix et al., 1999). In such situations, changes in nitrogen metabolism can be ascribed to several hormonal, metabolic, and behavioural changes. Briefly, amino acids are released by muscle proteolysis and provide substrates for the synthesis of proteins associated with inflammation. In such a situation, changes in the patterns of free amino acids are observed in plasma and urine.

The examples cited above show that amino acid metabolism can be affected by various nutritional and/or pathological situations, with two major consequences: a large variation in blood amino acid concentrations and a negative nitrogen balance. In these situations, individuals have to adjust several of their physiological functions involved in the defence/adaptation to amino acid limitation by regulating numerous genes. In the next section the specific role of amino acids in the adaptation to two different amino acids deficient diets will be considered.

2.2. Specific examples of the role of amino acids in the adaptation to protein deficiency

Protein undernutrition

Prolonged feeding on a low protein diet causes a fall in the plasma level of most essential amino acids. Protein undernutrition has its most devastating consequences during growth. For example, leucine and methionine concentrations can be reduced from about 100–150 μM and 18–30 μM to 20 μM and 5μM, respectively, in plasma of children affected by kwashiorkor (Grimble and Whitehead, 1970; Baertl et al., 1974). It follows that individuals have to adjust several physiological functions in order to adapt to this amino acid deficiency. One of the main consequences of feeding a low protein diet is the dramatic inhibition of growth. Growth is controlled by a complex interaction of genetic, hormonal and nutritional factors. A large part of this control is due to growth hormone (GH) and insulin-like growth factors (IGFs). The biological activities of the IGFs are modulated by the IGF-binding proteins (IGFBPs) that specifically bind IGF-I and IGF-II (Lee et al., 1993). Of the six IGFBPs, IGFBP-1 is the only one that displays rapid dynamic regulation *in vivo* with serum levels varying ten-fold or more depending on the nutritional state (Straus, 1994). Strauss et al., 1993) demonstrated that a dramatic overexpression of IGFBP-1 was responsible for growth inhibition, in response to prolonged feeding of a low protein diet. Known regulators of IGFBP-1 expression are GH, insulin or glucose. However, the high IGFBP-1 levels found in response to a protein deficient diet cannot be explained by these factors. It has been demonstrated that a fall in the amino acid concentration was directly responsible for IGFBP-1 induction (Straus et al., 1993; Jousse et al., 1998). Therefore, amino acid limitation, as occurs during dietary protein deficiency, participates in the down-regulation of growth through the induction of IGFBP-1.

Imbalanced diet

The ability to synthesize protein is essential for survival, and protein synthesis is dependent on the simultaneous supply of the 20 precursor amino acids. Because mammals cannot synthesize all of the amino acids, the diet must provide the remaining ones. Thus, in the event of a deficiency in one of the indispensable amino acid, the remaining amino acids are catabolized and lost and body protein are broken down to provide the limiting amino acid (Munro, 1976). It follows that mammals (with the exception of the ruminants) need mechanisms that provide for selection of a balanced diet. The capacity to distinguish balance from imbalance among the amino acids in the diet and to select for the growth limiting essential amino acid provides an adaptive advantage to animals.

After eating an amino acid imbalanced diet, animals first recognise the amino acid deficiency and then develop a conditioned taste aversion. Recognition and anorexia resulting from an amino acid imbalanced diet takes place very rapidly (Rogers and Leung, 1977; Gietzen et al., 1986). The mechanisms that underlie the recognition of protein quality must act by the way of the amino acids resulting from intestinal digestion of proteins. It has been observed that a marked decrease in the blood concentration of the limiting amino acid can become apparent as early as few hours after feeding an imbalanced diet. The anorectic response is correlated with a decreased concentration of the limiting amino acid in the plasma. Several lines of evidence suggest that the fall in the limiting amino acid concentration is detected in the brain. Gietzen (1993, 2000) reviews the evidence that a specific brain area, the anterior piriform cortex (APC), can sense the amino acid concentration. This recognition phase is associated with localised decreases in the concentration of the limiting amino acid and with important changes in protein synthesis rate and gene expression. Subsequent to recognition of the deficiency the second step, development of anorexia, involves another part of the brain.

These two examples demonstrate that a variation in blood amino acid concentration can activate, in target cells, several control processes that can specifically regulate the expression of target genes. Although the role of the amino acids considered to be regulators of genes expression is understood in only a few nutritional situations, recent progress has been made in understanding the mechanisms by which amino acid limitation controls the expression of several genes.

3. Amino acid control of gene expression

In mammalian cells, a few examples of specific mRNA types that are induced following amino acid deprivation have been reported (Fafournoux et al., 2000; Straus, 1994). However, the impact of amino acid availability on the control of gene expression is just beginning to be investigated extensively. At the molecular level, most results

have been obtained by studying the up-regulation of C/EBP homologous protein (*CHOP*), asparagine synthetase (*AS*) and the cationic amino acid transporter (*Cat-1*) gene expression by a limitation in amino acid availability. It was shown that the regulation of CHOP and AS expression by amino acid concentration has both transcriptional and post-transcriptional components.

3.1. Post-transcriptional regulation of genes expression by amino acid availability

For most of the amino acid regulated genes that have been studied, it has been shown that the mRNA is stabilized in response to an amino acid depletion. However, the molecular mechanisms involved in this process have not been identified yet (Gong et al., 1991; Bruhat et al., 1997; Aulak et al., 1999).

Recently, it was shown that the translation rate of specific genes could be regulated by amino acid availability. Hatzoglou and collaborators demonstrated that amino acid depletion initiates molecular events that specifically activate translation of the CAT-1 gene. They have shown the presence of an Internal Ribosome Entry Site (IRES) located within the 5′UTR of the Cat-1 mRNA (Fernandez et al., 2000). This IRES is involved in the amino acid control of translation of the Cat-1 transcript. Under conditions of amino acid starvation, translation from this IRES is stimulated whereas the cap-dependent protein synthesis is decreased. Another example of translation induced by amino acid starvation was recently reported for the branched-chain α-ketoacid dehydrogenase kinase, but the mechanism of translational control was not studied (Doering and Danner, 2000). This mechanism of compensatory response allows translation of major proteins despite the inhibition of the cap-dependent translational apparatus.

3.3. Transcriptional activation of mammalian genes by amino acid starvation

It was also established that the increase in CHOP or AS mRNA following amino acid starvation was mainly due to increased transcription. By first identifying the genomic *cis*-elements and then the corresponding transcription proteins responsible for regulation of these specific target genes, it is anticipated that one can progress backwards up the signal transduction pathway to understand the individual steps required.

CHOP

CHOP is a stress-inducible nuclear protein that dimerizes with members of the C/EBP family of transcription factors (Ron and Habener, 1992). Although CHOP was initially shown to function as a dominant negative inhibitor of gene transcription by forming stable heterodimers with C/EBPs and preventing them from binding to DNA, later studies showed that CHOP-C/EBP heterodimers are capable of recognizing novel DNA target sequences and thereby of activating gene transcription (Ubeda et al., 1999). The *CHOP* gene is regulated tightly by a wide variety of stresses in mammalian cells (Luethy and Holbrook, 1992; Sylvester et al., 1994; Schmitt-Ney and Habener, 2000) including many conditions that are known to induce an endoplasmic reticulum stress response (ERSR) (Wang et al., 1996). The ERSR, also known as the unfolded protein response (UPR), is an intracellular signaling pathway to remedy the accumulation of unfolded protein in the ER. Transcriptional control of *CHOP* by the ERSR involves the binding of ATF-6 in the presence of NF-Y to the *cis*-acting ERSR element (ERSE) located between nt −75 and −93 (Fig. 13.1) (Yoshida et al., 2000).

Amino acid starvation regulates *CHOP* transcription through a specific amino acid response (AAR) pathway that is distinct from the ERSR signaling cascade (Jousse et al., 1999). Transcriptional control elements used by the AAR pathway are contained within nucleotides −313 to −295 of the *CHOP* promoter (Bruhat et al., 2000). This short sequence can regulate a basal promoter in response to starvation of several individual amino acids and therefore was called the AAR element (AARE). Mutations affecting a stretch of nine nucleotides (called the "AARE core"; 5′-ATTGCATCA-3′) in the AARE result in a loss of amino acid responsiveness.

The sequence of the *CHOP* AARE region shows some homology with the specific binding sites of the C/EBP and ATF/CREB transcription factor families. We have shown that the transcription factors ATF-2 and C/EBPβ bind the *CHOP* AARE *in vitro* in starved and unstarved conditions. However, when knockout cell lines for these two proteins were tested, amino acid-dependent expression of *CHOP* was blocked in ATF-2-deficient cells, but not in cells lacking C/EBPβ expression, demonstrating that only ATF-2 is essential for the transcriptional activation of *CHOP* by leucine starvation. This result was supported by the observation that expression of a dominant negative form of ATF-2 suppressed the starvation-dependent transcription from a *CHOP* promoter/luciferase reporter construct (Bruhat et al., 2000).

ATF-2 contains a DNA binding domain consisting of a cluster of basic amino acids and a leucine zipper region (b-ZIP domain). Through its leucine zipper region, ATF-2 can form heterodimers with other b-ZIP proteins. It is likely that ATF-2 could bind to the *CHOP* AARE as an heterodimer with an unknown transcription factor (referred to as X in Fig. 13.1) and then activate transcription in response to amino acid starvation. It is well known that the transactivating capacity of ATF-2 is activated *via* phosphorylation of the N-terminal residues Thr-69, Thr-71 and Ser-90 (Gupta et al., 1995; Livingstone et al., 1995). There are two lines of evidence suggesting that ATF-2 phosphorylation belongs to the AAR pathway leading to the transcriptional activation of *CHOP* by amino acids: (i) Leucine starvation induces ATF-2 phosphorylation in human cell lines (unpublished data), and (ii) an ATF-2 dominant negative mutant (Sano et al., 1999), in which the three residues cannot be phosphorylated, inhibits the *CHOP* promoter activity enhanced by leucine starvation. Taken together these data suggest that the specific AAR pathway that leads to the transcriptional activation of *CHOP* may involve a phosphorylation of prebound ATF-2 rather than an increase in ATF-2 binding. However, the identity of the kinases involved in ATF-2 phosphorylation by amino acid starvation remains to be discovered (Fig. 13.2).

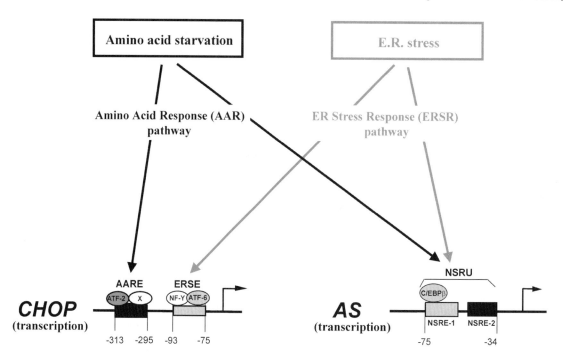

Fig. 13.2. *Cis*-acting elements required for induction of *CHOP* and *AS* genes following amino acid starvation or ER stress pathway. The *cis*-acting elements required for induction of the *CHOP* gene following amino acid starvation (AARE) or the ERSR pathway (ERSE) are located in sequences separated by several hundred base pairs. Transcriptional control of *CHOP* by ER stress involves the binding of ATF-6 in the presence of NF-Y to the *cis*-acting ER stress response element (ERSE) located between nt –75 and –93. The Amino Acid Response Element (AARE) used by the AAR pathway is contained within nucleotides –313 to –295. ATF-2 binds to the AARE and is essential for the transcriptional activation of *CHOP* by leucine starvation. The *AS* NSRU sequence (nt –75 to 34) is required for activation of the gene following either amino acid limitation or activation of the UPR pathway. C/EBPβ binds *in vitro* to the NSRE-1 sequence and is involved in the activation of the gene by AAR and ERSR pathways.

Asparagine synthetase

Asparagine synthetase (AS) is expressed in most mammalian cells as a housekeeping enzyme responsible for the biosynthesis of asparagine from aspartate and glutamine (Andrulis et al., 1987). The levels of AS mRNA increases not only in response to asparagine starvation but also after leucine, isoleucine and glutamine deprivation (Gong et al., 1991; Hutson and Kilberg, 1994; Hutson et al., 1997). AS is transcriptionally regulated by both amino acid and glucose deprivation of cells.

Guerrini et al. (1993) were the first to document the presence of a genomic element within the human AS promoter that mediated amino acid-dependent regulation of transcription. Those authors identified a region from nt –70 to –62 that functioned as an amino acid response element (AARE). Subsequently, using *in vivo* footprinting

and single nucleotide mutagenesis, Barbosa-Tessmann et al. (1999, 2000) demonstrated that the promoter sequence 5′-TGATGAAAC-3′ nt –68 to –60, the region first identified by Guerrini et al. (1993), was also responsible for the induction of the *AS* gene following glucose starvation involving the activation of the ERSR pathway. The ERSR activation demonstrates that this sequence serves in a broader capacity than simply as an AARE. To reflect this broader substrate detecting capability, the group of M. Kilberg has labeled this sequence the Nutrient Sensing Response Element-1 (NSRE-1) (Fig. 13.2). The AS NSRE-1 sequence differs by only two nucleotides from the *CHOP* AARE core sequence. A second element (5′-GTTACA-3′; nt –48 to –43), eleven nucleotides downstream, is also absolutely required for activation of *AS* by AAR and ERSR pathways and is called the NSRE-2 (Sui et al 2001). The term

Nutrient Sensing Response Unit (NSRU) has been coined by M. Kilberg to describe the collective function of these sequences.

Recently, Siu et al., 2001 have documented that C/EBPβ binds *in vitro* to the NSRE-1 *cis*-element within the AS promoter and is involved in the activation of the gene by the AAR and ERSR pathways (Fig. 13.2). Previous *in vivo* footprinting experiments showed that nutrient limitation caused an increase in protein binding at NSRE-1 (Barbosa-Tessmann et al., 2000). Consistent with these results, C/EBPβ binding is increased by amino acid and glucose starvation. It also appears that C/EBPβ itself, or a protein at a prior step in the pathway is synthesized *de novo* following activation of either the AAR or ERSR pathways.

From these data, it appears that the *CHOP* AARE and *AS* NSRE-1 sequences show some similarity. However, there are several lines of evidence suggesting that induction of *CHOP* and *AS* following amino acid starvation does not occur through a unique and common mechanism.

(i) The *cis*-acting elements required for induction of the *CHOP* gene following amino acid starvation (AARE) or the ERSR pathway (ERSE) are located in sequences separated by several hundred base pairs, whereas the *AS* NSRE-1 sequence is required for activation of the gene following either amino acid limitation or activation of the UPR pathway.

(ii) The amino acid specificity with regard to the degree of induction of these two genes is different (Jousse et al., 2000).

(iii) The region immediately following the *CHOP* AARE does not have a readily identifiable sequence that would correspond to NSRE-2.

(iv) ATF-2 binds *in vitro* to the *CHOP* AARE sequence and is essential for the transcriptional activation of *CHOP* by leucine starvation whereas Sui et al. (2001) have shown this transcription factor does not bind to the *AS* NSRE-1 sequence.

Collectively, the properties of the *CHOP* AARE and the *AS* NSRE-1 demonstrate that they are structurally related but functionally distinct.

ATF-4 and the amino acid signaling pathways
It appears that mammalian cells have more than one amino acid signaling pathway independent of the ERSR pathway. However, the individual steps required for these pathways are not well understood. Recently, Harding et al. (2000) revealed a signaling pathway for regulating gene expression in mammals that is homologous to the well characterized yeast general control response to amino acid deprivation. Its components include the mammalian homologue of the GCN2 kinase, the initiation factor eIF2α and ATF-4, a member of the ATF/CREB proteins. Like the yeast GCN4 transcript, the ATF-4 mRNA contains a uORF in its 5′-untranslated region that allows translation when cap-dependent translation is inhibited. The authors showed that GCN2 activation, phosphorylation of eIF2α and translational activation of ATF-4 are necessary but not sufficient for the induction of CHOP expression in response to leucine starvation (see Fig. 13.2).

Because ATF-4 belongs to the b-ZIP transcription factor family, it can interact and form a dimer with several transcription factors. In the case of amino acid starvation, it is possible that ATF-4 could bind to the *CHOP* AARE and/or to the *AS* NSRE-1 sequences (Fig. 13.3) and interact with other transcription factors (e.g. ATF-2 or C/EBPβ). However, the precise role of ATF-4 as well its interactions with transcription factors such as ATF-2 or C/EBPβ remain to be demonstrated.

Taken together, these results also show that at least two different pathways, that lead to ATF-2 phosphorylation and to ATF-4 expression (Fig. 13.3), are necessary to induce CHOP expression in response to one stimulus (leucine starvation). The different steps involved in these pathways remains to be discovered, particularly the kinases that lead to ATF2 phosphorylation as well as the mechanisms that cause ATF-4 overexpression. This section shows that increased gene transcription in response to amino acid starvation is linked to the translation of regulatory factors that depend on the presence of IRES or uORF in their messenger 5′-noncoding region. In this situation, cap dependent translation is inhibited. The role of amino acids in the regulation of cap-dependent protein synthesis and proteolysis will be discussed in the next section.

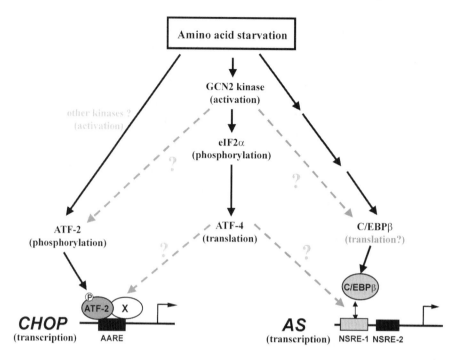

Fig. 13.3. Possible role of ATF-4 in the AAR pathway. A signaling pathway for regulating gene expression in mammals has been identified that is homologous to the well-characterized yeast general control response to amino acid deprivation. Its components include the mammalian homologue of the GCN2 kinase, the initiation factor eIF2α and ATF-4. It is possible that ATF-4 could bind to the *CHOP* AARE (and could represent X) or to the *AS* NSRE-1 sequences and interact with ATF-2 or C/EBPβ, respectively. However, the direct role of ATF-4 in the transcriptional regulation of *CHOP* and *AS* by amino acid availability remains to be demonstrated.

4.   Amino acid control of protein metabolism

It is now well established that amino acids can be direct regulators of protein turn-over as are growth factors and hormones like insulin. Their role has been demonstrated in many tissues including pancreas, skeletal muscle, liver and adipocytes. There is now a better knowledge of the regulatory amino acids and the molecular mechanisms by which they control protein breakdown and protein synthesis. Those aspects are developed below together with the specific role of the protein kinase, mTOR (for mammalian Target Of Rapamycin) in the mediation of the amino acid response. Indeed, among the protein kinase(s)/phosphatase(s) that participate in the mediation of amino acids effects on protein turnover, mTOR is a key player as it has the potential to co-regulate, in an opposite manner, protein synthesis and breakdown.

4.1.   Regulation of protein degradation by amino acids

Protein degradation is a highly controlled process that is performed by elaborate proteolytic systems located in multiple cellular compartments. Most cytosolic and nuclear proteins are degraded by the proteasome, a multisubunit proteolytic complex composed of a central multicatalytic core, the 20S proteasome, that can be associated with different regulatory complexes, especially the PA700 (19S regulator) to form the 26S proteasome which recognizes ubiquitinated proteins (for review: Hershko and Ciechanover, 1998; Voges et al., 1999; Pickart, 2000). A fraction of intracellular proteins can also be sequestered in autophagic vacuoles and delivered for their degradation to the lysosome, enriched in multiple catalytic enzymes, including the endopeptidases cathepsins (for review see Turk et al., 2001). Autophagic vacuoles could be generated from portions of the endo-plasmic reticulum free of ribosomes or specific organelles (Stromhaug et al., 1998). The comple-mentation of yeast autophagy-defective mutants allowed the cloning of a series of genes regulating autophagy, named Apg (for Autophagy genes) (Tsukada and Ohsumi, 1993). In particular, a new conjugation system involved in the formation of autophagosomes has been characterized. It is very similar to the ubiquitin conjugation system also encoded by the yeast genome, but involves dis-tinct, non-homologous proteins (Mizushima et al., 1998). These proteins are conserved in higher eukaryotes and expressed in many tissues (for reviews Kim and Klionsky, 2000; Klionsky and Emr, 2000).

Role of amino acids in protein breakdown

An indication of the role of amino acids as signal transducers in the regulation of protein breakdown was obtained from studies performed on perfused rat liver: amino acid withdrawal increased the rate of protein degradation, and correlated with an increased number of larger autophagic vacuoles in the cytosol (macroautophagy). This response was reversed by perfusion with a group of eight amino acids (Mortimore et al., 1989; Blommaart et al., 1997). Among them, leucine was the most potent at inhibiting the formation of autophagosomes, followed by phenylalanine and tyrosine, and gluta-mine. Proline, methionine, histidine and trypto-phan also regulate protein breakdown. Alanine, although exhibiting no direct suppressor effect on macroautophagy, has an essential co-regulatory role for the action of the active amino acids.

As opposed to the high level of autophagy in liver, intracellular protein breakdown in skeletal muscle is mainly achieved via non-lysosomal path-ways, especially by the proteasome (50% of total protein breakdown). Fasting has been shown to activate both lysosome and proteasome-dependent proteolysis in skeletal muscle. However, the acti-vation of proteasome mainly results from increased circulating glucocorticoids (Wing and Goldberg, 1993). The lack of a direct effect of amino acid starvation on proteasome activity and on the level of mRNAs encoding proteasome

subunits has been confirmed in C2C12 myotubes. In those cells it was clearly shown that leucine or total amino acids starvation increases the rate of protein breakdown by inducing macroautophagy. Amino acid starvation does not modify mRNA levels encoding some components of the lyso-somal pathways (Mordier et al., 2000) nor the lysosomal cathepsin B and L proteolytic activities (unpublished observation). Thus, amino acids reg-ulate protein breakdown mainly by controlling the accession of substrates, by the process of auto-phagy, for their degradation in the lysosome.

Role of mTOR in amino acid mediated protein degradation

The possible involvement of mTOR in the regula-tion of protein breakdown mediated by amino acids was first supported by the study of Bloommart et al. (1995). In isolated hepatocytes, they found that inhibition of autophagic proteolysis following amino acid re-supplementation was lin-early correlated with an increase in the phosphory-lation of protein ribosomal S6. This led the authors to conclude that phosphorylation of S6 was inhibi-tory for autophagy. S6 phosphorylation is depend-ent on p70s6k activity, itself directly dependent on mTOR (Brown et al., 1995). Since then, the genetic studies performed in yeast clearly estab-lished that specific signaling proteins not related to S6, downstream mTOR, regulate macroauto-phagy. In yeast, inactivation of TOR by rapamycin induces autophagy in nutrient rich conditions. The same result is obtained when a yeast strain carrying a temperature sensitive mutant of TOR is cultured at a non-permissive temperature. Rapamycin treat-ment of individual Apg mutants (deficient for auto-phagy) has no more positive effect on autophagy induction, suggesting that the site of action of all APG proteins is downstream of the site of action of TOR (Noda and Ohsumi, 1998). In particular, Apg13 is a direct substrate for TOR. Apg13 under-goes dephosphorylation following rapamycin treatment or nitrogen deprivation. A consequence is its association with Apg1 that encodes a Ser/Thr kinase and whose activity is increased upon starva-tion and is necessary for macroautophagy induc-tion (Kamada et al., 2000; Scott et al., 2000).

Although the above-mentioned studies demonstrate the involvement of mTOR in the regulation of amino acid starvation-induced autophagy, one study performed on C12C12 myotubes, shows that upon leucine starvation, the induction of autophagy and the associated increase in the rate of protein breakdown does not result of mTOR inactivation (Mordier et al., 2000). This observation suggests that amino acids could also act on an mTOR independent signaling pathway for the regulation of protein breakdown, especially in muscle cells.

4.2.   Regulation of protein synthesis by amino acids

Translation of mRNA in mammals is a GTP and ATP energy-dependent process. The clamping of the ribosomal subunits 40S and 60S (initiation) and their progression along the mRNA sequence (elongation) requires different sets of accessory proteins (eukaryotic Initiation Factors eIFs and eukaryotic Elongation Factors eEFs) interacting with the mRNA or the ribosome itself (for review, Proud and Denton, 1997) (Fig. 13.4). Amino acids can regulate the cell translational capacity, especially the branched chain amino acids, as well as additional amino acids like methionine or histidine. However, the branched chain amino acid leucine seems to be the most potent amino acid in regulating cell protein synthesis capacity.

Amino acids and the regulation of 43S pre-initiation complex formation
The initiator (Met)-tRNA has to be coupled with eIF2-GTP for its association with the 40S ribosomal subunit and then its binding to the mRNA. The generation of eIF2-GTP is dependent on the activity of the guanine–nucleotide exchange factor eIF2B. eIF2B is composed of 5 subunits (α to ε) and mediates the release of GDP from eIF2 to regenerate the active eIF2-GTP complex. eIF2B activity is lowered when specific residues of the α subunit of eIF2 and/or the ε subunit of eIF2B are phosphorylated (Fig. 13.4 (1)).

In L6 myoblasts and skeletal muscle, the reduced protein synthesis capacity, associated with amino acid deficiency, is correlated with a decrease in eIF2B activity (Kimball et al., 1998; Vary et al., 1999). In L6 muscle cells, the increase in eIF2B phosphorylation in association with leucine depletion has been shown to be mediated by a transient activation of the glycogen synthase kinase 3 (GSK-3) whose activity is regulated by mTOR (Peyrollier et al., 2000). Among the five characterized phosphorylation sites in eIF2B, Ser535 is the substrate for GSK-3 (Wang et al., 2001b).

Amino acid starvation also induces an imbalanced tRNA charging that leads to the activation of an eIF2 α kinase; mGCN2 (the mammalian equivalent of the yeast GCN2) is able to increase the phosphorylation state of eIF2 at Ser51. When phosphorylated, eIF2 is converted into a competitive inhibitor of eIF2B (Pain, 1994; Gray and Wickens, 1998).

Amino acids and the regulation of the eIF4F translation initiation complex formation
The cap-dependent translation initiation requires fixation of the eIF4F complex for subsequent recruitment of the 60S ribosomal subunit. Formation of the eIF4F complex, composed of the association of the cap binding protein eIF4E and the RNA helicase eIF4A proteins to the scaffold eIF4G, is regulated by the availability of eIF4E (Fig. 13.4 (2)). eIF4E can be sequestered by the low phosphorylated form of 4E-BP1 (4E Binding Protein 1). 4E-BP1 is phosphorylated on many residues by the action of mTOR and other kinases whose identity remains unknown (Beretta et al., 1996).

Under conditions of appropriate amino acid supply, 4E-BP1 is phosphorylated and does not sequester eIF4E. Upon amino acid starvation, there is a reduction in the association of eIF4E with eIF4G that correlates with a decreased phosphorylation of 4E-BP1 in L6 myoblasts (Kimball et al., 1999), CHO-IR cells (Hara et al., 1998), skeletal muscle, pancreatic cells and adipocytes (Vary et al., 1999; Xu et al., 1998; Fox et al., 1998a). This effect is abolished by amino acid re-addition but this reversal is blocked by the mTOR inhibitor, rapamycin, indicating an effect of amino acids on the mTOR kinase activity.

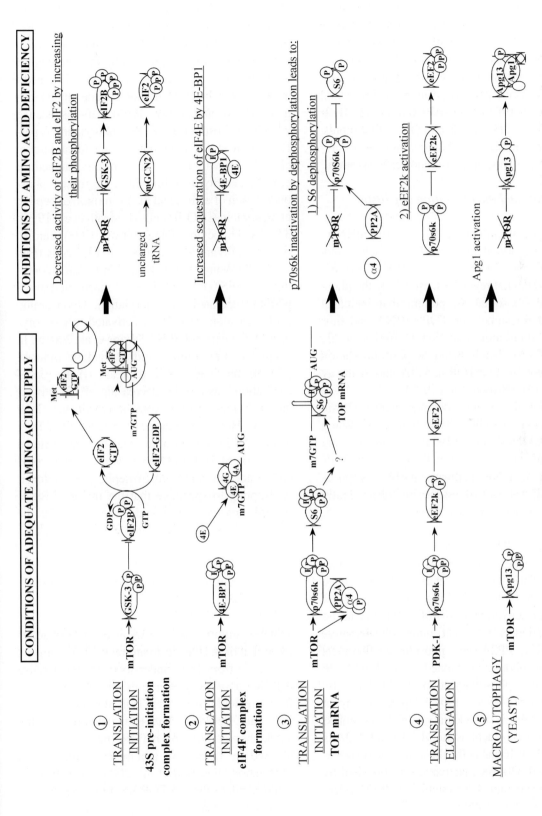

Fig. 13.4. Regulation of protein turn-over by amino acids: role of mTOR. This scheme shows the major phosphoproteins involved in the regulation of protein turnover at different levels, translation initiation (1), (2), (3), translation elongation (4), and macroautophagy (5). The transition from adequate amino acid supply to amino acid deficiency inactivates mTOR and modifies the activity of protein kinases/phosphatases, including GSK-3, mGCN2, p70S6k, pp2A, and eEF2k, that leads to modifications of protein-protein or protein/RNA interactions. The consequence is a reduction in synthesis capacity and the induction of macroautophagy.

The number and identity of the phosphorylation sites of 4E-BP1 required for the release of eIF4E is not fully known. Rapamycin treatment strongly inhibits the phosphorylation of 4E-BP1 on residues Ser65 and Thr70 in HEK 293 cells stimulated with serum. However, phosphorylation of Ser65 and Thr70 does not disrupt 4E-BP1 binding to eIF4E (Gingras et al., 2001), indicating that other inputs, in addition to mTOR inactivation, must be involved to get the full response to amino acid starvation.

Amino acids and the regulation of p70s6k activity

The translation initiation of specific mRNAs encoding ribosomal and translation factors that contain 5′ pyrimidine tracts (5′-TOP mRNAs), requires additional regulators, especially the phosphorylated ribosomal S6 protein that increases affinity of ribosomes for TOP mRNA and thus facilitates translation initiation (Jefferies et al., 1994, 1997) As already stated, S6 phosphorylation is dependent on the mTOR and S6kinase pathway (Fig. 13.4 (3)) (Brown et al., 1995).

Amino acid starvation down-regulates p70s6k activity in many cell lines (Hara et al., 1998; Wang et al., 1998; Iiboshi et al., 1999; Shigemitsu et al., 1999; Kimball et al., 1999) through an mTOR dependent signaling pathway. p70s6k is phosphorylated on at least 10 residues by independently regulated protein kinases (Weng et al., 1995, 1998). mTOR is responsible for phosphorylation at three residues: Thr229, Thr389, Ser404, but the principal target of rapamycin induced p70s6k inactivation is Thr389 (Pearson et al., 1995). Evidence has been provided, showing that the mTOR dependent inactivation of p70s6K following amino acid starvation is not direct: under physiological conditions, a PP2A-type phosphatase interacts with a phosphoprotein named α 4 (Tap42p in yeast) whose phosphorylation is regulated by mTOR (Murata et al., 1997). Upon rapamycin treatment, there is a dissociation of the complex, consecutive to α 4 dephosphorylation and PP2A is then active. In Jurkat cells, the amino acid starvation induced p70S6K inactivation is abolished by addition of calyculin A, an inhibitor of PP2A activity (Peterson et al., 1999).

The elongation factors (eEFs) are used for ribosome progression on all mRNAs and thus control global protein synthesis. Two factors have been characterized: eEF-1 promoting the binding of aminoacyl-tRNA to the ribosome and eEF-2 mediating the translocation step of peptide chain elongation (Proud and Denton, 1997). eEF-2 is inactive when it is phosphorylated. Phosphorylation is performed by a Ca^{2+}/calmodulin-dependent kinase: eEF-2k. Growth factors like insulin can reduce eEF-2k activity resulting in a decreased phosphorylation of eEF-2 and the promotion of peptide elongation in CHO-IR cells (Redpath et al., 1996), 3T3-L1 adipocytes (Diggle et al., 1998), cardiac myocytes (Wang et al., 2000), and ES cells (Fig. 13.4 (4)) (Wang et al., 2001a). Under basal conditions, in ES cells, eEF-2k is inactivated by a p70S6k mediated phosphorylation. Upon amino acid starvation, eEF-2k is activated, phosphorylates and inactivates eEF-2 (Wang et al., 2001a). In summary, the amino acid starvation induced p70S6k inactivation will decrease (i) the global translation capacity by decreasing eEF2 activity, and (ii) the specific translation of the TOP mRNAs. However, because a recent report of Tang et al. (2001) demonstrates that the translational repression of TOP mRNAs in amino acid-starved cells is not correlated with S6 dephosphorylation, it is tempting to speculate that the major effect of p70S6k inactivation would be to decrease the capacity for peptide chain elongation during translation.

4.3. Regulation of mTOR kinase activity by amino acids

Whether amino acids directly affect mTOR activity or that of mTOR upstream regulatory elements is still far from fully understood mainly because mTOR's phosphorylation sites and substrate specificity are not known. An mTOR autophosphorylation site, Ser2481, has been identified. This autokinase activity of mTOR, however, is not blocked by rapamycin treatment or amino acid deprivation (Peterson et al., 2000). Amino acids could regulate some mTOR upstream elements of the insulin-signaling pathway, especially Akt (or

PKB). Akt can phosphorylate mTOR on Ser2448. Amino acid starvation does inhibit Akt-dependent mTOR Ser2448 phosphorylation and this identifies a possible mechanism by which amino acid starvation could repress mTOR activity. The consequence is also the establishment of insulin resistance (Nave et al., 1999). However, in most other studies, amino acid starvation or re-supplementation does not seem to regulate Akt activity (Wang et al., 1998; Hara et al., 1998; Iiboshi et al., 1999; Peyrollier et al., 2000). In C2C12 myoblasts, leucine starvation induces a decrease in S6 phosphorylation with no significant decrease in the level of phospho-mTOR Ser2448 (S. Mordier, unpublished).

Amino acids could exert their effect on mTOR by affecting the t-RNA charging state. In Jurkat cell lines, the addition of many different amino acids alcohols (that inhibits their corresponding t-RNA synthetase) has the same effect as amino acid starvation: the inhibition of p70s6k activity. This result has been confirmed with temperature-sensitive mutants of histidine tRNA synthetase (Iiboshi et al., 1999). This may represent a general response as most individual essential amino acids seem to affect tRNA charging. However, in skeletal muscle or adipocytes, most of the effects of amino acids on mTOR signaling are mediated by leucine and to a lesser extent the other branched-chained amino acids (Fox et al., 1998b; Anthony et al., 2000) suggesting that a tRNA charging-independent mechanism must be involved. Especially, in adipocytes, amino acids use an amino acid alcohol-insensitive mechanism to regulate mTOR (Pham et al., 2000; Lynch, 2001).

5. Conclusion

In mammals, the plasma amino acid concentration shows striking alterations as a function of nutritional or pathological conditions. Amino-acidemia can arise after a protein rich meal, whereas, under poor nutritional conditions, the organism can experience limitations in the supply of essential amino acids. In such a situation, an adaptive response takes place by adjusting several physiological functions involved in defence/adaptation to amino acid limitation. The cellular machinery compensates for a deficiency in amino acids by both (i) decreasing the cap-dependent protein translation initiation and the capacity for peptide chain elongation, and (ii) increasing the rate of degradation of the resident proteins by induction of macroautophagy. These two processes contribute to the restoration of the free amino acid pool. At the same time, cells have the capacity to specifically increase both (i) the translation of a set of proteins independently of the cap-dependent process, and (ii) the transcription of specific genes.

The molecular mechanisms involved in the cellular response to amino acid availability have just begun to be discovered. Some components of the pathway(s) involved in the regulation of gene expression in response to amino acid limitation have been identified: the role of mGCN2 associated with tRNA charging and the possible role of the transcription factors ATF-2, ATF-4 and/or C/EBPβ. However, the precise cascade of molecular events by which the cellular concentration of an individual amino acid regulates gene expression has not been identified yet.

The mechanisms involved in the amino acid regulation of protein turnover are more precisely known. Leucine is the most potent amino acid that exerts control on protein turnover. Amino acids regulate signaling pathway(s) that converge at the kinase mTOR. Bifurcating pathways downstream of mTOR control, in an opposite manner, protein synthesis and protein breakdown. Some of the signalling molecules involved in the regulation of macroautophagy downstream of mTOR remain to be identified in higher eukaryotes. Although mTOR is considered to be an intracellular checkpoint control of amino acid availability, a recent report (Talloczy et al., 2002) demonstrates that, in yeast, the gcn2 kinase also regulates starvation induced macroautophagy. Gcn2 is then the second kinase identified as a co-regulator of protein synthesis and breakdown in yeast.

Defining the molecular steps by which individual amino acids can regulate gene expression and protein turnover will be an important contribution to our understanding of metabolite control in mammalian cells. The molecular basis for gene regulation by dietary protein intake is important with respect to the regulation of physiological functions of individuals living under conditions of restricted or excessive food intake.

References

Andrulis, I.L., Chen, J. and Ray, P.N. (1987). Isolation of human cDNAs for asparagine synthetase and expression in Jensen rat sarcoma cells. Mol. Cell. Biol. 7, 2435–2443.

Anthony, J.C., Yoshizawa, F., Anthony, T.G., Vary, T.C., Jefferson, L.S. and Kimball, S.R. (2000). Leucine stimulates translation initiation in skeletal muscle of postabsorptive rats via a rapamycin-sensitive pathway. J. Nutr. 130, 2413–2419.

Aoki, T.T., Brennan, M.F., Muller, W.A., Soeldner, J.S. Alpert, J.S., Saltz, S.B., Kaufmann, R.L., Tan, M.H. and Cahill, G.F. (1976). Amino acid levels across normal forearm muscle and splanchnic bed after a protein meal. Am. J. Clin. Nutr. 29, 340–350.

Attaix, D., Combaret, L., Tilignac, T. and Taillandier, D. (1999). Adaptation of the ubiquitin-proteasome proteolytic pathway in cancer cachexia. Mol. Biol. Rep. 26, 77–82.

Aulak, K.S., Mishra, R., Zhou, L., Hyatt, S.L., de Jonge, W., Lamers, W., Snider, M. and Hatzoglou, M. 1999. Post-transcriptional regulation of the arginine transporter Cat-1 by amino acid availability. J. Biol. Chem. 274, 30424–30432.

Baertl, J.M., Placko, R.P. and Graham, G.G. (1974). Serum proteins and plasma free amino acids in severe malnutrition. Am. J. Clin. Nutr. 27, 733–742.

Barbosa-Tessmann, I.P., Chen, C., Zhong, C., Schuster, S.M., Nick, H.S. and Kilberg, M.S. (1999). Activation of the unfolded protein response pathway induces human asparagine synthetase gene expression. J. Biol. Chem. 274, 31139–31144.

Barbosa-Tessmann, I.P., Chen, C. Zhong, C., Siu, F., Schuster, S.M., Nick, H.S. and Kilberg, M.S. (2000). Activation of the human asparagine synthetase gene by the amino acid response and the endoplasmic reticulum stress response pathways occurs by common genomic elements. J. Biol. Chem. 275, 26976–26985.

Beretta, L., Gingras, A.C., Svitkin, Y.V., Hall, M.N. and Sonenberg, N. (1996). Rapamycin blocks the phosphorylation of 4E-BP1 and inhibits cap-dependent initiation of translation. EMBO J. 15, 658–664.

Blommaart, E.F., Luiken, J.J. Blommaart, P.J., van Woerkom, G.M. and Meijer, A.J. (1995). Phosphorylation of ribosomal protein S6 is inhibitory for autophagy in isolated rat hepatocytes. J. Biol. Chem. 270, 2320–2326.

Blommaart, E.F., Luiken, J.J. and Meijer, A.J. (1997). Autophagic proteolysis, control and specificity. Histochem. J. 29, 365–385.

Brown, E.J., Beal, P.A., Keith, C.T., Chen, J., Shin, T.B. and Schreiber, S.L. (1995). Control of p70 s6 kinase by kinase activity of FRAP *in vivo*. Nature 377, 441–446.

Bruhat, A. and Fafournoux, P. (2001). Recent advances on molecular mechanisms involved in amino acid control of gene expression. Curr. Opin. Clin. Nutr. Metab. Care 4, 439–443.

Bruhat, A., Jousse, C., Carraro, V., Reimold, A.M., Ferrara, M. and Fafournoux, P. (2000). Amino acids control mammalian gene transcription, activating transcription factor 2 is essential for the amino acid responsiveness of the CHOP promoter. Mol. Cell. Biol. 20, 7192–7204.

Bruhat, A., Jousse, C., Wang, X.Z., Ron, D., Ferrara, M. and Fafournoux, P. (1997). Amino acid limitation induces expression of CHOP, a CCAAT/enhancer binding protein-related gene, at both transcriptional and post-transcriptional levels. J. Biol. Chem. 272, 17588–17593.

Cynober, L., Le Boucher, J. and Vasson, M.P. (1995). Arginine metabolism in mammals J. Nutr. Biochem. 6, 402–412.

Diggle, T.A., Redpath, N.T., Heesom, K.J. and Denton, R.M. (1998). Regulation of protein-synthesis elongation-factor-2 kinase by cAMP in adipocytes. Biochem. J. 336, 525–529.

Doering, C.B. and Danner, D.J. (2000). Amino acid deprivation induces translation of branched-chain alpha-ketoacid dehydrogenase kinase. Am. J. Physiol. Cell Physiol. 279, C1587–C1594.

Duplus, E., Glorian, M. and Forest, C. (2000). Fatty acid regulation of gene transcription. J. Biol. Chem. 275, 30749–30752.

Everson, W.V., Flaim, K.E., Susco, D.M., Kimball, S.R. and Jefferson, L.S. (1989). Effect of amino acid deprivation on initiation of protein synthesis in rat hepatocytes. Am. J. Physiol. 256, C18–C27.

Fafournoux, P., Remesy, C. and Demigne, C. (1990). Fluxes and membrane transport of amino acids in rat liver under different protein diets Am. J. Physiol. 259, E614–E625.

Fafournoux, P., Bruhat, A. and Jousse, C. (2000). Amino acid regulation of gene expression. Biochem. J. 351, 1–12.

Fernandez, J., Yaman, I., Mishra, R., Merrick, W.C., Snider, M.D., Lamers, W.H. and Hatzoglou, M. (2001). Internal ribosome entry site-mediated translation of a mammalian mRNA is regulated by amino acid availability. J. Biol. Chem. 276, 12285–12291.

Foufelle, F., Girard, J. and Ferre, P. (1998). Glucose regulation of gene expression. Curr. Opin. Clin. Nutr. Metab. Care 1, 323–328.

Fox, H.L., Kimball, S.R., Jefferson, L.S. and Lynch, C.J.

(1998a). Amino acids stimulate phosphorylation of p70S6k and organization of rat adipocytes into multicellular clusters. Am. J. Physiol. 274, C206–C213.

Fox, H.L., Pham, P.T., Kimball, S.R., Jefferson, L.S. and Lynch, C.J. (1998b). Amino acid effects on translational repressor 4E-BP1 are mediated primarily by L-leucine in isolated adipocytes. Am. J. Physiol. 275, C1232–C1238.

Gietzen, D.W. (1993). Neural mechanisms in the responses to amino acid deficiency. J. Nutr. 123, 610–625.

Gietzen, D.W. (2000). Amino acid recognition in the central nervous system. In: Neural and Metabolic Control of Macronutrient Intake (Berthoud, H.R. and Seeley, R.J., Eds.), pp. 339–357. CRC Press, Boca-Raton, FL.

Gietzen, D.W., Leung, P.M.B., Castonguay, T.W., Hartmann, W.J. and Rogers, G.R. (1986). Time course of food intake and plasma and brain amino acid concentrations in rat fed amino acid-imbalanced or -deficient diet. In: Interaction of the Chemical Senses with Nutrition. pp. 415–456. Academic Press, New York.

Gingras, A.C., Raught, B., Gygi, S.P., Niedzwiecka, A., Miron, M., Burley, S.K., Polakiewicz, R.D., Wyslouch-Cieszynska, A., Aebersold, R. and Sonenberg, N. (2001). Hierarchical phosphorylation of the translation inhibitor 4E-BP1. Genes Dev. 15, 2852–2864.

Gong, S.S., Guerrini, L. and Basilico, C. (1991). Regulation of asparagine synthetase gene expression by amino acid starvation. Mol. Cell Biol. 11, 6059–6066.

Gray, N.K. and Wickens, M. (1998). Control of translation initiation in animals. Annu. Rev. Cell. Dev. Biol. 14, 399–458.

Grimaldi, P.A. (2001). Fatty acid regulation of gene expression. Curr. Opin. Clin. Nutr. Metab. Care 4, 433–437.

Grimble, R.F., and Whitehead, R.G. (1970). Fasting serum-amino acid patterns in kwashiorkor and after administration of different levels of protein. Lancet 1, 918–920.

Guerrini, L., Gong, S.S., Mangasarian, K. and Basilico, C. (1993). *Cis*- and *trans*-acting elements involved in amino acid regulation of asparagine synthetase gene expression. Mol. Cell. Biol. 13, 3202–3212.

Gupta, S., Campbell, D., Derijard, B. and Davis, R.J. (1995). Transcription factor ATF2 regulation by the JNK signal transduction pathway. Science 267, 389–393.

Hara, K., Yonezawa, K., Weng, Q.P., Kozlowski, M.T., Belham, C. and Avruch, J. (1998). Amino acid sufficiency and mTOR regulate p70 S6 kinase and eIF-4E BP1 through a common effector mechanism. J. Biol. Chem. 273, 14484–14494.

Harding, H.P., Novoa, I.I., Zhang, Y., Zeng, H., Wek, R., Schapira, M. and Ron, D. (2000). Regulated translation initiation controls stress-induced gene expression in mammalian cells. Mol. Cell. 6, 1099–1108.

Hershko, A. and Ciechanover, A. (1998). The ubiquitin system. Annu. Rev. Biochem. 67, 425–479.

Hutson, R.G. and Kilberg, M.S. (1994). Cloning of rat asparagine synthetase and specificity of the amino acid-dependent control of its mRNA content. Biochem. J. 304, 745–750.

Hutson, R.G., Kitoh, T., Moraga Amador, D.A., Cosic, S., Schuster, S.M. and Kilberg, M.S. (1997). Amino acid control of asparagine synthetase, relation to asparaginase resistance in human leukemia cells. Am. J. Physiol. 272, C1691–C1699.

Iiboshi, Y., Papst, P.J., Kawasome, H. Hosoi, H., Abraham, R.T., Houghton, P.J. and Terada, N. (1999). Amino acid-dependent control of p70(s6k). Involvement of tRNA aminoacylation in the regulation. J. Biol. Chem. 274, 1092–1099.

Jeejeebhoy, K.N. (1981). Protein nutrition in clinical practice. Br. Med. Bull. 37, 11–17.

Jeevanandam, M., Horowitz, G.D., Lowry, S.F. and Brennan, M.F. (1984). Cancer cachexia and protein metabolism. Lancet 1, 1423–1426.

Jefferies, H.B., Fumagalli, S., Dennis, P.B., Reinhard, C., Pearson, R.B. and Thomas, G. (1997). Rapamycin suppresses 5′TOP mRNA translation through inhibition of p70s6k. EMBO J. 16, 3693–3704.

Jefferies, H.B., Reinhard, C., Kozma, S.C. and Thomas, G. (1994). Rapamycin selectively represses translation of the "polypyrimidine tract" mRNA family. Proc. Natl. Acad. Sci. USA 91, 4441–4445.

Jousse, C., Bruhat, A., Ferrara, M. and Fafournoux, P. (1998). Physiological concentration of amino acids regulates insulin-like-growth-factor-binding protein 1 expression. Biochem. J. 334, 147–153.

Jousse, C., Bruhat, A., Ferrara, M. and Fafournoux, P. (2000). Evidence for multiple signaling pathways in the regulation of gene expression by amino acids in human cell lines. J. Nutr. 130, 1555–1560.

Jousse, C., Bruhat, A., Harding, H.P., Ferrara, M. Ron, D. and Fafournoux, P. (1999). Amino acid limitation regulates CHOP expression through a specific pathway independent of the unfolded protein response. FEBS Lett. 448, 211–216.

Kamada, Y., Funakoshi, T., Shintani, T., Nagano, K., Ohsumi, M. and Ohsumi, Y. (2000). Tor-mediated induction of autophagy via an Apg1 protein kinase complex. J. Cell. Biol. 150, 1507–1513.

Kilberg, M.S., Hutson, R.J. and Laine, R.O. (1994). Amino acid-regulated gene expression in eukaryotic cells. FASEB J. 8, 13–19.

Kim, J., and Klionsky, D.J. (2000). Autophagy, cytoplasm-to-vacuole targeting pathway, and pexophagy in yeast and mammalian cells. Annu. Rev. Biochem. 69, 303–342.

Kimball, S.R., Horetsky, R.L. and Jefferson, L.S. (1998). Implication of eIF2B rather than eIF4E in the regulation of global protein synthesis by amino acids in L6 myoblasts. J. Biol. Chem. 273, 30945–30953.

Kimball, S.R., Shantz, L.M., Horetsky, R.L. and Jefferson, L.S. (1999). Leucine regulates translation of specific mRNAs in L6 myoblasts through mTOR-mediated changes in availability of eIF4E and phosphorylation of ribosomal protein S6. J. Biol. Chem. 274, 11647–11652.

Klionsky, D.J., and Emr, S.D. (2000). Autophagy as a regulated pathway of cellular degradation. Science 290, 1717–1721.

Lee, P.D., Conover, C.A. and Powell, D.R. (1993). Regulation and function of insulin-like growth factor-binding protein-1. Proc. Soc. Exp. Biol. Med. 204, 4–29.

Livingstone, C., Patel, G. and Jones, N. (1995). ATF-2 contains a phosphorylation-dependent transcriptional activation domain. EMBO J. 14, 1785–1797.

Luethy, J.D. and Holbrook, N.J. (1992). Activation of the gadd153 promoter by genotoxic agents, a rapid and specific response to DNA damage. Cancer Res. 52, 5–10.

Lynch, C.J. (2001). Role of leucine in the regulation of mTOR by amino acids, revelations from structure-activity studies. J. Nutr. 131, 861S–865S.

Mizushima, N., Noda, T., Yoshimori, T., Tanaka, Y., Ishii, T., George, M.D., Klionsky, D.J., Ohsumi, M. and Ohsumi, Y. 1998. A protein conjugation system essential for autophagy. Nature 395, 395–398.

Mordier, S., Deval, C., Bechet, D., Tassa, A. and Ferrara, M. (2000). Leucine limitation induces autophagy and activation of lysosome-dependent proteolysis in C2C12 myotubes through a mammalian target of rapamycin-independent signaling pathway. J. Biol. Chem. 275, 29900–29906.

Mortimore, G.E., Poso, A.R. and Lardeux, B.R. (1989). Mechanism and regulation of protein degradation in liver. Diabetes Metab. Rev. 5, 49–70.

Munro, H.N. (1970). Mammalian Protein Metabolism. Academic Press, New York.

Munro, H.N. (1976). Second Boyd Orr Memorial Lecture. Regulation of body protein metabolism in relation to diet. Proc. Nutr. Soc. 35, 297–308.

Murata, K., Wu, J. and Brautigan, D.L. (1997). B cell receptor-associated protein alpha4 displays rapamycin-sensitive binding directly to the catalytic subunit of protein phosphatase 2A. Proc. Natl. Acad. Sci. USA 94, 10624–10629.

Nave, B.T., Ouwens, M., Withers, D.J., Alessi, D.R. and Shepherd, P.R. (1999). Mammalian target of rapamycin is a direct target for protein kinase B, identification of a convergence point for opposing effects of insulin and amino-acid deficiency on protein translation. Biochem. J. 344, 427–431.

Noda, T., and Ohsumi, Y. (1998). Tor, a phosphatidylinositol kinase homologue, controls autophagy in yeast. J. Biol. Chem. 273, 3963–3966.

Ozalp, I., Young, V.R., Nagchaudhuri, J., Tontisirin, K. and Scrimshaw, N.S. (1972). Plasma amino acid response in young men given diets devoid of single essential amino acids. J. Nutr. 102, 1147–1158.

Pain, V.M. (1994). Translational control during amino acid starvation. Biochimie 76, 718–728.

Pearson, R.B., Dennis, P.B., Han, J.W., Williamson, N.A., Kozma, S.C., Wettenhall, R.E. and Thomas, G. (1995). The principal target of rapamycin-induced p70s6k inactivation is a novel phosphorylation site within a conserved hydrophobic domain. EMBO J. 14, 5279–5287.

Pégorier, J.P. (1998). Regulation of gene expression by fatty acids. Curr. Opin. Clin. Nutr. Metab. Care 1, 329–334.

Peng, Y., and Harper, A.E. (1970). Amino acid balance and food intake, effect of different dietary amino acid patterns on the plasma amino acid pattern of rats. J. Nutr. 100, 429–437.

Peterson, R.T., Beal, P.A., Comb, M.J. and Schreiber, S.L. (2000). FKBP12-rapamycin-associated protein (FRAP) autophosphorylates at serine 2481 under translationally repressive conditions. J. Biol. Chem. 275, 7416–7423.

Peterson, R.T., Desai, B.N., Hardwick, J.S. and Schreiber, S.L. (1999). Protein phosphatase 2A interacts with the 70-kDa S6 kinase and is activated by inhibition of FKBP12-rapamycin-associated protein. Proc. Natl. Acad. Sci. USA 96, 4438–4442.

Peyrollier, K., Hajduch, E., Blair, A.S., Hyde, R. and Hundal, H.S. (2000). L-leucine availability regulates phosphatidylinositol 3-kinase, p70 S6 kinase and glycogen synthase kinase-3 activity in L6 muscle cells, evidence for the involvement of the mammalian target of rapamycin (mTOR) pathway in the L-leucine-induced up-regulation of system A amino acid transport. Biochem. J. 350, 361–368.

Pham, P.T., Heydrick, S.J., Fox, H.L., Kimball, S.R., Jefferson, L.S. and Lynch, C.J. (2000). Assessment of cell-signaling pathways in the regulation of mammalian target of rapamycin (mTOR) by amino acids in rat adipocytes. J. Cell. Biochem. 79, 427–441.

Pickart, C.M. (2000). Ubiquitin in chains. Trends Biochem. Sci. 25, 544–548.

Proud, C.G. and Denton, R.M. (1997). Molecular mechanisms for the control of translation by insulin. Biochem. J. 328, 329–341.

Redpath, N.T., Foulstone, E.J. and Proud, C.G. (1996). Regulation of translation elongation factor-2 by insulin via a rapamycin-sensitive signalling pathway. EMBO J. 15, 2291–2297.

Rogers, Q.R., and Leung, P.M.B. (1977). The control of food intake, when and how are amino acids involved? In: The Chemical Senses and Nutrition (Kare, M.R. and Maller, O., Eds.), pp. 213–249. Academic Press, New York.

Ron, D. and Habener, J.F. (1992). CHOP, a novel developmentally regulated nuclear protein that dimerizes with transcription factors C/EBP and LAP and functions as a dominant-negative inhibitor of gene transcription. Genes Dev. 6, 439–453.

Sano, Y., Harada, J., Tashiro, S., Gotoh-Mandeville, R., Maekawa, T. and Ishii, S. (1999). ATF-2 is a common nuclear target of Smad and TAK1 pathways in transforming growth factor-beta signaling. J. Biol. Chem. 274, 8949–8957.

Schmitt-Ney, M. and Habener, J.F. (2000). CHOP/GADD153 gene expression response to cellular stresses inhibited by prior exposure to ultraviolet light wavelength band C (UVC). Inhibitory sequence mediating the UVC response localized to exon 1. J. Biol. Chem. 275, 40839–40845.

Scott, S.V., Nice, D.C., Nau, J.J., Weisman, L.S., Kamada, Y., Keizer-Gunnink, I., Funakoshi, T., Veenhuis, M., Ohsumi, Y. and Klionsky, D.J. (2000). Apg13p and Vac8p are part of a complex of phosphoproteins that are required for cytoplasm to vacuole targeting. J. Biol. Chem. 275, 25840–25849.

Shigemitsu, K., Tsujishita, Y., Hara, K., Nanahoshi, M., Avruch, J. and Yonezawa, K. (1999). Regulation of translational effectors by amino acid and mammalian target of rapamycin signaling pathways. Possible involvement of autophagy in cultured hepatoma cells. J. Biol. Chem. 274, 1058–1065.

Siu, F., Chen, C. Zhong, C. and Kilberg, M.S. (2001). CCAAT/Enhancer-binding protein-beta is a mediator of the nutrient-sensing response pathway that activates the human asparagine synthetase gene. J. Biol. Chem. 276, 48100–48107.

Straus, D.S. (1994). Nutritional regulation of hormones and growth factors that control mammalian growth. FASEB J. 8, 6–12.

Straus, D.S., Burke, E.J. and Marten, N.W. 1993. Induction of insulin-like growth factor binding protein-1 gene expression in liver of protein-restricted rats and in rat hepatoma cells limited for a single amino acid. Endocrinology 132, 1090–1100.

Stromhaug, P.E., Berg, T.O., Fengsrud, M. and Seglen, P.O. (1998). Purification and characterization of autophagosomes from rat hepatocytes. Biochem. J. 335, 217–224.

Svanberg, E., Jefferson, L.S., Lundholm, K. and Kimball, S.R. (1997). Postprandial stimulation of muscle protein synthesis is independent of changes in insulin. Am. J. Physiol. 272, E841–E847.

Sylvester, S.L., ap Rhys, C.M., Luethy-Martindale, J.D. and Holbrook, N.J. (1994). Induction of GADD153, a CCAAT/enhancer-binding protein (C/EBP)-related gene, during the acute phase response in rats. Evidence for the involvement of C/EBPs in regulating its expression [published erratum appears in J. Biol. Chem. 1995; 270, 14842]. J. Biol. Chem. 269, 20119–20125.

Talloczy Z., Jiang, W., Virgin, H.W., Leib, D.A., Scheuner, D., Kaufman, R.J., Eskelinen, E.-L. and Levine, B. (2002). Regulation of starvation and virus-induced autophagy by the eIF2α kinase signaling pathway. Proc. Natl. Acad. Sci. 99, 190–195.

Tang, H., Hornstein, E., Stolovich, M., Levy, G., Livingstone, M., Templeton, D., Avruch, J. and Meyuhas, O. (2001). Amino acid-induced translation of TOP mRNAs is fully dependent on phosphatidylinositol 3-kinase-mediated signaling, is partially inhibited by rapamycin, and is independent of S6K1 and rpS6 phosphorylation. Mol. Cell. Biol. 21, 8671–8683.

Towle, H.C. (1995). Metabolic regulation of gene transcription in mammals. J. Biol. Chem. 270, 23235–23238.

Tsukada, M., and Ohsumi, Y. (1993). Isolation and characterization of autophagy-defective mutants of *Saccharomyces cerevisiae*. FEBS Lett. 333, 169–174.

Turk, V., Turk, B. and Turk, D. (2001). Lysosomal cysteine proteases, facts and opportunities. EMBO J. 20, 4629–4633.

Ubeda, M., Vallejo, M. and Habener, J.F. (1999). CHOP enhancement of gene transcription by interactions with Jun/Fos AP-1 complex proteins. Mol. Cell. Biol. 19, 7589–7599.

Vary, T.C., Jefferson, L.S. and Kimball, S.R. (1999). Amino acid-induced stimulation of translation initiation in rat skeletal muscle. Am. J. Physiol. 277, E1077–E1086.

Vaulont, S., Vasseur-Cognet, M. and Kahn, A. (2000). Glucose regulation of gene transcription. J. Biol. Chem. 275, 31555–31558.

Voges, D., Zwickl, P. and Baumeister, W. (1999). The 26S proteasome, a molecular machine designed for controlled proteolysis. Annu. Rev. Biochem. 68, 1015–1068.

Wang, L., Wang, X. and Proud, C.G. (2000). Activation of mRNA translation in rat cardiac myocytes by insulin involves multiple rapamycin-sensitive steps. Am. J. Physiol. 278, H1056–H1068.

Wang, X., Campbell, L.E., Miller, C.M. and Proud, C.G. (1998). Amino acid availability regulates p70 S6 kinase and multiple translation factors [published erratum appears in Biochem. J. 1998; 335, 711]. Biochem. J. 334, 261–267.

Wang, X., Li, W., Williams, M., Terada, N., Alessi, D.R. and Proud, C.G. (2001a). Regulation of elongation factor 2 kinase by p90(RSK1) and p70 S6 kinase. EMBO J. 20, 4370–4379.

Wang, X., Paulin, F.E., Campbell, L.E., Gomez, E., O'Brien, K., Morrice, N. and Proud, C.G. (2001b). Eukaryotic initiation factor 2B, identification of multiple phosphorylation sites in the epsilon-subunit and their functions in vivo. EMBO J. 20, 4349–4359.

Wang, X.Z., Lawson, B., Brewer, J.W., Zinszner, H., Sanjay, A., Mi, L.J., Boorstein, R., Kreibich, G., Hendershot, L.M. and Ron, D. (1996). Signals from the stressed endoplasmic reticulum induce C/EBP-homologous protein (CHOP/GADD153). Mol. Cell. Biol. 16, 4273–4280.

Weng, Q.P., Andrabi, K., Kozlowski, M.T., Grove, J.R. and

Avruch, J. (1995). Multiple independent inputs are required for activation of the p70 S6 kinase. Mol. Cell. Biol. 15, 2333–2340.

Weng, Q.P., Kozlowski, M., Belham, C., Zhang, A., Comb, M.J. and Avruch, J. (1998). Regulation of the p70 S6 kinase by phosphorylation in vivo. Analysis using site-specific anti-phosphopeptide antibodies. J. Biol. Chem. 273, 16621–16629.

Wing, S.S., and Goldberg, A.L. (1993). Glucocorticoids activate the ATP-ubiquitin-dependent proteolytic system in skeletal muscle during fasting. Am. J. Physiol. 264, E668–E676.

Wolfe, R.R., Jahoor, F. and Hartl, W.H. (1989). Protein and amino acid metabolism after injury. Diabetes Metab. Rev. 5, 149–164.

Xu, G., Kwon, G., Marshall, C.A., Lin, T.A., Lawrence, J.C. and McDaniel, M.L. (1998). Branched-chain amino acids are essential in the regulation of PHAS-I and p70 S6 kinase by pancreatic beta-cells. A possible role in protein translation and mitogenic signaling. J. Biol. Chem. 273, 28178–28184.

Yoshida, H., Okada, T., Haze, K., Yanagi, H., Yura, T., Negishi, M. and Mori, K. (2000). ATF6 activated by proteolysis binds in the presence of NF-Y (CBF) directly to the *cis*-acting element responsible for the mammalian unfolded protein response. Mol. Cell. Biol. 20, 6755–6767.

Yoshizawa, F., Endo, H., Ide, H., Yagasaki, K. and Funabiki, R. (1995). Translational regulation of protein synthesis in the liver and skeletal muscle of mice in response to refeeding. Nutr. Biochem. 6, 130–136.

Young, V.R., El-Khoury, A.E., Melchor, S. and Castillo, L. (1994). The biochemistry and physiology of protein and amino acid metabolism, with reference to protein nutrition. In: Protein Metabolism during Infancy (Niels, C.R.R., Ed.), Vol. 33, pp. 1–28. Nestec Ltd., Vevey/Raven Press, Ltd., New York.

Ziegler, T.R., Gatzen, C. and Wilmore, D.W. (1994). Strategies for attenuating protein-catabolic responses in the critically ill. Annu. Rev. Med. 45, 459–480.

Sensing, Signaling and Cell Adaptation. Edited by K.B. Storey and J.M. Storey
© *2002 Elsevier Science B.V. All rights reserved.*

CHAPTER 14

Amino Acid-dependent Signal Transduction

Peter F. Dubbelhuis and Alfred J. Meijer
Department of Biochemistry, Academic Medical Center, University of Amsterdam, Amsterdam, The Netherlands

1. Introduction: amino acid stimulation of S6 phosphorylation

Amino acids are the precursors of a variety of intracellular N-containing molecules. In addition, they can be used for oxidation to generate energy. Amino acids can also function as regulators of enzyme activities and can control fluxes through metabolic pathways. A textbook example is that of the inhibition of liver pyruvate kinase by alanine which is considered to be relevant in fasting in order to prevent simultaneous operation of gluconeogenesis and glycolysis in hepatocytes. Other examples are the stimulation by amino acids of glucose-driven glycogen synthesis (Baquet et al., 1990) and the inhibition of autophagy by amino acids (Blommaart et al., 1997b, for review). The latter two properties of amino acids represent typical insulin-like actions, and we have been interested in the mechanisms underlying these phenomena.

Stimulation of glycogen synthesis by certain amino acids (glutamine, proline, alanine) is caused by cell swelling due to concentrative, Na$^+$-dependent, amino acid transport across the plasma membrane and by the intracellular production of impermeant catabolites such as glutamate and aspartate (Baquet et al., 1990; Plomp et al., 1990). In response to the initial swelling, cells undergo "regulatory volume decrease" when they try to restore their original volume by releasing KCl. The decrease in intracellular chloride concentration can be considerable and activates glycogen synthase phosphatase because chloride inhibits the enzyme

at concentrations usually found in cells (Meijer et al., 1992).

By contrast, autophagy is inhibited by amino acids such as leucine, tyrosine and phenylalanine that are transported by H$^+$-dependent mechanisms and are not concentrated to a major extent inside the cells (Blommaart et al., 1997b). Autophagic protein degradation in freshly isolated hepatocytes is not affected by cell swelling *per se*, but an increase in cell volume promotes the ability of leucine, tyrosine and phenylalanine to inhibit proteolysis (Blommaart et al., 1995).

Clearly, the reciprocal regulation of glycogen synthesis and of autophagic proteolysis by amino acids shares a common element, cell swelling, but differences also exist. Thus, leucine, tyrosine and phenylalanine do not affect glycogen synthesis (Baquet et al., 1990; Luiken et al., 1994).

The potency of leucine as an inhibitor of autophagic proteolysis (Blommaart et al., 1997b, for review) was of interest because for a long time leucine had also been known as an effective stimulator of protein synthesis, not only in the liver but also in other cell types (Tischler et al., 1982). We then discovered that the addition of a complete mixture of amino acids at physiological concentration to isolated rat hepatocytes strongly and rapidly stimulated the phosphorylation of a protein of 31 kDa that we identified as ribosomal protein S6 (Luiken et al., 1994; Blommaart et al., 1995). S6 has five phosphorylation sites, is a component of the 40S ribosomal subunit and its phosphorylation is required for the translation of the terminal oligopyrimidine ("TOP") family of mRNA molecules, containing an oligopyrimidine tract upstream of their transcription-initiation site (Duffner and Thomas, 1999); these mRNA molecules encode proteins belonging to the protein-translation machinery. Amino acid-stimulated phosphorylation of S6 was completely eliminated by rapamycin, indicating that the serine/threonine protein kinase mTOR (m̲ammalian t̲arget o̲f r̲apamycin) was on the pathway of amino acid-stimulated S6 phosphorylation. The stimulation of S6 phosphorylation by amino acids resembled that by insulin (Sturgill and Wu, 1991, for review). It must be stressed that in our experiments with amino acids, insulin was not present. Moreover, when insulin was added in the absence of amino acids, no effect on S6 phosphorylation was observed (Blommaart et al., 1995). However, although ineffective on its own, insulin did increase the effectiveness of amino acids to induce S6 phosphorylation in that lower concentrations of amino acids were needed to induce maximal S6 phosphorylation. The reason for the effectiveness of insulin in stimulating S6 phosphorylation in other cell types, as reported in the earlier literature, was the presence of amino acids in the culture media. Another important observation we made was that hypo-osmotically-induced cell swelling also increased the effectiveness of amino acids, such as leucine, tyrosine and phenylalanine, to promote S6 phosphorylation. Furthermore, under a wide variety of conditions we found a linear relationship between the degree of S6 phosphorylation and the percentage of inhibition of proteolysis as measured in the presence of cycloheximide to inhibit simultaneous protein synthesis (Blommaart et al., 1995). Of great significance was that rapamycin addition could partially, albeit not completely, release the inhibition of autophagic proteolysis by amino acids under these conditions. In our experiments, in the absence of cycloheximide, protein synthesis was also partially inhibited by rapamycin. It was concluded that the same mechanism is perhaps involved in the reciprocal control of protein synthesis and degradation which would be extremely efficient from the point of view of metabolic regulation (Blommaart et al., 1995). Interestingly, rapamycin addition to yeast cells also

stimulates autophagy (Noda and Ohsumi, 1998), indicating conservation of this control mechanism in evolution.

Amino acid-induced S6 phosphorylation was not only inhibited by rapamycin but also by wortmannin and LY294002, two structurally unrelated inhibitors of phosphatidylinositol 3-kinase (PI 3-kinase), and it was concluded that PI 3-kinase is another component of the amino acid-dependent signaling pathway (Blommaart et al., 1997a). It should be noted that the PI 3-kinase that is activated by insulin and other growth factors is, in fact, PI 3-kinase class I, which produces PtdIns $(3,4)P_2$ and PtdIns$(3,4,5)P_3$ (Vanhaesebroeck and Waterfield, 1999). By contrast, PI 3-kinase class III is probably not affected by insulin, and produces PtdIns(3)P. It has recently been shown that PtdIns(3)P is essential for autophagic protein degradation while PtdIns$(3,4)P_2$ and PtdIns$(3,4,5)P_3$ are inhibitory (Petiot et al., 2000).

2.   Amino acids and p70S6 kinase activation

The existence of amino acid-dependent signaling was confirmed a few years later by several groups, although in most studies the degree of phosphorylation of p70S6 kinase, the enzyme responsible for S6 phosphorylation in the intact cell (Dufner and Thomas, 1999), and its *in vitro* activity, were analyzed. p70S6 kinase is located downstream of mTOR and is presumably directly phosphorylated by mTOR (Burnett et al., 1998).

Amino acid-dependent signaling appeared not to be unique for hepatocytes, and amino acid-induced, rapamycin-sensitive, p70S6 kinase phosphorylation was found in many insulin-sensitive cell types, including muscle cells, adipocytes, hepatoma cells, CHO cells and pancreatic β-cells (Hara et al., 1998; Wang et al., 1998; Fox et al., 1998; Patti et al., 1998; Kimball et al., 1998; Xu et al., 1998a). The involvement of mTOR in the amino acid response was also supported by other experiments. Thus, in CHO-IR cells a rapamycin-resistant mutant of p70S6 kinase could be phosphorylated in the presence of insulin in a wortmannin-sensitive manner at Thr 412, critical for enzyme activity, irrespective of the presence of amino acids (Hara et al., 1998). Conversely, in human rhabdomyosarcoma Rh30 cells harbouring a rapamycin-resistant mutant of mTOR, amino acids stimulated p70S6 kinase activity in a rapamycin-insensitive manner (Iiboshi et al., 1999).

As in hepatocytes, amino acids and insulin also acted synergistically in other cell types (Hara et al., 1998; Patti et al., 1998; Xu et al., 1998a; Campbell et al., 1999; Tremblay and Marette, 2001) and among the various amino acids, leucine was the most effective (Hara et al., 1998; Wang et al., 1998; Patti et al., 1998; Kimball et al., 1998; Xu et al., 1998a; Shigemitsu et al., 1999b; Lynch et al., 2000; Xu et al., 2001). Insulin alone did not induce p70S6 kinase activation. In cases where it did stimulate on its own this could be ascribed to amino acids produced by autophagy (Shigemitsu et al., 1999a). The data also showed that leucine alone cannot completely mimic the effect of a mixture of all amino acids. It is likely, therefore, that other amino acids act in concert with leucine to elicit full activation of p70S6 kinase. A possible explanation is that amino acids which are transported together with Na^+ are concentrated against the concentration gradient. The ensuing increase in cell volume may then be responsible for a potentiation of the leucine effect, as discussed above for hepatocytes (see Introduction). This may also explain why glutamine is so potent in stimulating the effect of leucine in perfused liver (Shah et al., 1999) because glutamine potently increases cell volume (Baquet et al., 1990). Control experiments (not shown) carried out in our laboratory indicated that cell swelling does not affect plasma membrane leucine transport.

3. Amino acid stimulation of 4E-BP1 phosphorylation

4E-BP1 (also known as PHAS-1), in addition to p70S6 kinase, is another substrate of mTOR (Burnett et al., 1998). 4E-BP1 has several phosphorylation sites that are targets of different protein kinases and phosphorylation of the protein results in dissociation of the eIF4E.4E-BP1 complex so that eIF4E becomes available for cap-dependent mRNA translation-initiation (Gingras et al., 2001). Phosphorylation of 4E-BP1, like that of p70S6 kinase, was greatly increased in the presence of amino acids in a rapamycin-sensitive manner (Hara et al., 1998; Wang et al., 1998; Fox et al., 1998; Patti et al., 1998; Xu et al., 1998b; Shigemitsu et al., 1999a; Iiboshi et al., 1999), with leucine again being most effective (Hara et al., 1998; Fox et al., 1998; Xu et al., 1998a). Insulin was not required for this effect. In the absence of amino acids, insulin was unable to stimulate 4E-BP1 phosphorylation (Hara et al., 1998; Xu et al., 1998b).

4. Amino acid stimulation of eEF2kinase

Eukaryotic elongation factor 2 (eEF2), which mediates the translocation step of elongation, is inactive when phosphorylated at Thr56. Phosphorylation of eEF2 is under the control of eEF2 kinase. This kinase becomes inhibited when phosphorylated at Ser366 by p70S6 kinase so that the same factors (e.g., amino acids) that control p70S6 kinase activity also control eEF2 kinase phosphorylation, and thus the activity of eEF2 (Wang et al., 2001). Interestingly, eEF2 kinase can also be phosphorylated by p90[RSK]; in this case, eEF2 dephosphorylation is insensitive to rapamycin but sensitive to inhibitors of the MEK/Erk signaling pathway (Wang et al., 2001).

5. Amino acid stimulation of eIF2α

The eukaryotic initiation factor eIF2 recruits charged initiator tRNA to the 40S ribosomal subunit and becomes inactivated when Ser51 of the α subunit eIF2 is phosphorylated in the absence of amino acids. Dephosphorylation and activation are effected by amino acids, leucine in particular (Kimball et al., 1998). In yeast, a single kinase, GCN2 (general control non-depressible), is responsible for eIF2α phosphorylation. This kinase is activated by uncharged tRNAs because its

C-terminus structurally resembles histidyl-tRNA synthetase and other amino acyl-tRNA synthetases (Hinnebusch, 1997). Uncharged tRNA has been proposed as a sensor of amino acid starvation (see Section 14).

6. Involvement of PI 3-kinase and protein kinase B in amino acid-dependent signaling? Amino acid/insulin synergy

The mechanism responsible for the synergy between amino acids and insulin in stimulating p70S6 kinase, S6 and 4E-BP1 phosphorylation is of great interest because it may provide information on the mechanism by means of which amino acids stimulate signaling.

A simple explanation for the synergy would be that insulin promotes plasma membrane amino acid transport. In perfused muscle, however, insulin stimulated protein synthesis that was accompanied by decreased, not increased, intracellular amino acid concentrations (Shah et al., 2000a).

As discussed, insulin alone, in the absence of amino acids, did not affect phosphorylation of p70S6 kinase, S6 and 4E-BP1. However, the hormone did stimulate the activity of PI 3-kinase and protein kinase B (Hara et al., 1998; Patti et al., 1998; Campbell et al., 1999; Shigemitsu et al., 1999a), components upstream of mTOR in insulin-dependent signal transduction (Gingras et al., 2001). Phosphorylation of p70S6 kinase, S6 and of 4E-BP1 in the presence of amino acids alone, in the absence of insulin, can be prevented by inhibitors of PI 3-kinase (Blommaart et al., 1997a; Hara et al., 1998; Wang et al., 1998; Fox et al., 1998; Patti et al., 1998; Shigemitsu et al., 1999b). This suggests that PI 3-kinase is located upstream of p70S6 kinase in the signaling pathway. However, controversy exists on whether or not amino acids are able to stimulate PI 3-kinase activity, because direct attempts to show such activation have either failed (Hara et al., 1998; Patti et al., 1998; Shigemitsu et al., 1999a) or were successful, albeit that activation was only transient (Krause et al., 1996; Peyrollier et al., 2000). There is general agreement, however, that amino acids do not affect protein kinase B

activity (Hara et al., 1998; Wang et al., 1998; Patti et al., 1998; Campbell et al., 1999; Iiboshi et al., 1999; Peyrollier et al., 2000). This finding clearly eliminates protein kinase B as a component of the amino acid-dependent signaling pathway.

In order to account for the apparent contradiction between the inhibitory effects of PI 3-kinase inhibitors on amino acid-dependent signaling and the lack of effect of amino acids on PI 3-kinase activity, at least in some studies, we previously proposed (van Sluijters et al., 2000) that amino acids may not affect PI 3-kinase itself but rather PTEN, the lipid phosphatase that is responsible for the removal of the 3-phosphate of PtdIns $(3,4)P_2$ and PtdIns$(3,4,5)P_3$. This idea remains to be explored, however.

It has been suggested that inhibition of amino acid-dependent signaling by the PI 3-kinase inhibitors may, in fact, be ascribed to direct inhibition of mTOR (Hara et al., 1998; Patti et al., 1998). We consider this unlikely for three reasons. Firstly, inhibition of *in vitro* mTOR activity by wortmannin occurs at concentrations much higher than required for inhibition of *in vitro* PI 3-kinase activity (Brunn et al., 1996; cf. Blommaart et al., 1997a). Secondly, cells overexpressing protein kinase delta showed increased phosphorylation of 4E-BP1 which was rapamycin-sensitive but wortmannin-insensitive (Kumar et al., 2000). This is difficult to explain if mTOR would be inhibited by wortmannin. Thirdly, involvement of PI 3-kinase in amino acid-dependent signaling was also indicated by studies with PDK1 –/– cells (Wang et al., 2001) or with cells with overexpression of either PTEN or of a dominant-negative regulatory subunit p85 of PI 3-kinase class I (Tang et al., 2001), showing that these genetic manipulations strongly interfered with amino acid signaling.

Because amino acids may not directly activate PI 3-kinase, it was proposed that PI 3-kinase is on a pathway parallel to that of amino acids and that both the activation of PI 3-kinase (by insulin) and of mTOR or another kinase (by amino acids) are required for full activation of p70S6 kinase (Gingras et al., 2001; Wang et al., 2001). One possibility is that the phosphatidylinositol lipids are required for membrane anchoring of one or more kinases, the activity of which is regulated by amino acids (Tang et al., 2001). Another possibility arose from the above-mentioned studies showing that amino acid-dependent p70S6 kinase activation was abrogated in PDK1 –/– cells; however, amino acids were still able to increase phosphorylation of 4E-BP1 in a rapamycin-sensitive manner, indicating that activation of mTOR function in the presence of amino acids was maintained in these PDK1 deficient cells (Wang et al., 2001). Using p70S6 kinase phospho-specific antibodies it was concluded that activation of p70S6 kinase requires two separate inputs: one through PDK1, which results in phosphorylation of Thr229, and another through mTOR (PI 3-kinase and PDK1-independent) resulting in phosphorylation of Thr389 (Note: Thr229 and Thr389 are equivalent to Thr252 and Thr412 of the long splice variant of p70S6kinase) (Wang et al., 2001) (Fig. 14.1).

In order to account for the ability of high concentrations of amino acids to activate p70S6 kinase in the absence of insulin by a mechanism that is wortmannin- or LY294002-sensitive, however, one has to assume that either basal activity of PI 3-kinase, or only a slight stimulation of PI 3-kinase by amino acids (or inhibition of PTEN), may be sufficient for Thr389 phosphorylation in p70S6 kinase.

If correct, the scheme of Fig. 14.1 satisfactorily accounts for the synergy between amino acids and insulin regarding p70S6 kinase activation. It may even be speculated that cell swelling mimics the effect of insulin on PI 3-kinase (Krause et al., 1996) so that the effect of high concentrations of amino acids on p70S6 kinase consists of two components: one due to amino acids coupled to Na+ transport, giving rise to increased cell volume resulting in PI 3-kinase and PDK1 activation, and another component (e.g., leucine), required for mTOR activation (see also section on Amino acids and p70S6 kinase activation). This would explain why high concentrations of amino acids are able to activate p70S6 kinase in the absence of insulin.

7. Amino acids and mTOR activation

The mechanism by which amino acids activate mTOR is still unclear. *In vitro* kinase activity towards 4E-BP1 of mTOR, immunoprecipitated from either rapamycin-treated or amino acid-depleted CHO-IR or PC12 cells, was not different from mTOR immunoprecipitated from control cells (Hara et al., 1998; Kleijn and Proud, 2000). This suggests that mTOR activity changes were lost during isolation and may not be due to phosphorylation of mTOR but perhaps due to an

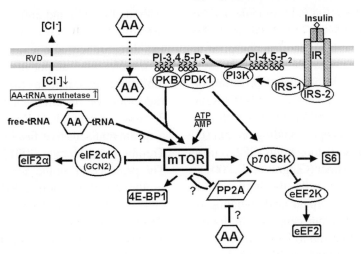

Fig. 14.1. Amino acid-dependent signal transduction. IR, insulin receptor; IRS, insulin receptor substrate; PI3K, phosphatidylinositol 3-kinase; PDK1, phosphoinositide-dependent kinase 1; PKB, protein kinase B; AA, amino acids; RVD, regulatory volume decrease. For other abbreviations, see text.

allosteric effect on mTOR itself or on an unknown protein kinase located upstream of mTOR (cf. Fig. 14.1). Other studies, however, did show stable changes in mTOR activity. Thus, mTOR isolated from amino acid-stimulated Jurkat cells could phosphorylate the protein phosphatase PP2A *in vitro* (Peterson et al., 1999). In HEK293 cells, amino acid addition in the absence of insulin increased phosphorylation of Ser2448 of mTOR. Moreover, *in vitro*, mTOR could be phosphorylated by protein kinase B, but only when mTOR was immunopurified from amino acid-treated cells (Navé et al., 1999). Although these experiments show that mTOR can undergo stable changes in phosphorylation in response to amino acid addition, the relevance of Ser2448 phosphorylation for mTOR activity for transmission of signals to p70S6 kinase and 4E-BP1 has been questioned on the basis of experiments with an mTOR mutant in which Ser2448 was replaced by alanine (Gingras et al., 2001).

8.   Amino acids and protein phosphatases

It has been suggested that amino acids may act as inhibitors of a protein phosphatase rather than as stimulators of a protein kinase. In yeast, for example, the rapamycin-sensitive TOR proteins are known to affect PP2A activity, by modulating the association of PP2A with the Tap42 protein: in the presence of nutrients Tap42 becomes phosphorylated and associates with PP2A which then becomes inhibited, while nutrient deprivation or rapamycin addition reverses these events (Gingras et al., 2001; for review). The mammalian ortholog of Tap42 is the B-cell receptor-binding protein α4 and this phosphoprotein, too, binds PP2A in a rapamycin-sensitive manner (Gingras et al., 2001).

So far, in mammalian cells, the issue of the involvement of PP2A in amino acid signaling is controversial. Thus, in brain cells and in Jurkat cells, p70S6 kinase appeared to be tightly associated with protein phosphatase 2A (Westphal et al., 1999; Peterson et al., 1999); in cells carrying the p70S6 kinase mutant that is resistant to rapamycin and to amino acid depletion, the association with

PP2A was lost (Peterson et al., 1999). Moreover, in some cell types, phosphorylation, and activation, of p70S6 kinase, S6 and of 4E-BP1 in the presence of rapamycin or in the absence of amino acids can be induced by PP2A inhibitors (Peterson et al., 1999; Parrott et al., 1999; Tang et al., 2001). Likewise, dexamethasone-induced dephosphorylation of p70S6 kinase and of 4E-BP1 in L6 myoblasts in the presence of amino acids can be corrected by these PP2A inhibitors (Shah et al., 2000b). In other cell types, however, PP2A inhibitors had no effect (Hara et al., 1998; Westphal et al., 1999). In our own experiments, carried out with isolated rat hepatocytes, the PP2A inhibitor calyculin, but not okadaic acid, induced rapamycin-insensitive hyperphosphorylation of p70S6 kinase, which was additive with the rapamycin-sensitive phosphorylation induced by amino acid addition (Fig. 14.2). The calyculin-induced hyperphosphorylation did not affect p70S6 kinase activity (not shown). This lack of effect on p70S6 kinase activity is in agreement with the lack of effect of phosphatase inhibitors on amino acid-induced phosphorylation of S6 (Blommaart et al., 1995).

9.   Negative feedback by amino acid signaling on insulin signaling

Although insulin and amino acids synergize with regard to their effects on mTOR mediated signaling, there are now also several reports showing that in muscle cells, adipocytes and hepatoma cells, amino acids cause a time-dependent downregulation of insulin-mediated activation of PI 3-kinase, protein kinase B and glucose transport in a rapamycin-sensitive fashion (Patti et al., 1998; Tremblay and Marette, 2001; Takano et al., 2001). The presence of amino acids results in increased ser/thr phosphorylation of IRS1, decreased binding of the p85 regulatory subunit of PI 3-kinase to IRS1 followed by increased, presumably proteasomal, degradation of IRS-1 (Tremblay and Marette, 2001; Takano et el., 2001). It has been proposed that this mechanism may underlie diminished glucose consumption during high-protein feeding (Patti et al., 1998; Tremblay and Marette,

A

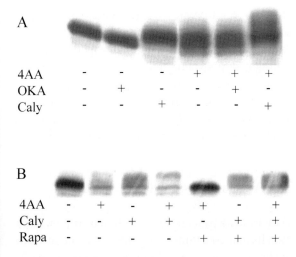

4AA	-	-	-	+	+	+
OKA	-	+	-	-	+	-
Caly	-	-	+	-	-	+

B

4AA	-	+	-	+	+	-	+
Caly	-	-	+	+	-	+	+
Rapa	-	-	-	-	+	+	+

Fig. 14.2. Effect of PP2A inhibitors on amino acid-dependent p70S6 kinase phosphorylation. Hepatocytes were incubated with 20 mM glucose for 1 h at 37°C under the conditions indicated. p70S6 kinase was analyzed by a bandshift assay. After electrophoresis (SDS-PAGE, 10% polyacrylamide), proteins were western blotted onto a PVDF membrane and the p70S6 kinase was probed with a polyclonal p70S6 kinase antibody, and visualized by enhanced chemiluminescence. Increased phosphorylation results in a lowered electrophoretic mobility of the p70S6 kinase protein. This is reflected in an upward shift of the protein band in the gel. In (A), it is shown that calyculin, but not okadaic acid, stimulated phosphorylation of p70S6 kinase and that the effect was additive with that caused by amino acid addition. In the experiment of (B), carried out with a different hepatocyte preparation, but under otherwise identical conditions, it is demonstrated that rapamycin prevented the phosphorylation of p70S6 kinase caused by amino acid addition but not that caused by calyculin. Abbreviations: 4AA, a complete mixture of amino acids, each amino acid being present at concentrations four times that found in the portal vein of a fasted rat; OKA, okadaic acid (20 nM); caly, calyculin (40 nM); rapa, rapamycin (100 nM) (Blommaart et al., 1995, for further experimental details).

2001). In this context, the amino acid composition of dietary protein may be important and the relatively low leucine content of fish protein protects against high-fat induced insulin resistance (Tremblay and Marette, 2001).

Apparently, and paradoxically, amino acids are required for insulin-mediated activation of mTOR and its downstream targets but they inhibit the initial part of the insulin signaling pathway. The paradox lies in the fact that PI 3-kinase activity is essential for amino acid-induced activation of mTOR and its downstream targets (see above). Down-regulation of PI 3-kinase activation by amino acids would be counter-productive, therefore, and would eventually lead to diminished protein synthesis and increased autophagic protein breakdown. This is highly unlikely. In order to resolve this problem one has to assume that part of the activation of PI 3-kinase by insulin proceeds independently from IRS-1. There is, indeed, evidence that the pathway via IRS-2 may escape feedback inhibition by amino acids (Takano et al., 2001). In our view, this residual, IRS-2-mediated, activation of PI 3-kinase would then be sufficient for amino acid-induced activation of mTOR and its downstream targets. mTOR may thus be considered as a metabolic switch that integrates both nutritional- and insulin-mediated signals (Takano et al., 2001). Amino acids would then simultaneously decrease transport and utilisation of glucose by insulin-sensitive tissues and at the same time increase protein synthesis and decrease autophagic proteolysis, thus contributing to stimulation of cell growth. The fact that an increase in cAMP, a catabolic signal, decreases mTOR activity (Scott and Lawrence, 1998), further supports a role of mTOR as a nutritional sensor and nicely accounts for the glucagon/insulin antagonism we previously observed with regard to S6 phosphorylation in hepatocytes (Blommaart et al., 1995).

10. mTOR as an ATP sensor. Involvement of AMP-dependent protein kinase?

In a recent study with HEK293 cells yet another function of mTOR was proposed in that this protein kinase may not only act as a sensor of amino acids but also of intracellular ATP (Dennis et al., 2001). It was noted that, among various protein kinases, the K_m of ATP for mTOR was high, about 1 mM, and in the physiological range of ATP concentrations. Moreover, by inhibiting either glycolytic or mitochondrial ATP production a correlation was found between the intracellular ATP concentration and the degree of phosphorylation of p70S6 kinase or 4E-BP1, as indicators of *in situ*

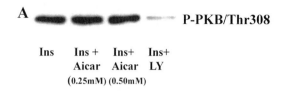

A   **P-PKB/Thr308**

Ins   Ins +   Ins+   Ins+
      Aicar   Aicar   LY
    (0.25mM) (0.50mM)

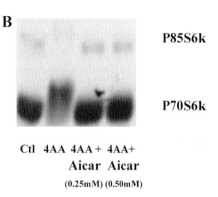

B   **P85S6k**

                       P70S6k

Ctl   4AA   4AA +   4AA+
           Aicar   Aicar
        (0.25mM) (0.50mM)

Fig. 14.3. Effect of AICAriboside on protein kinase B and p70S6 kinase phosphorylation. (A). Hepatocytes were preincubated for 20 min with 20 mM glucose. AICAriboside (if present) was then added at the concentrations indicated and incubation was continued for another 15 min, followed by addition of insulin (10^{-7} M) for 4 min. LY294002 (100 μM), if present, was added 2 min before insulin. (B). As in A, except that a complete mixture of amino acids (if present) was added after the 15 min exposure to AICAriboside. Incubation with amino acids was for 30 min. Protein kinase B was analyzed with a specific antibody against phospho-Thr308. p70S6 kinase was analyzed by bandshift assay with a polyclonal antibody (cf. Fig. 14.2). The data in (A) and (B) were obtained with the same hepatocyte preparation, and show that AICAriboside, which activates AMP-dependent protein kinase, prevented the stimulation by amino acids of p70S6 kinase phosphorylation (B) but had no effect on insulin-stimulated protein kinase B phosphorylation (A). p85S6 kinase is the nuclear form of p70S6 kinase and contains an N-terminal extension of 23 amino acids (Dufner and Thomas, 1999). The changes in phosphorylation of p85S6 kinase were similar to that of p70S6 kinase (B).

mTOR activity (Dennis et al., 2001). Because inhibition of ATP production also increases intracellular AMP levels via adenylate kinase (Hardie et al., 1998), we considered that AMP-dependent protein kinase may also contribute to mTOR inhibition when energy production is compromised. In agreement with this interpretation we observed in hepatocytes that AICAriboside—a compound which, after conversion to its phosphorylated

derivative, activates AMP-dependent protein kinase (Hardie et al., 1998)—inhibited p70S6 kinase phosphorylation without affecting insulin-stimulated phosphorylation of protein kinase B (Fig. 14.3). AICAriboside did not affect intracellular ATP (not shown). On the basis of these experiments with hepatocytes we propose that mTOR may also be a sensor of AMP.

11. Amino acid signaling in β-cells

Amino acid signaling is also found in pancreatic β-cells and constitutes a fascinating feedback loop in the regulation of whole body metabolism. In these cells, too, amino acids stimulated p70S6 kinase and 4E-BP1 phosphorylation, and the process was sensitive to inhibition by rapamycin and wortmannin. As in other cells, insulin alone, whether produced by the β-cells themselves (by glucose addition) or added externally was not effective unless amino acids are also present (Xu et al., 1998a; 1998b). Strikingly, in these cells among the various amino acids, leucine was again most effective (Xu et al., 1998a; 2001). While amino acids, via increased signaling, promote β-cell proliferation, it has been proposed that cytosolic glutamate in β-cells can directly promote exocytosis of insulin presumably by causing swelling of the insulin-containing granules (Maechler and Wolheim, 1999). Although attractive, this idea was refuted by data showing that with glutamine present, intracellular glutamate was extremely high, yet insulin release remained low (MacDonald and Fahien, 2000). Interestingly, glutamine and leucine (in the absence of glucose) acted synergistically, and in the presence of these two amino acids alone, insulin production was as high as observed in the presence of glucose alone, and the traditional view was maintained that allosteric activation of glutamate dehydrogenase by leucine provides α-oxoglutarate to the citric acid cycle (MacDonald and Fahien, 2000). Data by Xu et al. (2001) showed that the combination of glutamine plus leucine was also particularly effective in stimulating p70S6 kinase phosphorylation and that the ability of leucine analogues to stimulate glutamate dehydrogenase was closely associated with their ability to stimulate p70S6 kinase phosphorylation. Moreover, inhibition of the mitochondrial respiratory chain, with glycolysis being the only source of ATP production, eliminated the ability of glutamine plus leucine to stimulate p70S6 kinase. It was concluded that the same mitochondrial events that generate signals for leucine-stimulated exocytosis of insulin are required to activate the amino acid signaling pathway and that activation of protein synthesis by amino acid signaling contributes to enhanced β-cell function (Xu et al., 2001).

If, as discussed earlier (see Section 10), mTOR is indeed a sensor of ATP (or perhaps of AMP), a mechanistic explanation for the coupling between mitochondrial function and insulin release may be found. The key question to be answered then is the nature of the coupling between the amino acid-dependent signal transduction pathway and the process of insulin release. An attractive hypothesis would be that the closure of the K[+] channel in the plasma membrane of the β-cell is perhaps controlled by mTOR-mediated phosphorylation.

12. Amino acid signaling *in vivo*

Although studies on amino acid-dependent signal transduction have mainly been carried out in isolated cells, there is now ample evidence that amino acid signaling also plays an important role under physiological conditions. Thus, the protein anabolic response after a protein meal in man, rats and mice occurred in the absence of changes in insulin concentration and was accompanied by increased phosphorylation of 4E-BP1 and p70S6 kinase in muscle and liver (Shah et al., 2000a; Long et al., 2000; Balage et al., 2001). In the rat, inhibition of insulin production by diazoxide eliminated the effect of amino acids suggesting that, as in isolated cells, insulin and amino acids are also both required *in vivo* to induce a positive nitrogen balance (Balage et al., 2001).Likewise, in man, leucine and insulin synergized with respect to their ability to stimulate p70S6 kinase phosphorylation in muscle while insulin, but not leucine, increased

protein kinase B phosphorylation (Greiwe et al., 2001).

13. Amino acid signaling as a function of age

In both liver and muscle cells rapamycin-sensitive p70S6 kinase phosphorylation and activation decreased with age (Liu et al., 1998; Dardevet et al., 2000). This was not caused by a decrease in p70S6 kinase protein but, at least in muscle, by a decrease in the affinity for leucine (Dardevet et al., 2000). It is thought that this decline in signaling contributes to net loss of protein with age.

14. Mechanisms

From all these studies, even though some of the data obtained in the last few years are conflicting, the picture emerges that amino acids somehow directly activate mTOR activity. For this activation, PI 3-kinase may not be required; however, for stimulation of phosphorylation of downstream targets of mTOR (e.g., p70S6 kinase), PI 3-kinase activity is required and this explains why amino acid effects on these downstream targets are sensitive to inhibition by PI 3-kinase inhibitors (Fig. 14.1). It also explains the synergy between amino acids and insulin.

The question still to be answered is the mechanism by which amino acids can activate mTOR. Apart from the possibility that amino acids may stimulate a protein kinase acting on mTOR as substrate, an attractive mechanism is also that amino acids, indeed, inhibit a protein phosphatase (or a combination of both). The simplest mechanism would be if mTOR, in analogy with yeast, would be a direct substrate for PP2A, although this remains to be proven (Navé et al., 1999). Association of PP2A with mTOR downstream targets is also possible (cf. Fig. 14.1). However, the data with PP2A inhibitors are not conclusive and depend on the cell type studied. Possibly, another type of protein phosphatase is involved which is not inhibited by these compounds.

If, indeed, amino acids stimulate a protein kinase or inhibit a protein phosphatase, one possibility is that they do so by a direct, allosteric, effect on these proteins. Another possibility is that the plasma membrane contains a specific amino acid receptor. The existence of a receptor was proposed on the basis of the specific binding to the hepatocyte plasma membrane of Leu$_8$-Map, a small cell-impermeant globular peptide with eight leucine residues on the outside of the molecule; moreover, the peptide effectively inhibited autophagy and replacement of the leucine residues by isoleucine rendered the peptide inactive (Miotto et al., 1994). However, Leu$_8$Map did not affect amino acid signaling (van Sluijters et al., 2000; Lynch et al., 2000). Moreover, the effect on autophagy could be ascribed to the degradation of the peptide to free leucine (van Sluijters et al., 2000).

Evidence that amino acids may not act via a surface receptor was provided by the demonstration that inhibition of plasma amino acid transport inhibited the activation of p70S6 kinase (Iiboshi et al., 1999). This indicates that the direct target for amino acids must be located intracellularly. An, as yet, hypothetical mechanism is one in which the cell responds to changes in the charging of tRNAs. This hypothesis is based on data in yeast showing that on amino acid starvation free, uncharged, tRNA strongly binds to the protein kinase GCN2 which then becomes activated and phosphorylates the α subunit of the initiation factor eIF2 (see section Amino acid stimulation of eIF2α). This is then followed by increased synthesis of the transcription factor GCN4 which, in turn, is responsible for increased transcription of a large number of genes involved in amino acid synthesis and other metabolic processes needed under these conditions, including genes encoding proteins required for autophagy (Hinnebusch, 1997; Natarajan et al., 2001). Whether or not free tRNA, indeed, controls amino acid signaling in mammalian cells is controversial. Thus, in one study with T-lymphoblastoid Jurkat cells, inhibition of amino acid-tRNA synthetase with amino acid alcohols indeed prevented amino acid-induced activation of p70S6 kinase (Iiboshi et al., 1999). This was not confirmed, however, in another study using freshly isolated rat adipocytes (Lynch et al., 2000; Pham et al., 2000). In HEK-293 cells amino acid deprivation did not affect aa-tRNA levels (Dennis et al., 2001). The latter two studies suggest that intracellular amino acid pools rather than the degree of aa-tRNA charging controls amino acid signaling. Although these differences in results may be due to the different cell types used, we consider it highly unlikely that the amino acid-sensing mechanism would be cell type dependent. Further studies are clearly required to resolve this issue.

If it turns out that, indeed, tRNA is always fully charged with amino acids, a corollary is that protein synthesis is never substrate-limited, even under amino acid-deprived conditions, and that the rate of protein synthesis is exclusively determined by the amino acid concentration dependence of amino acid signaling.

If, on the other hand, tRNA acts as an amino sensor like in yeast and tRNA charging does affect amino acid signaling, a possible mechanism underlying the ability of cell swelling to potentiate amino acid-dependent signaling can be provided. Thus, during regulatory volume decrease, when intracellular chloride falls, amino acid-tRNA synthetases may become activated because chloride ions inhibit these enzymes, in analogy with the situation in certain bacteria (van Sluijters et al., 2000) (cf. Fig. 14.1).

Previously we postulated that amino acid signaling, ultimately leading to S6 phosphorylation, provides an efficient mechanism by which both autophagic protein degradation and protein synthesis could be oppositely controlled (cf. Introduction). We also proposed a mechanism by which S6 phosphorylation may contribute to the reciprocal control of protein synthesis and degradation. We hypothesized that S6 phosphorylation may promote binding of ribosomes to the endoplasmic reticulum (ER) and enhance ER-linked protein synthesis (Blommaart et al., 1995). Interestingly, in hepatocytes, synthesis of export protein, but not of housekeeping protein, declines after amino acid deprivation (Tanaka and Ichihara, 1983). Ribosome binding to the ER would reduce the availability of ribosome-free regions of the ER which are the source of the autophagosomal membrane

(Dunn, 1990). Thus, a common mechanism would stimulate ER-linked protein synthesis while at the same time inhibiting proteolysis. Removal of ribosomes by autophagy (Lardeux and Mortimore, 1987) is thus prevented. We still think such a mechanism is possible. In this context it is important to stress that activation of PI 3-kinase class I also simultaneously stimulates protein synthesis and inhibits autophagic protein degradation, as discussed earlier in this review.

15. Conclusions

Amino acids can no longer be considered as intermediates in metabolism only. They also strongly stimulate signal transduction, presumably by activation of mTOR. They synergize with insulin with regard to the activation of downstream targets of mTOR and in this way promote protein synthesis and inhibit autophagic protein degradation by the same signaling pathway. This is efficient from the point of view of metabolic regulation. The fact that insulin production in pancreatic β-cells is also promoted by amino acid-dependent signaling further adds to the anabolic and anti-catabolic effects of amino acids.

Acknowledgement

The authors are grateful to the Dutch Diabetes Fund for financial support of their work (Grant 96.604).

References

Balage, M., Sinaud, S., Prod'homme, M., Dardevet, D., Vary, T.C., Kimball, S.R., Jefferson L.S. and Grizard, J. (2001). Amino acids and insulin are both required to regulate assembly of the eIF4E. eIF4G complex in rat skeletal muscle. Am. J. Physiol. 281, E565–E574.

Baquet, A., Hue, L., Meijer, A.J., van Woerkom, G.M. and Plomp, P.J.A.M. (1990). Swelling of rat hepatocytes stimulates glycogen synthesis. J. Biol. Chem. 265, 955–959.

Blommaart, E.F.C., Luiken, J.J.F.P., Blommaart, P.J.E., van Woerkom, G.M. and Meijer, A.J. (1995). Phosphorylation of ribosomal protein S6 is inhibitory for autophagy in isolated rat hepatocytes. J. Biol. Chem. 270, 2320–2326.

Blommaart, E.F.C., Krause, U., Schellens, J.P., Vreeling-Sindelárová, H. and Meijer, A.J. (1997a). The phosphatidylinositol 3-kinase inhibitors wortmannin and LY294002 inhibit autophagy in isolated rat hepatocytes. Eur. J. Biochem. 243, 240–246.

Blommaart, E.F.C., Luiken, J.J.F.P and Meijer, A.J. (1997b). Autophagic proteolysis: control and specificity. Histochem. J. 29, 365–385.

Brunn, G.J., Williams, J., Sabers, C., Wiederrecht, G., Lawrence, J.C. and Abraham, R.T. (1996). Direct inhibition of the signaling functions of the mammalian target of rapamycin by the phosphoinositide 3-kinase inhibitors, wortmannin and LY294002. EMBO J. 15, 5256–5267.

Burnett, P.E., Barrow, R.K., Cohen, N.A., Snyder, S.H. and Sabatini, D.M. (1998). RAFT1 phosphorylation of the translational regulators p70 S6 kinase and 4E-BP1. Proc. Natl. Acad. Sci. USA 95, 1432–1437.

Campbell, L.E., Wang, X.M. and Proud, C.G. (1999). Nutrients differentially regulate multiple translation factors and their control by insulin. Biochem. J. 344, 433–441.

Dardevet, D., Sornet, C., Balage, M. and Grizard, J. (2000). Stimulation of in vitro rat muscle protein synthesis by leucine decreases with age. J. Nutr. 130, 2630–2635.

Dennis, P.B., Jaeschke, A., Saitoh, M., Fowler, B., Kozma, S.C. and Thomas, G. (2001). Mammalian TOR: a homeostatic ATP sensor. Science, 294, 1102–1105.

Dufner, A. and Thomas, G. (1999). Ribosomal S6 kinase signaling and the control of translation. Exp. Cell Res. 253, 100–109.

Dunn, W.A. (1990). Studies on the mechanisms of autophagy: formation of the autophagic vacuole. J. Cell Biol. 110, 1923–1933.

Fox, H.L., Kimball, S.R., Jefferson, L.S. and Lynch, C.J. (1998). Amino acids stimulate phosphorylation of p70 S6 kinase and organization of rat adipocytes into multicellular clusters. Am. J. Physiol. 274, C206–C213.

Gingras, A.C., Raught, B. and Sonenberg, N. (2001). Regulation of translation initiation by FRAP/mTOR. Genes Dev. 15, 807–826.

Greiwe, J.S., Kwon, G., McDaniel, M.L. and Semenkovich, C.F. (2001). Leucine and insulin activate p70 S6 kinase through different pathways in human skeletal muscle. Am. J. Physiol. 281, E466–E471.

Hara, K., Yonezawa, K., Weng, Q.P., Kozlowski, M.T., Belham, C. and Avruch, J. (1998). Amino acid sufficiency and mTOR regulate p70 S6 kinase and eIF-4E BP1 through a common effector mechanism. J. Biol. Chem. 273, 14484–14494.

Hardie, D.G., Carling, D. and Carlson, M. (1998). The AMP-activated/SNF1 protein kinase subfamily: metabolic sensors of the eukaryotic cell? Annu. Rev. Biochem. 67, 821–855.

Hinnebusch, A.G. (1997). Translational regulation of yeast GCN4. A window on factors that control initiator-tRNA binding to the ribosome. J. Biol. Chem. 272, 21661–21664.

Iiboshi, Y., Papst, P.J., Kawasome, H., Hosoi, H., Abraham, R.T., Houghton, P.J. and Terada, N. (1999). Amino acid-dependent control of p70(s6k). Involvement of tRNA aminoacylation in the regulation. J. Biol. Chem. 274, 1092–1099.

Kimball, S.R., Horetsky, R.L. and Jefferson, L.S. (1998). Implication of eIF2B rather than eIF4E in the regulation of global protein synthesis by amino acids in L6 myoblasts. J. Biol. Chem. 273, 30945–30953.

Kleijn, M. and Proud, C.G. (2000). Glucose and amino acids modulate translation factor activation by growth factors in PC12 cells. Biochem J. 347, 399–406.

Krause, U., Rider, M.H. and Hue, L. (1996). Protein kinase signaling pathway triggered by cell swelling and involved in the activation of glycogen synthase and acetyl-CoA carboxylase in isolated rat hepatocytes. J. Biol. Chem. 271, 16668–16673.

Kumar, V., Pandey, P., Sabatini, D., Kumar, M., Majumder, P.K., Bharti, A., Carmichael, G., Kufe, D. and Kharbanda, S. (2000). Functional interaction between RAFT1/FRAP/mTOR and protein kinase Cδ in the regulation of cap-dependent initiation of translation. EMBO J. 19, 1087–1097.

Lardeux, B.R. and Mortimore, G.E. (1987) Amino acid and hormonal control of macromolecular turnover in perfused rat liver. Evidence for selective autophagy. J. Biol. Chem. 262, 14514–14519.

Liu, Y., Gorospe, M., Kokkonen, G.C., Boluyt, M.O., Younes, A., Mock, Y.D., Wang, X., Roth, G.S. and Holbrook, N.J. (1998). Impairments in both p70 S6 kinase and extracellular signal-regulated kinase signaling pathways contribute to the decline in proliferative capacity of aged hepatocytes. Exp. Cell Res. 240, 40–48.

Long, W., Saffer, L., Wei, L. and Barrett, E.J. (2000). Amino acids regulate skeletal muscle PHAS-I and p70 S6-kinase phosphorylation independently of insulin. Am. J. Physiol. 279, E301–E306.

Luiken, J.J.F.P., Blommaart, E.F.C., Boon, L., van Woerkom, G.M. and Meijer, A.J. (1994). Cell swelling and the control of autophagic proteolysis in hepatocytes: involvement of phosphorylation of ribosomal protein S6? Biochem. Soc. Trans. 22, 508–511.

Lynch, C.J., Fox, H.L., Vary, T.C., Jefferson, L.S. and Kimball, S.R. (2000). Regulation of amino acid-sensitive TOR signaling by leucine analogues in adipocytes. J. Cell. Biochem. 77, 234–251.

MacDonald, M.J. and Fahien, L.A. (2000). Glutamate is not a messenger in insulin secretion. J. Biol. Chem. 275, 34025–34027.

Maechler, P. and Wollheim, C.B. (1999). Mitochondrial glutamate acts as a messenger in glucose-induced insulin

exocytosis. Nature 402, 685–689.

Meijer, A.J., Baquet, A., Gustafson, L., van Woerkom, G.M. and Hue, L. (1992). Mechanism of activation of liver glycogen synthase by swelling. J. Biol. Chem. 267, 5823–5828.

Miotto, G., Venerando, R., Marin, O., Siliprandi, N. and Mortimore, G.E. (1994). Inhibition of macroautophagy and proteolysis in the isolated rat hepatocyte by a non-transportable derivative of the multiple antigen peptide Leu$_8$-Lys$_4$-Lys$_2$-Lys-βAla. J. Biol. Chem. 269, 25348–25353.

Natarajan, K., Meyer, M.R., Jackson, B.M., Slade, D., Roberts, C., Hinnebusch, A.G. and Marton, M.J. (2001). Transcriptional profiling shows that Gcn4p is a master regulator of gene expression during amino acid starvation in yeast. Mol. Cell. Biol. 21, 4347–4368.

Navé, B.T., Ouwens, D.M., Withers, D.J., Alessi, D.R. and Shepherd, P.R. (1999). Mammalian target of rapamycin is a direct target for PKB: identification of a convergence point for opposing effects of insulin and amino-acid deficiency on protein translation. Biochem. J. 344, 427–431.

Noda, T. and Ohsumi, Y. (1998). Tor, a phosphatidylinositol kinase homologue, controls autophagy in yeast. J. Biol. Chem. 273, 3963–3966.

Parrott, L.A. and Templeton, D.J. (1999). Osmotic stress inhibits p70/85 S6 kinase through activation of a protein phosphatase. J. Biol. Chem. 274, 24731–24736.

Patti, M.E., Brambilla, E., Luzi, L., Landaker, E.J. and Kahn, C.R. (1998). Bidirectional modulation of insulin action by amino acids. J. Clin. Invest. 101, 1519–1529.

Peterson, R.T., Desai, B.N., Hardwick, J.S. and Schreiber, S.L. (1999). Protein phosphatase 2A interacts with the 70-kDa S6 kinase and is activated by inhibition of FKBP12-rapamycin associated protein. Proc. Natl. Acad. Sci. USA 96, 4438–4442.

Petiot, A., Ogier-Denis, E., Blommaart, E.F.C., Meijer, A.J. and Codogno, P. (2000). Distinct classes of phosphatidylinositol 3′-kinases are involved in signaling pathways that control macroautophagy in HT-29 cells. J. Biol. Chem. 275, 992–998.

Peyrollier, K., Hajduch, E., Blair, A.S., Hyde, R. and Hundal, H.S. (2000). L-leucine availability regulates phosphatidylinositol 3-kinase, p70 S6 kinase and glycogen synthase kinase-3 activity in L6 muscle cells: evidence for the involvement of the mammalian target of rapamycin (mTOR) pathway in the L-leucine-induced up-regulation of system A amino acid transport. Biochem. J. 350, 361–368.

Pham, P.T., Heydrick, S.J., Fox, H.L., Kimball, S.R., Jefferson, L.S. Jr and Lynch, C.J. (2000). Assessment of cell-signaling pathways in the regulation of mammalian target of rapamycin (mTOR) by amino acids in rat adipocytes. J. Cell. Biochem. 79, 427–441.

Plomp, P.J.A.M., Boon, L., Caro, L.H.P., van Woerkom,

G.M. and Meijer, A.J. (1990). Stimulation of glycogen synthesis in hepatocytes by added amino acids is related to the total intracellular content of amino acids. Eur. J. Biochem. 191, 237–243.

Scott, P.H. and Lawrence, J.C. (1998). Attenuation of mammalian target of rapamycin activity by increased cAMP in 3T3-L1 adipocytes. J. Biol. Chem. 273, 34496–34501.

Shah, O.J., Antonetti, D.A., Kimball, S.R. and Jefferson, L.S. (1999). Leucine, glutamine, and tyrosine reciprocally modulate the translation initiation factors eIF4F and eIF2B in perfused rat liver. J. Biol. Chem. 274, 36168–36175.

Shah, O.J., Anthony, J.C., Kimball, S.R. and Jefferson, L.S. (2000a). 4E-BP1 and S6K1: translational integration sites for nutritional and hormonal information in muscle. Am. J. Physiol. 279, E715–E729.

Shah, O.J., Kimball, S.R. and Jefferson, L.S. (2000b). Glucocorticoids abate p70(S6k) and eIF4E function in L6 skeletal myoblasts. Am. J. Physiol. 279, E74–E82.

Shigemitsu, K., Tsujishita, Y., Hara, K., Nanahoshi, M., Avruch, J. and Yonezawa, K. (1999a). Regulation of translational effectors by amino acid and mammalian target of rapamycin signaling pathways. Possible involvement of autophagy in cultured hepatoma cells. J. Biol. Chem. 274, 1058–1065.

Shigemitsu, K., Tsujishita, Y., Miyake, H., Hidayat, S., Tanaka, N., Hara, K. and Yonezawa, K. (1999b). Structural requirement of leucine for activation of p70 S6 kinase. FEBS Lett. 447, 303–306.

Sturgill, T.W. and Wu, J. (1991). Recent progress in characterization of protein kinase cascades for phosphorylation of ribosomal protein S6. Biochim. Biophys. Acta 1092, 350–357.

Takano, A., Usui, I., Haruta, T., Kawahara, J., Uno, T., Iwata, M. and Kobayashi, M. (2001). Mammalian target of rapamycin pathway regulates insulin signaling via subcellular redistribution of insulin receptor substrate 1 and integrates nutritional signals and metabolic signals of insulin. Mol. Cell. Biol. 21, 5050–5062.

Tanaka, K. and Ichihara, A. (1983). Different effects of amino acid deprivation on syntheses of intra- and extracellular proteins in rat hepatocytes in primary culture. J. Biochem. (Tokyo) 94, 1339–1348.

Tang, H., Hornstein, E., Stolovich, M., Levy, G., Livingstone, M., Templeton, D., Avruch, J. and Meyuhas, O. (2001). Amino acid-induced translation of TOP mRNAs is fully dependent on phosphatidylinositol 3-kinase-mediated signaling, is partially inhibited by rapamycin, and is independent of S6K1 and rpS6 phosphorylation. Mol. Cell. Biol. 21, 8671–8683.

Tischler, M.E., Desautels, M. and Goldberg, A.L. (1982). Does leucine, leucyl-tRNA, or some metabolite of leucine regulate protein synthesis and degradation in skeletal and cardiac muscle? J. Biol. Chem. 257, 1613–1621.

Tremblay, F. and Marette, A. (2001). Amino acid and insulin signaling via the mTOR/p70 S6 kinase pathway. A negative feedback mechanism leading to insulin resistance in skeletal muscle cells. J. Biol. Chem. 276, 38052–38060.

Vanhaesebroeck, B. and Waterfield, M.D. (1999). Signaling by distinct classes of phosphoinositide 3-kinases. Exp. Cell Res. 253, 239–254.

van Sluijters, D.A., Dubbelhuis, P.F., Blommaart, E.F.C. and Meijer A.J. (2000). Amino-acid-dependent signal transduction. Biochem J. 351, 545–550.

Wang, X., Campbell, L.E., Miller, C.M. and Proud, C.G. (1998). Amino acid availability regulates p70 S6 kinase and multiple translation factors. Biochem. J. 334, 261–267.

Wang, X., Li, W., Williams, M., Terada, N., Alessi, D.R. and Proud, C.G. (2001) Regulation of elongation factor 2 kinase by p90(RSK1) and p70 S6 kinase. EMBO J. 20, 4370–4379.

Westphal, R.S., Coffee, R.L., Marotta, A., Pelech, S.L. and Wadzinski, B.E. (1999). Identification of kinase–phosphatase signaling modules composed of p70 S6 kinase–protein phosphatase 2A (PP2A) and p21-activated kinase-PP2A. J. Biol. Chem. 274, 687–692.

Xu, G., Kwon, G., Cruz, W.S., Marshall, C.A. and McDaniel M.L. (2001). Metabolic regulation by leucine of translation initiation through the mTOR-signaling pathway by pancreatic β-cells. Diabetes 50, 353–360.

Xu, G., Kwon, G., Marshall, C.A., Lin, T.A., Lawrence, J.C. and McDaniel, M.L. (1998a). Branched-chain amino acids are essential in the regulation of PHAS-I and p70 S6 kinase by pancreatic beta-cells. A possible role in protein translation and mitogenic signaling. J. Biol. Chem. 273, 28178–28184.

Xu, G., Marshall, C.A., Lin, T.A., Kwon, G., Munivenkatappa, R.B., Hill, J.R., Lawrence, J. C. and McDaniel, M.L. (1998b). Insulin mediates glucose-stimulated phosphorylation of PHAS-I by pancreatic beta cells. An insulin-receptor mechanism for autoregulation of protein synthesis by translation. J. Biol. Chem. 273, 4485–4491.

Sensing, Signaling and Cell Adaptation. Edited by K.B. Storey and J.M. Storey
© *2002 Elsevier Science B.V. All rights reserved.*

CHAPTER 15

Signal Transduction Pathways Involved in the Regulation of Drug Metabolizing Enzymes

Vidya Hebbar and A.-N. Tony Kong*

Department of Pharmaceutics, Ernest Mario School of Pharmacy and the Environmental and Occupational Health Sciences Institute (EOHSI), Rutgers, The State University of New Jersey, Piscataway, NJ 08854, USA

1. Introduction

Responses of cells to external stimuli are the focus of numerous investigative studies. The stimuli come in the form of chemicals, pollutants, ultraviolet light (UV), drugs that are prescribed and food ingredients. These foreign agents or xenobiotics induce events that lead to various cellular responses including homeostasis, proliferation, differentiation, apoptosis and necrosis. An important aspect of the cellular responses are the enzymes responsible for the metabolism of the xenobiotics, namely, the drug metabolizing enzymes (DMEs), which include the phase I and phase II DMEs. Phase I DMEs introduce a hydrophilic moiety to the xenobiotic and the phase II enzymes significantly aid in the conjugation of the xenobiotics hence greatly increasing the hydrophilicity.

This chapter reviews the cellular mechanisms involved in the expression of DMEs when cells are exposed to xenobiotics. Phase I DMEs, which include an important group, the cytochrome P450 enzymes (CYP) metabolize most of the clinically available drugs and other xenobiotics. The induction of some of these enzymes is mediated by the action of receptors, for example, the aryl hydrocarbon receptor (AhR) and other xeno-sensors such

as, the orphan nuclear receptors. Phase II drug metabolizing reactions catalyzed by enzymes, such as NAD(P)H: quinone oxidoreductase (NQO1), glutathione S-transferase (GST) and UDP-glucuronosyltransferase (UGT), are most often preceded by phase I biotransformation. These phase II enzymes are induced by a cascade of signal transduction events eventually leading to the action of transcription factors, for example NF-E2 related factor 2 (Nrf2) on a *cis*-acting DNA response element, the antioxidant response element (ARE) which is found in the promoter region of the phase II genes. Nrf2 in association with other proteins binds to the ARE, and mediates the expression of phase II genes. These downstream events are regulated and triggered by upstream signaling kinases. Mitogen-activated protein kinases (MAPKs) and their upstream kinases are activated by chemical signals and other extracellular stimuli. Different classes of compounds have activated the MAPK pathways and these studies have led to useful information regarding the individual kinases activated or inhibited as well as the transcription factors involved and the genes that they control. These compounds are phenolic antioxidants, for example, the green tea polyphenols and the isothiocyanate class of naturally occurring chemopreventive agents. These pathways have given us important insights into the mechanisms involved in phase II enzyme induction.

---

*Corresponding author

2. Drug Metabolizing Enzymes (DMEs)

2.1. General considerations

The human body is constantly exposed to xenobiotics, which are found very readily in the environment. These xenobiotics are present in chemicals, pesticides, environmental pollutants and also, in the food that we ingest, including fruits and vegetables. It is highly imperative that the body has an efficient mechanism of eliminating some of these potentially toxic xenobiotics, in order to protect the cells from the harmful effects of these agents. Xenobiotics, especially drugs, are highly lipophilic in nature, which make them excellent candidates for quick absorption into the body. Thus, to achieve efficient elimination of these agents, they have to undergo biotransformation into hydrophilic compounds, which make it easier for them to be eliminated in the urine or feces.

In the metabolism of xenobiotics, certain enzymes help in the process of biotransformation of drugs to less harmful compounds and subsequently aid in the detoxification of the compound. Xenobiotic biotransforming enzymes or DMEs include two major groups, phase I and phase II enzymes. The cytochrome P450s (CYP), the most important class of the phase I DMEs, catalyze the introduction or exposure of functional groups such as, $-OH$, $-NH_2$, $-SH$ or $-COOH$, rendering the xenobiotics slightly hydrophilic. Thus, phase I enzymes are involved in hydrolysis, reduction and oxidation reactions, for example, hydroxylation of aromatic and aliphatic carbons, heteroatom oxidation (N-, S-oxidation) and hydrolysis of ester and amide groups. The other important members that constitute phase I enzymes are the flavin-containing monooxygenases, esterases and amidases. Phase II enzymes are involved in reactions that make the xenobiotics substantially more hydrophilic and hence make the metabolites highly excretable. The phase II reactions almost always follow the phase I reactions, and include reactions such as, glucuronidation, conjugation with glutathione, sulfation, acetylation, methylation and conjugation with amino acids.

2.2. Phase I DMEs

2.2.1. CYP enzymes

The CYP enzymes are a group of enzymes that are very active in metabolizing xenobiotics and most current clinically approved drugs. The term "cytochrome P450" was first used to describe a pigment in the liver cell (Omura and Sato, 1962). There are 17 families among the CYPs, which are further divided into subfamilies and contain several enzymes within each subfamily. The enzymes responsible for drug/xenobiotic metabolism belong to a small group of CYP families, namely, CYP1, 2, 3 and 4.

A very important consideration for all DMEs, in particular the CYP enzymes, is that some of them exhibit genetic polymorphisms. These can lead to occurrence of two phenotypes, high or low metabolizers, which make individuals at high risk from exposure to xenobiotics. These genetic polymorphisms exist when different alleles occur in the population and encode proteins that have activities, which differ significantly from the normal range. Once these differences are known and studied, it would be possible to alter or administer drugs based on an individual's activity of the metabolizing enzyme.

The CYP1A family of enzymes metabolizes drugs that fall under the class of polycyclic aromatic hydrocarbons in addition to a wide range of xenobiotics and cancer causing agents such as benzo[a]pyrene. The induction of this enzyme is mediated mainly through the AhR. The CYP2 family is another major class that metabolizes several xenobiotic agents used presently. For example, CYP2B6 that is induced by phenobarbital, metabolizes compounds like bupropion. This is an enzyme that is not highly expressed in the liver, unlike CYP2E1 that metabolizes acetaminophen and ethanol, and comprises 7% of the total hepatic CYP content. CYP2C9, an important member of this family of CYP enzymes, is polymorphic and is responsible for the metabolism of several non-steroidal anti-inflammatory drugs (NSAIDs). Similarly, enzyme CYP2C19 also displays polymorphisms in the population and is responsible for the metabolism of several important xenobiotics such as omeprazole, diazepam and propranolol (Wrighton and Stevens, 1992).

CYP3A4, an enzyme responsible for the metabolism of about 60% of the total xenobiotics administered, constitutes around 30% of the total hepatic CYP content. Because of its importance in xenobiotic metabolism and its high diversity for substrates, this enzyme has been extensively studied. CYP2, 3 and 4 enzymes are induced by the action of the appropriate ligand binding to the orphan receptors and subsequently lead to the expression of phase I genes.

2.2.2. *Aryl hydrocarbon receptor*

The CYP1A genes are induced by polycyclic aromatic hydrocarbons (PAHs) and dioxins (Whitlock, 1999). These compounds bind to a cytoplasmic receptor, AhR, dissociating it from a complex formed with heat shock protein (hsp90). This enables the receptor to be phosphorylated by tyrosine kinase thus activating the AhR. Upon entering the nucleus the activated AhR binds to the aryl hydrocarbon nuclear translocator (ARNT) protein. This heterodimer along with transcription factors such as NF1 and Sp1, then binds to the xenobiotic responsive element (XRE) present in the promoter region of several of the CYP1A gene family and induces the expression of the CYP1A genes. The CYP1A1 gene is extremely important in the metabolism of benzo[a]pyrene (BaP) which is found in cigarette smoke (Barouki and Morel, 2001). Two successive epoxidation reactions produce the highly carcinogenic diol epoxide form of BaP. This diol epoxide form is further detoxified by phase II DMEs such as GSTs and UGTs. Hence the ratio between CYP1A1 and that of the phase II DMEs is extremely important, to overcome the harmful effects of the highly reactive and harmful intermediates.

2.2.3. *Orphan nuclear receptors*

Orphan nuclear receptors belong to a superfamily of nuclear receptors, which include the steroid hormone receptors. These nuclear receptors act as ligand-activated transcription factors and they control diverse functions in the cell including lipid metabolism and xenobiotic metabolism. Nuclear

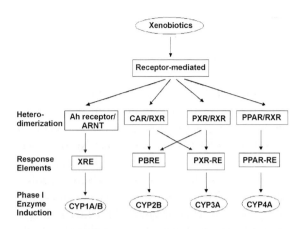

Fig. 15.1. A schematic of the phase I drug metabolizing enzyme pathways.

receptors share structural similarity but vary widely in the ligands that they are sensitive to. These proteins contain an NH_2-terminal domain that contains a ligand-independent transcriptional activation function (AF-1), a DNA binding region that contain two zinc finger motifs, a hinge region required for flexibility between dimerization with other proteins and DNA binding, and an extensive COOH-terminal domain that incorporates functions of dimerization, ligand-binding and a ligand-dependent activation function (AF-2). Once a ligand binds, the receptor dissociates from corepressors and recruits coactivators to undergo its transcriptional role (McKenna et al., 1999). The orphan nuclear receptors include among others, peroxisome proliferator-activated receptors (PPARs), steroid xenobiotic receptor/pregnane X receptor (SXR/PXR) and constitutive androstane receptor (CAR). These receptors form heterodimers with retinoid X receptor (RXR), whose endogenous ligand is 9-*cis* retinoic acid (Mangelsdorf and Evans, 1995). RXR typically does not act alone. A schematic of the phase I DME pathways is depicted in Fig. 15.1.

(i) Peroxisome proliferator-activated receptors

Peroxisome proliferator-activated receptors (PPARs) (α, γ, δ) act to regulate a large number of genes involved in differentiation and lipid metabolism. This is in response to compounds collectively termed as peroxisome proliferators. These compounds are a fibrate family of hypolipidemic drugs, herbicides, pesticides and natural and synthetic fatty acids. Once activated PPARs heterodimerize with RXRα and bind to *cis*-acting regulatory elements called peroxisome proliferator response elements (PPRE). Induction of PPRE leads to transcriptional activation of genes encoding for the peroxisomal β-oxidation system and CYP4A isoforms, CYP4A1 and CYP4A3 (Reddy et al., 1986; Johnson et al., 1996). It has also been observed that in the liver with peroxisome proliferation, the CYP4A family of enzymes is significantly induced. This generates H_2O_2 thus leading to intracellular oxidative stress (Yeldandi et al., 2000).

(ii) Steroid xenobiotic receptor/pregnane X receptor (SXR/PXR) and constitutive androstane receptor (CAR)

CYP3A and CYP2B are important members of the CYP family of enzymes, responsible for the metabolism of a large number of pharmaceutical drugs and herbal medicines. The two nuclear receptors that function in the metabolism of xenobiotics are CAR and PXR/SXR. In the presence of a ligand, receptors such as the human steroid xenobiotic receptor (SXR) and its rodent homolog, pregnane X receptor (PXR) bind to the 5′ regulatory region of the CYP3A4 gene and stimulate gene expression (Blumberg et al., 1998).

CAR forms a heterodimer with RXR and binds to the retinoid acid response element (β-RARE) and also the phenobarbital response element (PBRE), which is found upstream of the CYP2B genes (Trottier et al., 1995). It has recently been shown that there is cross-talk between SXR/PXR and CAR and their response elements. SXR/PXR, in addition to binding to the CYP3A4 DNA binding region, also binds to the response element upstream of the CYP2B gene and activates the PBRE. CAR, in addition to activating the PBRE and regulating the CYP2B genes, also binds to the CYP3A4 response element and regulates gene expression (Xie et al., 2000).

Thus the complexity of the response and the regulatory network is an important consideration since these response elements are involved in the metabolism of a diverse number of xenobiotics.

This cross-talk between the nuclear receptors and the response elements possibly affords protection to the poor metabolizer by improving the capacity of the xenobiotic response system, however it also has the potential to increase drug–drug interactions.

2.3. Phase II DMEs

Phase II DMEs catalyze reactions such as sulfation, glucuronidation, acetylation, methylation and conjugation with glutathione or amino acids. The co-factors react with groups that are either added onto the xenobiotic by phase I biotransformation reactions or can react with groups already present on the chemical entity. These reactions result in large increases in xenobiotic hydrophilicity thus facilitating and hastening excretion of the xenobiotic. Phase II DMEs are located primarily in the cytosol but UGTs occur in the microsomes. A list of phase II reactions and the enzymes involved is provided in Table 15.1.

Phase II enzymes such as GST, UGT and NQO1 are induced by several classes of compounds for example, (1) planar aromatic compounds (polycyclic, azo and halogenated) which also activate phase I enzymes and hence are bifunctional inducers and (2) monofunctional inducers such as diphenols, isothiocyanates and thiocarbamates (Talalay et al., 1987).

3. The antioxidant response element (ARE)

The 5′ promoter region of rodent GSTYa subunit genes has shown the presence of several DNA binding elements, which include the xenobiotic response element (XRE), hepatocyte nuclear factor-1 (HNF-1), activator protein 1 (AP-1) and AP-1-like elements. In addition to these enhancers, the identification of a new enhancer, called the ARE or the electrophile response element (EpRE), was an important and ground-breaking discovery (Rushmore, et al. 1990; Friling et al., 1990). This enhancer was localized between the nucleotides −722 to −682 of the 5′ flanking region. This 5′ region containing the ARE was the DNA element

Table 15.1. Summary of the phase II drug metabolizing enzymes and their reactions

Reaction	Phase II Enzyme and Co-factor	Drugs/Classes of Drugs Metabolized
1 Glucuronidation	UDP-glucuronosytransferases (UGTs). Co-factor: Uridine diphosphate glucuronic acid (UDPGA)	Hormones, bile acids, acetaminophen, benzo(a)pyrene
2 Sulfation	Sulfotransferases. Co-factor: 3'-phosphoadenosine-5'- phosphosulfate	Ethanol, polyethylene glycols, cholesterol, acetaminophen, dopamine
3 Methylation	Phenol and catechol O-methyl transferase (POMT, COMT), N-methyl transferases, thiopurine methyltransferase, thiol methyl transferase. Co-factor: S-adenosylmethionine	Phenol, catechol, nicotinamide, histamine, 6-mercaptopurine
4 Acetylation	N-acetyl transferase (NAT1 and NAT2 in humans). Co-factor: Acetyl coenzyme A	Aromatic amines, hydrazines, hydrazides
5 Amino acid conjugation	Conjugation with glycine, glutamine, taurine, serine and proline	Benzoic acid, aromatic hydroxylamines
6 Glutathione conjugation	Glutathione S-transferase (GST)	Electrophilic compounds
7 Quinone reduction	NAD(P)H:quinone oxidoreductase (NQO1, QR, DT-diaphorase)	Quinones and quinoid compounds

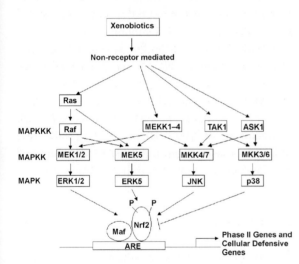

Fig. 15.2. A schematic of the mitogen-activated protein kinase pathways leading to ARE- mediated phase II detoxifying enzyme induction.

responsible for basal and inducible expression of the GSTYa subunit gene, by planar aromatic compounds such as β-naphthoflavone, 3-methylcholanthrene, 2,3,7,8-tetrachlorodibenzo-p-dioxin, and electrophilic compounds such as tert-butylhydroquinone (tBHQ).

The ARE was also detected upstream of another phase II gene, namely, rat and human NQO1 gene

(Favreau and Pickett, 1991; Li and Jaiswal, 1992). The ARE core sequence was assigned to the following essential nucleotides, GTGACNNNGC, where N is any nucleotide. In addition to the NQO1 and GST genes, the ARE has been found in cis-active sequences of the heavy and light chain genes encoding the γ-glutamylcysteine synthetase holoenzyme (GCS) and hemoxygenase-1 (HO-1). GCS and glutathione synthetase are required for the formation of the tripeptide gluthathione that is required in glutathione conjugation reactions (refer to Table 15.1). The enzyme GCS is responsible for the rate-limiting step in the production of glutathione (GSH). Thus the discovery of ARE seems to be extremely important when we consider that the phase II enzymes protect cells from oxidative stress. A scheme of the pathways leading to ARE-mediated phase II gene expression is depicted in Fig. 15.2.

4.   Transacting factors regulating the expression of Phase II enzymes

Nuclear proteins that are found to bind to the ARE have been identified and include c-Jun, Jun-B, Jun-D, c-Fos, Fra1, Nrf1, Nrf2, YABP, ARE-BP1.

The core sequence of the ARE is similar to the binding sequence of the erythroid transcription factor 2 (NF-E2) (Andrews et al., 1993). An important member of this group of transacting factors that binds to the ARE was identified as the NF-E2 related factor-2 (Nrf2). It is a member of the Cap 'n' Collar (CNC) subfamily of basic region-leucine zipper proteins (bZip) (Moi et al., 1994). There is a high degree of homology between the bZip family of transcription factors in the basic region and leucine zipper domains as well as in the CNC domains. Other members of the CNC bZip family of transcription factors are Nrf1, Nrf3, Bach1 and Bach2. Nrf2 has a 589 amino acid open reading frame and is a powerful activator of RNA Polymerase II. The N-terminal region of Nrf2 is high in glutamic and aspartic acids and is believed to be the transactivation domain. It belongs to a subfamily of bZip factors containing leucine zippers that renders it incapable of self-dimerization. It dimerizes with other bZip proteins and this interaction is believed to facilitate binding to the ARE. It was intriguing that the expression of Nrf2 was significantly higher in tissues that express high levels of phase II enzymes, for example, the liver, lung, kidney, heart and muscle.

Gene knockout experiments have emphasized the role that Nrf2 plays in protection against potential toxicants. In targeted gene knockout experiments of Nrf2 (Nrf2 –/–), it is observed that the mice are normal, but the expression of NQO1 is impaired (Itoh et al., 1997). When Nrf2 (–/–) mice were treated with the food preservative, butylated hydroxytoluene, there was a high mortality rate among the knockout mice as compared to the wild type. The mice died of acute lung failure and the levels of GCS mRNA were greatly reduced (Chan and Kan, 1999). Overexpression of Nrf2 cDNA restored levels of GSH in the Nrf2 (–/–) fibroblasts and also increased levels of GCS transcripts. The Nrf2 knockout mice experiments have provided useful information regarding the role that it plays in the regulation of phase II enzymes and these mice may serve as a tool for the screening of potentially harmful xenobiotics. Since oxidative damage leads to DNA breakage, which ultimately leads to the initiation of cancers, the Nrf2 knockout mice could also be used to screen potential cancer causing agents.

Proteins associated with Nrf2: Several proteins that are associated with Nrf2 have been isolated. These proteins are diverse and can function as partners, coactivators, repressors or inhibitory proteins. Some of the proteins that are associated with Nrf2, are found to be partners of the CNC-bZip family of transcription factors, whereas other proteins have been isolated by specific binding to Nrf2 from yeast two-hybrid experiments. Here we will discuss a small but nevertheless important group of proteins associated with Nrf2.

Nrf1 and Nrf2 do not form heterodimers with each other but dimerize with other bZip partners. One of the first Nrf2 partners to be identified, were the small Maf proteins (Maf G, Maf K and Maf F). The erythroid transcription factor NF-E2, which plays a critical role in erythroid and megakaryocyte gene expression is a heterodimer made up of two bZip proteins, one subunit p45 is a member of the CNC family and the p18 subunit, is a member of the small Maf protein family. Since Nrf2 belongs to the same family as NF-E2, the small Mafs were proposed to form heterodimers with Nrf2 as well. The Maf family is divided into the large Mafs (vMaf, cMaf, MafB and Nrl) and the small Mafs (Fujiwara et al., 1993; Blank et al., 1997). The small Mafs differ from the large Maf proteins in that they do not contain a transactivation domain but only the basic region and leucine zipper domains which mediate DNA binding and dimerization.

Small Mafs bind to a TRE-type Maf recognition element (T-MARE), a sequence that resembles ARE quite strikingly. In fact, ARE is thought to form a subset of T-MARE regulatory sequences (Motohashi et al., 1997). Studies performed on the small Mafs suggest that an optimal balance between the small Mafs and CNC proteins is required for promoting transcriptional activation. Thus the small Maf proteins can form hetero or homodimers and either positively or negatively regulate transcription (Dhakshinamoorthy and Jaiswal, 2000).

Another important coactivator protein of Nrf2 isolated because of its virtue of binding to other nuclear transcription factors, is the cAMP-response element-binding protein (CREB)-binding protein (CBP). Both CBP and its homologous protein, p300 serve as coactivators for nuclear factors. These coactivators have histone acetyltransferase activity, which makes the chromatin more relaxed. In the context of NF-E2, CBP acetylated Maf G in its basic region and this enhanced its DNA binding ability. Hence CBP might serve two functions, to modify chromatin structure and secondly, to mediate transcription factor activity (Hung et al., 2001). Recently it was shown that, CBP/p300 plays a role in mediating ARE activation of phase II genes via the Nrf2/Maf heterodimer (Zhu and Fahl, 2001). In another independent study, the regions that bind to CBP on Nrf2 were assigned to the Neh4 and Neh5 domains in the transactivation domain of Nrf2 protein (Katoh et al., 2001). Thus the coactivator CBP/p300 exerts its action in two possible ways, to acetylate proteins that possess histone acetyltransferase activity, thus promoting chromatin relaxation and secondly to bridge the distance between the transcription factors and the DNA binding element.

One of the associated proteins of Nrf2, isolated by specific binding obtained from yeast two-hybrid experiments, is a cytoskeletal-associated protein called Keap1 (Kelch-like ECH-associated protein 1). A proposed mechanism for Nrf2 is that it is bound in the cytosol to Keap1 and in the presence of a stress signal, Nrf2 is released from Keap1 in the cytosol and then translocates to the nucleus, where it exerts its action (Itoh et al., 1999).

Besides the small Mafs there are other proteins that can interact with Nrf2 and mediate transcription of phase II genes. Activating transcription factor 4 (ATF4), a member of the activating transcription factor/cAMP responsive element-binding protein family of transcription factors, was found to dimerize with Nrf1. Recently, ATF4 was also identified as an Nrf2 interacting protein from yeast two hybrid experiments. This study emphasized the role of ATF4 in basal and cadmium-induced regulation of the HO-1 gene (He et al., 2001). The HO-1 gene is activated by stress responses including heme, heavy metals, tumor promoters, UV and inflammatory cytokines. ATF4 is now known to be

a stress response protein and its interaction with Nrf2 further reiterates its role in regulating stress responses via the HO-1 gene.

5. Mitogen-activated protein kinases (MAPKs)

The Nrf2-ARE complex is responsive to antioxidants as well as oxidative stress thus mediating downstream expression of phase II genes. It is important to understand the upstream events that lead to this downstream induction of phase II genes. MAPKs belong to a group of serine/threonine kinases and play important roles in cellular differentiation, cell proliferation and cell death upon activation by various extracellular stimuli (Robinson and Cobb, 1997). Three MAPK members are well studied and they are (1) extracellular signal-regulated kinases (ERK), (2) c-jun NH_2-terminal kinases (JNK, also called as stress-activated protein kinases, SAPKs) and (3) p38 (Gomez and Cohen, 1991; Hibi et al., 1993; Kyriakis et al., 1994). These MAPKs are phosphorylated at their threonyl and tyrosyl residues and are induced by the action of an upstream kinase, MAP kinase kinase (MAPKK). The tripeptide motif at which the phosphorylation occurs is TXY. This MAP kinase kinase is in turn induced by phosphorylation at Ser/Thr sites by an upstream kinase called, MAP kinase kinase kinase (MAPKKK) (Marshall, 1994).

Raf-1, the first MAPKKK identified activates ERK through MEK1/2 but has little effect on JNK and p38 (Robinson and Cobb, 1997). The ERKs are strongly activated by insulin, EGF, PDGF and FGF. JNK and p38 are stress-activated MAPK members and are induced by heat shock, oxidative stress, ionizing radiation and other stress conditions (Derijard et al., 1994; Kyriakis and Avruch, 1996). Another class of stress-activated MAPK called ERK5 or Big MAP kinase-1 (BMK1) is also induced by oxidative stress (peroxide) and osmotic shock (Abe et al., 1996). The MAPK kinases, MKK4 and MKK7 can phosphorylate and activate all three isoforms of JNK. MKK3 and MKK6, two MAPKKs phosphorylate p38 and have no effect on

JNK or ERK (Raingeaud et al., 1996). ERK5 (BMK1) is activated by MEK5 and is the only known substrate for MEK5 (Chao et al., 1999).

MAPKKKs (MAP3Ks) are broadly classified into three categories, the MEK kinases (MEKKs), the mixed lineage kinases (MLKs) and the thousand and one kinases (TAOs). Among the mammalian MEKKs, the common feature is a conserved catalytic domain, and they include MEKKs1-4, apoptosis signal-regulating kinase-1 (ASK1), TFG-β-activated kinase-1 (TAK1) and Tpl-2 (the product of cot protooncogene) (Gerwins et al., 1997; Ichijo et al., 1997). The kinase domain of MEKK1 has a binding site for Ras and it also can bind Rho family GTPases. MEKK1 can also activate MEK1/2 leading to the activation of ERKs, similar to Raf-1 and activates MKK7 and MKK4. MEKK1 itself is activated by TNF, oxidant stress and receptor tyrosine kinase agonists (Baud et al., 1999; Fanger et al., 1997). TAK1 and ASK1 can activate JNKs and p38 via MKK4, MKK3 and MKK6.

6. MAPK activation by phenolic compounds and isothiocyanates: induction of Phase II enzymes

We have demonstrated that phenolic antioxidants such as, BHA and green tea polyphenols (GTPs) potently activate MAPKs (JNK1 and ERK2) in a wide spectrum of mammalian cell lines, such as, human cervical squamous carcinoma HeLa cells, human hepatoma HepG2 cells and murine hepatoma Hepa1c1c7 cells (Yu et al., 1997a,b). The demethylated and active metabolite of BHA, tBHQ, that is another potent inducer of phase II genes, can stimulate ERK2 at concentrations ranging from 50–100 μM but stimulated JNK1 only to a small extent (Yu et al., 1997b, 1999). GTPs are known to induce expression of immediate early genes, c-jun and c-fos and activate ARE chloramphenicol acetyltransferase (CAT) reporter gene expression. We further investigated the activities of GTPs and indeed found that, (−)-epigallocatechin-3-gallate (EGCG) and (−)-epicatechin-3-gallate (ECG) induced ARE-Luc reporter gene activity. These specific GTPs also potently activated the MAPKs, ERK and p38, and in addition, JNK activity was increased by EGCG (Chen et al., 2000).

Recent observations have shown that ERK2 plays a direct and positive role in the activation of ARE and the induction of the phase II enzyme, NQO1 (Yu et al., 1999). By employing an ERK inhibitor, PD98059, which blocked the activation of ERK and a dominant-negative mutant of ERK, it was observed that ARE induction after tBHQ treatment was greatly reduced. Since tBHQ did not stimulate JNK1, it is concluded that it may not be a major contributor to the induction of ARE and the phase II enzymes. We have observed that both BHA and tBHQ altered p38 MAPK in a time and dose dependent manner (Yu et al., 2000b). It appeared that p38 is a negative regulator of ARE-mediated induction of NQO1. This was observed not only with an inhibitor of p38, but also, with a dominant-negative mutant of p38 or its upstream activator, MKK3. Taken together, these data show that the delicate balance of the MAPK cascades by various agents may be central in the regulation of phase II genes.

Natural as well as synthetic cancer chemopreventive isothiocyanates such as phenethyl-isothiocyanate (PEITC), from cruciferous vegetables and sulforaphane, found in broccoli induce the expression of phase II DMEs (Prestera et al., 1993). PEITC and sulforaphane induced MAPK activities at concentration 5–10 μM. PEITC stimulated only JNK and ERK, whereas sulforaphane activated only ERK and inhibited both basal as well as UV-induced JNK activities (Yu et al., 1999). The differences in chemical structure between these isothiocyanates might play a crucial role in determining the MAPKs that are actively involved. Sulforaphane-induced ARE induction was abolished by the use of ERK dominant-negative mutant as well as the ERK inhibitor. The upstream activator Ras, did not have an effect on ARE activation when cells were treated with sulforaphane or tBHQ. This indicates that the effect of the isothiocyanates or the phenolic antioxidant, tBHQ is most likely by a Ras-independent mechanism. On the other hand, Raf-1 kinase activity was stimulated by sulforaphane and tBHQ, which shows that the induction of ARE and phase II DMEs may involve a direct activation of Raf-1 kinase by the isothiocyanate, sulforaphane and the phenolic compound, tBHQ.

7. MAPK activation and ARE-mediated gene expression via Nrf2-dependent mechanism

Recent studies from our laboratory have shown that Nrf2 induces the activation of ARE in a dose dependent manner (Yu et al., 2000a). The dominant negative mutant of Nrf2 (deletion of N-terminal transactivation domain), blocks the activation of ARE and also blocked activation by inducers such as BHA and tBHQ. Expression of MEKK1, TAK1 and ASK1 in HepG2 cells activated Nrf2 transcriptional activity of ARE-Luc reporter gene and this activity was blocked by the use of the dominant negative mutants of the upstream kinases. Hence it is tempting to speculate that the transcriptional activity of Nrf2 on ARE-dependent genes is dependent on upstream kinases such as MEKK1, TAK1 and ASK1. These observations indicate that phenolic antioxidants play a role in the early signaling events thus leading to Nrf2 transcriptional activation of ARE-mediated phase II genes.

8. Concluding remarks

In conclusion, we and others have demonstrated the importance of the ARE in the regulation of phase II detoxifying enzymes. The upstream events that lead to this induction have a significant impact on protection conferred on the cells by these enzymes. The protective function of the phase II detoxifying enzymes can be used to our advantage, by studying drugs that will induce these enzymes. A chemopreventive agent is the term used for a drug or natural substance, which will not cause any harm to the body but that would halt the initiation of cancer or its progression. The use of drugs that can function as chemopreventive agents is being widely studied and although progress is slow, this is a very important aspect in relation to cancer management and its treatment. Monofunctional inducers, that induce only the phase II enzymes, would serve as very good candidates for this purpose. Our aim should be to understand the signaling events leading to phase II enzyme induction and utilize this knowledge to decipher the mechanism of known drugs and to develop more promising chemopreventive agents.

Acknowledgements

We thank Chi Chen, Dr. Bok-Ryang Kim, Dr. Owuor and all the members of Dr. Tony Kong's laboratory for helpful contributions. Supported in part by the National Institutes of Health grants RO1-CA73674 and RO1-CA94828 to A.-N.T. Kong.

References

Abe, J., Kusuhara, M., Ulevitch, R.J., Berk, B.C. and Lee, J.D. (1996). Big mitogen-activated protein kinase 1 (BMK1) is a redox-sensitive kinase. J. Biol. Chem. 271, 16586–16590.

Andrews, N.C., Erdjument-Bromage, H., Davidson, M.B., Tempst, P. and Orkin, S.H. (1993). Erythroid transcription factor NF-E2 is a haematopoietic-specific basic-leucine zipper protein. Nature 362, 722–728.

Barouki, R. and Morel, Y. (2001). Repression of cytochrome P450 1A1 gene expression by oxidative stress: mechanisms and biological implications. Biochem. Pharmacol. 61, 511–516.

Baud, V., Liu, Z.G., Bennett, B., Suzuki, N., Xia, Y. and Karin, M. (1999). Signaling by proinflammatory cytokines: oligomerization of TRAF2 and TRAF6 is sufficient for JNK and IKK activation and target gene induction via an amino-terminal effector domain. Genes Dev. 13, 1297–1308.

Blank, V., Kim, M.J. and Andrews, N.C. (1997). Human MafG is a functional partner for p45 NF-E2 in activating globin gene expression. Blood 89, 3925–3935.

Blumberg, B., Sabbagh, W., Juguilon, H., Bolado, J., van Meter, C.M., Ong, E.S. and Evans, R.M. (1998). SXR, a novel steroid and xenobiotic-sensing nuclear receptor. Genes Dev. 12, 3195–3205.

Chan, K. and Kan, Y.W. (1999). Nrf2 is essential for protection against acute pulmonary injury in mice. Proc. Natl.

Acad. Sci. USA 96, 12731–12736.

Chao, T.H., Hayashi, M., Tapping, R.I., Kato, Y. and Lee, J.D. (1999). MEKK3 directly regulates MEK5 activity as part of the big mitogen-activated protein kinase 1 (BMK1) signaling pathway. J. Biol. Chem. 274, 36035–36038.

Chen, C., Yu, R., Owuor, E.D. and Kong, A.N. (2000). Activation of antioxidant-response element (ARE), mitogen-activated protein kinases (MAPKs) and caspases by major green tea polyphenol components during cell survival and death. Arch. Pharm. Res. 23, 605–612.

Derijard, B., Hibi, M., Wu, I.H., Barrett, T., Su, B., Deng, T., Karin, M. and Davis, R.J. (1994). JNK1: a protein kinase stimulated by UV light and Ha-Ras that binds and phosphorylates the c-Jun activation domain. Cell 76, 1025–1037.

Dhakshinamoorthy, S. and Jaiswal, A.K. (2000). Small maf (MafG and MafK) proteins negatively regulate antioxidant response element-mediated expression and antioxidant induction of the NAD(P)H:Quinone oxidoreductase1 gene. J. Biol. Chem. 275, 40134–40141.

Fanger, G.R., Johnson, N.L. and Johnson, G.L. (1997). MEK kinases are regulated by EGF and selectively interact with Rac/Cdc42. Embo J. 16, 4961–4972.

Favreau, L.V. and Pickett, C.B. (1991). Transcriptional regulation of the rat NAD(P)H:quinone reductase gene. Identification of regulatory elements controlling basal level expression and inducible expression by planar aromatic compounds and phenolic antioxidants. J. Biol. Chem. 266, 4556–4561.

Friling, R.S., Bensimon, A., Tichauer, Y. and Daniel, V. (1990). Xenobiotic-inducible expression of murine glutathione S-transferase Ya subunit gene is controlled by an electrophile-responsive element. Proc. Natl. Acad. Sci. USA 87, 6258–6262.

Fujiwara, K.T., Kataoka, K. and Nishizawa, M. (1993). Two new members of the maf oncogene family, mafK and mafF, encode nuclear b-Zip proteins lacking putative trans-activator domain. Oncogene 8, 2371–2380.

Gerwins, P., Blank, J.L. and Johnson, G.L. (1997). Cloning of a novel mitogen-activated protein kinase kinase kinase, MEKK4, that selectively regulates the c-Jun amino terminal kinase pathway. J. Biol. Chem. 272, 8288–8295.

Gomez, N. and Cohen, P. (1991). Dissection of the protein kinase cascade by which nerve growth factor activates MAP kinases. Nature 353, 170–173.

He, C.H., Gong, P., Hu, B., Stewart, D., Choi, M.E., Choi, A.M. and Alam, J. (2001). Identification of activating transcription factor 4 (ATF4) as an Nrf2-interacting protein. Implication for heme oxygenase-1 gene regulation. J. Biol. Chem. 276, 20858–20865.

Hibi, M., Lin, A., Smeal, T., Minden, A. and Karin, M. (1993). Identification of an oncoprotein- and UV-responsive protein kinase that binds and potentiates the c-Jun activation domain. Genes Dev. 7, 2135–2148.

Hung, H.L., Kim, A.Y., Hong, W., Rakowski, C. and Blobel, G.A. (2001). Stimulation of NF-E2 DNA binding by CREB-binding protein (CBP)-mediated acetylation. J. Biol. Chem. 276, 10715–10721.

Ichijo, H., Nishida, E., Irie, K., ten Dijke, P., Saitoh, M., Moriguchi, T., Takagi, M., Matsumoto, K., Miyazono, K. and Gotoh, Y. (1997). Induction of apoptosis by ASK1, a mammalian MAPKKK that activates SAPK/ JNK and p38 signaling pathways. Science 275, 90–94.

Itoh, K., Chiba, T., Takahashi, S., Ishii, T., Igarashi, K., Katoh, Y., Oyake, T., Hayashi, N., Satoh, K., Hatayama, I., Yamamoto, M. and Nabeshima, Y. (1997). An Nrf2/ small Maf heterodimer mediates the induction of phase II detoxifying enzyme genes through antioxidant response elements. Biochem. Biophys. Res. Commun. 236, 313–322.

Itoh, K., Wakabayashi, N., Katoh, Y., Ishii, T., Igarashi, K., Engel, J.D. and Yamamoto, M. (1999). Keap1 represses nuclear activation of antioxidant responsive elements by Nrf2 through binding to the amino-terminal Neh2 domain. Genes Dev. 13, 76–86.

Johnson, E.F., Palmer, C.N. and Hsu, M.H. (1996). The peroxisome proliferator-activated receptor: transcriptional activation of the CYP4A6 gene. Ann. NY Acad. Sci. 804, 373–386.

Katoh, Y., Itoh, K., Yoshida, E., Miyagishi, M., Fukamizu, A. and Yamamoto, M. (2001). Two domains of Nrf2 cooperatively bind CBP, a CREB binding protein, and synergistically activate transcription. Genes Cells 6, 857–868.

Kyriakis, J.M. and Avruch, J. (1996). Sounding the alarm: protein kinase cascades activated by stress and inflammation. J. Biol. Chem. 271, 24313–24316.

Kyriakis, J.M., Banerjee, P., Nikolakaki, E., Dai, T., Rubie, E.A., Ahmad, M.F., Avruch, J. and Woodgett, J.R. (1994). The stress-activated protein kinase subfamily of c-Jun kinases. Nature 369, 156–160.

Li, Y. and Jaiswal, A.K. (1992). Regulation of human NAD(P)H:quinone oxidoreductase gene. Role of AP1 binding site contained within human antioxidant response element. J. Biol. Chem. 267, 15097–15104.

Mangelsdorf, D.J. and Evans, R.M. (1995). The RXR heterodimers and orphan receptors. Cell 83, 841–850.

Marshall, C.J. (1994). MAP kinase kinase kinase, MAP kinase kinase and MAP kinase. Curr. Opin. Genet. Dev. 4, 82–89.

McKenna, N.J., Lanz, R.B. and O'Malley, B.W. (1999). Nuclear receptor coregulators: cellular and molecular biology. Endocr. Rev. 20, 321–344.

Moi, P., Chan, K., Asunis, I., Cao, A. and Kan, Y.W. (1994). Isolation of NF-E2-related factor 2 (Nrf2), a NF-E2-like basic leucine zipper transcriptional activator that binds to the tandem NF-E2/AP1 repeat of the beta-globin locus control region. Proc. Natl. Acad. Sci. USA 91, 9926–9930.

Motohashi, H., Shavit, J.A., Igarashi, K., Yamamoto, M. and Engel, J.D. (1997). The world according to Maf. Nucleic Acids Res. 25, 2953–2959.

Omura, T. and Sato, R. (1962). A new cytochrome in liver microsomes. J. Biol. Chem. 237, 1375–1376.

Prestera, T., Holtzclaw, W.D., Zhang, Y. and Talalay, P. (1993). Chemical and molecular regulation of enzymes that detoxify carcinogens. Proc. Natl. Acad. Sci. USA 90, 2965–2969.

Raingeaud, J., Whitmarsh, A.J., Barrett, T., Derijard, B. and Davis, R.J. (1996). MKK3- and MKK6-regulated gene expression is mediated by the p38 mitogen-activated protein kinase signal transduction pathway. Mol. Cell. Biol. 16, 1247–1255.

Reddy, J.K., Goel, S.K., Nemali, M.R., Carrino, J.J., Laffler, T.G., Reddy, M.K., Sperbeck, S.J., Osumi, T., Hashimoto, T., Lalwani, N.D. et al. (1986). Transcription regulation of peroxisomal fatty acyl-CoA oxidase and enoyl-CoA hydratase/3-hydroxyacyl-CoA dehydrogenase in rat liver by peroxisome proliferators. Proc. Natl. Acad. Sci. USA 83, 1747–1751.

Robinson, M.J. and Cobb, M.H. (1997). Mitogen-activated protein kinase pathways. Curr. Opin. Cell Biol. 9, 180–186.

Rushmore, T.H., King, R.G., Paulson, K.E. and Pickett, C.B. (1990). Regulation of glutathione S-transferase Ya subunit gene expression: identification of a unique xenobiotic-responsive element controlling inducible expression by planar aromatic compounds. Proc. Natl. Acad. Sci. USA 87, 3826–3830.

Talalay, P., De Long, M.J. and Prochaska, H.J. (1987). Molecular mechanisms in protection against carcinogenesis. In: Cancer Biology and Therapeutics (Cory, J.G. and Szentivanyi, A., Eds.), pp. 197–216. Plenum, NY.

Trottier, E., Belzil, A., Stoltz, C. and Anderson, A. (1995). Localization of a phenobarbital-responsive element (PBRE) in the 5′-flanking region of the rat CYP2B2 gene. Gene 158, 263–268.

Whitlock, J.P. (1999). Induction of cytochrome P4501A1. Annu. Rev. Pharmacol. Toxicol. 39, 103–125.

Wrighton, S.A. and Stevens, J.C. (1992). The human hepatic cytochromes P450 involved in drug metabolism. Crit. Rev. Toxicol. 22, 1–21.

Xie, W., Barwick, J.L., Simon, C.M., Pierce, A.M., Safe, S., Blumberg, B., Guzelian, P.S. and Evans, R.M. (2000). Reciprocal activation of xenobiotic response genes by nuclear receptors SXR/PXR and CAR. Genes Dev. 14, 3014–3023.

Yeldandi, A.V., Rao, M.S. and Reddy, J.K. (2000). Hydrogen peroxide generation in peroxisome proliferator-induced oncogenesis. Mutat. Res. 448, 159–177.

Yu, R., Chen, C., Mo, Y.Y., Hebbar, V., Owuor, E.D., Tan, T.H. and Kong, A.N. (2000a). Activation of mitogen-activated protein kinase pathways induces antioxidant response element-mediated gene expression via a Nrf2-dependent mechanism. J. Biol. Chem. 275, 39907–39913.

Yu, R., Jiao, J.J., Duh, J.L., Gudehithlu, K., Tan, T.H. and Kong, A.N. (1997a). Activation of mitogen-activated protein kinases by green tea polyphenols: potential signaling pathways in the regulation of antioxidant-responsive element-mediated phase II enzyme gene expression. Carcinogenesis 18, 451–456.

Yu, R., Lei, W., Mandlekar, S., Weber, M.J., Der, C.J., Wu, J. and Kong, A.T. (1999). Role of a mitogen-activated protein kinase pathway in the induction of phase II detoxifying enzymes by chemicals. J. Biol. Chem. 274, 27545–27552.

Yu, R., Mandlekar, S., Lei, W., Fahl, W.E., Tan, T.H. and Kong, A.T. (2000b). p38 mitogen-activated protein kinase negatively regulates the induction of phase II drug-metabolizing enzymes that detoxify carcinogens. J. Biol. Chem. 275, 2322–2327.

Yu, R., Tan, T.H. and Kong, A.T. (1997b). Butylated hydroxyanisole and its metabolite tert-butylhydroquinone differentially regulate mitogen-activated protein kinases. The role of oxidative stress in the activation of mitogen-activated protein kinases by phenolic antioxidants. J. Biol. Chem. 272, 28962–28970.

Zhu, M. and Fahl, W.E. (2001). Functional characterization of transcription regulators that interact with the electrophile response element. Biochem. Biophys. Res. Commun. 289, 212–219.

Sensing, Signaling and Cell Adaptation. Edited by K.B. Storey and J.M. Storey

CHAPTER 16

Biological Actions of Infrared Radiation

Lee Laurent-Applegate and Stéphanie Roques
Laboratory of Oxidative Stress and Aging, University Hospital, Lausanne, Switzerland

1. Introduction

1.1. Infrared radiation spectrum

Terrestrial solar radiation is composed primarily of visible (55%) and infrared (40%) radiation with the remaining 5% being in the ultraviolet region of the spectrum. Interactions of matter and radiation are involved in all events in the universe. Apart from the force of gravity, the electromagnetic field whose quantum is the photon of light is a force experienced in the everyday world. Electromagnetic photons have a perhaps infinite range of energies but only a very narrow band including ultraviolet, visible, and near-infrared light interacts with matter primarily by the excitation of electrons in atoms or molecules. Electronic excitation comprises the majority of the matter–energy interactions that have been evolved in the creation and maintenance of living systems on earth. Life on earth would not be as it is without solar radiation as it feeds us through photosynthesis and infrared radiation provides the warmth that we need to heat the earth's surface so that it is not otherwise frozen.

When skies are clear and there are no clouds, two-thirds of solar energy arrives at the earth's surface and only 1% of this incident energy is recuperated by photosynthetic organisms via their pigments, such as chlorophyll. The energy that is re-emitted by the earth's surface depends largely on the surface topography (i.e., water, vegetation, rocks, desert) and the earth also absorbs a large part of the solar spectrum wavelengths. Principally, only IR is not re-emitted as these wavelengths are absorbed by water, carbon dioxide, ozone and molecules in general with more than three atoms (Fig. 16.1). The "Greenhouse effect" is in fact the heating of the atmosphere by IR radiation re-emitted by the Earth (Phillips, 1985).

It was indeed the warming effects of the IR component that brought attention to these wavelengths within the solar spectrum already described in 1801 by Sir William Herschel. The spectrum of IR was then oriented to the near-, mid- and far-IR regions by arbitrary subdivision, much like that done for the ultraviolet radiation spectrum.

Infrared radiation is the invisible portion of the electromagnetic spectrum following the long or red end of the visible light region and extending all the way to the microwave range. Similar to the UV portion of the sun's spectrum, the activity of infrared radiation is not uniform across the spectrum. As the wavelength of electromagnetic radiation increases, the energy level per photon decreases. Therefore, if equal absorption occurs, fewer thermal effects can be induced with the longer wavelengths. Generally, X-rays ionise atoms, UV and visible light induce photochemical reactions and IR radiation is thought to induce only molecular vibrations that manifest as an increase in temperature. However, wavelengths up to 800–900 nm in the near-IR region are capable of inducing photochemical reactions, too. The mid- and far-IR radiation induces biological effects similar to those seen in hyperthermia. For the most part, we will concentrate in this chapter on near-IR radiation effects that have photochemical interactions that produce reactive oxygen species (ROS).

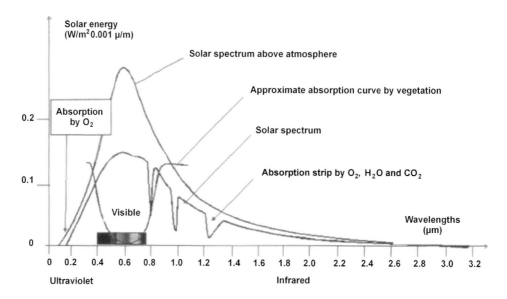

Fig. 16.1. Electromagnetic spectrum showing the solar spectrum above the atmosphere and at the Earth's surface. (Figure in part from http://perso.libertysurf.fr/pst/svtiufm/atmosphr.htm)

Heat can act in two different measures: (1) by increasing tissue temperature which accelerates all metabolic reactions non-specifically. or (2) by initiating specific responses to heat stimuli that are dependent on the type, intensity, duration and region of the body to which the stimulus is applied. Infrared-A applied locally has sufficient penetration properties where heat transfer and conduction by the bloodstream can have therapeutic effects on tissues at some distance from the site of direct irradiation (Vaupel et al. 1991).

1.2. Penetration properties in tissue of infrared radiation

The depth of transmission of electromagnetic radiation in tissue is generally proportional to the wavelength and the longer wavelengths penetrate deeper. Infrared radiation can penetrate from 0.7 to 30 mm into tissue which is sufficient for transmission through the chest wall. This aspect is very important to the therapeutic uses of IR discussed later in this chapter.

Infrared, which comes to the earth's surface in moderate climate zones (not in deserts where there is no significant content of water vapour in the atmosphere), is considered as water-filtered infrared and the major component is water-filtered infrared A (wIRA). Concerning the depth of transmission of infrared radiation (especially infrared A, B and parts of C) in biological tissues, the importance of reflection (away from the tissue surface) and diffraction (into the tissue) is quite small compared to the effect of absorption in the tissue and this absorption and therefore penetration is very different depending on the wavelength. The so-called water bands within the infrared A are highly absorbed in tissues (contributing to a surface heating of the skin within the first ~250–500 micrometers) and therefore have only a poor penetration ability in tissue. Infrared B and those parts of infrared C near to infrared B are highly absorbed in tissue (contributing to a surface heating of the skin within the first approximately 250–500 micrometers) and therefore have only a poor penetration ability in tissue. If the water bands within the infrared A and the whole infrared B and those parts of infrared C near to infrared B are subtracted from a mixture of infrared A, B and those parts of infrared C near to infrared B then the result is water-filtered infrared A (wIRA = infrared without water bands within the infrared A and without infrared B and without infrared C). wIRA has a low absorption coefficient and therefore a high penetration ability in tissue without much surface heating of the skin within the first ~250–500 micrometers.

A conventional infrared radiator has no water filtering system and delivers a mixture of all parts of infrared A, infrared B and those parts of infrared C near to infrared B. These sources include wavelengths with high absorption and therefore poor penetration ability in tissue and produce surface heating in the upper layer of the skin. It is advisable to have the delivery of a quality of radiation which shows small absorption coefficients and good penetration ability. Therefore, the different types of IR can be clinically defined more correctly as IR for basic infrared, IRA for infrared A, wIRA for water filtered infrared and laser IR for monochromatic forms of infrared radiation.

2. Electromagnetic radiation and human health

2.1. Reactive oxygen species and their role in bio-regulation

Traditionally it has been accepted that particular physiological reactions in living cells are evoked by specialized molecules such as hormones, cytokines and neurotransmitters. There are now data showing that reactive oxygen species (ROS) participate in most, if not all, cellular reactions (Khan and Wilson, 1995; Saran et al., 1998; Gamaley and Klybin, 1999). Living cells normally react to external stimuli in specific ways such as: (a) they are induced to perform their specialized function, (b) they change their specialization by either differentiation or de-differentiation, (c) they enter into mitotic cycles and proliferate, or (d) they proceed into apoptosis or "programmed cell death". ROS themselves imitate the action of many bio-molecular signals upon cellular functions including regulation of genomic activity although the mechanism of action remains puzzling for conventional biochemical paradigms.

2.2. Free radical reactions and light production

Very weak radiation emission from cells is thought to result from radical reactions which can be produced by biological events such as lipid peroxidation. In studies of microsomal lipid peroxidation (Cadenas, 1984; Wright et al., 1979), it has been shown that the amount of malonaldehyde produced and the intensity of emitted light are related to each other. Based on these studies, Inaba and co-workers (Devaraj et al., 1991) proposed that oxygen dependent light emission in rat liver nuclei was caused most likely by lipid peroxidation in the nuclear membrane. As discussed in detail by Cadenas and Sies (1982), free radical decomposition of lipid hydroperoxides leads to the formation of excited chemiluminescent species by the self-reaction of secondary lipid peroxyradicals, producing either singlet molecular oxygen or excited carbonyl groups.

A very interesting model by Nagl and Popp (1983) suggests that there is a negative feedback loop in living cells that couples together states of coherent weak photon emission and the conformational state of the cellular DNA. Photon transfer or non-radiation chemical pumping from the cytoplasmic metabolism could result in changes in DNA conformation via exciplex/eximer formation. Their model is supported by experimental data reviewed by Birks (1975) who also suggested that excimers act as precursors of the pyrimidine photodimers which play a key role in the radiation damage of DNA and have been most recently proposed as biomolecular signals in communication (Applegate et al., 1999). It seems that an effective intracellular mechanism of photon trapping in normal human cells exists and that this type of light-trapping mechanism in UV-induced pyrimidine dimers could be responsible for influencing metabolic and cellular events such as were shown in photodimer-induced melanogenesis (Eller et al., 1994).

3. Cellular interactions with IR radiation

3.1. Cellular recognition and migration

Certain cells seem to have a kind of infrared "vision. In a study by Albrecht-Buehler (1991) using a specially designed phase-contrast light microscope with an infrared spot illuminator, it was shown that mammalian cells were able to extend pseudopodis towards the infrared sources nearby. The strongest responses were observed following IR irradiation of cells in the range of 800–900 nm with intermittent pulses rates of 30–60/min. Cells were able to sense specific infrared wavelengths and to determine the direction of individual sources. The results suggest the existence of cellular receptor pigments that can absorb in the spectral range of 800–900 nm. There are indeed several kinds of bacterial chlorophylls within this range and other pigments also absorb in the far red and infrared such as plant chlorophyll with its absorption peak at 680 nm (Fuller et al., 1963; Darnell et al., 1990) and the infrared receptors in the facial pits of rattle snakes which absorb in the range of 2000–3000 nm and longer (Bullock and Cowles, 1952). Albrecht-Buehler suggests that pseudopod migration could have an association with IR absorption by a cellular component such as centrioles in the localization of infrared light sources in cells which would not be unreasonable considering that retinal cells have been shown to detect single photons of visible light.

3.2. Sources of low dose infrared radiation

In the above study by Albrecht-Buehler (1991), the biological effect was independent of temperature increase. This is an area of confusion in many studies that use IR radiation as it has been described that heat is equivalent to IR radiation. The majority of studies in the literature claim to show biological effects of IR when there are serious increases in the temperature of tissues caused by either infrared light bulbs, hot water baths, heated air and even heating coils. The fact that there is an increase in temperature does not mean that the same biological events will take place, especially if the source to increase the temperature includes photon light energy. It is true that a dominant property noticed from IR radiation is its thermal capacity and the possibility to induce hyperthermia but IR radiation from the solar spectrum or light sources can also have independent effects from those of temperature increase related to the photonic capacity or the interaction of photons with matter.

This phenomenon, termed "biostimulation", is related to the stimulatory effects of low-energy light on various biological systems which is independent of temperature. More recently, water-filtered near-IR devices have become available that successfully eliminate the over-heating effects for IR radiation sources. However, it should be noted that the output of these devices still includes red light. Although near IR radiation by definition finishes at the far end of the visible light region, the most commonly used sources of IR are lamps that emit red light and this should be taken into account when assessing the studies.

3.3. Sources of high dose infrared radiation

Lasers are a common source of coherent monochromatic radiation used as high-energy devises to generate tissue temperatures from 50 to several thousand degrees Celsius. They are used for tissue coagulation, vaporization and cutting applications (Colver, 1986, 1989). Many of the lasers emit in the infrared range and are being investigated for hyperthermia induction as an adjunct to both radiation and Photodynamic Therapy (PDT) for the treatment of malignancies. However, these IR lasers can be used at much lower doses for such purposes if photosensitizers are used such as in PDT. IR radiation would provide an interesting source for the biostimulation and excitation of fluorochromes. In this therapy, ROS are activated following photosensitization reactions. Cellular responses are therefore similar to those of signals induced following oxidative stress (Granville et al., 1999). IR could be an important aid to vascular photosensitizers as a deeper penetration of wavelengths is necessary and lower photosensitivity of surrounding tissue would be emphasized.

The biological effects of IR not associated with heat generation have been highly neglected mainly due to the lack of appropriate filtration of IR radiation sources. Reports of the induction of heat shock proteins by IR is most likely associated with the heat generated rather than with the IR radiation itself (Maytin et al., 1993). Protective effects of IR radiation on cultured skin cells that have been previously exposed to UVB and UVA radiation could be related to a direct induction by IR-A radiation of other protective proteins as we have seen in recent studies with ferritin in human skin *in vivo* (Applegate et al., 1998). Subsequent protection of cultured cells has been shown to be specifically related to ferritin induction following UVA radiation (Vile and Tyrrell, 1993). It is interesting to note that if we take into account the total amount of DNA damage to human skin cells following solar irradiation and the overall capacity of cells for repair of the induced damage, the incidence of skin cancer in the human population is much lower than what would be expected (Sutherland, 1996). It seems logical that evolution has assured that overall sun exposure is not too hazardous for non-haired creatures. UVB and UVA radiation exposure is very damaging to human skin being the major factors implicated in photoaging and carcinogenesis. Of the solar UV, 90% is UVA radiation which is less mutagenic and carcinogenic. However, UVA given monochromatically can induce skin cancers including melanoma in an experimental fish model (Setlow, 1995) and epidemiologically there is a causal relationship for artificial UVA exposure used for medical or cosmetic reasons and malignant melanoma induction (Westerdahl, 1994). These and most studies have used monochromatic or narrow wavebands of each spectra. As overall sun exposure does not seem to induce as much long-term damage to human skin as would be expected from extrapolating acute damage, perhaps IR radiation does indeed play a very important role in the natural protection of human skin, just as we have seen experimentally where IR exposure protects against cell killing and can induce ferritin levels (Applegate et al., 1998, 1999, 2000).

4. Gene and protein induction

Menezes et al. (1998) have shown that a pre-irradiation of human skin fibroblasts with IR at wavelengths of 700–2000 nm and at a temperature of 25°C can protect the cells against a subsequent cytotoxicity induced by UVA and UVB radiation. The protective effect provided by the pre-irradiation with IR did not seem to be associated with a cell cycle stimulation or act through protein synthesis. The protective effect was cumulative and could last up to 24 hours following the initial dose of IR. The results from this study strongly demonstrate the cellular effects of IR radiation without the thermal qualities and the authors conclude that there is an extremely effective role for IR radiation in the prevention of damage to skin cells by UV radiation.

4.1. Survival curves of human skin fibroblasts following IR-A radiation

In our early studies, we have shown that dermal fibroblasts of six individuals subjected to Infrared-A irradiation were resistant for doses up to 10.2×10^6 J/m^2 IR-A radiation as assessed by trypan blue exclusion. Survival curves were established for human skin fibroblasts from a 29-year-old male irradiated at 22–25°C using two different IR-A radiation sources. We have found that even very high doses of IR-A were non-lethal to this cell type and IR-A could even protect against cell killing induced by the long period of cell contact in phosphate buffered saline (PBS) (Fig. 16.2).

4.2. Induction of protective proteins

Very little information is available on the induction of stress genes in human skin by infrared radiation although protective effects of IR have been seen on

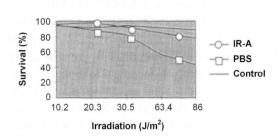

Fig. 16.2. Cellular protection provided by IR-A radiation. Percent survival of human skin fibroblasts as a function of increasing doses of IR-A radiation. An average of four experiments is shown with controls (solid line), PBS alone (squares) or cells treated with IR-A radiation (circles).

cultured cells irradiated with other wavelengths (UVB and UVA) of solar radiation (Menezes et al., 1998).

Because the skin is in continual contact with the external environment, its ability to provide protection against oxidative damage due to toxic chemicals and radiation is a primordial function. There is a vast repertoire of antioxidant defenses in human skin cells including antioxidant enzymes (e.g., superoxide dismutase, catalase, glutathione peroxidase, glutathione reductase) as well as lipophilic (tocopherol and ubiquinol) and hydrophilic (glutathione, ascorbate) antioxidants (Yohn et al., 1991; Shindo et al., 1994). In addition to the constitutive defense system in human skin cells, the heme oxygenase (HO-1) gene is strongly induced in cultured human skin fibroblasts by oxidants including UVA, providing the first example of a marked gene induction by an oxidizing carcinogen. Induction of HO-1 in fibroblasts was shown to correlate with an enhancement of cellular ferritin levels (Vile and Tyrrell, 1993; Vile et al., 1994). Ferritin constitutes the major storage site for non-metabolized intracellular iron and therefore plays a critical role in regulating the availability of iron to catalyze such harmful reactions as the peroxidation of lipids and the production of hydroxy radicals. In cultured skin fibroblasts, it was shown that the increase in ferritin levels constitutes an adaptive response that serves to protect these cells from subsequent oxidative damage (Vile and Tyrrell, 1993; Vile et al., 1994). Furthermore, the studies by Vile et al. (1994) have clearly shown that cellular sensitivity to UVA radiation is increased by treatment of human skin fibroblasts with HO-1 antisense oligonucleotides or desferoxamine, both of which eliminate UVA-mediated increases in cellular ferritin. Ferritin levels, which increase following UVA radiation exposure, would result in an enhancement of cellular iron-sequestering capacity and confer increased resistance to oxidative stress (Vile and Tyrrell, 1993; Vile et al., 1994). In cultured epidermal keratinocytes, ferritin levels are 3–7 fold higher than in dermal fibroblasts (Applegate and Frenk, 1995a) and it has been suggested that these high levels of ferritin are a constitutive response. In this respect, the keratinocyte, as clearly the primary

target for oxidative stress generated by solar irradiation, would benefit from the continuous protection provided by high levels of ferritin (Applegate et al., 1995a and b).

This defense mechanism involving ferritin has previously been defined *in vitro* following UVA I radiation and the induction of this protein is related to cell type and wavelength of radiation (Applegate et al., 1998). Infrared radiation has been seen to have positive effects on cultured cells relating to their proliferation and protection against subsequent oxidative stress which is unrelated to the induction of Heat Shock Proteins. Importantly, we have recently seen that a putative defense mechanism involving ferritin is induced following IR-A exposure to human skin. The induction of this important protective mechanism is seen at doses of IR-A radiation that do *not* produce alterations involving DNA damage, oxidative stress molecules and proteases involved in degrading processes in aged skin.

Recently, a Russian group has shown that pulsed low dose IR-A radiation from lasers is able to induce proteins responsible for cell adhesion (very important in wound healing) (Karu et al., 2001a). This cell adhesion process was shown to be dependent on modulation of the activity of the respiratory chain (Karu et al., 2001b) and it was demonstrated that free radical and redox processes are also involved in the cell matrix interaction (Karu et al., 2001a, 2001c).

4.3. Quantitative expression of ferritin in vitro following IR radiation

Dermal fibroblasts of six individuals were subjected to Infrared-A irradiation and ferritin levels were assessed by ELISA. No cell killing was seen for doses up to 10.2×10^6 J/m^2 IR-A radiation as assessed by trypan blue exclusion. In previous studies our laboratory has seen that young cells are exceptionally resistant to oxidative stress or UV radiation. Both cells from very young (11–13 week fetal stage) and adult were resistant to the highest doses of IR-A radiation employed and responded in a similar manner with ferritin induction of up to 133% over basal cell levels (Fig. 16.3).

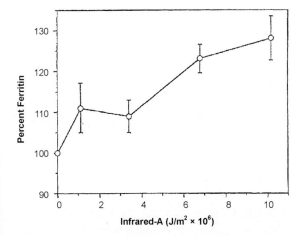

Fig. 16.3. Percent increase in ferritin in human skin cells in vitro following IR-A radiation. Ferritin was measured as a function of increasing doses of IR-A radiation. Experiments with each of six cells lines were repeated 2–4 times and data points are the overall average of the experiments (±SD).

4.4. Expression of ferritin in human skin in vivo following infrared-A radiation

In the epidermis, ferritin staining was shown to increase 24 hours following infrared-A radiation (Applegate et al., 1998). There was a definite increase in epidermal ferritin staining following the two highest doses of infrared-A (3.4 and 10.2×10^6 J/m^2) and the suprabasal layer and spinous keratinocytes also became lightly stained following these doses of infrared-A radiation (Fig. 16.4). Increased staining for ferritin following these doses of infrared-A was also observed in hair follicles.

No changes in ferritin staining were seen in both the papillary and reticular dermis, vessel walls, muscles or sebaceous glands. Overall epidermal ferritin levels changed by 120–220% (Fig. 16.4).

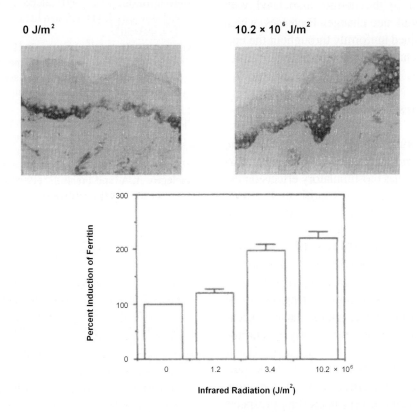

Fig. 16.4. Ferritin immunostaining in human skin *in vivo* 24 h following infrared-A radiation. Skin biopsies taken at 24 h following each dose were frozen, cut at 5 μm and stained with ferritin antibodies using the avidin:biotin enhancement immunostaining and counterstained with Papanicolaou. Precise ferritin staining is seen in the basal cell layer of the epidermis (dark coloration) of normal non-irradiated skin. Increased basal and suprabasal staining for ferritin is seen following IR-A radiation. Video image analysis shows a dose-dependent increase for ferritin induction following IR-A exposure (bar graph).

4.5. Expression of DNA damage, oxidative stress molecules and proteases of human skin in vivo following infrared-A radiation

Using the same doses of infrared-A that gave a clear induction of ferritin, we could detect no change in nuclear damage to DNA as assessed by the induction of pyrimidine dimers or by p53 expression when compared to non-irradiated control biopsies (1–2 nuclei stained for p53 or pyrimidine dimers for 4 mm of skin). In addition, no protease induction (collagenase, stromelysin, gelatinase) involved in dermal degradation could be detected nor were levels of stress molecules (heme oxygenase, heat shock protein, nitric oxide) altered even following the highest dose (10.2×10^6 J/m^2) of IR-A radiation. Staining was seen throughout the epidermal tissue for all stress molecules and was not altered quantitatively. Heme oxygenase seen in the dermis associated with blood vessels did not change. Proteases which were lightly stained uniformly throughout the epidermis did not change quantitatively.

5.   Perspectives

There is a growing volume of evidence both *in vitro* and *in vivo* that low energy red and near-infrared radiation has biostimulatory effects on tissues. Most of the evidence is specific to dermatology for the beneficial effects on wound healing. *In vitro*, there is extensive evidence with skin cells in culture for the protective effects by IR-A radiation. In the laboratory, low energy IR-A has been shown to have effects on DNA, RNA and protein synthesis related to cell proliferation and protection of human cells. In the clinic, studies have concentrated on wound healing, particularly in chronic wounds such as leg ulcers (Bihari and Mester, 1989; Sugrue et al., 1990). Because of the profound penetration properties of IR-A radiation, these wavelengths have very interesting prospects for use as therapeutic treatments (Bowker et al., 1981; Gartner, 1986; Stelian et al., 1992). Whatever the molecular target and mechanisms involved in the biological effects of IR-A radiation,

Mother Nature has provided a particular protection induced by IR-A for cellular defense against cytotoxic events, such as UV radiation.

Acknowledgements

We are indebted to the Erwin Braun Foundation for support of our studies and in particular to Dr. Hoffmann for discussions and help with all of the literature on the subject and to Prof. Gebbers with his continual encouragement. We would also like to thank Corinne Scaletta for technical help. In addition, a part of these studies were also supported by the Swiss League Against Cancer (KFS 695-7-1998).

References

Albrecht-Buehler, G. (1991). Surface extensions of 3T3 cells towards distant infrared light sources. J. Cell Biol. 114, 493–502.

Applegate, L.A., Goldberg, L.H., Ley, R.D. and Ananthaswamy, H.N. (1990). Hypersensitivity of skin fibroblasts from basal cell nevus syndrome patients to killing by ultraviolet B but not by ultraviolet C radiation. Cancer Res. 50, 637–641.

Applegate, L.A. and Frenk, E. (1995a). Oxidative defense in cultured human skin fibroblasts and keratinocytes from sun-exposed and non-exposed skin. Photoderm. Photoimmunol. Photomed. 11, 95–101.

Applegate, L.A. and Frenk, E. (1995b). Cellular defense mechanisms of the skin against oxidant stress and in particular UVA radiation. Eur. J. Dermatol. 5, 97–103.

Applegate, L.A., Scaletta, C., Panizzon, R. and Frenk, E. (1998). Evidence that ferritin is UV inducible in human skin: Part of a putative defense mechanism. J. Invest. Dermatol. 111, 159–163.

Applegate, L.A., Scaletta, C., Niggli, H., Panizzon, R. and Frenk, E. (1999). *In vivo* induction of pyrimidine dimers in human skin by UVA radiation : initiation of cell damage and/or intercellular communication? Int. J. Mol. Med. 3, 467–472.

Applegate, L.A., Scaletta, C., Panizzon, R., Frenk, E., Hohlfeld, P. and Schwarzkopf, S. (2000). Induction of the putative protective protein ferritin by infrared radiation: Implication in skin repair. Int. J. Mol. Med. 5, 247–251.

Bihari, I. and Mester, A.R. (1989). The biostimulative effect of low level laser therapy of long-standing crural ulcers using helium neon laser, helium neon plus infrared lasers, and noncoherent light: Preliminary report of a randomized double blind comparative study. Laser Therapy 1, 97–98.

Birks, J.B. (1975) Excimers. Rep. Progr. Phys. 38, 903–974.

Bowker, P., Martin, C.J., Fulton, D. and Muir, I.F. (1981). The use of infrared radiation to reduce heat loss in patients: experiments with a phantom. Clin. Phys. Physiol. Meas. 2, 257–270.

Bullock, T.H. and Cowles, G.J. (1952). Physiology of an infrared receptor: the facial pit of pit vipers. Science 115, 541–543.

Cadenas, E. and Sies, H. (1982). Low level chemiluminescence of liver microsomal fractions initiated by terbutyl hydroperoxide. Eur. J. Biochem. 124, 349–356.

Cadenas, E. (1984). Biological chemiluminescence. Photochem. Photobiol. 40, 823–830.

Colver G.B., Jones, R.L., Cherry, G.W., Dawber, R.P. and Ryan, T.J. (1986). Precise dermal damage with an infrared coagulator. Br. J. Dermatol. 114, 603–608.

Colver, G.B. (1989). The infrared coagulator in dermatology. Dermatol. Clin. 7, 155–167.

Devaraj, B., Scott, R.Q., Roschger, P. and Inaba, H. (1991). Ultraweak light emission from rat liver nuclei. Photochem. Photobiol. 54, 289–293.

Darnell, J., Lodish, H. and Baltimore, D. (1990). Molecular Cell Biology, 2nd edition. W.H. Freeman, New York.

Eller, M.S., Yaar, M. and Gilchrest, B.A. (1994). DNA damage and melanogenesis, Nature 372, 413–414.

Fuller, R.C., Conti, S.F. and Mellin, D.B. (1963). The structure of the photosynthetic apparatus in the green and purple sulfur bacteria. In: Bacterial Photosynthesis (Gest, H., San Pietro, A. and Vernon, L.P., Eds.), pp. 71–87. The Antioch Press, Yellow Springs, OH.

Gamaley, I.A. and Klybin, I.V., (1999) Roles of ROS: signalling and regulation of cellular functions. Int. Rev. Cytol. 188, 203–255.

Gartner, C. (1986). Pain treatment in seronegative spondylarthritis (spa) with infrared-lasers. Z. Rheumatologie 45, 232–233.

Granville D.J., Shaw, J.R., Leong, S., Carthy, C.M., Margaron, P., Hunt, D.W. and McManus, B.M. (1999). Release of cytochrome c, Bax migration, Bid cleavage, and activation of caspases 2, 3, 6, 7, 8, and 9 during endothelial cell apoptosis. Am. J. Pathol. 155, 1021–1025.

Kahn, A.U. and Wilson, T. (1995) Reactive oxygen species as second messengers. Chem. Biol. 2, 437–445.

Karu, T.I., Pyatibrat, L.V. and Kalendo, G.S. (2001a). Cell attachment to extracellular matrices is modulated by pulsed radiation at 820 nm and chemicals that modify the activity of enzymes in the plasma membrane. Lasers Surg. Med. 29, 274–281.

Karu, T.I., Pyatibrat, L.V. and Kalendo, G.S. (2001b). Donors of NO and pulsed radiation at lambda=820 nm exert effects on cell attachment to extracellular matrices. Toxicol. Lett. 121, 57–61.

Karu, T.I., Pyatibrat, L.V. and Kalendo, G.S. (2001c). Cell attachment modulation by radiation from a pulsed light diode (lambda=820 nm) and various chemicals. Lasers Surg. Med. 28, 227–236.

Kummer, U, Valeur, K.R., Baier G., Wegmann, K., Olsen, L.F. (1996) Oscillations in the peroxidase-oxidase reaction: a comparison of different peroxidases. Biochim. Biophys. Acta. 1289, 397–403.

Maytin, E.V., Murphy, L.A. and Merrill, M.A. (1993). Hyperthermia induces resistance to ultraviolet light B in primary and immortalized murine keratinocytes. Cancer Res. 53, 4952–4959.

Menezes, S., Coulomb, B., Lebreton, C. and Dubertret, L. (1998). Non-coherent near infrared radiation protects normal human dermal fibroblasts from solar ultraviolet toxicity. J. Invest. Dermatol. 111, 629–633.

Nagl, W. and Popp, F.A. (1983). A physical (electromagnetic) model of differentiation. Basic considerations. Cytobios 37, 45–62.

Phillips, M. (1985). Electromagnetic radiation. In: The New Encyclopaedia Britannica. 15th ed., Vol. 6., pp. 644–645. Encyclopaedia Britannica Inc., Chicago.

Saran, M., Michel, C. and Bors, W. (1998). Radical function in vivo: a critical review of current concepts and hypothesis. Z. Naturforsch. 53, 210–227.

Setlow, R.B. (1996). Relevance of in vivo models in melanoma skin cancer. Photochem. Photobiol. 63, 410–412.

Shindo, Y., Witt, E., Han, D., Epstein, W. and Packer, L. (1994). Enzymatic and non-enzymatic antioxidants in epidermis and dermis of human skin. J. Invest. Dermatol. 102, 122–124.

Stelian, J., Gil, I., Habot, B., Rosenthal, M., Abramovici, I., Kutok, N. and Khahil, A. (1992). Improvement of pain and disability in elderly patients with degenerative osteoarthritis of the knee treated with narrow-band light therapy. J. Am. Geriatr. Soc. 40, 232–233.

Sugrue, M.E., Carolan, J., Leen, E.J., Feeley, T.M., Moore, D.J. and Shanik, G.D. (1990). The use of infrared laser therapy in the treatment of venous ulceration. Ann. Vasc. Surg. 4, 179–181.

Sutherland, B.M. (1996). Mutagenic lesions in photocarcinogenesis : induction and repair of pyrimidine dimers. Photochem. Photobiol. 63, 375–377.

Vaupel, P., Stohrer, M., Groebe, K. and Rzeznik, J. (1991). Localized hyperthermia in superficial tumors using water-filtered infrared-A-radiation : evaluation of temperature distribution and tissue oxygenation in subcutaneous rat tumors. Strahlenther Onkol. 167, 353–354.

Vile, G.F. and Tyrrell, R.M. (1993). Oxidative stress resulting from ultraviolet A irradiation of human skin fibroblasts leads to a heme oxygenase-dependent increase in ferritin. J. Biol. Chem. 268, 14678–14681.

Vile, G.F., Basu-Modak, S., Waltner, C. and Tyrrell, R.M. (1994). Haem oxygenase-1 mediates an adaptive response to oxidative stress in human skin fibroblasts. Proc. Natl. Acad. Sci. USA 91, 2607–2610.

Westerdahl, J., Olsson, H. and Masbäck, A. (1994). Use of sunbeds and sunlamps and malignant melanoma in Southern Sweden. Am. J. Epidemiol. 140, 691–699.

Wright, J.R., Rumbaugh, R.C., Colby, H.D. and Miles, P.R. (1979). The relationship between chemiluminescence and lipid peroxidation in rat hepatic microsomes. Arch. Biochem. Biophys. 192, 344–351.

Yohn, J.J., Norris, D.A., Yrastorza, D.G., Buno, I.J., Leff, J.A. and Hake, S.S. (1991). Disparate antioxidant enzyme activities in cultured human cutaneous fibroblasts, keratinocytes, and melanocytes. J. Invest. Dermatol. 97, 405–409.

Sensing, Signaling and Cell Adaptation. Edited by K.B. Storey and J.M. Storey
© *2002 Elsevier Science B.V. All rights reserved.*

CHAPTER 17

Energy Sensing and Photostasis in Photoautotrophs

Norman P.A. Huner[1], Alexander G. Ivanov[1], Kenneth E. Wilson[2], Ewa Miskiewicz[1], Marianna Krol[1] and Gunnar Öquist[3]

[1]*Department of Plant Sciences, University of Western Ontario, London, Canada*
[2]*Department of Molecular Biology, University of Geneva, Geneva, Switzerland*
[3]*Department of Plant Physiology, Umeå University, Umeå, Sweden*

1. Introduction

Erwin Schrödinger defined life as negative entropy (Schrödinger, 1944). Life is an endergonic process characterized by structural and functional order, the maintenance of which reflects homeostasis. The complex, integrated metabolic pathways characteristic of all living cells not only represent the mechanism by which cells bioysnthesize and degrade a multitude of cellular constituents but also represent the mechanism by which cells regulate and transform energy within the cell. The ultimate source of this energy for the biosphere is the sun. Photoautotrophs link all other organisms to the sun through their ability to use visible light as their sole energy source and CO_2 as their sole source of carbon. The mechanism for the reduction CO_2 derived though photosynthesis, in turn, sustains all living organisms with respect to most of their bioenergetic and structural needs. Thus, photosynthesis must be considered the single most important process on earth. In this review, we focus our discussion on the role of energy sensing through redox reactions associated with electron transport in the detection and subsequent acclimation of organisms to changes in their environment with particular emphasis on plants, green algae and cyanobacteria. We show through specific examples how energy sensing through modulation of photosynthetic electron transport is used by photoautotrophs to sense changes in light intensity,

temperature, nutrient availability and water availability. Thus, the general concept of energy sensing is applicable to photoautotrophic as well as heterotrophic organisms.

2. Energy sensing and redox status

The regulation of cellular energy required to maintain homeostasis is essential for the growth and survival of all living cells. The most comprehensive research regarding cellular energy sensing has been performed in heterotrophic bacteria. Since Anton van Leeuwenhoek first observed the movement of living creatures through his microscope in 1676, microbiologists have discovered that bacteria are able to detect a variety of energy-related signals to guide their movement from environments that supply minimal energy resources to those that provide optimal energy resources. Energy taxis is the term used to describe such movements (Taylor et al., 1999), representing an important component to explain the following behaviours: (1) aerotaxis, the movement of bacteria to seek optimal oxygen concentrations and (2) phototaxis, the movement of unicellular, motile photosynthetic prokaryotes and eukaryotes to an optimal light environment. Although bacteria move in response to perceived chemical gradients in their environments, chemotaxis occurs independent of the uptake and metabolism of the chemical stimulus (Taylor et al.,

1999). Thus, the binding of the external effector molecule to the chemotaxis receptor is sufficient to induce the chemotactic response.

How do cells sense changes in their energy status? In the case of aerotaxis, heterotrophic bacteria such as *E. coli* sense a decrease in cellular energy levels when O_2 becomes limiting for respiration (Laszlo and Taylor, 1981; Taylor, 1983). Furthermore, in photosynthetic bacterial species such as *Rhodospirillum sphaeroides* and *Rhodospirillum centenum*, aerotaxis as well as phototaxis represents a response to changes in the redox status of the bacterial electron transport chain (Grishanin et al., 1997; Romagnoli and Armitage, 1999). Ferguson et al. (1987) and Taylor et al. (1999) have suggested a general model that links energy sensing to the redox status of the electron transport chain. We have integrated these concepts into the model presented in Fig. 17.1.

Photosynthetic and respiratory electron transport represent the principal mechanisms to transform energy in living cells. Both energy transforming systems exhibit similar basic requirements: first, an energy source is required in the form of reducing power derived either from light and the subsequent photo-oxidation of photosystem reaction centers (P) in the case of photosynthesis or from reductants generated by carbon oxidation in the case of respiration (Fig. 17.1). Second, electron sinks, which may be represented, for example, by CO_2 in the case of photosynthesis and O_2 in the case of respiration are required to accept these electrons. The electron sources and electron sinks are connected by an electron transport chain which is characterized by an intervening, redox-sensitive, quinone pool (Q) pool in bacterial, mitochondrial and chloroplastic electron transport systems. This quinone is represented by ubiquinone in the inner membrane of mitochondria and plastoquinone in the thylakoid membrane of chloroplasts. The chemiosmotic principle couples the oxidation/reduction of the electron transport components to the generation of a proton motive force (pmf) whose essential components include a membrane potential ($\Delta\Psi$) and a proton gradient (ΔpH). This is, in turn, coupled to the synthesis of ATP through the H^+-dependent ATP synthase.

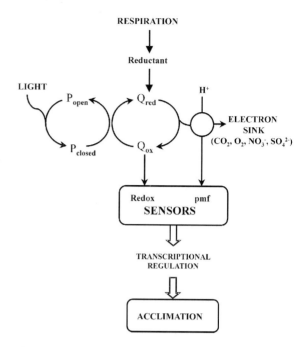

Fig. 17.1. Model of cellular energy sensing illustrating the possible relationships between electron transport, redox sensing and acclimation in photoautotrophic and heterotrophic organisms. In photosynthetic organisms, light is the ultimate source of energy. It is converted to redox potential energy through closing of reaction centers (P) photochemically by the absorption of light and the subsequent opening of reaction centers by the quinone pool (Q) and intersystem electron transport. A myriad of electron transport acceptors can are used as electron sinks. Since the oxidation of the Q pool is considered to be diffusion dependent, the rate-limiting step in photosynthetic as well as respiratory electron transport in eukaryotes and prokaryotes is the oxidation of this Q pool. As a consequence, various environmental factors will modulate the redox state of the Q pool. Thus, the energy balance between energy available in the form of either light in the case of photosynthetic electron transport or reductant in the form of NADH for respiratory electron transport versus energy consumed through metabolic electron sinks may be sensed by changes in either the redox status of the quinone pool and / or the proton motive force (pmf) generated by the electron transport chain. Modulation of either the redox status of the electron transport chain or the pmf generated act as signals, which are transduced through some unknown signal transduction pathway to regulate gene transcription and ultimately acclimation to the new environment.

The precise mechanism for energy sensing still remains a paradox. However, ATP levels do not appear to represent the primary signal in bacterial systems. Environmentally induced changes in

ATP levels appear to be too slow to account for regulatory changes involved in energy conservation (Shioi et al., 1982). The suggestion that modulation of chloroplastic ATP levels is important in sensing/signalling mechanisms associated with photoacclimation in higher plants and green algae remains equivocal (Melis et al., 1985). Alternatively, it is proposed that energy sensors may monitor changes in the redox status of electron carriers, in particular, the redox status of the quinone pool and/or one of the components of the pmf ($\Delta\Psi$ or ΔpH) (Ferguson et al., 1987; Allen et al., 1995; Huner et al., 1998; Taylor et al., 1999; Bauer et al., 1999). Generally, the rate of electron transport may be limited either by the availability of electron donors at the source or by the availability of a sufficient supply of electron acceptors at the sinks. Modulation of either source or sink capacity will modulate the relative redox status of the intervening electron transport chain as well as alter the pmf. The sensors would then transduce the electron transport signal, that is, either the redox status or the pmf, into biochemical signals that regulate the transcription of genes involved in acclimation (Fig. 17.1).

Taylor and Zhulin (1999) propose that PAS domains are important signalling modules within sensory proteins that detect these environmental-induced changes in cellular energy status. These domains have been identified in proteins across all kingdoms of living organisms and are important modules in proteins as varied as photoreceptors, phytochromes, clock proteins as well as oxygen and redox sensors (Taylor and Zhulin, 1999). The PAS domains are located in the cytoplasm and an investigation of completed microbial genomes indicates a strong correlation between the total number of PAS domains present and the number of photosynthetic and respiratory components present. This is consistent with the hypothesis that PAS domains are primarily involved in sensing changes in O_2 concentration, redox potential and light (Zhulin et al., 1997). However, Georgellis et al. (2001) reported recently that redox signal transduction by the well-known ArcB sensor kinase of the bacterium, *Haemophilus influenzae* does not exhibit a PAS domain.

In prokaryotes, PAS domains are present almost exclusively in the sensor of two-component signal transduction systems, which consist of a transmembrane sensor protein, and a cytoplasmic protein called the response regulator. These two-component systems not only sense environmental change but also transduce the specific environmental signals to light quality, redox potential, oxygen concentration, nitrogen assimilation, nutrient levels and osmotic potential (Hoch and Silhavy, 1995; Urao et al., 2000). The mechanism by which the signal is transduced from the sensor to the response regulator involves protein phosphorylation–dephosphorylation. Activation of the sensor by the external environment induces an autokinase activity that uses ATP to autophosphorylate a specific histidine residue in the cytoplasmic domain of the sensor. Hence, all two-component systems are histidine sensor kinases. This phosphohistidine residue serves as a substrate for the soluble, cytoplasmic response regulator, which catalyzes the transfer of the phosphate group from the histidine residue of the sensor to an aspartyl residue in the response regulator. The response regulator typically is a transcriptional factor capable of triggering a cellular response through binding to the promoter region of a specific gene to affect its expression. Two-component signal transduction systems were initially thought to be associated with prokaryotes only. Recently, however, two-component systems have been discovered in yeast as well as higher plants (Urao et al., 2000).

3. Photosynthetic electron transport

Plants, green algae and cyanobacteria have evolved a very specialized thylakoid membrane system, which enables them to transform light energy into useable chemical energy in the form of ATP and NADPH for the reduction of CO_2 to carbohydrate. To accomplish this, extremely fast photophysical processes and photochemical reactions are integrated with much slower biochemical reactions. Photosystem II (PSII) and PSI are integral thylakoid membrane protein complexes each consisting of numerous distinct polypeptides (Fig.

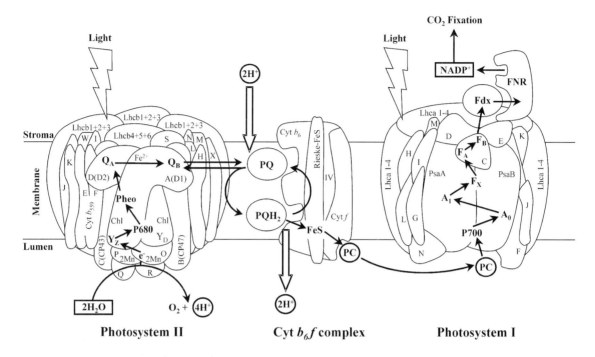

Fig. 17.2. Model of the photosynthetic membrane demonstrating the major polypeptide constituents and electron transport components. Photosystem II: D1 (PsbA) and D2 (PsbB) are the hydrophobic subunits of the PSII reaction center binding the photochemical Chl P680, Y_Z, the primary electron acceptor Pheo, and the bound quinones Q_A and Q_B. The major (Lhcb1–3) and minor (Lhcb 4–6) light harvesting polypeptides are also presented. Chlorophyll a – containing CP43 (PsbC) and CP47 (PsbD) function as core antenna. The hydrophobic subunit S (PsbS) has a major role in photoprotection. The minor polypeptides PsbH-PsbX and Cyt b_{559} represent hydrophobic subunits (PsbE and PsbF) with unknown function. The lumen-oriented hydrophilic subunits PsbO, PsbP and PsbQ are involved in water oxidation. The function of PsbR is unknown. Photosystem I: The reaction center heterodimer of PSI (PsaA/PsaB) binds the special photochemically active Chl pair (P700), the primary electron acceptor (A_0), phylloquinone (A_1), and 4Fe-4S cluster (F_X). The stromal-oriented PsaC polypeptide carries two 4Fe-4S clusters (F_A, F_B). Light harvesting polypeptides of PSI (Lhca1-4) are also shown. Lumenal-oriented PsaF subunit functions as a plastocyanin (PC) docking site. Several hydrophilic subunits exhibit stromal orientation: the ferredoxin (Fdx) docking site (PsaD); the cyclic electron transport component (PsaE) as well as the Lhca1-4 linker (PsaH). PsaG, PsaI, PsaJ, PsaK, PsaL, PsaM and PsaN represent subunits with unknown function. FNR = ferredoxin:NADP$^+$ reductase. Cyt b_6/f complex: Cyt$f$, Cyt b_6, Rieske-FeS and Subunit IV are the major polypeptides of the Cyt b_6/af complex transferring electrons from reduced plastoquinone (PQ) to oxidized PC. Minor polypeptide subunits (PetG, PetM, and PetL) of Cyt b_6/f complex with unknown functions are not shown. The transfer of electrons from water to NADP$^+$ in the linear electron transport (Z-scheme) is illustrated with solid arrows. Open arrows illustrate the translocation of protons.

17.2). The bulk of the chlorophyll and carotenoid present within the chloroplast thylakoid membrane is bound to the Lhcb and Lhca family of light harvesting polypeptides associated with PSII and PSI respectively, the PSII core antenna polypeptides of PsbB (CP47) and PsbC (CP43), the PSII reaction center polypeptides PsbA (D1), PsbD (D2) and the PSI reaction center polypeptides PsaA and PsaB (Green and Durnford, 1996).

The primary photochemical reactions of photosynthesis are dependent upon the photophysical processes of light absorption and energy transfer

within the light harvesting antenna pigment bed and the subsequent excitation migration from the light harvesting pigments, through the core antenna to reaction center pigments, P680 and P700 of PSII and PSI, respectively (Fig. 17.2). Excitation energy transfer occurs on a time scale of femtoseconds to picoseconds (10^{-15} to 10^{-12}s) which makes this probably the fastest process in all of biology. When the excitation energy reaches either a PSII or a PSI reaction center, photo-oxidation of PSII (P680 $\rightarrow$ P680$^+$) and PSI (P700 $\rightarrow$ P700$^+$) occurs on a time scale of nanoseconds to microseconds (10^{-9} to 10^{-6} s). The electrons generated by the photo-oxidation of P700 are used to reduce $NADP^+$ to NADPH via ferredoxin (Fdx) and the enzyme, FNR. Photosynthetic reducing power is consumed in a variety of biosynthetic reactions including CO_2 fixation through the reductive pentose phosphate cycle (RPPC), lipid biosynthesis as well as nitrate and sulfate assimilation.

However, to process excitation energy from the light harvesting and core antenna pigments on a continuous basis, a cycle of photo-oxidation followed by reduction of the reaction centers of PSI and PSII must also occur on a continuous basis. Electrons from Q_A^- are transferred via Q_B and convert plastoquinone (PQ) to plastoquinol (PQH_2) which is subsequently oxidized by the Cyt b_6f complex. This is considered to be the rate-limiting step in photosynthetic electron transport and occurs on the time scale of milliseconds (10^{-3} s) (Haehnel, 1984). $P700^+$ oxidizes the Cyt b_6f complex via plastocyanin (PC) which converts $P700^+$ back to P700. The oxidation of PQH_2 via the Cyt b_6f complex occurs concomitantly with the vectorial transport of protons from the stroma to the thylakoid lumen (Fig. 17.2). In the case of PSII, $P680^+$ exhibits a sufficiently positive reduction potential to oxidize H_2O resulting in the reduction of $P680^+$ back to P680, the release of O_2 and the release of protons into the thylakoid lumen (Fig. 17.2). The proton gradient generated by photosynthetic electron transport is used for the chemiosmotic synthesis of ATP by the chloroplast H^+-dependent ATP synthase.

4. Excitation pressure and redox sensing

As discussed above, the process of photosynthesis represents an integration of extremely rapid, temperature-insensitive photochemical reactions with relatively slow, temperature-dependent biochemical reactions. In the conversion of light energy into redox potential energy, the rate-limiting step is considered to be the conversion of "closed" PSII reaction centers to "open" PSII reaction centers through the oxidation of Q_A^- by PQ and the Cyt b_6/f complex of the intersystem electron transport chain (see Fig. 17.2):

$$[Y_Z \, P680 \, Pheo \, Q_A \, Q_B]_{open} + photon \rightarrow [Y_Z \, P680^+ \, Pheo \, Q_A^- \, Q_B]_{closed}$$

$$[Y_Z \, P680^+ \, Pheo \, Q_A^- \, Q_B]_{closed} + (PQ)_{ox} + H_2O \rightarrow [Y_Z \, P680 \, Pheo \, Q_A \, Q_B]_{open} + (PQ)_{red} + \tfrac{1}{2}O_2 + 2H^+$$

$$(PQ)_{red} + (Cyt \, b_6/f)_{ox} \rightarrow (PQ)_{ox} + (Cyt \, b_6/f)_{red}$$

Since the oxidation of $(PQ)_{red}$ by $(Cyt \, b_6/f)_{ox}$ is diffusion limited (Haehnel, 1984), the "closure" of PSII reaction centers through photochemistry will always be faster than the subsequent "opening" of PSII reaction centers through intersystem electron transport. Thus, exposure to excess light causes the over-reduction of the PQ pool which, in turn, increases the probability that PSII reaction centers remain closed. As a consequence, any environmental condition which exacerbates the difference between the rates at which PSII reaction centers are closed through photochemistry versus the rates at which they are opened through redox biochemistry will induce the over-reduction of the PQ pool and affect the relative redox state of Q_A in PSII reaction centers such that $[Q_A^-]/[Q_A] + [Q_A^-] > [Q_A]/[Q_A] + [Q_A^-]$. This is defined as excitation pressure which is a measure of the relative reduction state of Q_A in PSII reaction centers (Huner et al., 1998).

The rate of energy absorption by PSII is proportional to $\sigma_{PSII} \cdot I$ where σ_{PSII} is the functional absorption cross-sectional area of PSII and I is the absorbed photon flux. This product is, by and large, insensitive to temperature in the biologically significant range. Under light saturating conditions, the rate of utilization of the absorbed light through temperature-sensitive photosynthetic electron transport and the ultimate use of these photosynthetic electrons to reduce carbon, oxygen, nitrogen and sulfur may be expressed as $n \cdot \tau^{-1}$ where n is the number of electron sinks and τ^{-1} is the turnover rate of these sinks (Durnford and Falkowski, 1997). Accordingly, photoautrophs are exposed to excitation pressure whenever $\sigma_{PSII} \cdot I > n \cdot \tau^{-1}$ (Huner et al., 1998).

According to this inequality, excitation pressure may be induced by changes in several different environmental parameters. For example, increasing growth irradiance at a constant temperature would cause an increase in the relative reduction state of Q_A due to an increase in I, and thus an increase in $\sigma_{PSII} \cdot I$, assuming no changes in the capacity to utilize the absorbed energy, that is no change in $n \cdot \tau^{-1}$. Theoretically, a similar increase in the relative reduction state of Q_A could be created by maintaining the same irradiance but decreasing the temperature. The lower temperature would decrease the rate of the biochemical redox reactions that utilize the absorbed energy, that is, would decrease $n \cdot \tau^{-1}$, with no change in $\sigma_{PSII} \cdot I$. Similarly, exposure to drought or the lack of specific essential nutrients would also cause a decrease in $n \cdot \tau^{-1}$ due to limitations in the availability of electron acceptors such as CO_2, NO_3^- or SO_4^{2-}.

Chlorophyll a fluorescence is a property exhibited by all photosynthetic organisms due to its essential role in the structure and function of the photosynthetic apparatus. Typically, less than 3% of the absorbed light is ever re-emitted as chlorophyll fluorescence and, at room temperature, most of the chlorophyll a fluorescence emanates from PSII. Quantification of chlorophyll a fluorescence induction has proved to be an extremely useful tool to assess the structure and function of PSII and the overall process of photosynthesis (Krause and Weis, 1991). Excitation pressure can be estimated as the relative reduction state of Q_A, that is $[Q_A^-]$ / $[Q_A] + [Q_A^-]$, which can be conveniently measured *in vivo* as $1 - qP$ using pulse amplitude modulated chlorophyll fluorescence where qP is the photochemical quenching parameter (Huner et al., 1998).

5. Photostasis

Any change in environmental conditions such as light, temperature, water and nutrient availability may modulate the photochemical reactions of photosynthesis to a different extent than the biochemical reactions involved in carbon reduction cycle, photorespiration, and nitrogen and sulfur assimilation. Consequently, these environmental changes will modulate the relative reduction state of Q_A and hence will modulate excitation pressure. Excitation pressure reflects an imbalance between energy absorbed through photochemistry and energy utilized through the consumption of photo-

synthetically generated reducing power ($\sigma_{PSII} \cdot I > n \cdot \tau^{-1}$ (Huner et al., 1998). The predisposition of photosynthetic organisms to attain a balance in energy budget is defined as photostasis. This would be attained whenever energy absorption equals energy utilization through photochemistry plus nonphotochemical dissipation of excess energy (Fig. 17.3). Due to the photophysical and photochemical nature of light absorption, energy transfer and charge separation, photostasis is rarely, if ever, attained under natural environmental conditions. In fact, the light absorbed usually exceeds the energy that can be consumed through metabolism.

The inequality illustrated above also provides insights into the possible mechanisms by which photosynthetic organisms may respond to the imbalance in energy budget in an attempt to attain photostasis. Figure 17.3 illustrates the possible fates of absorbed light energy through the photosynthetic apparatus. Energy balance could be attained by either reducing σ_{PSII} by reducing light harvesting antenna size and/or reducing the effective absorption cross-sectional area of PSII by dissipating energy nonphotochemically as heat (Krause and Weis, 1991; Horton et al., 1996). Alternatively, photostasis also could be attained by increasing sink capacity ($n \cdot \tau^{-1}$). This may be accomplished by elevating the levels of Calvin cycle enzymes and enzymes involved in cytosolic sucrose biosynthesis, which would increase the capacity for CO_2 assimilation relative to the capacity for photosynthetic electron transport. Clearly, photoautotrophs may exploit any one or a combination of these mechanisms to attain photostasis in an environment which exhibits daily as well as seasonal changes in irradiance, temperature, water availability and nutrient status.

6. Excitation pressure, photoprotection and photostasis

According to the discussion above, exposure of plants to either high light, low temperature, nutrient limitations or drought may induce high excitation pressure and create an energy imbalance whereby $\sigma_{PSII} \cdot I > n \cdot \tau^{-1}$. Chronic exposure to high

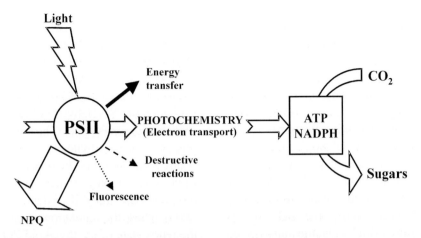

Fig. 17.3. A model illustrating the possible fates of absorbed light energy through the photosynthetic apparatus. Since the oxidation of PSII through intersystem electron transport is assumed to be the rate limiting step in photosynthetic electron transport (Haehnel, 1984), PSI is not included. Normally, excitation energy transferred to PSII is consumed through useful photochemistry, which drives photosynthetic electron transport. The energy is ultimately converted to ATP and NADPH needed for the reduction of CO_2 to complex carbohydrates. When the energy absorbed by PSII exceeds that which can used for useful photochemistry, the excess energy can be dissipated as heat through non-photochemical quenching (NPQ). Alternatively, excess energy absorbed by PSII can be redistributed through energy transfer to PSI. This energy transfer occurs through state transitions. Whenever the light energy absorbed through PSII exceeds the capacity for both photochemistry and non-photochemical quenching, PSII reaction centers are subjected to destructive reactions, which lead to irreversible photoinhibition. Typically less than 3% of the light absorbed by PSII at room temperature is lost as chlorophyll fluorescence. Changing rates of photochemistry and non-photochemical quenching modulates this chlorophyll fluorescence signal. As a consequence, chlorophyll fluorescence is a very sensitive, non-invasive probe of PSII structure and function.

excitation pressure may lead to photoinhibition of photosynthesis. Photoinhibition is defined as the light dependent decrease in photosynthetic rate which may occur whenever the photon flux is in excess of that required for photosynthesis (Long et al., 1994), that is, whenever $\sigma_{PSII} \cdot I > n \cdot \tau^{-1}$. Rapidly reversible photoinhibition is a consequence of an increase in thermal energy dissipation (NPQ, non-photochemical quenching) which leads to a decrease in the effective σ_{PSII} and a down regulation of PSII activity (Öquist et al., 1992). This process or any other process (Fig. 17.3) which protects PSII from over-excitation in the absence of protein synthesis is referred to as "photoprotection" and typically involves xanthophyll cycle pigments (Demmig-Adams and Adams, 1996). If, however, the absorbed energy exceeds the capacity both to utilize this energy through photochemistry as well as exceeds the capacity for NPQ, the result is irreversible photoinhibition or "photodamage" due to the greater rate of destruction versus repair of the D1 PSII reaction center protein (Melis,

1999). To protect themselves from photodamage, photosynthetic organisms must attempt to decrease excitation pressure by balancing the energy absorbed through photochemistry versus the energy either utilized through metabolism or dissipated through NPQ. It follows that the acclimation response to these environmental conditions should reflect the mechanisms by which photosynthetic organisms attempt to reduce excitation pressure in order to re-establish a balance in energy budget, that, is, re-establish photostasis. Below we discuss briefly specific examples of acclimation in green algae, cyanobacteria and higher plants, which support this hypothesis.

6.1. Photoacclimation

Adjustments to the structure and function of the photosynthetic apparatus in response to changes in growth irradiance are called photoacclimation. One mechanism of photoacclimation involves the modulation of the size and composition of the light harvesting antennae of PSI and PSII (Melis, 1998). Generally, there is an inverse relationship between growth irradiance and light harvesting antenna size. Thus, low growth light promotes large PSI and PSII light harvesting antenna size whereas growth at high light generates a small photosynthetic unit size. Recently, Escoubas et al. (1995) showed that the modulation of the size of LHCII is the consequence of regulation of the nuclear *Lhcb* gene family by the redox state of the PQ pool in the chloroplast thylakoid membrane in *Dunaliella tertiolecta*. When the PQ pool is reduced by exposure to high light, the transcription of the *Lhcb* genes is downregulated, which results in a decrease in the size of LHCII. This photoprotective mechanism is consistent with the notion that photostasis in response to high light may be attained through modulation of σ_{PSII}.

It is now established that the xanthophyll, zeaxanthin, is involved in the nonphotochemical dissipation of excess light (Demmig-Adams and Adams, 1996). The xanthophyll cycle consists of the light-dependent conversion of the light harvesting xanthophyll, violaxanthin, to the energy quenching xanthophylls, antheraxanthin and zeaxanthin (Yamamoto and Bassi, 1999). There is now a consensus that a close relationship exists between an increase in the capacity for NPQ, the extent of the thylakoid ΔpH and the increase in xanthophyll cycle activity (Gilmore, 1997). The capacity for NPQ is closely related to the expression of *PsbS*, a gene regulating NPQ in *Arabidopsis thaliana* (Li et al., 2000). Acclimation to high light appears to result in a persistent engagement of the xanthophyll cycle quenching of excess energy through NPQ (Demmig-Adams and Adams, 1996) which leads to photostasis via a decrease in the functional absorption cross-section of PSII (σ_{PSII}) even though the physical size of the light harvesting complex has not changed.

The spectral distribution of the solar radiation reaching the earth is attenuated due to filtering through either aquatic environments (Falkowski, 1983), crop and forest canopies (Bjorkman and Ludlow, 1972), or through a single leaf (Vogelmann et al., 1996). Such attenuation inevitably results in an imbalance in the absorption of light

between PSII and PSI, which causes a decreased efficiency of linear electron transport. In terrestrial plants and green algae, light absorbed preferentially by PSII relative to PSI (state 2) leads to an over-reduction of the PQ pool whereas preferential excitation of PSI relative to PSII (state 1) results in oxidation of the PQ pool. The redox state of the PQ pool regulates a thylakoid protein kinase which controls the phosphorylation state of the peripheral Lhcb antenna polypeptides and energy transfer between PSII and PSI (Allen and Pfannschmidt, 2000). Thus, the regulation of energy transfer by the redox state of the thylakoid PQ pool reflects a photoprotective mechanism to counteract the potential for uneven absorption of light by PSI and PSII by adjustment of σ_{PSII} to maintain photostasis and to ensure maximum photosynthetic efficiency on a short-term basis.

It was first proposed by Fujita et al. (1994) that modulation of photosystem stoichiometry is a response to changes in the redox state of the intersystem electron transport chain to ensure equal rates of electron flow through both PSI and PSII (Allen and Pfannschmidt, 2000). Pfannschmidt et al. (1999) have shown that the transcription of the chloroplast encoded *psbA* which codes for D1 and *psaAB* genes which code for the PSI reaction center polypeptides are controlled by the redox state of the PQ pool. Over-reduction of the PQ pool by the preferential excitation of PSII not only favours energy transfer from PSII to PSI through phosphorylation of LHCII but also favours the activation of *psaAB* transcription and the concomitant repression of *psbA*. Conversely, oxidation of the PQ pool by preferential excitation of PSI not only favours de-phosphorylation of LHCII but also the activation of transcription of *psbA* and the repression of *psaAB* (Pfannschmidt et al., 1999; Allen and Pfannschmidt, 2000). Thus, PQ, the redox sensor that controls state transitions, also appears to be the sensor that regulates chloroplast photosystem stoichiometry.

6.2. Cold acclimation

As predicted, photosynthetic adjustment during cold acclimation of the unicellular green algae

Chlorella vulgaris and *Dunaliella salina* by growth at low temperature and moderate irradiance $5°C/150$ μmol m^{-2} s^{-1} (5/150) mimics photoacclimation of these algal species grown at high light and moderate temperatures (27/2200) (Huner et al., 1998). Cells grown at 5/150 are indistinguishable from those grown at 27/2200 with respect to photosynthetic efficiency, photosynthetic capacity, pigmentation, Lhcb content and sensitivity to photoinhibition. These results are explained on the basis that cultures grown at either 5/150 or 27/2200 indeed are exposed to comparable excitation pressure measured as $1-qP$ (Huner et al., 1998). Similar conclusions regarding the role of excitation pressure have been reported for thermal and photoacclimation of *Laminaria saccharina* (Machalek et al., 1996) and cold acclimation of the filamentous cyanobacterium, *Plectonema boryanum* (Miskiewicz et al., 2000). These results are consistent with the thesis that exposure to low temperature creates a similar imbalance in energy budget as exposure to high light.

These green algal and cyanobacterial species are unable to up-regulate carbon metabolism and thus are unable to adjust the capacity of electron-consuming sinks during growth and development at low temperature (Savitch et al., 1996; Miskiewicz et al., 2000). As a consequence, these organisms are unable to adjust $n \cdot \tau^{-1}$ significantly with respect to changes in growth temperature. Thus, to attain photostasis, these organisms adjust $\sigma_{PSII} \cdot I$ through a reduction in the size of PSII light-harvesting complex coupled with an increased capacity for NPQ which result in a decrease in σ_{PSII}. The redox state of the PQ pool acts as a chloroplastic sensor regulating the expression of the nuclear encoded *Lhcb* genes whereas the transthylakoid ΔpH acts as a sensor regulating xanthophyll activity and hence NPQ (Wilson and Huner, 2000). In contrast, the redox sensor for *Plectonema boryanum* appears to be located downstream of the PQ pool and Cyt b_6/f complex (Miskiewicz et al., 2000). In addition, *Plectonema boryanum* modulates I by accumulating the carotenoid, myxoxanthophyll, in the cell membrane. This carotenoid is a nonphotosynthetic pigment and acts as a natural sunscreen to protect the photosynthetic apparatus from excess light (Miskiewicz et al., 2000).

Cold temperate conifers such as Lodgepole pine (*Pinus contorta* L.) and herbaceous cereals such winter wheat (*Triticum aestivum* L.) and winter rye (*Secale cereale* L.) are representative of some of the most cold tolerant plants that retain their foliage during the autumn and winter (Levitt, 1980). This capacity to cold acclimate is an essential requirement for the development of maximum freezing tolerance, which allows them to survive the freezing temperatures during the winter. However, these two groups of plants exhibit quite different strategies for the utilization of light energy during growth and cold acclimation (Öquist et al., 2001; Savitch et al., 2002). Cold acclimation of conifers induces the cessation of primary growth in contrast to winter cereals, which require continued growth and development during the cold acclimation period to attain maximum freezing tolerance (Fowler and Carles, 1979). In the context of these different growth strategies, the requirement for photosynthetic assimilates also differs considerably. Conifers exhibit a decreased requirement for photosynthetic assimilates upon the induction of dormancy and cold acclimation. In contrast, overwintering cereals maintain a high demand for photoassimilates during cold acclimation.

As a consequence of the decreased sink demand for photoassimilates, that is, a decrease in $n \cdot \tau^{-1}$, conifers exhibit feedback inhibition of CO_2 assimilation (Savitch et al., 2002). To attain photostasis under these conditions, conifers adjust their capacity and efficiency to absorb light by decreasing the content of PSII reaction centers as well as the content of Lhcb polypeptides and their associated pigments. In addition, conifers increase their capacity for NPQ through the up-regulation of *PsbS* and the xanthophyll cycle with the concomitant aggregation of the major light harvesting pigment–protein complexes (Ottander et al., 1995; Savitch et al., 2002). Energetically, this results in a highly quenched state. Since conifers exhibit the capacity to recover fully from this quenched state with the onset of spring (Ottander et al., 1995), this capacity to down-regulate photosynthesis during cold acclimation is an important mechanism for the successful establishment of evergreen conifers in cold temperate and subarctic climates.

In contrast, winter cereals maintain maximum efficiency and capacity for light absorption through the light harvesting complexes and reaction centers of PSII and PSI with a minimum investment in the expression of *PsbS* and the capacity for nonphotochemical quenching of absorbed light energy (Huner et al., 1998; Savitch et al., 2002). This maximizes $\sigma_{PSII} \cdot I$ which should lead to maximum excitation pressure at low temperature. However, excitation pressure is moderated due to an increased capacity for CO_2 assimilation through the up-regulation of transcription and translation of genes coding for Rubisco, the rate-limiting enzyme for photosynthetic CO_2 fixation, as well as regulatory enzymes of cytosolic sucrose and vacuolar fructan biosynthesis (Hurry et al., 1996). This reprogramming of carbon metabolism to match the continued absorption of light energy has a dual function: it not only maximizes the chemical energy and carbon pool necessary for the renewed growth in the spring but the accumulation of photosynthetic end-products such as sucrose provides cryoprotectants to stabilize the cells during freezing events during the winter (Hurry et al., 1996). The response of cereals to excitation pressure extends beyond photosynthesis to include the regulation of plant morphology and freezing tolerance (Huner et al., 1998).

6.3. Nutrient limitations

Growth of higher plants such as sugar beet under Fe deficiency induces a significant decrease in photosynthetic capacity as a consequence of a co-ordinated decrease in the content of LHCII, electron transport components and Rubisco (Terry, 1983; Winder and Nishio, 1995). In addition to the decrease in the apparent size of LHCII, growth under Fe deficient conditions increases the capacity for NPQ (Abadía et al., 2000). Thus, these observations are consistent with the thesis that plants grown under Fe deficient conditions attempt to attain photostasis by decreasing σ_{PSII} by lowering the LHCII content and increasing the xanthophyll cycle activity. Furthermore, the observation that

the content of Rubisco, the CO_2 fixing enzyme, is reduced during growth under Fe deficient conditions ensures that the rate of light absorption ($\sigma_{PSII} \cdot I$) matches the capacity for CO_2 assimilation ($n \cdot \tau^{-1}$).

In the green alga, *Chlamydomonas reinhardtii*, both phosphate and sulfate limitations independently result in a decrease in photosynthetic efficiency ($\sigma_{PSII} \cdot I$) (Wykoff et al., 1998). This is due to the combined effects of increased capacity for xanthophyll cycle-dependent NPQ plus an enhanced energy transfer from PSII to PSI during exposure to nutrient-limited conditions. Thus, the photoacclimation response induced by nutrient limitations in this green alga, at least in part, appears to involve an adjustment of σ_{PSII}.

6.4. Drought

Photorespiration is the light-dependent evolution of CO_2. The photorespiratory pathway is initiated by the fixation of O_2 by Rubisco producing phosphoglycolate, which is metabolized in the photorespiratory pathway to form CO_2 and NH_3. Although it appears that photorespiration could be a wasteful process, photorespiration could serve as an important energy sink ($n \cdot \tau^{-1}$) through its consumption of ATP and NADPH thus preventing the over-reduction of the PQ pool when leaf intercellular CO_2 concentrations are limiting. The protective role of photorespiration may be especially important under stress conditions such as drought when leaf stomates close in the light in order to prevent excess water loss (Wingler et al., 1999). Drought-stress stimulated photorespiratory activity in barley (*Hordeum vulgare* L.) with a concomitant increase in xanthophyll-cycle dependent NPQ (Wingler et al., 1999). Thus, it appears that drought-stressed barley may maintain photostasis through a combination of enhanced sink capacity ($n \cdot \tau^{-1}$) due to the stimulation of photorespiration coupled with a decrease in σ_{PSII}.

7. Summary

A common characteristic of all photoautrophic organisms exposed to changes in environmental factors such as light intensity, temperature, nutrient status or water availability is that they exhibit an imbalance between the energy absorbed through photochemistry relative to the energy utilized through electron transport and cellular metabolism. Regardless of the environmental change, such an imbalance leads to an increased reduction of the PQ pool which reflects increased excitation pressure which can be estimated *in vivo* by pulse amplitude modulated chlorophyll fluorescence spectroscopy. Excitation pressure can be detected whenever $\sigma_{PSII} \cdot I > n \cdot \tau^{-1}$. The predisposition of photosynthetic organisms to attain a balance in energy budget is defined as photostasis. This minimizes excitation pressure and prevents irreversible, photoinhibitory damage. At least two sensing mechanisms related to the PQ pool of the thylakoid membranes appear to be involved in detecting changes in excitation pressure and hence energy imbalance: (1) the redox state of the PQ pool regulates both nuclear and chloroplastic photosynthetic gene expression and (2) the transthylakoid ΔpH, and hence chloroplastic pmf, regulates the xanthophyll cycle and non-photochemical quenching of excess light energy. Both sensing mechanisms operate coincidentally to protect the photosynthetic apparatus from chronic photodamage. Thus, the photosynthetic apparatus appears to have a dual function: not only does it function as the traditional energy transformer but it also functions as primary environmental sensor. This is consistent with the notion of a "grand design" for photosynthesis (Arnon, 1982; Anderson et al., 1995). Furthermore, the concept of energy sensing through electron transport represents a mechanism by which both heterotrophic and photoautotrophic organisms may acclimate to a changing environment (Fig. 17.1).

Acknowledgements

NPAH and GÖ are grateful for financial support from NSERCC and NFR respectively and acknowledge the support of the Swedish Foundation for International Cooperation in Research and Higher Education.

References

Abadía, J., Morales, F. and Abadía, A. (2000). Photosystem II efficiency in low chlorophyll, iron deficient leaves. Plant and Soil 215, 183–192.

Allen, J.F., Alexciev, K. and Hakansson, G. (1995). Regulation of redox signalling. Current Biol. 5, 869–872.

Allen, J.F. and Pfannschmidt, T. (2000). Balancing the two photosystems: photosynthetic electron transfer governs transcription of reaction center genes in chloroplasts. Phil. Trans. Roy. Soc. Lond. B 355, 1351–1359.

Anderson, J.M., Chow, W.S. and Park, Y.-I. (1995). The grand design of photosynthesis: acclimation of the photosynthetic apparatus to environmental cues. Photosynth. Res. 46, 129–139.

Arnon, D.I. (1982). Sunlight, earth life: the grand design of photosynthesis. The Sciences 22, 22–27.

Bauer, C.E., Elsen, S. and Bird, T.H. (1999). Mechanism for redox control of gene expression. Ann. Rev. Microbiol. 53,495–523.

Björkman, O. and Ludlow, M.M. (1972). Characterization of the light climate on the floor of a Queensland rainforest. Carnegie Inst. Washington Year Book 71, 85–94.

Demmig-Adams, B. and Adams, W.W. (1996). The role of xanthophyll cycle carotenoids in the protection of photosynthesis. Trends Plant Sci. 1, 21–26.

Durnford, D.G. and Falkowski, P.G. (1997). Chloroplast redox regulation of nuclear gene transcription during photoacclimation. Photosynth. Res. 53, 229–241.

Escoubas, J.-M., Lomas, M., LaRoche, J. and Falkowski, P.G. (1995). Light intensity regulates cab gene transcription via the redox state of the plastoquinone pool in the green alga, *Dunaliella tertiolecta*. Proc. Nat. Acad. Sci. USA 92, 10237–10241.

Falkowski, P.G. (1983). Light–shade adaptation and vertical mixing of marine phytoplankton: a comparative field study. J. Mar. Res. 41, 215–237.

Ferguson S.J., Jackson, J.B., and McEwan, A.G. (1987). Anaerobic respiration in the Rhodospirillaceae: characterization of pathways and evaluation of roles in redox balancing during photosynthesis. FEMS Microbiol. Rev. 46, 117–143.

Fowler, D.B. and Carles, R.J. (1979). Growth, development and cold tolerance of fall acclimated cereal grains. Crop Sci. 19, 915–922.

Fujita, Y., Murakami, A., Aizawa, K. and Ohki, K. (1994). Short-term and long-term adaptation of the photosynthetic apparatus: homeostatic properties of thylakoids. In: Advances in Photosynthesis, Vol. 1, The Molecular Biology of Cyanobacteria. (Bryant, D.A., Ed.), pp. 677–692. Kluwer Academic, Dordrecht.

Georgellis, D., Kwon, O., Lin, E.C.C., Wong, S.M. and Akerley, B.J. (2001). Redox signal transduction by the ArcB sensor kinase of *Haemophilus influenza* lacking the PAS domain. J. Bacteriol. 183, 7206–7212.

Gilmore, A.M. (1997). Mechanistic aspects of xanthophyll cycle-dependent photoprotection in higher plant chloroplasts and leaves. Physiol. Plant. 99, 197–209.

Green, B.R. and Durnford, D.G. (1996). The chlorophyll-carotenoid proteins of oxygenic photosynthesis. Ann. Rev. Plant Physiol. Plant Mol. Biol. 47, 685–714.

Grishanin, R.N., Gauden, D.E. and Armitage, J.P. (1997). Photoresponses in *Rhodobacter sphaeroides*: role of photosynthetic electron transport. J. Bacteriol. 179, 24–30.

Haehnel, W. (1984). Photosynthetic electron transport in higher plants. Ann. Rev. Plant Physiol. 35, 659–693.

Hoch, J.A. and Silhavy, T.J. (1995). Two-component signal transduction. ASM Press, Washington, D.C.

Horton, P., Ruban, A. and Walters, R.G. (1996). Regulation of light harvesting in green plants. Ann. Rev. Plant Physiol. Plant Mol. Biol. 47, 655–684.

Huner, N.P.A., Öquist, G. and Sarhan, F. (1998). Energy balance and acclimation to light and cold. Trends Plant Sci. 3, 224–230.

Hurry, V., Huner, N.P.A., Selstam, E., Gardestrom, P. and Öquist, G. (1996). Photosynthesis at low temperatures. In: Photosynthesis: a Comprehensive Treatise. (Raghavendra, A.S., Ed.), pp. 238–249. Cambridge University Press, Cambridge.

Krause, G.H. and Weis, E. (1991). Chlorophyll fluorescence and photosynthesis: the basics. Ann. Rev. Plant Physiol. Plant Mol. Biol. 42, 313–349.

Laszlo, D.J. and Taylor, B.L. (1981). Aerotaxis in *Salmonella typhimurium*: role of electron transport. J. Bacteriol. 145, 990–1001.

Levitt, J. (1980). Responses of Plants to Environmental Stresses. Academic Press, New York.

Long, S.P., Humphries, S. and Falkowski, P.G. (1994). Photoinhibition of photosynthesis in nature. Ann. Rev. Plant Physiol. Plant Mol. Biol. 45, 633–662.

Machalek, K.M., Davison, I.R. and Falkowski, P.G. (1996). Thermal acclimation and photoacclimation of photosynthesis in the brown alga *Laminaria saccharina*. Plant Cell Environ. 19, 1005–1016.

Melis, A. (1998). Photostasis in plants. In: Photostasis and Related Phenomena. (Williams, E. and Thistle, R., Eds.), pp. 207–220. Plenum Press, New York.

Melis, A. (1999). Photosystem-II damage and repair cycle in chloroplasts: what modulates the rate of photodamage *in vivo*? Trends Plant Sci. 4, 130–135.

Melis, A., Manodori, A., Glick, R.E., Ghirardi, M.L., McCauley, S.W. and Neale, P.J. (1985). The mechanism of photosynthetic membrane adaptation to environmental stress conditions: a hypothesis on the role of electron-transport capacity and of ATP/NADPH pool in the regulation of thylakoid membrane organization and function. Physiol. Vég. 23, 757–765.

Miskiewicz, E., Ivanov, A.G., Williams, J.P., Khan, M.U., Falk, S. and Huner, N.P.A. (2000). Photosynthetic accli-

mation of the filamentous cyanobacterium, *Plectonema boryanum* UTEX 485, to temperature and light. Plant Cell Physiol. 41, 767–775.

Öquist, G., Chow, W.S. and Anderson, J.M. (1992). Photoinhibition of photosynthesis represents a mechanism for long term regulation of photosystem II. Planta 186, 450–460.

Ottander, C., Campbell, D. and Öquist, G. (1995). Seasonal changes in photosystem II organization and pigment composition in *Pinus sylvestris*. Planta 197, 176–183.

Pfannschmidt, T., Nilsson, A. and Allen, J.F. (1999). Photosynthetic control of chloroplast gene expression. Nature 397, 625–628.

Romagnoli, S. and Armitage, J.P. (1999). The role of the chemosensory pathways in transient changes in swimming speed of *Rhodobacter sphaeroides* induced by changes in photosynthetic electron transport. J. Bacteriol. 181, 34–39.

Savitch, L.V., Leonardos, E.D., Krol, M., Jansson, S., Grodzinski, B., Huner, N.P.A. and Öquist, G. (2002). Two different strategies for light utilization in photosynthesis in relation to growth and cold acclimation. Plant Cell Environ., in press.

Savitch, L.V., Maxwell, D.P. and Huner, N.P.A. (1996). Photosystem II excitation pressure and photosynthetic carbon metabolism in *Chlorella vulgaris*. Plant Physiol 111, 127–136.

Shioi, J., Galloway, R.J., Niwano, M., Chinock, R.E. and Taylor, B.L. (1982). Requirement of ATP in bacterial chemotaxis. J. Biol. Chem. 257, 7969–7975.

Schrödinger, E. (1944). What is Life? Cambridge University Press, Cambridge.

Taylor, B.L. (1983). Role of proton motive force in sensory transduction in bacteria Role of proton motive force in sensory transduction in bacteria. Ann. Rev. Microbiol. 37, 551–573.

Taylor, B.L. and Zhulin, I.B. (1999). PAS domains: internal sensors of oxygen, redox potential and light. Microbiol. Mol. Biol. Rev. 63, 479–506.

Taylor, B.L. and Zhulin, I.B. and Johnson, M.S. (1999). Aerotaxis and other energy-sensing behaviour in bacteria. Ann. Rev. Microbiol. 53, 103–128.

Terry, N. (1983). Limiting factors in photosynthesis. IV. Iron stress mediated changes in light harvesting and electron transport capacity and its effects on photosynthesis in vivo. Plant Physiol. 71, 855–860.

Urao, T., Yamaguchi-Shinozaki, K. and Shinozaki, K. (2000). Two-component systems in plant signal transduction. Trends Plant Sci. 5, 67–74.

Wilson, K.E. and Huner, N.P.A. (2000). The role of growth rate, redox-state of the plastoquinone pool and the trans-thylakoid ΔpH in photoacclimation of *Chlorella vulgaris* to growth irradiance and temperature. Planta 212, 93–102.

Winder, T.L. and Nishio, J. (1995). Early iron deficiency

stress response in leaves of sugar beet. Plant Physiol. 108, 1487–1494.

Wingler, A., Quick, W.P., Bungard, R.A., Bailey, K.J., Lea, P.J. and Leegood, R.C. (1999). The role of photorespiration during drought stress: an analysis utilizing barley mutants with reduced activities of photorespiratory enzymes. Plant Cell Environ. 22, 361–373.

Wykoff, D.D., Davies, J.P., Melis, A. and Grossman, A.R. (1998). The regulation of photosynthetic electron transport during deprivation in *Chlamydomonas reinhardtii*. Plant Physiol. 117, 129–139.

Yamamoto, H.Y. and Bassi, R. (1996). Carotenoids: localization and function. In: Advances in Photosynthesis, Vol 4, Oxygenic Photosynthesis: the Light Reactions.

(Ort, D.R. and Yocum, C.F., Eds.), pp. 539–563. Kluwer Academic, Dordrecht.

Yu, L., Zhao, J., Mühlenhoff, U., Bryant, D.A. and Golbeck, J.H. (1993). PsaE is required for *in vivo* cyclic electron transport flow around photosystem I in the cyanobacterium *Synechocystis* sp. PCC 7002. Plant Physiol. 103, 171–180.

Zhang, H., Whitelegge, J.P. and Cramer, W.A. (2001). Ferredoxin:NADP$^+$ oxidoreductase is a subunit of the chloroplast cytochrome *b6f* complex. J. Biol. Chem. 276, 38159–38165.

Zhulin, I.B., Taylor, B.L. and Dixon, R. (1997). PAS domain S-boxes in Archaea bacteria and sensors for oxygen and redox. Trends Biochem. Sci. 22, 331–333.

Sensing, Signaling and Cell Adaptation. Edited by K.B. Storey and J.M. Storey
© 2002 Elsevier Science B.V. All rights reserved.

CHAPTER 18

The Uncoupling Proteins Family: From Thermogenesis to the Regulation of ROS

Marie-Clotilde Alves-Guerra, Claire Pecqueur, Alison Shaw[1], Elodie Couplan, Maria Del Mar Gonzalez Barroso, Daniel Ricquier, Frédéric Bouillaud and Bruno Miroux*
CEREMOD (UPR 9078 CNRS), Meudon, and Institut de Recherches Necker-Enfants Malades, Paris France
[1]*EMBL, Heidelberg Germany*

1. Introduction

Over the last 20 years the mitochondrial uncoupling proteins have been regarded as candidate genes for nonshivering thermogenesis, diet-induced thermogenesis, resting metabolic rate, and body mass control. The first uncoupling protein to be discovered, UCP1, indeed plays a crucial role in cold adaptation of mammals and in the body mass control of rodents. The discovery of UCP1 homologues named UCP2, UCP3 and the plant stUCP, strongly stimulated research in the field of mitochondrial bioenergetics. It is still unclear whether the new UCP1 homologues have a physiological uncoupling activity and therefore contribute to the proton leak observed in all mitochondria. However, recent data led to the proposal that the uncoupling proteins may have an additional function: the regulation of mitochondrial reactive oxygen species. This chapter presents the uncoupling protein family and reviews data suggesting a role for the UCPs in oxidative stress situations.

2. UCP1 belongs to the mitochondrial anion carriers family

2.1. Historical background

Mitochondria contain two compartments bounded

by the inner and the outer membranes. Due to nonspecific pore proteins, the outer membrane is permeable to many small metabolites. On the other hand the inner membrane needs a strict control of its permeability since the maintenance of the high electrochemical gradient created by mitochondrial respiratory chain is necessary for energy conservation and ATP synthesis in mitochondria. Therefore, the inner membrane does not contain porins and metabolite transport occurs through specialized mitochondrial carriers operating in an energy conservative way. They transport anion substrates such as ADP, ATP, phosphate, oxoglutarate, citrate, glutamate and malate. The transport of these metabolites is often coupled to protons probably because the high membrane potential generated by the respiratory chain (negative inside) opposes anion transport into mitochondria. These proteins play an important role in several metabolic pathways including gluconeogenesis, urea synthesis, and the citric acid cycle (Fig. 18.1) (for review, Kramer and Palmieri, 1989). The ADP/ATP carrier was first sequenced in 1982 by Klingenberg's laboratory (Aquila et al., 1982) and a few years later the amino acid sequence of the uncoupling protein was obtained both by protein sequencing (Aquila et al., 1985) and by isolation of UCP1 cDNA in our laboratory (Bouillaud et al., 1986). The peptide sequence analysis of both carriers revealed a threefold repeat structure of a hundred

---

*Corresponding author.

amino acids (Saraste and Walker, 1982), each repeat containing a conserved energy sequence signature (PROSITE accession number PS00215). The complete sequencing of the yeast genome led to the identification of at least 35 members of the family, most of them of unknown function (el Moualij et al., 1997). All members of the family share about 30% amino acid identity, are organized in six membrane spans (Miroux et al., 1993), two per repeat, and probably act as a dimer (Schroers et al., 1998). Among these carriers, the uncoupling proteins form a subfamily of mitochondrial carriers that are not expressed in all mitochondria and that seem to uncouple oxidative phosphorylation by facilitating the re-entry of protons into the mitochondrial matrix.

2.2. UCP1: the uncoupling protein of brown adipose tissue

UCP1 is uniquely expressed in brown adipose tissue (BAT), a thermogenic organ present mainly in the interscapular area but also in the peri-renal and peri-aortic region of all rodents and hibernating mammals. In other mammals, including humans, the BAT and UCP1 protein disappear shortly after birth. In response to cold, mammals can produce heat by two separate mechanisms: shivering by their skeletal muscles or metabolic thermogenesis which is assessed by the rate of oxygen consumption. Brown adipocytes are specialized for metabolic thermogenesis which is also called non-shivering thermogenesis. As opposed to white adipocytes, brown adipocytes exhibit considerable metabolic activity that is reflected by a very high number of mitochondria and many small drops of triglycerides. Under exposure to the cold, sympathetic nerves activate brown adipocyte specific β3 adrenergic receptors which in turn stimulate the transcription of the UCP1 gene and the lysis of triglycerides into free fatty acids. Free fatty acids are converted to acyl-CoA and reach the mitochondrial matrix with the help of the carnitine/acyl-carnitine transporter. The β-oxidation of acyl- CoA produces acetyl-CoA that is fed into the tricarboxylic acid cycle. The overall process generates reduced cofactors like NADH that are re-oxidized by complex I of the respiratory chain. Electron transport through the respiratory complexes II, III, and IV is coupled to the formation of an electrochemical proton gradient across the inner membrane. This proton gradient is used by some transporters but primarily by the FoF1-ATPase for ATP synthesis (chemio-osmotic theory of Peter Mitchell). In conventional cells, catabolic pathways such as β-oxidation of fatty acids or glycolysis are under the control of ATP and cofactor levels. In brown adipocytes, UCP1 acts as a proton carrier, dissipates the electrochemical proton gradient and uncouples the re-oxidation of cofactors by the respiratory chain from ATP synthesis. Once UCP1 is activated, lipid oxidation is no longer repressed by the levels of NADH or ATP. The blood flow through brown fat increases, brown adipocytes collect substrates and oxygen at a high rate from blood, and the energy derived from the intense catabolic activity in BAT is not converted into ATP but is, instead, dissipated as heat in the blood of the animal (Fig. 18.1). Hibernators use metabolic thermogenesis in order to raise their body temperature to 37°C in just a few minutes as they arouse from hibernation. BAT activity also controls the body mass of non-hibernating rodents which decreases in winter when cold adaptation requires metabolic thermogenesis. The efficiency of the respiratory uncoupling by UCP1 relies on its abundance in the inner membrane. UCP1 can reach up 5% of the total mitochondrial protein of cold-adapted rats. The protein is also tightly regulated. Purine nucleotides are inhibitors whereas free fatty acids and retinoïc acid strongly activate the uncoupling activity of UCP1 (Rial et al., 1999; Rial and Nicholls, 1983). Finally, genetic ablation of brown adipose tissue and deletion of the UCP1 gene in mice confirmed the essential role of BAT and of UCP1 in cold-adapted thermogenesis (Enerback et al., 1997; Lowell et al., 1993). In humans, UCP1 protein disappears shortly after birth but is often detected in pathologies such pheochromocytoma or hibernoma tumours. However, given its influence on whole body energy expenditure, UCP1 remains an interesting therapeutic target for the treatment of obesity (for a complete review see Gonzalez-

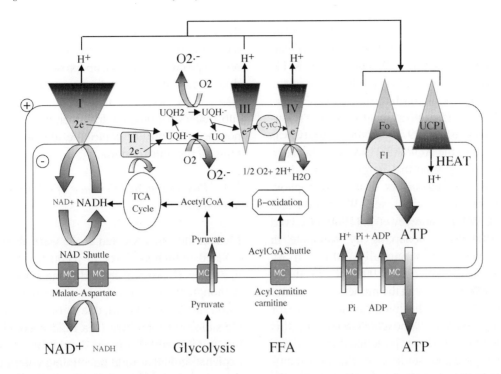

Fig. 18.1. A simple view on mitochondrial metabolism. Mitochondria play a central role in cofactor regeneration and ATP production. NADH and FADH are oxidized by the complexes I and II of the respiratory chain, respectively. Electron transport through the respiratory chain leads to the formation of superoxide anions ($O_2{}^{\cdot -}$) at the level of the semi-ubiquinone ($UQH^{\cdot -}$). By pumping out protons, complexes I, III and IV of the respiratory chain generate an electrochemical proton gradient ($\Delta\mu H^+$) which is used by the F1Fo ATP synthase to produce ATP, or in brown adipocytes, by uncoupling protein 1 to produce heat. Mitochondria also influence, via the oxidation of cofactors, the ATP produced by glycolysis. Mitochondrial carriers (MC) transport key metabolites for glycolysis, the oxidation of free fatty acids (FFA), and ATP synthesis in and out of the mitochondria.

Barroso et al., 2000). Figure 18.1 illustrates the role of UCP1 and of other mitochondrial transporters in mitochondria. In this simple view, the role of the mitochondria is to make ATP, mitochondrial carriers help ATP synthesis by transporting adequate substrates and UCP1 generates a proton leak in brown adipocytes for a clear physiological purpose: cold-adapted thermogenesis.

2.3. The mitochondrial carriers: a multifunctional protein family?

Over the last ten years many experimental data coming from various laboratories have changed our conventional view of mitochondria and especially of the mitochondrial carriers family. First of all, it appeared that mitochondria are not only devoted to ATP synthesis but that they also actively participate in apoptosis or programmed cell death (for review see Loeffler and Kroemer, 2000). For instance, Petit et al. (1995) found that a drop of mitochondrial membrane potential is an early event of dexamethasone-induced apoptosis in thymocytes and that it occurs before the fragmentation of DNA. This is due to the formation of the mitochondrial transition pore (MTP) (for review see Crompton, 1999). The exact composition of the MTP is still a matter of debate but the ADP/ATP carrier is believed to be a part of it because carboxyatractyloside (CAT) and bongkrekic acid, two highly specific inhibitors of the ADP/ATP carrier, regulate the opening of the MTP. Thus, a substrate-specific exchanger like the ADP/ATP carrier can be converted into a nonspecific pore inside a large protein complex. Brustovetsky and Klingenberg (1996) also showed that in proteoliposomes, the ADP/ATP carrier could be converted into a large channel by addition of calcium which is also an activator of the MTP. Intriguingly, deletion of nine amino acids in the conserved motif of the sixth α-helix of UCP1 also converted the protein into a nonspecific pore (Gonzalez-Barroso et al., 1997). Another activity was also assigned to the ADP/ATP carrier. It has been known for a long time that long chain free fatty acids (FFA) are uncouplers of oxidative phosphorylation (for review see Wojtczak and Schonfeld, 1993). Andreyev et al. (1989) observed that mitochondrial uncoupling by palmitate in skeletal muscle was partially inhibited by bongkrekic acid. Skulachev (1991) proposed that fatty acids cycling through the mitochondrial inner membrane could be a physiological mechanism for uncoupling oxidative phosphorylation. In this model, free fatty acids are protonated in the intermembrane space and rapidly cross the inner membrane due to their long hydrophobic carbon tails. The basic pH of the matrix side facilitates the de-protonation of free fatty acids but in their anionic form, FFA cannot cross the inner membrane again without the help of anion transporters such as the ADP/ATP carrier. Schonfeld et al. (1996) were able to confirm this model and the involvement of the ADP/ATP carrier by using azido derivatives of long chain fatty acids. Brustovetsky and Klingenberg (1994) showed that in proteoliposomes, the ADP/ATP carrier mediates a FFA dependent uncoupling activity. It was later shown that the dicarboxylate carrier (Wieckowski and Wojtczak, 1997), the aspartate/glutamate antiporter, and the phosphate carrier (Zackova et al., 2000) also trigger fatty acid mediated uncoupling. The groups of Jezeck and Garlid proposed that the protonophoric activity of UCP1 is uniquely due to its fatty acid transport activity (Garlid et al., 1996), a hypothesis that is still matter of debate (Gonzalez-Barroso et al., 1998; Jaburek et al., 1999). Overall, these data suggest that all the mitochondrial carriers may be able to work in three different modes: (i) as a specific anion substrate transporter, (ii) as a nonspecific long chain free fatty acid transporter, (iii) as a nonspecific pore (for review see Bouillaud et al., 2001). Depending of the cell type,

the ratio between FFA and their specific substrates, the calcium concentration, the status of the cell (at rest, in proliferation, or in apoptosis) the mitochondrial carriers may switch from one conformation to another. By doing so, they would either provide substrates for mitochondrial metabolism, act as a nonspecific pore, or contribute to a mild uncoupling of the mitochondria.

2.4. Physiological relevance of a "mild uncoupling"

Most of the data reviewed above were obtained on isolated mitochondria or in reconstituted proteoliposomes (Brustovetsky and Klingenberg, 1994). Although there is no doubt about the utility of anion substrate transport, one can wonder what the physiological relevance of a mild uncoupling by the mitochondrial carriers could be. A simple explanation is that mild uncoupling contributes to the proton leak observed in all mitochondria. Proton leak is defined as a non-energy coupled proton conductance and accounts for 20–40% of the standard metabolic rate of endothermic and ectothermic vertebrates and invertebrates (for review see Stuart et al., 2001). Although proton leak does not explain endothermia *per se*, a strong increase in proton conductance such as the one mediated by UCP1 in BAT ultimately leads to production of heat. Similarly, the thermogenic effect of hyperthyroidism has been at least partially explained by an increase in ADT/ATP carrier expression and, consequently, an increase in FFA sensitive proton conductance (Schonfeld et al., 1997). Another important physiological role of mitochondrial uncoupling was first postulated by Skulachev (1996) as a means of preventing an excessive production of reactive oxygen species (ROS).

3. The discovery of the uncoupling proteins homologues

3.1. Historical background

In recent years, we and other laboratories have observed signals in Western blots which suggested that UCP1 was expressed in tissues other than BAT such as liver (Shinohara et al., 1991) and muscle (Yoshida et al., 1998). These results turned out to be misleading (Ricquier et al., 1992). Given that consensus sequences exist for all mitochondrial carriers and given the size of the family, it is not surprising to observe cross-reactivity in Western and Northern blot experiments. Nevertheless, some authors still think that UCP1 is specifically expressed in tissues other than BAT (Nibbelink et al., 2001). For other laboratories, these conflicting results strongly suggested the existence of UCP1 homologues. In 1997 our laboratory found the first mammalian UCP homologue, UCP2 (Fleury et al., 1997) and collaborated on the characterisation of a cold-induced plant UCP homologue (Laloi et al., 1997). The same year, UCP3 was also discovered (Boss et al., 1997; Vidal-Puig et al., 1997). All these proteins share about 60% sequence similarity with UCP1 but only 30% sequence similarity with the other mitochondrial carriers (for a complete review see Bouillaud et al., 2001). Interestingly, UCP2 and UCP3 genes are only separated by a few kilobases in the genome suggesting that they arose by a duplication event (Pecqueur et al., 1999). The discovery of the UCP homologues was acknowledged with great excitement because they were seen as important pathways for the regulation of energy expenditure in man.

3.2. In vivo *distribution of the UCP homologues*

In contrast to UCP3 which is expressed only in muscle and brown adipose tissue, expression of UCP2 mRNA seems to be almost ubiquitous. It is highly abundant in spleen, thymocyte, lung, stomach, white adipose tissue (WAT), and macrophages. Small amounts of mRNA are also found in muscle, kidney, brain, testis, liver and heart (for review see Ricquier and Bouillaud, 2000). However, at the protein level, the situation is more complicated. Initially, immunological data obtained with anti-UCP1 antibodies in these tissues were ascribed to UCP2 or UCP3 (Larrouy et al., 1997; Liu et al., 1998; Negre-Salvayre et al., 1997). However, it gradually appeared that none of the

numerous commercially available UCP2 antipeptide antibodies were able to reproduce the same pattern of UCP2 protein expression. In order to circumvent this problem, we expressed the full length UCP2 in *E. coli*, purified the protein and injected it into rabbits. Mitochondria isolated from UCP2(–/–) mice allowed us to assess the specificity of all anti-UCP2 antibodies. We showed that all UCP2 antibodies cross-react with other more abundant carriers, and that only the antibody against the full length UCP2 was sensitive enough to detect the protein "in vivo" (Pecqueur et al., 2001). At present, UCP2 protein is easily detected in spleen, lung, intestine, stomach, macrophages, pancreas and thymus. It is now clear that both UCP2 and UCP3 are expressed at extremely low levels compared with the expression of UCP1 in BAT. For instance the UCP2 content in spleen is 0.6% of the amount of UCP1 in BAT (Pecqueur et al., 2001). Similarly, Harper and colleagues found that the level of UCP3 in muscle is between 0.1-0.5% of the level of UCP1 in warm adapted hamster BAT mitochondria (Harper et al., 2002). Interestingly, both UCP2 and UCP3 are expressed in tissues where ATP mainly comes from glycolysis and not from mitochondrial respiration. For instance UCP3 seems to be more highly expressed in anaerobic type 2b muscle fibers than in aerobic fibers (Hesselink et al., 2001) and UCP2 is expressed in cells and tissues that do not contain a high amount of mitochondria.

3.3. *The uncoupling activity of the UCP1 homologues in heterologous expression systems*

3.3.1. *Expression in the yeast* Saccharomyces cerevisiae

The putative uncoupling activity of both UCP2 and UCP3 protein is still a matter of controversy and debate. It was first advocated that, despite their high sequence homology with UCP1, they were lacking two conserved histidines residues essential for the uncoupling activity of UCP1 (Bienengraeber et al., 1998). However, when overexpressed in yeast UCP2 decreases the membrane potential of intact yeast cells (Fleury et al., 1997) and stimulates the oxygen consumption of isolated yeast mitochondria when stimulated by retinoic acid (Rial et al., 1999). Arguing the lack of UCP2 inhibition by GDP, Martin Brand and colleagues suggested that, depending of the level of expression of all UCPs in yeast, the uncoupling of mitochondria by the recombinant protein could be an artifact (Harper et al., 2002; Stuart et al., 2001; Stuart et al., 2001). It is certainly true that expression of membrane proteins at high levels can lead to inclusion body formation or to a protein misfolded into the membrane as shown for UCP3 (Heidkaemper et al., 2000). Nevertheless, with the appropriate expression vector and growth conditions, expression in yeast of UCP1 and UCP2 led to the discovery of highly specific activators (Rial et al., 1999). Therefore, the yeast expression system might be suitable for high throughput screening of UCPs regulatory ligands.

3.3.2. Reconstitution into liposomes

Since both UCP2 and UCP3 are expressed at very low levels in vivo, they could not be purified in the same way as UCP1 from BAT mitochondria. Therefore several groups over-expressed the UCPs in *E. coli* with the objective to restore their uncoupling activity into liposomes. Jaburek and coworkers demonstrated first that UCP2 and UCP3 behave identically to UCP1 in liposomes except for the regulation by nucleotides (Jaburek et al., 1999). In agreement with the results obtained on yeast mitochondria they found that nucleotides had a very low affinity for UCP2 and UCP3. On the other hand, Echtay and colleagues recently showed that coenzyme Q (ubiquinone) co-purifies with UCP1, and is an essential cofactor for UCP1 proton transport activity (Echtay et al., 2000). By adding ubiquinone to the purified protein from *E. coli*, they also demonstrated that UCP1, UCP2 and UCP3 have almost the same protonophoric uncoupling activity and, in contrast to the preceding studies, show very similar regulation by purine nucleotides (Echtay et al., 2001).

3.3.3. Overexpression of the UCPs in mammalian cell lines and transgenic mice

In order to determine the influence of UCPs on the

cell metabolism, several groups used adenovirus-mediated overexpression to produce either UCP2 or UCP3 in various cell types. Overexpression of UCP2 in normal rat islet decreased their ATP content by 50% and inhibited glucose stimulated insulin secretion (GSIS) (Chan et al., 1999, 2001) whereas in islets of Zucker diabetic rats UCP2 overexpression decreased ATP and ADP concentrations, increased the ATP/ADP ratio and improved GSIS (Wang et al., 1999). Both studies support the uncoupling effect of UCP2 on islet mitochondria which in turn stimulates glucose and fatty acid oxidation. The apparent contradiction of effects on GSIS between the two studies can be explained by the fact that in Zucker fatty rats the ATP level and ATP/ADP ratio are twice as high as compared with lean Zucker rats. Wang et al. (1999) also suggested that by stimulating glucose usage, UCP2 might lower the abnormal level of glucose metabolic products thereby restoring the sensitivity of the islet to extracellular glucose.

On the other hand, overexpression of UCP3 in muscle cells lowered the membrane potential, and surprisingly elevated the ATP/ADP ratio which in turn stimulated glucose and fatty acid oxidation (Garcia-Martinez et al., 2001; Huppertz et al., 2001). Overexpression of either UCP1 or UCP3 in muscle of transgenic mice also stimulated substrate oxidation and prevents diet-induced obesity (Clapham et al., 2000; Li et al., 2000). Finally, it appears that, when expressed at the level of UCP1 by genetic manipulations, both UCP2 and UCP3 are able to decrease the membrane potential, uncouple respiration from ATP synthesis and increase substrate oxidation. However, immunological studies showed that UCP2 and UCP3 probably never reach the level of UCP1 in BAT.

In a recent study, Fink and coworkers, measured the UCP2 dose-dependent proton leak in adenovirus infected INS-1 cells (Fink et al., 2002). The smallest proton leak they observed in isolated mitochondria still required a level of UCP2 protein that was far above its *in vivo* level. Therefore, the uncoupling activity observed after overexpression of UCP2 and UCP3 (Cadenas et al., 2002) might not be of physiological relevance.

4. Lessons from the UCP knockout mice

4.1. UCPs modify mitochondrial energy metabolism but not whole body energy expenditure

Mice lacking UCP1 are cold intolerant but not obese (Enerback et al., 1997) whereas brown adipose tissue ablated mice develop obesity (Lowell et al., 1993). This shows the importance of BAT in the energetic metabolism of the animals but also that UCP1 *per se* is not sufficient to unbalance body weight regulation. Therefore, it is not surprising that UCP2 and UCP3 knockout mice are not obese even after being fed a high fat diet. They are also resistant to exposure to the cold which confirms BAT as the major organ for cold-adapted thermogenesis. In contrast to the overexpression studies, the absence of a metabolic disorder in the knockout mice suggests that UCP2 and UCP3 proteins do not significantly contribute to the basal metabolic rate. However a reduced proton leak was observed in mitochondria isolated from *Ucp3(–/–)* skeletal muscle (Gong et al., 2000; Vidal-Puig et al., 2000) and in intact thymocytes cells of *Ucp2(–/–)* mice (Krauss et al., 2002), compared with wild type mice. Cline et al. (2001) also showed that under fasting conditions, when UCP3 protein is increased in skeletal muscle, the coupling of oxidative phosphorylation is 2-4 fold higher in skeletal muscle of *Ucp3(–/–)* mice compared with wild type mice. This is due to an increase in the ATP production rate and not to an increase in TCA cycle rate, suggesting that regeneration of NADH might have priority over ATP homeostasis (Cline et al., 2001). Finally, these data show that UCPs are able to alter mitochondrial functions but that none of them regulates the whole body energy expenditure. However, when working on gene deficient mice, compensatory mechanisms cannot be excluded as well as the effect of heterosis. For instance, Hoffman et al. (2001) showed in a very elegant series of experiments, that congenic C57BL/6J and 129/SvImj *Ucp1(–/–)* mice are cold sensitive whereas *Ucp1(–/–)* mice on the F1 hybrid background are cold resistant. Interestingly, whatever the genetic background of the *Ucp1(–/–)* mice, their mito-chondria isolated from BAT exhibit the same FFA mediated reduction in proton leak compared with the *Ucp1(+/+)* mice, showing that on the hybrid background, cold adaptation is achieved by an unknown compensatory mechanism (Hofmann et al., 2001). Since UCP2 and UCP3 deficient mice were also studied on a hybrid background, it is probably too early to draw definitive conclusions on their contribution to the whole body energy metabolism.

4.2. Improvement of β cells function in Ucp2(–/–) mice

The group of Brad Lowell found that *Ucp2(–/–)* mice have higher serum insulin and lower blood glucose levels compared with the wild type mice (Zhang et al., 2001). They also measured a small, but significant, increase in ATP levels in isolated pancreatic islets from *Ucp2(–/–)* mice which correlates with greater insulin secretion by the islets in response to glucose (GSIS). Similarly, *Ucp2(–/–)* mice showed a greater GSIS compared with wild type. Moreover, the lack of UCP2 in *ob/ob* mice increases the level of serum insulin, decreases blood glucose levels and significantly improves insulin secretion 15 minutes after glucose injection. The authors concluded that UCP2 is a negative regulator of insulin secretion and a major link between obesity, β cell dysfunction and type II diabetes (Zhang et al., 2001). Since insulin secretion is dependent on the activity of the ATP sensitive potassium channel, these important results strongly support the uncoupling activity of UCP2 in lowering the ATP/ADP ratio. However, given that UCP2 is expressed in many tissues, one cannot rule out other indirect effects of the absence of UCP2 on glucose tolerance. It should also be noted that although UCP2 is up-regulated in pancreatic islets of *ob/ob* mice (Zhang et al., 2001), the level of serum insulin is 33-fold higher in *ob/ob* mice compared with wild type.

4.3. Improvement of macrophage oxidative burst in Ucp2(–/–) mice

In collaboration with Sheila Collins (Duke

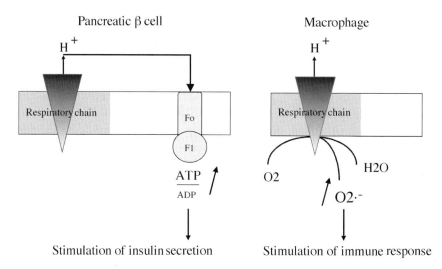

Fig. 18.2. Phenotypes observed in Ucp2(–/–) mice. In pancreatic β cells, Zhang et al. (2001) showed an increase in glucose stimulated insulin secretion which correlated with an increase in the ATP/ADP ratio. In macrophages, the lack of UCP2 led to an increase in the level of reactive oxygen species and improved the toxoplasmacidal function of the cell (Arsenijevic et al., 2000). Although both observations are consistent with a mild uncoupling function of UCP2, its transport activity is not fully understood.

University) and Denis Richard (Quebec) we also generated *Ucp2(–/–)* mice. Since UCP2 protein is expressed in organs of the immune system, *Ucp2(–/–)* mice were challenged for the responsiveness of their immune system. After infection with *Toxoplasma gondii*, an intracellular protozoan brain parasite, all *Ucp2(–/–)* mice survived over the 80-days duration of the experiment whereas their wild type littermates succumbed to infection within 28 to 51 days (Arsenijevic et al., 2000). It is known that ROS produced by macrophages have a potent toxoplasmacidal effect. Consistent with the phenotype of the mice, macrophages isolated from *Ucp2(–/–)* mice produced more ROS compared with wild type macrophages, when infected *in vitro* by *T. gondii*. Moreover, addition of L-histidine, a ROS quencher, abolished the *in vitro* toxoplasmacidal activity of UCP2(–/–) macrophages (Arsenijevic et al., 2000). These results suggest that UCP2 not only decreases the ATP level as described by the group of Lowell but also limits the production of ROS in mitochondria (Fig. 18.2). This finding also implies that, when secreting insulin, pancreatic β cells in which UCP2 content should be low, would then produce more ROS (for review see Langin, 2001).

5. UCP family, from thermogenesis to the regulation of ROS

5.1. UCP1, a misleading model

UCP1 was the first uncoupling protein to be isolated and was, over the last years, a conceptual model for studying the newly discovered UCPs. Unfortunately, the UCP1 model has been misleading for several reasons. Firstly, UCP1 is abundantly expressed in BAT, a thermogenic organ with a very high content of mitochondria. In contrast UCP2 and UCP3 are expressed at low levels and, at least for UCP2, in cells containing a small number of mitochondria. Secondly, results obtained after heterologous expression of UCP1 were mostly consistent with its behaviour *in vivo* whereas over-expression of UCP2 or UCP3 led to misleading conclusions, precisely because *in vivo* they are expressed at low levels. Thirdly, the concept of uncoupling mitochondrial respiration for heat production could not be applied to UCP2 and UCP3 and instead these proteins seem to have a more discrete and subtle function in cell metabolism. In fact, the only function that all the UCPs seem to share is their ability to influence the production of reactive oxygen species. Inhibition of

UCP1 by addition of GDP in BAT mitochondria results in an strong increase in ROS production (Negre-Salvayre et al., 1997) and macrophages from UCP2(–/–) mice and muscle mitochondria from UCP3(–/–) mice produce more ROS compared with wild type mice (Arsenijevic et al., 2000; Vidal-Puig et al., 2000).

5.2. UCP2 and UCP3: new players in the mitochondrial redox balance?

Reactive oxygen species such as O_2^{-}, OH^{-}, or $ONOO^{-}$ are oxygen derivative molecules containing a highly reactive free electron. As electron donor species, they cause peroxidation of lipids, damage DNA, or inactivate enzymes such as the aconitase by oxidizing the Fe-S reaction center. Even at moderate levels they are extremely toxic for the mitochondria because DNA damage results in a progressive dysfunction of the respiratory chain which in turn further increases the level of ROS. However, at a low level, they are considered to be signaling molecules that stimulate the transcription of genes and facilitate cell proliferation (Sauer et al., 2001). Mitochondria are an important source of ROS production at the level of complexes I (NADH/ubiquinone oxidoreductase) and III (ubiquinol/cytochrome c oxidoreductase) of the electron transport chain (for review see Turrens, 1997). When the electrochemical proton gradient ($\Delta\mu H^+$) increases, the entry of electrons into the respiratory chain slows down and the half-life of the semiquinone UQH^{-} increases. As an excellent electron donor, UQH^{-} reacts with oxygen to produce ROS probably on both sides of the inner membrane (Fig. 18.1). This probably occurs in state 4 respiration or during the state 3 to state 4 transition when the $\Delta\mu H^+$ increases above its ROS producing threshold value. Skulachev (1997) hypothesized that a mild uncoupling would prevent the unwanted increase of $\Delta\mu H^+$ and therefore the exponential rise of ROS. It is therefore tempting to explain the knockout phenotypes of the UCPs by the mild uncoupling hypothesis of Skulachev. Other data also support the regulatory function of UCP2 and UCP3 on ROS. For instance, injection of lipopolysaccharides (LPS) in mice,

which causes an oxidative stress mediated by neutrophil infiltration in lung, also strongly increases UCP2 protein in lung mitochondria (Pecqueur et al., 2001). In contrast, addition of LPS to isolated macrophages decreases the level of UCP2 mRNA (Arsenijevic et al., 2000) which is consistent with the increase in ROS observed in stimulated macrophages. Regarding UCP3, Huppertz et al. (2001) showed that stimulation of glucose uptake in cells over-expressing UCP3 occurs through a phosphoinositide 3-kinase-dependent (PI3K) mechanism. They suggested that the PI3K could be activated by ROS and not by a simple uncoupling of mitochondrial regulation. In addition, Echtay and co-workers (2002) showed that superoxide activates all three uncoupling proteins. These striking results reinforce the hypothesis that the UCPs have a protective function against ROS, a function that would be regulated directly by the level of ROS. Given the wide tissue distribution of the UCPs and the multiple roles of ROS in inflammatory response, diabetes, cell proliferation and aging (Finkel and Holbrook, 2000), it is likely that UCPs play a role in various physiological situations like acute lung injury, inflammatory bowel disease, and neuro-degenerative diseases.

Acknowledgements

Our work is supported by Centre National de la Recherche Scientifique, Institut National de la Santé et de la Recherche Médicale, Association de Recherches sur le Cancer (to F.B.), Institut de recherche Servier (to D.R.). M.C.A.G. and E.C. are supported by Nestlé and Servier respectively.

References

Andreyev, A., Bondareva, T.O., Dedukhova, V.I., Mokhova, E.N., Skulachev, V.P., Tsofina, L.M., Volkov, N.I. and Vygodina, T.V. (1989). The ATP/ADP-antiporter is involved in the uncoupling effect of fatty acids on mitochondria. Eur. J. Biochem. 182, 585–592.

Aquila, H., Link, T.A. and Klingenberg, M. (1985). The uncoupling protein from brown fat mitochondria is related to the mitochondrial ADP/ATP carrier. Analysis of sequence homologies and of folding of the protein in the membrane. EMBO J. 4, 2369–2376.

Aquila, H., Misra, D., Eulitz, M. and Klingenberg, M. (1982). Complete amino acid sequence of the ADP/ATP carrier from beef heart mitochondria. Hoppe Seylers Z. Physiol. Chem. 363, 345–349.

Arsenijevic, D., Onuma, H., Pecqueur, C., Raimbault, S., Manning, B.S., Miroux, B., Couplan, E., Alves-Guerra, M.C., Goubern, M., Surwit, R., Bouillaud, F., Richard, D., Collins, S. and Ricquier, D. (2000). Disruption of the uncoupling protein-2 gene in mice reveals a role in immunity and reactive oxygen species production. Nature Genet. 26, 435–439.

Bienengraeber, M., Echtay, K.S. and Klingenberg, M. (1998). H+ transport by uncoupling protein (UCP-1) is dependent on a histidine pair, absent in UCP-2 and UCP-3. Biochemistry 37, 3–8.

Boss, O., Samec, S., Paoloni-Giacobino, A., Rossier, C., Dulloo, A., Seydoux, J., Muzzin, P. and Giacobino, J.P. (1997). Uncoupling protein-3: a new member of the mitochondrial carrier family with tissue-specific expression. FEBS Lett. 408, 39–42.

Bouillaud, F., Couplan, E., Pecqueur, C. and Ricquier, D. (2001). Homologues of the uncoupling protein from brown adipose tissue (UCP1): UCP2, UCP3, BMCP1 and UCP4. Biochim. Biophys. Acta 1504, 107–119.

Bouillaud, F., Weissenbach, J. and Ricquier, D. (1986). Complete cDNA-derived amino acid sequence of rat brown fat uncoupling protein. J. Biol. Chem. 261, 1487–1490.

Brustovetsky, N. and Klingenberg, M. (1996). Mitochondrial ADP/ATP carrier can be reversibly converted into a large channel by Ca^{2+}. Biochemistry 35, 8483–8488.

Brustovetsky, N. and Klingenberg, M. (1994). The reconstituted ADP/ATP carrier can mediate H+ transport by free fatty acids, which is further stimulated by mersalyl. J. Biol. Chem. 269, 27329–27336.

Cadenas, S., Echtay, K.S., Harper, J.A., Jekabsons, M.B., Buckingham, J.A., Grau, E., Abuin, A., Chapman, H., Clapham, J.C. and Brand, M.D. (2002). The basal proton conductance of skeletal muscle mitochondria from transgenic mice overexpressing or lacking uncoupling protein-3. J. Biol. Chem. 277, 2773–2778.

Chan, C.B., De Leo, D., Joseph, J.W., McQuaid, T.S., Ha, X.F., Xu, F., Tsushima, R.G., Pennefather, P.S., Salapatek, A.M. and Wheeler, M.B. (2001). Increased uncoupling protein-2 levels in beta-cells are associated with impaired glucose-stimulated insulin secretion: mechanism of action. Diabetes 50, 1302–1310.

Chan, C.B., MacDonald, P.E., Saleh, M.C., Johns, D.C., Marban, E. and Wheeler, M.B. (1999). Overexpression of uncoupling protein 2 inhibits glucose-stimulated insulin secretion from rat islets. Diabetes 48, 1482–1486.

Clapham, J.C., Arch, J.R., Chapman, H., Haynes, A., Lister, C., Moore, G.B., Piercy, V., Carter, S.A., Lehner, I., Smith, S.A., Beeley, L.J., Godden, R.J., Herrity, N., Skehel, M., Changani, K.K., Hockings, P.D., Reid,

D.G., Squires, S.M., Hatcher, J., Trail, B., Latcham, J., Rastan, S., Harper, A.J., Cadenas, S., Buckingham, J.A., Brand, M.D. and Abuin, A. (2000). Mice overexpressing human uncoupling protein-3 in skeletal muscle are hyperphagic and lean. Nature 406, 415–418.

Cline, G.W., Vidal-Puig, A.J., Dufour, S., Cadman, K.S., Lowell, B.B. and Shulman, G.I. (2001). In vivo effects of uncoupling protein-3 gene disruption on mitochondrial energy metabolism. J. Biol. Chem. 276, 20240–20244.

Crompton, M. (1999). The mitochondrial permeability transition pore and its role in cell death. Biochem J. 341, 233–249.

Echtay, K.S., Roussel, D., St-Pierre, J., Jekabsons, M.B., Cadenas, S., Stuart, J.A., Harper, J.A., Roebuck, S.J., Morrison, A., Pickering, S., Clapham, J.C. and Brand, M.D. (2002). Superoxide activates mitochondrial uncoupling proteins. Nature 415, 96–99.

Echtay, K.S., Winkler, E., Frischmuth, K. and Klingenberg, M. (2001). Uncoupling proteins 2 and 3 are highly active H(+) transporters and highly nucleotide sensitive when activated by coenzyme Q (ubiquinone). Proc. Natl. Acad. Sci. USA 98, 1416–1421.

Echtay, K.S., Winkler, E. and Klingenberg, M. (2000). Coenzyme Q is an obligatory cofactor for uncoupling protein function. Nature 408, 609–613.

el Moualij, B., Duyckaerts, C., Lamotte-Brasseur, J. and Sluse, F.E. (1997). Phylogenetic classification of the mitochondrial carrier family of *Saccharomyces cerevisiae*. Yeast 13, 573–581.

Enerback, S., Jacobsson, A., Simpson, E.M., Guerra, C., Yamashita, H., Harper, M.E. and Kozak, L.P. (1997). Mice lacking mitochondrial uncoupling protein are cold-sensitive but not obese. Nature 387, 90–94.

Fink, B.D., Hong, Y.S., Mathahs, M.M., Scholz, T.D., Dillon, J.S. and Sivitz, W.I. (2002). UCP2-dependent proton leak in isolated mammalian mitochondria. J. Biol. Chem. 277, 3918–3925.

Finkel, T. and Holbrook, N.J. (2000). Oxidants, oxidative stress and the biology of aging. Nature 408, 239–247.

Fleury, C., Neverova, M., Collins, S., Raimbault, S., Champigny, O., Levi-Meyrueis, C., Bouillaud, F., Seldin, M.F., Surwit, R.S., Ricquier, D. and Warden, C.H. (1997). Uncoupling protein-2: a novel gene linked to obesity and hyperinsulinemia. Nature Genet. 15, 269–272.

Garcia-Martinez, C., Sibille, B., Solanes, G., Darimont, C., Mace, K., Villarroya, F. and Gomez-Foix, A.M. (2001). Overexpression of UCP3 in cultured human muscle lowers mitochondrial membrane potential, raises ATP/ADP ratio, and favors fatty acid vs. glucose oxidation. FASEB J. 15, 2033–2035.

Garlid, K.D., Orosz, D.E., Modriansky, M., Vassanelli, S. and Jezek, P. (1996). On the mechanism of fatty acid-induced proton transport by mitochondrial uncoupling protein. J. Biol. Chem. 271, 2615–2620.

Gong, D.W., Monemdjou, S., Gavrilova, O., Leon, L.R., Marcus-Samuels, B., Chou, C.J., Everett, C., Kozak, L.P., Li, C., Deng, C., Harper, M.E. and Reitman, M.L. (2000). Lack of obesity and normal response to fasting and thyroid hormone in mice lacking uncoupling protein-3. J. Biol. Chem. 275, 16251–16257.

Gonzalez-Barroso, M.D.M., Ricquier, D. and Cassard-Doulcier, A. (2000). The human uncoupling protein-1 gene (UCP1): present status and perspectives in obesity research. Obesity Rev. 1, 61–72.

Gonzalez-Barroso, M.M., Fleury, C., Bouillaud, F., Nicholls, D.G. and Rial, E. (1998). The uncoupling protein UCP1 does not increase the proton conductance of the inner mitochondrial membrane by functioning as a fatty acid anion transporter. J. Biol. Chem. 273, 15528–155232.

Gonzalez-Barroso, M.M., Fleury, C., Levi-Meyrueis, C., Zaragoza, P., Bouillaud, F. and Rial, E. (1997). Deletion of amino acids 261–269 in the brown fat uncoupling protein converts the carrier into a pore. Biochemistry 36, 10930–10935.

Harper, J.A., Stuart, J.A., Jekabsons, M.B., Roussel, D., Brindle, K.M., Dickinson, K., Jones, R.B. and Brand, M.D. (2002). Artifactual uncoupling by uncoupling protein 3 in yeast mitochondria at the concentrations found in mouse and rat skeletal muscle mitochondria. Biochem J. 361, 49–56.

Heidkaemper, D., Winkler, E., Muller, V., Frischmuth, K., Liu, Q., Caskey, T. and Klingenberg, M. (2000). The bulk of UCP3 expressed in yeast cells is incompetent for a nucleotide regulated H$^+$ transport. FEBS Lett. 480, 265–270.

Hesselink, M.K., Keizer, H.A., Borghouts, L.B., Schaart, G., Kornips, C.F., Slieker, L.J., Sloop, K.W., Saris, W.H. and Schrauwen, P. (2001). Protein expression of UCP3 differs between human type 1, type 2a, and type 2b fibers. FASEB J. 15, 1071–1073.

Hofmann, W.E., Liu, X., Bearden, C.M., Harper, M.E. and Kozak, L.P. (2001). Effects of genetic background on thermoregulation and fatty acid-induced uncoupling of mitochondria in UCP1-deficient mice. J. Biol. Chem. 276, 12460–12465.

Huppertz, C., Fischer, B.M., Kim, Y.B., Kotani, K., Vidal-Puig, A., Slieker, L.J., Sloop, K.W., Lowell, B.B. and Kahn, B.B. (2001). Uncoupling protein 3 (UCP3) stimulates glucose uptake in muscle cells through a phosphoinositide 3-kinase-dependent mechanism. J. Biol. Chem. 276, 12520–12529.

Jaburek, M., Varecha, M., Gimeno, R.E., Dembski, M., Jezek, P., Zhang, M., Burn, P., Tartaglia, L.A. and Garlid, K.D. (1999). Transport function and regulation of mitochondrial uncoupling proteins 2 and 3. J. Biol. Chem. 274, 26003–26007.

Kramer, R. and Palmieri, F. (1989). Molecular aspects of isolated and reconstituted carrier proteins from animal mitochondria. Biochim. Biophys. Acta 974, 1–23.

Krauss, S., Zhang, C.Y. and Lowell, B.B. (2002). A significant portion of mitochondrial proton leak in intact thymocytes depends on expression of UCP2. Proc. Natl. Acad. Sci. USA 99, 118–122.

Laloi, M., Klein, M., Riesmeier, J.W., Muller-Rober, B., Fleury, C., Bouillaud, F. and Ricquier, D. (1997). A plant cold-induced uncoupling protein. Nature 389, 135–136.

Langin, D. (2001). Diabetes, insulin secretion, and the pancreatic beta-cell mitochondrion. N. Engl. J. Med. 345, 1772–1774.

Larrouy, D., Laharrague, P., Carrera, G., Viguerie-Bascands, N., Levi-Meyrueis, C., Fleury, C., Pecqueur, C., Nibbelink, M., Andre, M., Casteilla, L. and Ricquier, D. (1997). Kupffer cells are a dominant site of uncoupling protein 2 expression in rat liver. Biochem. Biophys. Res. Commun. 235, 760–764.

Li, B., Nolte, L.A., Ju, J.S., Han, D.H., Coleman, T., Holloszy, J.O. and Semenkovich, C.F. (2000). Skeletal muscle respiratory uncoupling prevents diet-induced obesity and insulin resistance in mice. Nature Med. 6, 1115–1120.

Liu, Q., Bai, C., Chen, F., Wang, R., MacDonald, T., Gu, M., Zhang, Q., Morsy, M.A. and Caskey, C.T. (1998). Uncoupling protein-3: a muscle-specific gene upregulated by leptin in ob/ob mice. Gene 207, 1–7.

Loeffler, M. and Kroemer, G. (2000). The mitochondrion in cell death control: certainties and incognita. Exp. Cell. Res. 256, 19–26.

Lowell, B.B., S-Susulic, V., Hamann, A., Lawitts, J.A., Himms-Hagen, J., Boyer, B.B., Kozak, L. P. and Flier, J.S. (1993). Development of obesity in transgenic mice after genetic ablation of brown adipose tissue. Nature 366, 740–742.

Miroux, B., Frossard, V., Raimbault, S., Ricquier, D. and Bouillaud, F. (1993). The topology of the brown adipose tissue mitochondrial uncoupling protein determined with antibodies against its antigenic sites revealed by a library of fusion proteins. EMBO J. 12, 3739–3745.

Negre-Salvayre, A., Hirtz, C., Carrera, G., Cazenave, R., Troly, M., Salvayre, R., Penicaud, L. and Casteilla, L. (1997). A role for uncoupling protein-2 as a regulator of mitochondrial hydrogen peroxide generation. FASEB J. 11, 809–815.

Nibbelink, M., Moulin, K., Arnaud, E., Duval, C., Penicaud, L. and Casteilla, L. (2001). Brown fat UCP1 is specifically expressed in uterine longitudinal smooth muscle cells. J. Biol. Chem. 276, 47291–47295.

Pecqueur, C., Alves-Guerra, M.C., Gelly, C., Levi-Meyrueis, C., Couplan, E., Collins, S., Ricquier, D., Bouillaud, F. and Miroux, B. (2001). Uncoupling protein 2, in vivo distribution, induction upon oxidative stress, and evidence for translational regulation. J. Biol.

Chem. 276, 8705–8712.

Pecqueur, C., Cassard-Doulcier, A.M., Raimbault, S., Miroux, B., Fleury, C., Gelly, C., Bouillaud, F. and Ricquier, D. (1999). Functional organization of the human uncoupling protein-2 gene, and juxtaposition to the uncoupling protein-3 gene. Biochem. Biophys. Res. Commun. 255, 40–46.

Petit, P.X., Lecoeur, H., Zorn, E., Dauguet, C., Mignotte, B. and Gougeon, M.L. (1995). Alterations in mitochondrial structure and function are early events of dexamethasone-induced thymocyte apoptosis. J. Cell Biol. 130, 157–167.

Rial, E., Gonzalez-Barroso, M., Fleury, C., Iturrizaga, S., Sanchis, D., Jimenez-Jimenez, J., Ricquier, D., Goubern, M. and Bouillaud, F. (1999). Retinoids activate proton transport by the uncoupling proteins UCP1 and UCP2. EMBO J. 18, 5827–5833.

Rial, E. and Nicholls, D.G. (1983). The regulation of the proton conductance of brown fat mitochondria. Identification of functional and non-functional nucleotide-binding sites. FEBS Lett. 161, 284–288.

Ricquier, D. and Bouillaud, F. (2000). The uncoupling protein homologues: UCP1, UCP2, UCP3, StUCP and AtUCP. Biochem. J. 345 Pt 2, 161–179.

Ricquier, D., Raimbault, S., Champigny, O., Miroux, B. and Bouillaud, F. (1992). Comment to Shinohara et al. (1991) FEBS Lett. 293, 173–174. The uncoupling protein is not expressed in rat liver. FEBS Lett. 303, 103–106.

Saraste, M. and Walker, J.E. (1982). Internal sequence repeats and the path of polypeptide in mitochondrial ADP/ATP translocase. FEBS Lett. 144, 250–254.

Sauer, H., Wartenberg, M. and Hescheler, J. (2001). Reactive oxygen species as intracellular messengers during cell growth and differentiation. Cell. Physiol. Biochem. 11, 173–186.

Schonfeld, P., Jezek, P., Belyaeva, E.A., Borecky, J., Slyshenkov, V.S., Wieckowski, M.R. and Wojtczak, L. (1996). Photomodification of mitochondrial proteins by azido fatty acids and its effect on mitochondrial energetics. Further evidence for the role of the ADP/ATP carrier in fatty-acid-mediated uncoupling. Eur. J. Biochem. 240, 387–393.

Schonfeld, P., Wieckowski, M.R. and Wojtczak, L. (1997). Thyroid hormone-induced expression of the ADP/ATP carrier and its effect on fatty acid-induced uncoupling of oxidative phosphorylation. FEBS Lett. 416, 19–22.

Schroers, A., Burkovski, A., Wohlrab, H. and Kramer, R. (1998). The phosphate carrier from yeast mitochondria. Dimerization is a prerequisite for function. J. Biol. Chem. 273, 14269–14276.

Shinohara, Y., Shima, A., Kamida, M. and Terada, H. (1991). Uncoupling protein is expressed in liver mitochondria of cold-exposed and newborn rats. FEBS Lett. 293, 173–174.

Skulachev, V.P. (1991). Fatty acid circuit as a physiological mechanism of uncoupling of oxidative phosphorylation. FEBS Lett. 294, 158–162.

Skulachev, V.P. (1997). Membrane-linked systems preventing superoxide formation. Biosci. Rep. 17, 347–366.

Stuart, J.A., Cadenas, S., Jekabsons, M.B., Roussel, D. and Brand, M.D. (2001). Mitochondrial proton leak and the uncoupling protein 1 homologues. Biochim. Biophys. Acta 1504, 144–158.

Stuart, J.A., Harper, J.A., Brindle, K.M., Jekabsons, M.B. and Brand, M.D. (2001). A mitochondrial uncoupling artifact can be caused by expression of uncoupling protein 1 in yeast. Biochem. J. 356, 779–789.

Stuart, J.A., Harper, J.A., Brindle, K.M., Jekabsons, M.B. and Brand, M.D. (2001). Physiological levels of mammalian uncoupling protein 2 do not uncouple yeast mitochondria. J. Biol. Chem. 276, 18633–18639.

Turrens, J.F. (1997). Superoxide production by the mitochondrial respiratory chain. Biosci. Rep. 17, 3–8.

Vidal-Puig, A., Solanes, G., Grujic, D., Flier, J.S. and Lowell, B.B. (1997). UCP3: an uncoupling protein homologue expressed preferentially and abundantly in skeletal muscle and brown adipose tissue. Biochem. Biophys. Res. Commun. 235, 79–82.

Vidal-Puig, A.J., Grujic, D., Zhang, C.Y., Hagen, T., Boss, O., Ido, Y., Szczepanik, A., Wade, J., Mootha, V., Cortright, R., Muoio, D.M. and Lowell, B.B. (2000). Energy metabolism in uncoupling protein 3 gene knockout mice. J. Biol. Chem. 275, 16258–16266.

Wang, M. Y., Shimabukuro, M., Lee, Y., Trinh, K.Y., Chen, J.L., Newgard, C.B. and Unger, R.H. (1999). Adenovirus-mediated overexpression of uncoupling protein-2 in pancreatic islets of Zucker diabetic rats increases oxidative activity and improves beta-cell function. Diabetes 48, 1020–1025.

Wieckowski, M.R. and Wojtczak, L. (1997). Involvement of the dicarboxylate carrier in the protonophoric action of long-chain fatty acids in mitochondria. Biochem. Biophys. Res. Commun. 232, 414–417.

Wojtczak, L. and Schonfeld, P. (1993). Effect of fatty acids on energy coupling processes in mitochondria. Biochim. Biophys. Acta 1183, 41–57.

Zackova, M., Kramer, R. and Jezek, P. (2000). Interaction of mitochondrial phosphate carrier with fatty acids and hydrophobic phosphate analogs. Int. J. Biochem. Cell. Biol. 32, 499–508.

Zhang, C.Y., Baffy, G., Perret, P., Krauss, S., Peroni, O., Grujic, D., Hagen, T., Vidal-Puig, A.J., Boss, O., Kim, Y.B., Zheng, X.X., Wheeler, M.B., Shulman, G.I., Chan, C.B. and Lowell, B.B. (2001). Uncoupling protein-2 negatively regulates insulin secretion and is a major link between obesity, beta cell dysfunction, and type 2 diabetes. Cell 105, 745–755.

Sensing, Signaling and Cell Adaptation. Edited by K.B. Storey and J.M. Storey

CHAPTER 19

Regulation of Proliferation, Differentiation and Apoptosis of Brown Adipocytes: Signal Transduction Pathways Involved

Almudena Porras and Manuel Benito

Departamento de Bioquímica y Biología Molecular II (Instituto de Bioquímica, Centro Mixto del Consejo Superior de Investigaciones Científicas (C.S.I.C.) y de la Universidad Complutense de Madrid (U.C.M)), Facultad de Farmacia, Universidad Complutensede Madrid, Ciudad Universitaria, Madrid, Spain

1. Introduction

1.1. Function and characteristics of brown adipose tissue

Brown adipose tissue (BAT) is characterized by the presence of the uncoupling protein-1 (UCP-1) in the mitochondrial inner membrane, which allows the production of heat by the mechanism called "non-shivering thermogenesis" (Trayhurn and Milner, 1989).

Brown adipocytes are highly specialized in order to dissipate energy as heat. Like white adipocytes, they have the enzymatic machinery required for lipid synthesis, but additionally they present a high capacity for fatty acid β-oxidation and they express UCP-1. Thus, unlike white adipose tissue (WAT), BAT does not accumulate lipids just as a storage depot, but as a source of fatty acids to be oxidized in the mitochondria when BAT thermogenesis is activated in order to produce heat.

Morphologically, brown adipocytes are characterized by the presence of multiple lipid droplets and a high number of mitochondria, which express UCP-1 (Trayhurn and Milner, 1989).

BAT is activated under particular circumstances such as cold exposure, high fat diet feeding and during the perinatal period as a defense against cold or obesity (Trayhurn and Milner, 1989).

1.2. Uncoupling protein-1: function and activation

The function of UCP-1 is to uncouple the proton electrochemical gradient generated by the respiratory chain from ATP synthesis. UCP-1 allows the re-entry of protons and, as a consequence, energy liberated from substrate oxidation is dissipated as heat (Fig. 19.1). This mechanism is reversible, since it can be inhibited by purine nucleotides such as GDP (Trayhurn and Milner, 1989).

Acute activation of BAT thermogenesis upon cold-exposure or in other situations appears to be mediated by noradrenaline (NA) liberated from the sympathetic nervous system (Fig. 19.1). By acting through β-receptors, NA induces an increase in cAMP, which finally leads to activation of lypolysis. The resulting free fatty acids are used as substrates for mitochondrial respiration and as activators of UCP-1 (Trayhurn and Milner, 1989).

2. Development, differentiation and involution of brown adipose tissue

Activation of BAT under the physiological circumstances in which "non-shivering thermogenesis" is required, includes acute activation of pre-existing BAT and development of new highly differentiated brown adipocytes. This BAT

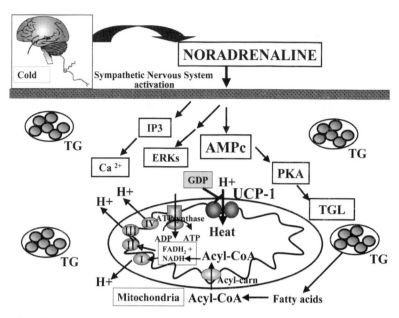

Fig. 19.1. Mechanism of cold-induced thermogenesis: activation of UCP-1. Cold is sensed by the brain and sympathetic nerves are activated, releasing NA. Acting through different types of adrenergic receptors, NA leads to activation of different signal transduction pathways such as ERKs and cAMP/PKA cascades. PKA phosphorylates triglyceride lipase (TGL) and fatty acids are released from triacylglyceride depots, which act as activators of UCP-1 and substrates for mitochondrial oxidation generating NADH. H^+ extruded by the respiratory chain reenter through UCP-1 instead of through ATP synthase and energy from the proton electrochemical gradient is released as heat. UCP-1 can be inhibited by purine nucleotides such as GDP.

development involves processes of proliferation of brown adipocyte precursor cells as well as their subsequent differentiation. In addition, pre-existing brown adipocytes are fully differentiated (Trayhurn and Milner, 1989).

On the other hand, when activation of BAT is not required, an involution of this tissue is produced. This process has not been well characterized, but it is very likely that induction of cell death by apoptosis could play a role. The absence of noradrenergic stimulation (Nisoli et al., 1997; Briscini et al., 1998) or the presence of TNF-α (Nisoli et al., 1997; Porras et al., 1997) are known to induce apoptosis in these cells. In addition, apoptosis seems also to contribute to the reduced activity of BAT observed in obesity (Nisoli et al., 2000).

3. Transcriptional control of brown adipose tissue differentiation

Differentiation of BAT includes thermogenic and adipogenic differentiation. Therefore, a parallel induction of the expression of *ucp-1* and adipogenic-related genes is produced.

3.1. Regulation of adipogenesis

Adipogenesis in brown adipocytes is essentially regulated by the same transcription factors as in white adipocytes, although there are some differences due to their distinct roles.

PPARγ (peroxisome proliferator-activated receptor-γ) is involved in the adipogenic and thermogenic differentiation of BAT. Treatment of brown adipocyte cell lines with PPARγ ligands and its administration to mice induces the differentiation and/or the accumulation of BAT (Tai et al., 1996). In addition, BAT development is retarded in PPARγ knockout mouse obtained through tetraploid rescue of the placenta defects (Barak et al., 1999). However, PPARγ is not required for the initial commitment of the BAT lineage, while being necessary for the complete development of the tissue. This means that PPARγ would be downstream of earlier adipogenic factors such as the CCAAT/

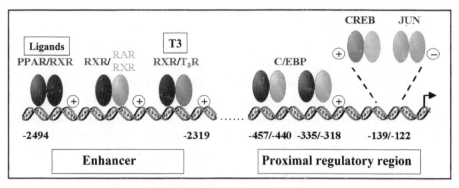

UCP-1 regulatory regions

Fig. 19.2. Proximal and distal regulatory regions of *ucp-1* gene. Regulatory regions of rat and mouse *ucp-1* gene include a distal enhancer with binding sites for PPAR, retinoid receptors (RAR and RXR) and T3 receptors (T3R); and a proximal regulatory region with two C/EBP sites and a CRE site, which can bind CREB (a positive regulator) and c-Jun (a negative regulator).

enhancer-binding protein (C/EBPβ) as has been proposed for WAT (Spiegelman and Flier, 1996).

A role for C/EBPβ and δ proteins in BAT adipogenesis is also evident based on the reduced lipid accumulation produced in brown adipocytes from mice lacking C/EBPβ and/or δ (Tanaka et al., 1997). However, the role of C/EBPα appears not to be so important for BAT differentiation, since in C/EBPα deficient mice, expressing C/EBPα in the liver in order to avoid hypoglycemia, a selective absence of WAT is produced, meanwhile mammary WAT and BAT are relatively unaffected (Linhart et al., 2001). These data indicate that C/EBPα is not required for the expression of several adipogenic genes in brown adipocytes and it is very likely that C/EBPδ, which is increased under these circumstances, can compensate for C/EBPα.

3.2. Regulation of UCP-1 expression

UCP-1 expression is mainly controlled through regulation of gene transcription, although other mechanisms are also involved. Different regulatory sequences have been identified in the proximal and distal regions of the UCP-1 promoter; of particular importance is the 220-bp enhancer element located approximately 2.4 kilobases upstream of the mouse and rat UCP-1 genes (Cassard-Doulcier et al., 1993; Kozak et al., 1994). This region (see Fig. 19.2) presents binding sites for PPARγ (Tai et al., 1996), thyroid hormone

receptors (Rabelo et al., 1996a) and retinoid acid receptors (Alvarez et al., 1995; 2000), which are positively regulated by their ligands and by other extracellular signals such as NA. In addition, it has been recently described that PPAR-α can also bind to the same region that PPARγ and up-regulate *ucp-1* gene expression (Barberá et al., 2001).

The action of these transcription factors is regulated by the coactivator, PGC1 (PPAR-γ-co-activator) (Puigserver et al., 1998). This factor is highly expressed in BAT, being induced upon cold exposure via β-adrenoreceptors. PGC1 activates UCP-1 expression as well as mitochondrial enzymes expression (Puigserver et al., 1998; Wu et al., 1999), which contributes to increase the thermogenic capacity of BAT.

On the other hand, in the proximal regulatory region of the mouse *ucp-1* gene, there are two C/EBP-responsive elements (Yubero et al., 1994) and a CRE (cAMP responsive element) site (Fig. 19.2), which is able to bind the positive regulator, CREB (CRE-binding protein), and the negative regulator, c-Jun (Yubero et al., 1998). In non-differentiated cells, c-Jun seems to repress the *ucp-1* gene due to the low levels of CREB along with the low transactivating activity exerted by C/EBPs, while in differentiated brown adipocytes CREB binds the CRE site leading to *ucp-1* expression. C/EBPs could also contribute to the increased UCP-1 expression under these conditions (Yubero et al., 1994; 1998).

The molecular mechanisms controlling the transcription of the human *ucp-1* gene are not so well characterized, although a 350-bp hormone response region with a high homology with the rat *ucp-1* enhancer has been identified (Gónzalez-Barroso et al., 2000). This region binds retinoid acid receptors, RARs and RXRs, CREB/ATF factors and PPARγ although there is no PPRE consensus sequence.

4. Regulation of proliferation of brown adipocytes: signals and signal transduction pathways involved

As mentioned above, the development and recruitment of BAT involves processes of proliferation and differentiation in order to increase the number of highly differentiated brown adipocytes. Proliferation of brown adipocytes is activated by different growth factors and hormones. Among these signals, NA, IGF-I and insulin are probably the most relevant. NA was the first mitogen identified by experiments carried out *in vivo* (Geloen et al., 1988; Geloen et al., 1992). It was shown that treatment of rats with NA or the β-adrenergic agonist, isoproterenol, induced the proliferation of brown adipocyte precursor cells (Geloen et al., 1988). These results suggested that under physiological conditions, cold exposure stimulates BAT growth by increasing the release of NA from sympathetic nerves, which activates proliferation of brown adipocyte precursor cells via β-adrenergic receptors. This was proved by experiments done with denervated and intact BAT pads of rats, showing that intact innervation was required for cold-induced proliferation of brown adipocyte precursor cells (Geloen et al., 1992). Moreover, NA infusion mimicked the effect of cold exposure, bypassing the requirement of sympathetic innervation.

The mitogenic effect of NA was also demonstrated using mouse brown adipocyte precursor cells maintained in primary culture (Bronnikov et al., 1992). In these cells, β_1-receptors seemed to mediate the proliferative effect of NA by increasing cAMP levels, while α_2-receptors exerted an opposing effect. However, the mechanisms

mediating the mitogenic effect of NA in rat fetal brown adipocytes appears to be different. ERKs (extracellular regulated kinases) activation is required for NA-induced proliferation (Fig. 19.3), while the cAMP/PKA pathway plays no role (Valladares et al., 2000b). Under these conditions, ERKs activation is mediated by β-, α_2- and α_1-adrenergic receptors through a mechanism dependent on MEK, but independent of cAMP/PKA. In contrast, in brown adipocytes from newborn rats (Shimizu et al., 1997) and from adult mice (Lindquist and Rehnmark, 1998) β-adrenergic receptors are the main mediators of ERKs activation with a low contribution of α_1-receptors and the cAMP/PKA cascade is necessary for this activation (Lindquist and Rehnmark, 1998; Lindquist et al., 2000; Shimizu et al., 1997). Therefore, depending on the system, the mechanism leading to ERKs activation appears to differ, but ERKs would be necessary for NA-induced proliferation in all cases.

All these discrepancies observed about the mechanism of ERKs activation using several model systems might be due to differences in the differentiation state and/or in the pattern of expression of adrenergic receptors subtypes. Regarding this, changes in the expression of β_3- versus β_1-receptors as well as in the coupling to cAMP accumulation occur during the differentiation of mouse brown adipocytes in culture (Bronnikov et al., 1999a). β_1-Receptors are present in proliferative precursor cells, whereas β_3-receptors are expressed in differentiated brown adipocytes.

In relation to the role played by α-receptors, α_2-receptors seem to inhibit proliferation (Bronnikov et al., 1992; Valladares et al.,2000b), while a positive (Rehmark et al., 1990) and a negative effect of α_1-receptors (Bronnikov et al., 1999b; Valladares et al., 2000b) has been described. In rat fetal brown adipocytes, α_1- and α_2-adrenergic receptors inhibit proliferation and the blockade of the ERKs cascade enhanced this antiproliferative effect (Fig. 19.3); therefore, other signaling pathways activated by these receptors might be inhibiting proliferation (Valladares et al., 2000b).

In addition to the direct role of NA inducing proliferation of brown adipocytes, some studies

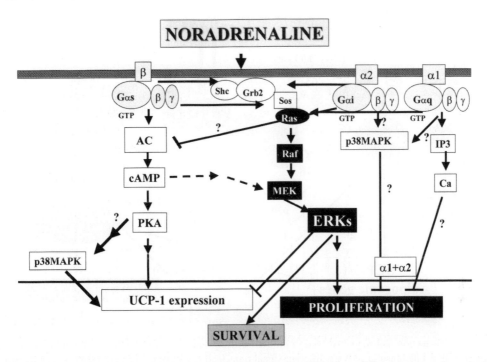

Fig. 19.3. Signal transduction pathways involved in the regulation of proliferation, differentiation and survival of brown adipocytes induced by noradrenaline. NA acting through α_1-, α_2-and β-receptors activates the Ras/ERKs cascade, which mediates proliferation and survival. cAMP/PKA cascade is also activated by β-receptors leading to UCP-1 expression, probably through p38MAPK activation. α_1-, α_2-Receptors could also activate other signal transduction pathways which might mediate inhibition of proliferation.

indicate that NA potentiates the mitogenic effect of different growth factors and neuropeptides such as EGF, PDGF, a/b-FGFs and vasopressin (García and Obregón, 1997). Moreover, it has also been proposed that the NA-induced expression of FGF-2 could contribute to mediate its proliferative effect (Yamashita et al., 1995).

IGF-I (Lorenzo et al, 1993; Porras et al., 1998) and insulin (Porras et al., 1998) are also important mitogens for rat fetal brown adipocytes maintained in primary culture with a potential important role in the development of BAT occurring during the fetal period. ERKs activation by these signals is required to induce proliferation of brown adipocytes (Porras et al., 1998), while the activation of PI3K (phosphoinositide-3-Kinase) plays no role (Valverde et al., 1997). Thus, the blockade of ERKs activation by the MEK1 specific inhibitor, PD98059, prevented IGF-I/insulin induced proliferation (Porras et al., 1998). IRS-1 is also a requirement for the mitogenic effect of these growth factors, and as a consequence, in brown adipocytes

from IRS-1 deficient mice insulin-induced mitogenesis and ERKs activation are impaired (Valverde et al., 2001). Therefore, the mitogenic pathway induced by IGF-I/insulin in brown adipocytes would require an intact IRS-1, which associates with Grb-2 (Valverde et al., 1998a; 2001) leading to the activation of the Ras-ERKs cascade, which in term mediates proliferation (Porras et al., 1998) (Fig. 19.4).

FGF-16, could also play a role in the proliferation of rat embryonic brown adipocytes since it is expressed coincident with activation of proliferation and is able to induce mitogenesis in rat fetal brown adipocytes maintained in primary culture (Konishi et al., 2000).

Regarding negative regulation of brown adipocytes proliferation, TNF-α has been identified as an inhibitor of basal and IGF-I-induced proliferation of rat fetal brown adipocytes maintained in primary culture (Porras et al., 1997; Valladares et al., 2000a). This antiproliferative effect of TNF-α appears to be mediated by p38MAPK (p38

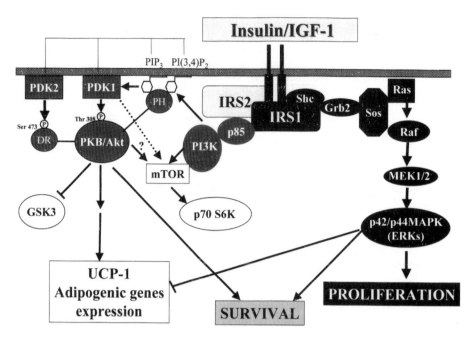

Fig. 19.4. Signal transduction pathways involved in the IGF-I/insulin-induced proliferation, differentiation and survival of brown adipocytes. IGF-I and insulin activate different signal transduction pathways such are ERKs and PI3K cascades. The IRS1-Grb2-Ras-ERKs cascade is required for proliferation, while the IRS1-PI3K-Akt is necessary for differentiation. In addition, both pathways mediate the survival effect.

Mitogen activated protein kinase), while ERKs activation plays an opposite role (Valladares et al., 2000a). Changes in the levels of the proliferating cellular nuclear antigen (PCNA) could account for the regulation exerted by these two pathways, while no regulation of cyclin D1 was produced (Fig. 19.5). Retinoic acid (RA) isomers (all-*trans*-RA and 9-*cis*-RA) also inhibit proliferation of brown adipocytes from mice maintained in primary culture (Puigserver et al., 1996).

5. Regulation of differentiation of brown adipocytes: signals and signal transduction pathways involved

Different signals have been identified as regulators of BAT thermogenic and/or adipogenic differentiation. Among them, the positive regulators are the best characterized and, in particular, those involved in the induction of the *ucp-1* gene.

NA released from the sympathetic nervous system is one of the most important positive regulators

of BAT and the first identified. As described above, NA was shown to be necessary for brown adipocytes hyperplasia upon cold exposure, β-receptors being the main mediators of this effect (Geloen et al., 1988; 1992). In addition, exposure of rats to cold also induced an increase in UCP-1 mRNA level, which can be mimicked by injection of animals with a selective β-agonist (Ricquier et al., 1986). This role of NA has been confirmed by studies done with brown adipocytes or brown adipocyte derived cell lines maintained in culture. NA increased the level and activity of UCP-1 in rat fetal brown adipocytes maintained in primary culture via β-receptors (Porras et al., 1989) and induced UCP-1 synthesis in mouse brown adipocytes differentiated in culture, playing a major role the β-adrenergic receptors-mediated cAMP increase (Kopecky et al., 1990), although a synergistic effect of α_1-receptors has also been proposed (Rehnmark et al., 1990). Similarly, in brown adipocytes from Syrian hamsters, thermogenic activity was stimulated by the β_3-adrenergic-mediated cAMP increase, being potentiated by the

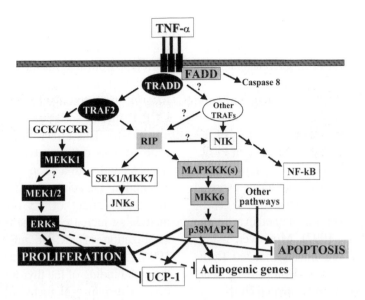

Fig. 19.5. Signal transduction pathways activated by TNF-α in brown adipocytes. Role played by MAPK superfamily in proliferation, differentiation and apoptosis. TNF-α activates ERKs, JNKs, p38MAPK, NF-κB and caspase 8. p38MAPK mediates TNF-α-induced apoptosis and inhibition of proliferation and plays a positive role in the regulation of UCP-1 and adipogenic genes expression. In contrast, ERKs are essential mediators of the inhibitory effect of TNF-α on UCP-1 expression, but are only partial mediators of its negative effect on adipogenic genes, other pathways being responsible for this. ERKs are also positive regulators of proliferation and survival.

α_1-receptors-mediated Ca^{2+} elevation (Zhao et al., 1997). In a brown adipocyte derived cell line obtained from a hibernoma, UCP-1 expression was also stimulated by NA, non-selective β-agonists or β_3-selective agonists (Ross et al., 1992).

Regarding the molecular mechanisms mediating the effect of NA on UCP-1 induction, the β_3-adrenergic-induced cAMP increase seems to play an important role (Kopecky et al., 1990; Rehnmark et al., 1990; Bronnikov et al., 1999a; 1999b) as well as activation of PKA (Valladares et al 2000b; Klein et al., 2000; Fredriksson et al., 2001). In addition, it has been recently shown that p38MAPK is also necessary as a downstream target of the cAMP/PKA cascade (Cao et al., 2001). In contrast, ERKs activation induced by NA (Valladares et al., 2000b) or by a β_3-agonist (Klein et al., 2000) plays a negative role in the regulation of UCP-1 expression (Fig. 19.3).

IGF-I and insulin are also important positive regulators of BAT differentiation, inducing adipogenic and/or thermogenic differentiation (Lorenzo et al., 1993; Guerra et al., 1994a; Teruel et al., 1996). IGF-I was shown to induce the expression of UCP-1 (Lorenzo et al., 1993; Guerra et al., 1994a) and adipogenic-related genes in rat fetal brown adipocytes maintained in primary culture (Lorenzo et al., 1993; Teruel et al., 1996). These data suggested that IGF-I could play a role in the differentiation of BAT in the fetal period. According to this, the expression of IGF-I and its receptor in rat BAT during the last days of the fetal period is coincident with an induction of the expression of UCP-1 and several adipogenic-related genes (Teruel et al., 1995). In addition, it is possible that IGF-I can also play a role in BAT upon cold-exposure, since IGF-I mRNA is transiently upregulated (Duchamp et al., 1997).

The action of insulin on BAT differentiation is preferentially exerted through induction of the expression of different adipogenic-related genes. Insulin up-regulates fatty acid synthase (FAS), glycerol-3-phosphate dehydrogenase (G3PD) and glucose transporter isoform 4 (Glut 4) mRNAs in rat fetal brown adipocytes maintained in primary culture and this is accompanied by an increase in the cellular lipid content (Teruel et al., 1996; Porras et al., 1998). Moreover, mice lacking insulin receptors in BAT show a progressive decrease in BAT weight and in the expression of adipogenic-related genes, while UCP-1 expression is increased (Guerra et al., 2001).

In relation to the molecular mechanisms involved in the actions of IGF-I and/or insulin on BAT differentiation, the relevance of the PI3K pathway should be noted. Thus, inhibition of PI3K with the chemical inhibitors, wortmannin or LY294002, prevented the IGF-I-induced expression of UCP-1 and some adipogenic-related genes in rat fetal brown adipocytes maintained in primary culture (Valverde et al., 1997). Moreover, using fetal brown adipocyte cell lines generated from IRS-1 deficient mice, it was demonstrated that IRS-1 was required for insulin-induced PI3K and Akt activation, as well as for the stimulation of lipid synthesis (Valverde et al., 1999). Ras proteins seem also to be involved in the IGF-I-induced UCP-1 expression (Porras et al., 1996) in brown adipocyte derived cell lines (Benito et al., 1993). In contrast, IGF-I/insulin-induced activation of ERKs plays a negative role in the regulation of the adipocytic and thermogenic differentiation in rat fetal brown adipocytes and, as a consequence, its inhibition enhanced IGF-I-induced UCP-1 and insulin-induced FAS mRNAs expression (Porras et al., 1998). Therefore, all these data indicate that the PI3K/Akt pathway is essential for the IGF-I/insulin-induced differentiation of brown adipocytes, while the ERKs pathway exerts an inhibitory effect (Fig. 19.4).

T3 and retinoic acid are also positive regulators of the thermogenic differentiation of BAT, which bind to their nuclear receptors (Guerra et al., 1994; Rabelo et al., 1996a; 1996b; Alvarez et al., 1995; 2000), leading to transactivation of the *ucp-1* gene (Fig. 19.2).

The direct effect of T3 on UCP-1 expression has not been well characterized until recently. First studies done *in vivo* with hypothyroid and euthyroid rats indicated that T3 *per se* was not able to induce UCP-1 expression, although it was required for the cold- and NA-induced UCP-1 mRNA expression (Silva, 1988; Bianco et al., 1988). Similarly, studies carried out with freshly dispersed brown adipocytes also indicated that T3

was required for maximal effect of cAMP on UCP-1 mRNA expression (Bianco et al., 1992). Different from this, in rat fetal brown adipocytes maintained in culture it was demonstrated for the first time that prolonged treatment with T3 induced UCP-1 mRNA and protein expression (Guerra et al., 1994b). This effect of T3 correlated with the expression of different types of thyroid receptors mRNAs and with a strong increase in the nuclear T3-binding capacity. The mechanisms mediating this direct effect of T3 on UCP-1 expression include an increase of gene transcription rate and stabilization of UCP-1 mRNA (Guerra et al., 1996). The activation of transcription is exerted through binding of the thyroid receptors to the two thyroid response elements (TRE) located in the *ucp-1* gene enhancer, that were later identified (Rabelo et al., 1996a).

Retinoic acid (RA) was also described as a transcriptional activator of the *ucp-1* gene, which acts through the RA-responsive region of the enhancer (Alvarez et al., 1995; Larose et al., 1996). Different types of retinoic acid receptors mediate this effect of RA *in vitro* (Alvarez et al., 1995; Larose et al., 1996; Rabelo et al., 1996b; Alvarez et al., 2000) and might also *in vivo* (Puigserver et al., 1996).

BAT differentiation and/or function can be also positively modulated by leptin, which is particularly important in relation to obesity due to its action activating diet-induced thermogenesis. Leptin administration to rats increases sympathetic activity in BAT, UCP-1 mRNA levels (Scarpace et al., 1997) and glucose utilization (Siegrist-Kaiser and Meier, 1997). In addition, leptin induces malic enzyme and lipoprotein lipase expression in brown adipocytes maintained in primary culture (Siegrist-Kaiser and Meier, 1997). On the other hand, leptin action also seems to be required for basal and cold-stimulated PGC-1 expression in BAT from rodents (Kakuma et al., 2000).

The negative regulation of BAT differentiation is not so well characterized, TNF-α being the most important signal. The first identified negative action of TNF-α on BAT was the induction of insulin resistance (Valverde et al., 1998b). Thus, pretreatment of rat fetal brown adipocytes with

TNF-α impaired insulin signaling and inhibited insulin-induced expression of several adipogenic differentiation mRNAs markers such as FAS and G3PD, as well as those for the transcription factors C/EBPα and β and PPARγ (Valverde et al., 1998b). This effect seemed to be due to the impairment of the association of IRS-2 with PI3K, which led to a decrease in the PI3K activity. However, recent data indicate that TNF-α is also a direct inhibitor of the thermogenic and adipogenic differentiation, acting through mechanisms mediated by TNF-α receptors, but not related with the induction of insulin resistance (Nisoli et al., 2000; Valladares et al., 2001). Thus, it has been demonstrated that TNF-α inhibits the expression of UCP-1 and adipocyte-specific genes in rat fetal brown adipocytes (Valladares et al., 2001). ERKs activation by TNF-α seems to mediate the decrease in UCP-1 expression, but not that of adipogenic genes (Valladares et al., 2001), although ERKs exerts a negative effect on these genes, too (Porras et al., 1998). In contrast, p38MAPK appears to play a positive role in thermogenic and adipogenic differentiation (Fig. 19.5), probably through up-regulation of C/EBP-α and β proteins (Valladares et al., 2001). Although C/EBP-α and β could be mediators of the inhibitory action of TNF-α because their levels were reduced upon TNF-α treatment and a partial reversion was observed when ERKs activation was blocked, other transcription factors might explain the different regulation of UCP-1 and adipogenic genes by ERKs (Valladares et al., 2001). According to this, recent data indicate that TNF-α induces PPARγ phosphorylation via ERKs activation, which could mediate inhibition of UCP-1 expression but not inhibition of adipogenesis (Porras et al., 2002). Thus, the negative effect of TNF-α on UCP-1 mRNA expression under basal conditions or in cells treated with the selective PPARγ agonist, rosiglitazone, can be reversed by inhibition of ERKs. In contrast, FAS and malic enzyme mRNAs downregulation or the decrease in lipid accumulation cannot be prevented (Porras et al., 2002).

The physiological relevance of TNF-α as a negative regulator of BAT differentiation has been evidenced by experiments performed in mice lacking TNF-α receptors (Nisoli et al., 2000). TNF-α was proposed to be responsible for the reduction in UCP-1 expression observed in BAT of obese (*ob/ob*) mice, because the absence of TNF receptors resulted in an increase in UCP-1 and β₃-adrenergic receptor expression in obese mice. Therefore, TNF-α appears to mediate the inhibition of differentiation occurring in BAT from obese mice and could also be a relevant signal regulating the involution of BAT after weaning.

6. Apoptosis of brown adipocytes: signals and signal transduction pathways involved in its regulation

Apoptosis is a specific process of cell elimination in normal tissues, that is an essential physiological mechanism for the development, maintenance and repair of tissues as well as for the regulation of many forms of cancer (Hengartner, 2000). Until very recently, almost nothing was known about the role of apoptosis in the regulation of adipose tissue. However, based on recent reports it appears to be an important mechanism regulating the loss of white and brown adipocytes under physiological and/or pathological conditions, TNF-α being a relevant positive regulator (Nissoli et al., 1997; Porras et al., 1997; Prins et al., 1997).

TNF-α induces apoptosis in rat fetal brown adipocytes maintained in primary culture (Porras et al., 1997) and brown adipocytes from young rats differentiated in culture (Nisoli et al., 1997). Although simultaneous treatment with TNF-α and cycloheximide enhances and accelerates the effect of TNF-α inducing apoptosis (Nisoli et al., 1997), cycloheximide is not required (Nisoli et al., 1997; Porras et al., 1997). Therefore, TNF-α would be the negative physiological signal inducing programmed cell death of brown adipocytes in order to control BAT homeostasis. In addition, TNF-α was also proposed to be responsible for the apoptosis produced in obesity, since in obese rats a parallel increase in TNF-α production and in the number of apoptotic brown adipocytes is produced (Nisoli et al., 1997). This idea was confirmed by experiments carried out with genetically obese (*ob/ob*) mice with targeted null mutation in the gene encoding p75 and/or p55 TNF-α receptor (TNFR) (Nisoli et al., 2000). In the *ob/ob* mice lacking p55 TNFR (*ob/ob*-p55–/–) or both TNFR (*ob/ob*-p55–/ – p75–/–) the number of apoptotic nuclei were significantly lower than in the wild type *ob/ob* and very close to control lean mice (Nisoli et al., 2000). In *ob/ob* mice lacking p75 TNFR the number of apoptotic cells was higher than in *ob/ob*-p55–/– or *ob/ob*-p55–/– p75–/–, but lower than in those seen in obese controls (Nisoli et al., 2000). Therefore, the activity of TNF-α is required for the obesity-induced BAT apoptosis, being p55 receptor the main mediator, although p75 TNFR seems to play a role, too. Similarly, TNF-α could also be an important physiological mediator of the regression of BAT after weaning by inducing apoptosis in these cells.

In relation with the possible molecular mechanisms responsible for the TNF-α induced apoptosis in brown adipocytes, p38 MAPK is a mediator of this process, while ERKs activation exerts an antiapoptotic effect and JNKs seems not to play any role (Fig. 19.5) (Valladares et al., 2000a). Activating transcription-1 (ATF1) is a potential downstream effector of p38MAPK upon TNF-α treatment, because its phosphorylation is extensively decreased by the p38MAPK inhibitor, SB203580. On the other hand, c-Jun could be a potential target for ERKs and p38MAPK, since TNF-α-induced c-Jun mRNA expression is mediated by ERKs, while p38MAPK exerts an inhibitory effect. However, TNF-α induced NF-κB DNA-binding activity seems to be independent of p38MAPK and ERKs. On the other hand, TNF-α-induced apoptosis in brown adipocytes does not involve changes in the expression of the antiapoptotic gene, bcl-2, and/or the pro-apoptotic gene, bax (Briscini et al., 1998).

Although TNF-α seems to be a relevant physiological signal inducing brown adipocytes apoptosis, other signals can also play a role. Thus, the expression of the oncogene H-ras (H-ras[Lys12]) was shown to induce apoptosis in brown adipocyte derived cell lines upon serum deprivation by a mechanism requiring Raf-1 association with phosphorylated Bcl-2, down-regulation of Bcl-2 and up-regulation of Bcl-xS expression (Navarro et al., 1999).

As opposed to TNF-α, NA and IGF-I are survival factors for brown adipocytes. IGF-I rescues from TNF-α-induced apoptosis (Porras et al., 1997) and from H-ras-induced apoptosis upon serum-deprivation by a mechanism depending on ERKs and PI3K (Fig. 19.4) (Navarro et al, 1998). NA also protects against TNF-α (Nisoli et al., 1997; 2001) and serum-deprivation induced apoptosis (Briscini et al., 1998). This effect of NA mimics the action of cold exposure (Nisoli et al., 1997; Briscini et al., 1998; Linquist and Rehmark, 1998), which induces the survival of brown adipocytes. Therefore, NA appears to be the mediator of the survival effect of cold exposure. NA can also play an important role in other situations such as diet-induced thermogenesis and as a consequence the reduced catecholaminergic activity produced in obesity could contribute to the induction of apoptosis (Nisoli et al., 1997).

Different mechanisms have been proposed to explain the protective effect of NA on brown adipocytes such as an increase in the Bcl-2/Bax ratio (Briscini et al., 1998) and in the expression of Hsp70 through NO generation (Nisoli et al., 2001). In addition, the activation of ERKs by NA through β- and α₁ adrenergic receptors-mediated cAMP and Ca^{2+} increases protected mouse brown adipocytes from apoptosis (Linquist and Rehmark, 1998). NA also induced the expression of basic fibroblast growth factor (bFGF), which appears to promote cell survival through a MEK-dependent activation of ERKs, too (Linquist and Rehmark, 1998). Thus, it is possible that NA released by the sympathetic nerves upon cold stimulation stimulates survival of brown adipocytes through direct activation of ERKs at short time periods (Fig. 19.3), maintaining this effect through an bFGF-induced ERKs activation.

Taken together all the data about the function of ERKs in apoptosis and survival, it is easy to understand that ERKs are relevant mediators of brown adipocytes survival upon NA, IGF-I and b-FGF activation or even when activated by a proapoptotic factor such as TNF-α. What is still unknown is the mechanism mediating this effect, although it is very likely that ERKs can promote survival by regulating the expression of anti-apoptotic and pro-apoptotic genes such as bcl-xS (Navarro et al, 1998) or other genes through the regulation of transcription factors such as c-Jun (Valladares et al., 2000a).

7. Conclusions

BAT development, function and involution is controlled by many extracellular signals, which regulate proliferation, differentiation and apoptosis of brown adipocytes. Some of them, such as T3 or retinoic acid bind to their nuclear receptors, trans-activating *ucp-1* gene, while others bind to membrane receptors activating different signaling pathways. Among these pathways, ERKs, PI3K, p38MAPK and the cAMP/PKA cascades are of particular relevance. Activation of ERKs is required for proliferation and survival induced by several signals, while playing a negative role in differentiation. In contrast, PI3K mediates IGF-I/insulin-induced differentiation and cAMP/PKA cascade is essential for NA-induced UCP-1 expression. On the other hand, p38MAPK has been recently identified as an important mediator of differentiation and apoptosis upon activation by different extracellular signals.

Acknowledgements

We are grateful to Margarita Fernández for comments on the manuscript. We apologize to the authors whose original work is not included in the references owing to space limitations.

References

Alvarez, R., de Andrés, J., Yubero, P., Viñas, O., Mampel, T., Iglesias, R., Giralt, M. and Villarroya, F. (1995). A novel regulatory pathway of brown fat thermogenesis. Retinoic acid is a transcriptional activator of the mitochondrial uncoupling protein gene. J. Biol Chem. 270, 5666–5673.

Alvarez, R., Checa, M.L., Brun, S., Viñas, O., Mampel, T., Iglesias, R., Giralt, M. and Villarroya, F. (2000). Both retinoic-acid-receptor- and retinoid-X-receptor-dependent signalling pathways mediate the induction of the brown-adipose-tissue-uncoupling-protein-1 gene by retinoids. Biochem. J. 345, 91–97.

Barak, Y., Nelson, M.C., Ong, E.S., Jones, Y.Z., Ruiz-Lozano, P., Chien, K.R., Koder, A. and Evans, R.M. (1999). PPARγ is required for placental, cardiac and adipose tissue development. Mol. Cell 4, 585595.

Barberá, M.J., Schlüter, A., Pedraza, N., Iglesias, R., Villarroya, F. and Giralt, M. (2001). Peroxisome proliferator-activated receptor α activates transcription of the brown fat uncoupling protein-1 gene. J. Biol. Chem. 276, 1486–1493.

Benito, M., Porras, A. and Santos, E. (1993). Establishment of permanent brown adipocyte cell lines achieved by transfection with SV40 large T antigen and ras genes. Exp. Cell Res. 209, 248–254.

Bianco, A.C., Sheng, X.Y. and Silva, J.E. (1988). Triiodothyronine amplifies norepinephrine stimulation of uncoupling protein gene transcription by a mechanism not requiring protein synthesis. J. Biol. Chem. 263, 18168–18175.

Bianco, A.C., Kieffer, J.D. and Silva, J.E. (1992). Adenosine 3′,5′-monophosphate and thyroid hormone control of uncoupling protein messenger ribonucleic acid in freshly dispersed brown adipocytes. Endocrinology 130, 2625–2633.

Briscini, L., Tonello, C., Dioni, L., Carruba, M.O. and Nisoli, E. (1998). Bcl-2 and Bax are involved in the sympathetic protection of brown adipocytes from obesity-linked apoptosis. FEBS Lett. 431, 80–84.

Bronnikov, G., Houstek, J. and Nedergaard, J. (1992). Beta-adrenergic, cAMP-mediated stimulation of proliferation of brown fat cells in primary culture. Mediation via beta 1 but not via beta 3 adrenoceptors. J Biol Chem. 267, 2006–2013.

Bronnikov, G., Bengtsson, T., Kramarova, L., Golozoubova, V., Cannon, B. and Nedergaard, J. (1999a). Beta1 to beta3 switch in control of cyclic adenosine monophosphate during brown adipocyte development explains distinct beta-adrenoceptor subtype mediation of proliferation and differentiation. Endocrinology 140, 4185–4197.

Bronnikov, G.E, Zhang, S.J., Cannon, B. and Nedergaard, J. (1999b). A dual component analysis explains the distinctive kinetics of cAMP accumulation in brown adipocytes. J Biol Chem. 274, 37770–37780.

Cao, W., Medvedev, A.V., Daniel, K.W. and Collins, S. (2001). β-Adrenergic activation of p38 MAP kinase in adipocytes: cAMP induction of the uncoupling protein 1 (UCP1) gene requires p38 MAP kinase. J Biol Chem. 276, 27077–27082.

Cassard-Doulcier, A.M., Gelly, C., Fox, N., Schrementi, J., Raimbault, S., Klaus, S., Forest, C., Bouillaud, F. and Ricquier, D. (1993). Tissue-specific and beta-adrenergic regulation of the mitochondrial uncoupling protein gene: control by cis-acting elements in the 5′-flanking region. Mol. Endocrinol. 7, 497–506.

Duchamp, C., Burton, K.A., Geloen, A. and Dauncey, M.J. (1997). Transient upregulation of IGF-I gene expression in brown adipose tissue of cold-exposed rats. Am. J. Physiol. 272, E453–460.

Fredriksson, J.M., Thonberg, H., Ohlson, K.B.E., Ohba, K., Cannon, B. and Nedergaard, J. (2001). Analysis of inhibition by H89 of UCP-1 gene expression and thermogenesis indicates protein kinase A mediation of β_3-adrenergic signalling rather than β_3-adrenoceptor antagonism by H89. Bichim. Biophys. Acta 1538, 206–217.

García, B. and Obregón, M.J. (1997). Norepinehrine potentiates the mitogenic effect of growth factors in quiescent brown preadipocytes: relationship with uncoupling protein messenger ribonucleic acid expression. Endocrinology 138, 4227–4233.

Geloen, A., Collet, A.J., Guay, G. and Bukowiecki, L.J. (1988). Beta-adrenergic stimulation of brown adipocyte proliferation. Am. J. Physiol. 254, C175–182.

Geloen, A., Collet, A.J and Bukowiecki, L.J. (1992). Role of sympathetic innervation in brown adipocyte proliferation. Am. J. Physiol. 263, R1176–R1181.

Gónzalez-Barroso, M.M., Pecqueur, C., Gelly, Ch., Sanchis, D., Alves-Guerra, M.C., Bouillaud, F., Ricquier, D. and Cassard-Doulcier, A.M. (2000). Transcriptional activation of the human *ucp-1* gene in a rodent cell line. J. Biol. Chem. 275, 31722–31732.

Guerra, C., Benito, M. and Fernández, M. (1994a). IGF-I induces the uncoupling protein gene expression in fetal rat brown adipocyte primary cultures: role of C/EBP transcription factors. Biochem. Biophys. Res. Commun. 201, 813–819.

Guerra, C., Navarro, P., Valverde, A.M., Arribas, M., Bruning, J., Kozak, L.P., Kahn, C.R. and Benito, M. (2001). Brown adipose tissue-specific insulin receptor knockout shows diabetic phenotype without insulin resistance. J. Clin. Invest. 108, 1205–1213.

Guerra, C., Porras, A., Roncero, C., Benito, M. and Fernández, M. (1994b). Triiodothyronine induces the expression of the uncoupling protein in long term fetal brown adipocyte primary cultures: Role of nuclear thyroid hormone receptor expression. Endocrinology 134, 1067–1074.

Guerra, C., Roncero, C., Porras, A., Fernández, M. and Benito, M. (1996). Triiodothyronine induces the transcription of the uncoupling protein gene and stabilizes its mRNA in fetal rat brown adipocyte primary cultures. J. Biol. Chem. 271, 2076–2081.

Hengartner, M.O. (2000). The biochemistry of apoptosis.
Nature 207, 769–776.

Kakuma, T., Wang, Z.W., Pan, W., Unger, R,H. and Zhou, Y.T. (2000). Role of leptin in peroxisome proliferator-activated receptor gamma coactivator-1 expression. Endocrinology 141, 4576–4582.

Klein, J., Fasshauer, M., Benito, M. and Kahn, C.R. (2000). Insulin and the β_3-adrenoceptor differentially regulate uncoupling protein-1 expression. Mol. Endocrinol. 14, 764–773.

Konishi, M., Mikami, T., Yamasaki, M., Miyake, A. and Itoh, N. (2000). Fibroblast growth factor-16 is a growth factor for embryonic brown adipocytes. J. Biol. Chem. 275, 12119–12122.

Kopecky, J., Baudysova, M., Zanitti, F., Janikova, D., Pavelka, S. and Houstek, J. (1990). Synthesis of mitochondrial uncoupling protein in brown adipocytes differentiated in cell culture. J. Biol. Chem. 265, 22204–22209.

Kozak, U.C., Kopecky, J., Teisinger, J., Enerback, S., Boyer, B. and Kozak, L.P. (1994). An upstream enhancer regulating brown-fat-specific expression of the mitochondrial uncoupling protein gene. Mol. Cell. Biol. 14, 59–67.

Larose, M., Cassard-Doulcier, A.M., Fleury, C., Serra, F., Champigny, O., Bouillaud, F. and Ricquier, D. (1996). Essential cis-acting elements in rat uncoupling protein gene are in an enhancer containing a complex retinoic acid response domain. J. Biol. Chem. 271, 31533–31542.

Lindquist, JM, Fredriksson, J.M., Rehnmark, S., Cannon, B. and Nedergaard, J. (2000). Beta 3- and alpha1-adrenergic Erk1/2 activation is Src- but not Gi-mediated in brown adipocytes. J. Biol. Chem. 275, 22670–22677.

Lindquist, J.M. and Rehnmark, S. (1998). Ambient temperature regulation of apoptosis in brown adipose tissue. Erk1/2 promotes norepinephrine-dependent cell survival. J. Biol. Chem. 273, 30147–30156.

Linhart, H.G., Ishimura-Oka, K., DeMayo, F., Kibe, T., Repka, D., Poindexter, B., Bick, R.J. and Darlington, G.J. (2001). C/EBPα is required for differentiation of white, but not brown, adipose tissue. Proc. Natl. Acad. Sci. 98, 12532–12537.

Lorenzo, M., Valverde, A.M., Teruel, T. and Benito, M. (1993). IGF-I is a mitogen involved in differentiation-related gene expression in fetal rat brown adipocytes. J. Cell Biol. 123, 1567–1575.

Navarro, P., Valverde, A.M., Benito, M. and Lorenzo, M. (1998). Insulin/IGF-I rescues immortalized brown adipocytes from apoptosis down-regulating Bcl-xS expression, in a PI 3-kinase- and map kinase-dependent manner. Exp. Cell Res. 243, 213–221.

Navarro, P., Valverde, A.M., Benito, M. and Lorenzo, M. (1999). Activated Ha-ras induces apoptosis by association with phosphorylated Bcl-2 in a mitogen-activated protein kinase-independent manner. J. Biol. Chem. 274,

18857–18863.

Nisoli, E., Briscini, L., Tonello, C., De Giuli-Morghen, C. and Carruba, M.O. (1997). Tumor necrosis factor-α induces apoptosis in rat brown adipocytes. Cell Death Differ. 4, 771–778.

Nisoli, E., Briscini, L., Giordano, A., Tonello, C., Wiesbrock, S.M., Uysal, K.T., Cinti, S., Carruba, M.O. and Hotamisligil, G.S. (2000). Tumor necrosis factor alpha mediates apoptosis of brown adipocytes and defective brown adipocyte function in obesity. Proc. Natl. Acad. Sci. USA 97, 8033–8038.

Nisoli, E., Regianini, L., Bulbarelli, A., Briscini, L., Breacale, R. and Carruba, M.O. (2001). Protective effects of noradrenaline against tumor necrosis factor-alpha-induced apoptosis in cultured rat brown adipocytes: role of nitric oxide-induced heat shock protein 70 expression. Int. J. Obes. 25, 1421–1430.

Porras, A., Álvarez, A.M., Valladares, A. and Benito, M. (1997). TNF-alpha induces apoptosis in rat fetal brown adipocytes in primary culture. FEBS Lett. 416, 324–328.

Porras, A., Álvarez, A.M., Valladares, A. and Benito, M. (1998). p42/p44 mitogen-activated protein kinases activation is required for the insulin-like growth factor-I/insulin induced proliferation, but inhibits differentiation, in rat fetal brown adipocytes. Mol. Endocrinol. 12, 825–834.

Porras, A., Fernández, M. and Benito, M. (1989). Adrenergic regulation of the uncoupling protein expression in foetal rat brown adipocytes in primary culture. Biochem. Biophys. Res. Commun. 163, 541–547.

Porras, A., Hernández, E.R. and Benito, M. (1996). Ras proteins mediate induction of uncoupling protein, IGF-I, and IGF-I receptor in rat fetal brown adipocyte cell lines. DNA Cell Biol. 15, 921–928.

Porras, A., Valladares, A., Álvarez, A.M., Roncero, C. and Benito, M. (2002). Differential role of PPARγ in the regulation of UCP-1 and adipogenesis by TNF-α in brown adipocytes. FEBS Lett. 520, 58–62.

Prins, J.B., Niesler, C.U., Winterford, C.M., Bright, N.A., Siddle, K., O'Rahilly, S., Walker, N.I. and Cameron, D.P. (1997). Tumor necrosis factor-alpha induces apoptosis of human adipose cells. Diabetes 46, 1939–1944.

Puigserver, P., Vázquez, F., Bonet, M.L., Picó, C. and Palou, A. (1996). *In vitro* and *in vivo* induction of brown adipocyte uncoupling protein (thermogenin) by retinoic acid. Biochem. J. 317, 827–833.

Puigserver, P., Wu, Z., Park, C.W., Graves, R., Wright, M. and Spiegelman, B.M. (1998). A cold-inducible coactivator of nuclear receptors linked to adaptive thermogenesis. Cell 92, 829–839.

Rabelo, R., Reyes, C., Schifman, A. and Silva, J.E. (1996a). Interactions among receptors, thyroid hormone response elements, and ligands in the regulation of the rat uncoupling protein gene expression by thyroid hormone. Endocrinology 137, 3478–3487.

Rabelo, R., Reyes, C., Schifman, A. and Silva, J.E. (1996b). A complex retinoic acid response element in the uncoupling protein gene defines a novel role for retinoids in thermogenesis. Endocrinology 137, 3488–3496.

Rehnmark, S., Nechad, M., Herron, D., Cannon, B. and Nedergaard, J. (1990). Alpha- and beta-adrenergic induction of the expression of the uncoupling protein thermogenin in brown adipocytes differentiated in culture. J. Biol. Chem. 265, 16464–16471.

Ricquier, D., Bouillaud, F., Toumelin, P., Mory, G., Bazin, R., Arch, J. and Penicaud, L. (1986). Expression of uncoupling protein mRNA in thermogenic or weakly thermogenic brown adipose tissue. Evidence for a rapid beta-adrenoreceptor-mediated and transcrptionally regulated step during activation of thermogenesis. J. Biol. Chem. 261, 13905–13910.

Ross, S.R., Choy, L., Graves, R.A., Fox, N., Solevjeva, V., Klaus, S., Ricquier, D. and Spiegelman, B.M. (1992). Hibernoma formation in transgenic mice and isolation of a brown adipocyte cell line expressing the uncoupling protein gene. Proc. Natl. Acad. Sci. USA 89, 7561–7565.

Scarpace, P.J., Matheny, M., Pollock, B.H. and Tumer, N. (1997). Leptin increases uncoupling protein expression and energy expenditure. Am. J. Physiol. 273, E226–E230.

Shimizu, Y., Tanishita, T., Minokoshi, Y. and Shimazu, T. (1997). Activation of mitogen-activated protein kinase by norepinephrine in brown adipocytes from rats. Endocrinology 138, 248–253.

Siegrist-Kaiser, C.A., Pauli, V., Juge-Aubry, C.E., Boss, O., Pernin, A., Chin, W.W., Cusin, I., Rohner-Jeanrenaud, F., Burger, A.G., Zapf, J. and Meie, C.A. (1997). Direct effects of leptin on brown and white adipose tissue. J. Clin. Invest. 100, 2858–2864.

Silva, J.E. (1988). Full expression of uncoupling protein gene requires the concurrence of norepinephrine and triiodothyronine. Mol. Endocrinol. 2, 706–713.

Spiegelman, B.M. and Flier, J.S. (1996). Adipogenesis and obesity: rounding out the big picture. Cell 87, 377–389.

Tai, T.A.C., Jennermann, C., Brown, K.K., Oliver, B.B., MacGinnitie, M.A., Wilkison, W.O., Brown, H.R., Lehmann, J.M., Kliewer, S. A., Morris, D.C. and Graves, R.A. (1996). Activation of the nuclear receptor peroxisome proliferator-activated γ promotes brown adipocyte differentiation. J. Biol. Chem. 271, 29909–29914.

Tanaka, T., Yoshida, N., Kishimoto, T. and Akira, S. (1997). Defective adipocyte differentiation in mice lacking the C/EBPβ and/or C/EBPδ gene. EMBO J. 16, 7432–7443.

Teruel, T., Valverde, A.M., Alvarez, A., Benito, M. and Lorenzo, M. (1995). Differentiation of rat brown adipocytes during late foetal development: role of insulin-like growth factor I. Biochem. J. 310, 771–776.

Teruel, T., Valverde, A.M., Benito, M. and Lorenzo, M. (1996). Insulin-like growth factor I and insulin induce adipogenic-related gene expression in fetal brown adipocyte primary cultures. Biochem. J. 319, 627–632.

Trayhurn, P. and Milner, RE. (1989). A commentary on the interpretation of *in vitro* biochemical measures of brown adipose tissue thermogenesis. Can. J. Physiol. Pharmacol. 67, 811–819.

Valladares, A., Álvarez, A.M., Ventura, J.J., Roncero, C., Benito, M. and Porras, A. (2000a). p38 mitogen-activated protein kinase mediates tumor necrosis factor-alpha-induced apoptosis in rat fetal brown adipocytes. Endocrinology 141, 4383–4395.

Valladares, A., Porras, A., Álvarez, A.M., Roncero, C. and Benito, M. (2000b) Noradrenaline induces brown adipocytes cell growth via beta-receptors by a mechanism dependent on ERKs but independent of cAMP and PKA. J. Cell. Physiol. 185, 324–330.

Valladares, A., Roncero, C., Benito, M. and Porras, A. (2001). TNF-alpha inhibits UCP-1 expression in brown adipocytes via ERKs. Opposite effect of p38MAPK. FEBS Lett. 493, 6–11.

Valverde, A.M, Kahn, C.R. and Benito M. (1999). Insulin signaling in insulin receptor substrate (IRS)-1-deficient brown adipocytes: requirement of IRS-1 for lipid synthesis. Diabetes 48, 2122–2131.

Valverde, A.M., Lorenzo, M., Navarro, P. and Benito, M. (1997). Phosphatidylinositol 3-kinase is a requirement for insulin-like growth factor I-induced differentiation, but not for mitogenesis, in fetal brown adipocytes. Mol. Endocrinol. 11, 595–607.

Valverde, A.M., Lorenzo, M., Pons, S., White, M.F. and Benito, M. (1998a). Insulin receptor substrate (IRS) proteins IRS-1 and IRS-2 differential signaling in the insulin/insulin-like growth factor-I pathways in fetal brown adipocytes. Mol. Endocrinol.12, 688–697.

Valverde, A.M., Mur, C., Pons, S., Álvarez, A.M., White, M.F., Kahn, C.R. and Benito, M. (2001). Association of insulin receptor substrate 1 (IRS-1) y895 with Grb-2 mediates the insulin signaling involved in IRS-1-deficient brown adipocyte mitogenesis. Mol Cell Biol. 21, 2269–2280.

Valverde, A.M., Teruel, T., Navarro, P., Benito, M. and Lorenzo, M. (1998b). Tumor necrosis factor-α causes insulin receptor substrate-2-mediated insulin resistance and inhibits insulin-induced adipogenesis in fetal brown adipocytes. Endocrinology 139, 1229–1238.

Wu, Z., Puigserver, P., Andersson, U., Zhang, Ch., Adelmant, G., Mootha, V., Troy, A., Cinti, S., Lowell, B., Scarpulla, R.C. and Spiegelman, B. (1999). Mechanisms controlling mitochondrial biogenesis and respiration trough the thermogenic coactivator PGC-1. Cell 98, 115–124.

Yamashita, H., Sato, N., Kizaki, T., Oh-ishi, S., Segawa, M., Saitoh, D., Ohira, Y. and Ohno, H. (1995). Norepinephrine stimulates the expression of fibroblast growth factor 2 in rat brown adipocyte primary culture. Cell Growth Diff. 6, 1457–1462.

Yubero, P., Barberá, M.J., Alvarez, R., Viñas, O., Mampel, T, Iglesias, R., Villarroya, F. and Giralt, M. (1998). Dominant negative regulation by c-Jun of transcription of the uncoupling protein-1 gene through a proximal cAMP-regulatory element: a mechanism for repressing basal and norepinephrine-induced expression of the gene before brown adipocyte differentiation. Mol. Endocrinol. 12, 1023–1037.

Yubero, P., Manchado, C., Cassard-Doulcier, A.M., Mampel, T., Viñas, O., Iglesias, R. Giralt, M. and Villarroya, F. (1994). CCAAT/enhancer binding proteins alpha and beta are transcriptional activators of the brown fat uncoupling protein gene promoter. Biochem. Biophys. Res. Commun. 198, 653–659.

Zhao, J., Cannon, B. and Nedergaard, J. (1997). α_1-Adrenergic stimulation potentiates the thermogenic action of β_3-adrenoreceptor-generated cAMP in brown fat cells. J. Biol. Chem. 272, 32847–32856.

Sensing, Signaling and Cell Adaptation. Edited by K.B. Storey and J.M. Storey
© *2002 Elsevier Science B.V. All rights reserved.*

CHAPTER 20

Control Analysis of Metabolic Depression

R. Keira Curtis, Tammie Bishop and Martin D. Brand
MRC Dunn Human Nutrition Unit, Hills Road, Cambridge, UK

1. Introduction

Metabolic control analysis has been used to look at many metabolic systems and so has relevance for those scientists that are researching metabolic depression. We provide an introduction to the theory and techniques used in metabolic control analysis, give several relevant examples of how it has been used to examine metabolism and signalling, and finally suggest how it could be used to investigate the mechanisms by which cells depress their metabolism.

2. What is control analysis?

Metabolic control analysis is a very powerful approach that is used to quantify the distribution of control over concentrations and fluxes in steady state metabolic pathways. It recognises that control is distributed among the many enzymes or steps in a given pathway or system, rather than being at a single site, and quantifies that control. The concept of the rate limiting enzyme is widely known, but if there were actually a single rate limiting step in a pathway, then changing the activity of that enzyme would be solely responsible for the change in the pathway flux, and altering any other step would have no effect at all. Frequently, control analysis finds that there is no single factor controlling the system flux, and instead control is distributed between several or all of the enzymes in the pathway.

Control analysis of metabolic systems was first described by Kacser and Burns (Kacser and Burns,

1973; Kacser et al., 1995) and by Heinrich and Rapoport (1974), more than 25 years ago. The essence of the approach is as follows. First, a metabolic system is described in terms of system variables (concentrations and fluxes) and parameters (enzymes, effectors such as hormones, inhibitors, etc). The system could be as simple as a short metabolic pathway or as large and complex as a multicellular organism. After the variables are measured in a reference state, the system is disturbed in some way. This could be a change in the activity of an enzyme, or a change in the concentration of a metabolite or effector. The two conditions are then compared in order to calculate various types of coefficients that describe the effect of small changes (perturbations) on the system. Alternatively, the kinetics of the enzymes may be measured in order to describe the behaviour of the system. Control analysis was originally described using an unbranched chain of enzymes as an example, but various theorems have expanded it to include more complex system structures such as branches or cycles (Fell and Sauro, 1985; Hofmeyr et al., 1986; Kacser, 1983).

3. Applications of control analysis

Control analysis can be used to investigate how the internal regulation of a system maintains homeostasis. The system's component enzymes regulate the flux, the intermediate concentrations and the enzyme rates, to balance precisely metabolite supply and demand. Control analysis has been used in the past to look at the internal regulation of glycolysis (Groen et al., 1986), mitochondrial respiration (Hafner et al., 1990; Rolfe and Brand, 1996), whole cell metabolism (Ainscow and Brand, 1999a; 1999c; Brand and Ainscow, 2000; Cornish-Bowden and Cardenas, 1999) and many other systems.

Metabolic control analysis has strong commercial potential. For example, it can suggest ways of manipulating a system to modify biosynthetic output (Fell, 1998; Fell and Thomas, 1995; Kacser and Acerenza, 1993; Kholodenko et al., 1998). In these cases, it is often useful to look at a system in terms of the supply and demand of metabolites or metabolic end products (Brand, 1996; Hafner et al., 1990). Metabolism is not just regulated by its component enzymes and intermediate concentrations, but also by changes in gene expression. This was addressed by ter Kuile and Westerhoff (2001), who extended control analysis to quantify the regulation of metabolism by these hierarchical mechanisms. Control analysis also has applications in genomics; it may be used to determine the function of novel genes, using an approach called FANCY (functional analysis by co-responses in yeast) (Oliver et al., 1998; Raamsdonk et al., 2001; Teusink et al., 1998).

Once the regulation within a system is understood, the analysis can be expanded to describe regulation by the external environment, the responses to external effectors such as hormones, drugs or poisons, or changes in parameters such as temperature or pH. This leads to another range of applications. Regulation analysis (Brand, 1997; Brand and Ainscow, 2000) can be used to find the sites of action of effectors and so has potential for finding drug targets (Kesseler and Brand, 1994b; Murphy, 2001). An effector may act directly on its target protein, process or enzyme, or indirectly, for example by acting through changes in metabolites that affect the target enzyme. The amount of direct and indirect response can be quantified using regulation analysis (Ainscow and Brand, 1999c). Enzyme kinetics in the presence and absence of effectors are compared to determine which change directly or indirectly as a result of the effector. For example, regulation analysis of potato tuber mitochondria located the sites of action of cadmium on

oxidative phosphorylation (Kesseler and Brand, 1994a–c). Control analysis has been applied to cell signalling, to investigate the relative importance of various signalling routes in transmitting a particular signal to its destination enzymes. Krauss and Brand investigated the signalling initiated by the mitogen Concanavalin A (Con A) in rat thymocytes and how this stimulated energy metabolism (Krauss et al., 1999; Krauss and Brand, 2000). Modular regulation analysis has also been applied to DNA microarray data to distinguish changes in gene expression that are important and involved in the physiological response from those that are not (Curtis and Brand, 2001; Brand and Curtis, 2002).

We propose a general method that can be used to investigate and quantify the signalling processes which initiate metabolic depression in response to an external stimulus.

4. Theory

4.1. Definition of coefficients

Control analysis uses coefficients to describe the distribution of control in a system. The coefficients are calculated using normalised changes and so are dimensionless, avoiding problems with units.

First are flux control coefficients (C). These describe how much control a particular enzyme, step or group of processes has over the flux. Flux control coefficients are calculated using Eq. 20.1 (the ratio of the fractional changes in the enzyme and in the flux) and depict how a small change in the activity of an enzyme affects the steady-state flux through the system.

$$C_{\text{enzyme}}^{\text{flux}} = \frac{\delta\text{flux}}{\text{flux}} \bigg/ \frac{\delta\text{enzyme}}{\text{enzyme}} \qquad (20.1)$$

A positive control coefficient means that an increase in the enzyme activity will result in an increase in the flux; a negative control coefficient means an increase in the enzyme activity will decrease the flux. Enzymes that have larger effects on the flux have higher flux control coefficients than those that only produce a small change in the system flux. An enzyme with a flux control coefficient of 0.5 over a system flux means that a 1% increase in enzyme activity results in a 0.5% increase in the system flux. There is a finite amount of control in the system, so the flux control coefficients for control over any particular flux sum to one (the summation theorem, Eq. 20.2).

$$\sum_{\text{all enzymes}} C_{\text{enzyme}}^{\text{flux}} = 1 \qquad (20.2)$$

The coefficients typically have values between zero and one, showing that control is distributed among the component enzymes, although some may have higher flux control coefficients than others. In systems containing branched pathways, moiety-conserved cycles or unusual kinetics, the coefficients may be more than one or negative (Fell, 1997; Hofmeyr et al., 1986; Kacser, 1983). Control coefficients act globally, meaning that after the small change is imposed the system is allowed to reach a new steady state before the new flux is measured. Because control coefficients are global, all the routes by which the change in enzyme activity affects the flux are included in the control coefficient and it is not necessary to know the precise mechanism or mechanisms of action.

An enzyme with a flux control coefficient close to one is responsible for most of the control over the flux. Often, enzymes in unbranched pathways are nearly all found to have small positive control coefficients, indicating that control over the flux is shared between many enzymes.

For many years, and in many biochemistry textbooks, phosphofructokinase (PFK) has been described as one of the key regulatory enzymes of glycolysis. Yet when this enzyme was overexpressed in a yeast system, there was no significant change in the glycolytic flux (Heinisch, 1986). This is probably because PFK is so closely regulated by its various allosteric effectors that the addition of more enzyme to the system has very little impact on the system flux. In control analysis terms, PFK has a low flux control coefficient over glycolysis. However, it is an important site of control over the concentrations of pathway metabolites and an important mediator of control exerted by steps elsewhere in the system that have strong control over glycolysis, such as ATP demand (Fell, 1997).

If the system under investigation consists of an unbranched chain of enzymes in steady state, then, whatever enzyme is measured, the flux through the system will be the same. But manipulating the activity of one enzyme in the pathway will force all the other enzymes to move to a new steady state where the rates of supply of each metabolite are equal to the demand. If the rate of one enzyme is increased, for example, its product may accumulate and stimulate the next enzyme in the pathway, and so on. In this way, enzymes also have flux control coefficients over the fluxes through other enzymes, and the control is mediated by the intermediates linking the enzymes.

Concentration control coefficients (also denoted C) describe how a small change in the activity of an enzyme, for example PFK, affects the concentration of a metabolite, perhaps ATP (Eq. 20.3).

$$C_{\text{enzyme}}^{\text{concentration}} = \frac{\delta\text{concentration}}{\text{concentration}} \Big/ \frac{\delta\text{enzyme}}{\text{enzyme}} \qquad (20.3)$$

Their definition and interpretation are very similar to that for flux control coefficients, but with the difference that they sum to zero for control over each metabolite (Eq. 20.4). This means that some concentration control coefficients will have positive values and some will be negative.

$$\sum_{\text{all enzymes}} C_{\text{enzyme}}^{\text{concentration}} = 0 \qquad (20.4)$$

Elasticity coefficients (denoted ε) describe the properties of individual processes or enzymes: how the kinetics of an enzyme are affected by metabolites and effectors. They describe how a small change in a variable (such as a metabolite or an effector concentration) affects the rate of an enzyme (Eq. 20.5), for example, how hexokinase is affected by changes in the glucose concentration.

$$\varepsilon_{\text{concentration}}^{\text{enzyme}} = \frac{\delta\text{enzyme}}{\text{enzyme}} \Big/ \frac{\delta\text{concentration}}{\text{concentration}} \qquad (20.5)$$

A positive elasticity means that an increase in the concentration will result in an increase in the rate; a negative elasticity means an increase in the concentration will decrease the rate. Elasticities are properties of isolated enzymes rather than the system, and are valid only for the conditions in which they are measured. Elasticities are local coefficients, as when they are measured, the whole system, apart from the imposed change, should be kept constant. The elasticity coefficient describes solely the direct action of the variable on the enzyme rate, hence a local effect. Every enzyme in a system will have one elasticity towards each substrate, product and effector. Later, we discuss how elasticity coefficients can be used to describe how a change in a signalling protein affects the metabolic rate, leading to a metabolically depressed state.

The connectivity theorems connect control coefficients to elasticity coefficients (Eq. 20.6, 20.7 and 20.8) (Kacser and Burns, 1973; Kacser et al., 1995).

$$\sum_{\text{all enzymes}} \left(C_{\text{enzyme}}^{\text{flux}} \cdot \varepsilon_{\text{concentration}}^{\text{enzyme}} \right) = 0 \qquad (20.6)$$

$$\sum_{\text{all enzymes}} \left(C_{\text{enzyme}}^{\text{concentration } a} \cdot \varepsilon_{\text{concentration } a}^{\text{enzyme}} \right) = -1 \qquad (20.7)$$

$$\sum_{\text{all enzymes}} \left(C_{\text{enzyme}}^{\text{concentration } a} \cdot \varepsilon_{\text{concentration } b}^{\text{enzyme}} \right) = 0 \qquad (20.8)$$

There is one connectivity theorem for each combination of a metabolite and a flux in the system. It is possible to obtain the control coefficients for a system mathematically from elasticity coefficients and fluxes by solving an appropriate set of simultaneous equations. Connectivity theorems are essential for this very useful calculation and were used in this way by Ainscow and Brand (1999d) to investigate rat hepatocyte metabolism.

4.2. Constructing the big picture

Once the elasticity and control coefficients for a system have been obtained, they can be put together to give a picture of how the whole system behaves. Response coefficients (R) describe the response of a concentration or a flux to a small change in a metabolite or an effector. They are the mathematical product of elasticity and control coefficients, for example, the elasticity of an enzyme to the effector, multiplied by how much that change in enzyme activity affects the output concentration or flux (i.e., the control coefficient of the enzyme activity over the concentration or flux) (Eq. 20.9).

$$^{\text{enzyme}} R_{\text{concentration}}^{\text{flux or concentration}} = C_{\text{enzyme}}^{\text{flux or concentration}} \cdot \varepsilon_{\text{concentration}}^{\text{enzyme}}$$

$$(20.9)$$

If there is a single primary target of the effector, a response coefficient is obtained. If there is more than one target, partial response coefficients are used and these divide the response into the sub-responses through each individual single enzyme. For example, the partial response of glycolysis to glucose, acting through hexokinase, could be obtained by multiplying how glucose affects the activity of hexokinase (elasticity coefficient) by how much control hexokinase has over the glycolytic flux (control coefficient). The total response of the flux or concentration to the change in metabolite or effector is simply the sum of the partial responses (Eq. 20.10).

$$R_{\text{concentration}}^{\text{flux or concentration}} = \sum_{\text{all enzymes}} {}^{\text{enzyme}} R_{\text{concentration}}^{\text{flux or concentration}}$$

$$(20.10)$$

Partial response coefficients are useful in a wide range of analyses, including locating the specific sites of action of effectors, finding the routes by which the response is effected, and measuring how much of the response is through each route (Brand, 1997). Partial external response coefficients are used to describe the response of a system to changes in its external environment. Partial internal response coefficients describe regulation within the system, for example how one metabolite concentration (e.g., glucose) controls the system flux (glycolysis) by its action on a particular enzyme (hexokinase).

Control analysis is strictly valid only for infinitely small changes, and assumes that the coefficients are linear over the observed range (Fell, 1997; Kacser and Burns, 1973; Kacser et al., 1995). Sometimes, it is not experimentally possible to make a small change in a parameter, instead a large step-wise change must be used. If this is this case, the elasticity coefficients are designated integrated elasticity coefficients (*Iε*), making it clear that they represent large step-changes (Eqs. 20.11 and 20.12) (Ainscow and Brand, 1999b).

$$I\varepsilon^{enzyme}_{parameter\ change} = \frac{\delta enzyme\ rate}{enzyme\ rate} \qquad (20.11)$$

$$^{enzyme}IR^{flux\ or\ concentration}_{parameter\ change} = C^{flux\ or\ concentration}_{enzyme} \cdot I\varepsilon^{enzyme}_{concentration}$$

$$(20.12)$$

The step-changes can also be incorporated into response coefficients: partial integrated response coefficients (*IR*) describe how a step-wise change in a parameter (or variable) affects a rate or concentration through its actions on each of a set of targets. We will use partial integrated response coefficients to represent how much of the observed metabolic depression is transmitted by different signalling molecules. Because there are different types of response coefficients, the control and regulation of many different systems can be described, and different types of information can be obtained (Brand, 1997), for example, which enzymes or intermediates are most important for the transmission of a response, how an enzyme rate or an intermediate concentration is regulated, or which are the direct and indirect effects of a drug.

Complex systems may be simplified by dividing them into a small number of modules or blocks. This is called the modular (or top-down) approach and was first used by Brand et al. (1988). Each block may contain several enzymes and metabolites, and it is simplest if the reaction blocks are connected to each other via single intermediates. The blocks do not have to be of equal size; one could contain a large metabolic pathway and another just a couple of enzymes and their single common metabolite. Elasticity, control and response coefficients apply in the same way to enzyme reaction blocks as to single enzymes. The most basic simplification of a system is into two reaction blocks and one metabolite intermediate: the first block is the supply block, producing the intermediate, and the second is the demand block,

consuming the intermediate. However, the modular method can cope with many more blocks and intermediates in a system. Modular control analysis is a very effective approach that simplifies a system into a manageable number of blocks so that the experimental complexity required is greatly reduced, yet the results obtained are of sufficient resolution to give a high quality picture of the behaviour of the system:

supply $\rightarrow$ metabolite intermediate $\rightarrow$ demand.

4.3. Calculation of coefficients

The coefficients of control analysis can be obtained in several different ways. The method of calculation of the coefficients depends on the types of experiments done, what information is gained from these experiments, and what the investigator is trying to find out.

The system can be perturbed in several different ways. Enzyme activity can be changed by altering the enzyme concentration, for example by adding purified enzyme to an *in vitro* system, or by changing the level of expression *in vivo*. Alternatively, specific or unspecific effectors or inhibitors may be used to change the activity of a particular enzyme or enzyme block. The concentrations of intermediates can be altered more directly by the addition of an extra branch to the system, to either produce or consume a particular metabolite.

Elasticity coefficients can be obtained directly, for example by titration of other parts of the system with inhibitors, as demonstrated by Hafner et al. (1990). Ainscow and Brand (1999a,c,d) used the multiple modulation method (Giersch, 1995), that we also use later, to get elasticity coefficients. This involves making several (combinations of) modulations to the system and measuring the fluxes and concentrations in each experiment. The multiple modulation method is very useful in situations where it is not possible to measure elasticities by keeping everything constant except the enzyme under investigation. It uses equations that describe the change in flux (J_i) in terms of the effects contributed by changes in each intermediate concentration (x) (Eq. 20.13).

$$\frac{\Delta J_i}{J_i} = \sum_{\text{all } x} \left(\varepsilon_x^i \cdot \frac{\Delta x}{x} \right) \qquad (20.13)$$

This is equivalent to holding the rest of the system constant, but mathematically. The same type of equation and method applies if the intermediates are enzymes, rather than the metabolite intermediates in Eq. 20.13. The different modulations are then used to generate a series of simultaneous equations that have the elasticities as the unknowns. Provided that the modulations of the system are independent of each other and enough manipulations are performed, there is sufficient information to solve the equations for the values of all the elasticities (Fell, 1997; Giersch, 1995; Kacser and Burns, 1979).

Flux and concentration control coefficients can either be obtained directly from experiments, or mathematically from experimentally obtained elasticities and fluxes. Experimentally, the activity of each enzyme in turn could be altered by adding more enzyme, or changing its expression and measuring the effect on the flux. The effect of each small, measurable change in enzyme activity on the flux or concentration under question provides the control coefficients (Eq. 20.1) and, according to the summation theorem (Eq. 20.2), these should sum to one. The mathematical method was used by both Hafner et al. (1990) and Ainscow and Brand (1999a,c,d). For a small system it may be simplest to obtain the control coefficients algebraically, as demonstrated by Hafner et al. (1990), using equations that are also listed by Brand (1996). These equations are derived from the summation and connectivity theorems (Eqs. 20.2, 20.4, 20.6, 20.7 and 20.8). Ainscow and Brand used matrix algebra to solve these equations for a more complex system (Hofmeyr et al., 1993; Reder, 1988; Sauro et al., 1987).

5. Examples

5.1. Control analysis of mitochondrial respiration in aestivating snails

Many studies on metabolic depression to date have involved investigating different processes (such as various glycolytic enzyme activities) to see whether or not they are down-regulated during metabolic depression. This is achieved by measuring the change in activity of a process in response to metabolic depression. Although knowing the elasticity of a process to a change in state such as metabolic depression indicates whether or not such a process plays a role in hypometabolism, the elasticities do not take into the account the control that each process has on overall outcomes such as respiration rate and glycolytic rate. So, for example, a process might have a large elasticity coefficient to metabolic depression (large change in the activity of the process upon metabolic depression) but if it has a low control coefficient over, for example glycolytic rate, even a large change in activity will not affect the overall glycolytic rate by much. Therefore, to quantify the roles of different processes in metabolic depression, both the elasticity and control coefficients need to be taken into account.

Bishop and colleagues measured elasticity and control coefficients to quantify the effects of metabolic depression on mitochondrial respiration in hepatopancreas cells from the snail *Helix aspersa* (Bishop and Brand, 2000; Bishop et al., 2002; Guppy et al., 2000). Snails undergo a metabolic depression called aestivation during the summer months in order to avoid desiccation. Mitochondrial respiration rates of cells from aestivating snails were found to be only a third of those of active snails. Mitochondrial respiration was divided into two blocks: mitochondrial membrane potential producers and mitochondrial membrane potential consumers, with membrane potential ($\Delta\Psi_m$) treated as the intermediate connecting the blocks. The respiration rates driving the $\Delta\Psi_m$ producers and consumers were then measured (as a function of $\Delta\Psi_m$) in hepatopancreas cells from active and aestivating snails. Integrated elasticity coefficients of $\Delta\Psi_m$ producers and consumers to metabolic depression were calculated as the change in these processes in response to aestivation. Control coefficients of $\Delta\Psi_m$ producers or consumers over mitochondrial respiration were calculated as the change in mitochondrial respiration induced by the changes in the $\Delta\Psi_m$ producers or consumers. By taking into account both the elasticity of the $\Delta\Psi_m$ producers and consumers to metabolic depression and the control coefficients of these processes over mitochondrial respiration, the effect of aestivation in decreasing mitochondrial respiration through both the $\Delta\Psi_m$ producers and consumers could be quantified. Approximately 75% of the decrease in mitochondrial respiration during aestivation could be attributed to changes in the $\Delta\Psi_m$ producers, with only 25% of the response occurring through a change in the $\Delta\Psi_m$ consumers. So, most of the metabolic depression induced by desiccating conditions is effected via a down-regulation of the $\Delta\Psi_m$ producers (the substrate oxidation block). Decreased supply of $\Delta\Psi_m$ leads to a secondary inhibition of the $\Delta\Psi_m$ consumers (ATP turnover and proton leak), and the system settles to a new steady state, which has a decreased membrane potential. The $\Delta\Psi_m$ consumers were also subdivided into a proton leak block and an ATP turnover block, and further analysis showed that the activity and kinetics of proton leak were unchanged, but the rate of proton leak was reduced, due to the decrease in membrane potential. This indicated that a primary shutdown of the proton leak is not a mechanism used to alter the rate of respiration during aestivation.

ATP consuming reactions, such as protein synthesis and the Na+/K+-ATPase, have been found to decrease during metabolic depression (Buck and Hochachka, 1993; Land et al., 1993) and much importance has been laid on decreasing the ATP consuming reactions to save energy during metabolic depression. If the results obtained in snails are proven to be a general principle of metabolic depression, this suggests instead that decreasing substrate oxidation or ATP producing reactions may be a more important cause of metabolic depression. This study demonstrates the importance of quantifying the effects of metabolic depression through various processes using metabolic control analysis.

5.2. Control analysis of whole cell metabolism

Ainscow and Brand (1999a,c,d) performed a modular kinetic analysis of the internal control of whole cell metabolism. Hepatocyte metabolism was divided into nine reaction blocks (glycogen breakdown, glucose release, glycolysis, lactate production, NADH oxidation, pyruvate oxidation, mitochondrial proton leak, mitochondrial phosphorylation, and ATP consumption) linked by five metabolite intermediates (intracellular glucose-6-phosphate, pyruvate and ATP levels, cytoplasmic NADH/NAD ratio, and mitochondrial membrane potential). The steady state fluxes and intermediate concentrations were measured in the presence and absence of various effectors of hepatocyte metabolism, e.g., oligomycin, an inhibitor of mitochondrial phosphorylation; cyanide, an effector of the pyruvate and NADH oxidation blocks; and glucose, an effector of glucose release and ATP consumption. Using the multiple modulation approach, various metabolic inhibitors were used in combinations as modulations of metabolism. In each modulation, the concentrations of the metabolite intermediates and the flux through each metabolic block were measured. Equation 20.13 was then used to calculate the elasticity coefficients of each reaction block to each intermediate. Next, the elasticity coefficients and steady state-fluxes through each block were used to calculate the control coefficients of the blocks over each other and over each intermediate, using Eqs. 20.2, 20.4, and 20.6–20.8 and matrix algebra. The appropriate control and elasticity coefficients were then multiplied to obtain partial internal response coefficients (Eq. 20.9). Flux control coefficients were partitioned to give partial flux control coefficients of each block over each other block via each intermediate, because the effects of one block on another are mediated by the intermediates in the system.

The system was then expanded to look at external regulation of metabolism by glucagon and adrenaline. The effects of the added hormones on the intermediates and fluxes were measured and combined with the coefficients from the previous elasticity analysis to locate the sites of action of these hormones and determine which pathways were used by glucagon and adrenaline to effect changes in metabolism. The greatest effects of adrenaline were observed to be a direct stimulation of glycogen breakdown, stimulation of glucose release via the intermediate glucose-6-phosphate (G6P), and inhibition of lactate production both directly and via G6P. The majority of the effects of glucagon were transmitted to the reaction blocks directly. This analysis is an excellent example of how modular control analysis can be used to distinguish the direct and indirect actions of effectors, the sites of action of effectors, and the relative importance of the different routes that transmit the response.

Ainscow and Brand (1998) then performed a Monte Carlo analysis of the data, to measure the statistical significance of the coefficients. Monte Carlo analysis involves using the mean and standard deviation of each experimentally obtained flux and concentration to generate a normal distribution of data points. Coefficients are calculated from each generated data set and this gives a distribution of coefficients that indicates the reliability of the analysis and which coefficients are particularly sensitive to experimental error. If 95% of the simulations give a coefficient that has a particular feature, for example is greater than another coefficient, then we have 95% confidence in that result.

5.3. Control analysis of cell signalling pathways

Krauss and Brand used the modular approach to quantify the effect of concanavalin A (Con A) on rat thymocyte metabolism (Krauss and Brand, 2000; Krauss et al., 1999). Con A is a mitogen that stimulates metabolism. First, the relative importance of the Con A signalling routes were quantified. The signalling pathways were divided into three blocks. One block contained protein kinase C (PKC) and PKC-like pathways, defined as those that were inhibited by bisindolylmaleimide I. The second was a calcineurin block, which contained the signalling routes that were inhibited by cypermethrin. The third was a block that contained the remaining unspecified unidentified signalling pathways. Respiration rates of isolated rat thymocytes were measured in the absence and

presence of Con A stimulation. Con A stimulated cells were treated with each inhibitor alone and then the two inhibitors combined and respiration rates were measured in each experiment. This quantified how much of the Con A stimulated increase in respiration rate was lost by inhibition of each signalling block. The PKC block was shown to transmit 54% of the Con A signal, the calcineurin block 30% and, by subtraction, the remaining 16% was attributed to unidentified signalling routes. The PKC block was further subdivided to allow a downstream MAPKK (mitogen activated protein kinase kinase) block of signalling reactions to be defined as those that were inhibited by PD98059. The MAPKK sub-block accounted for about 75% of the signal that passed through the PKC block.

The second phase of the analysis was to determine how each signalling pathway acted on the different blocks of mitochondrial respiration to give the overall change in metabolic rate (Krauss and Brand, 2000; Krauss et al., 1999). Oxidative phosphorylation was divided into two blocks, one containing mitochondrial membrane potential ($\Delta\Psi_m$) producers and another containing $\Delta\Psi_m$ consumers. As in the first example, $\Delta\Psi_m$ was treated as a single intermediate between the two blocks. The stimulated and unstimulated cells had similar membrane potentials, so any activations of the producer and consumer blocks were of equal magnitude. The kinetics of the $\Delta\Psi_m$ producing and consuming blocks were measured firstly in Con A stimulated cells without signalling inhibitor, secondly with the PKC block inhibitor, thirdly with the calcineurin block inhibitor, and finally in the presence of both inhibitors. The elasticities of each block to $\Delta\Psi_m$ were determined using Eq. 20.5 and titration with myxothiazol and oligomycin, which are inhibitors of the $\Delta\Psi_m$ producing and consuming reaction blocks, respectively. All the experimental data were then used to calculate the integrated elasticity of each reaction block to the effector, Con A, using Eq. 20.14 (Ainscow and Brand, 1999b). Finally, the response coefficients of the respiration rate to the effector, through each individual reaction block, were calculated using Eq. 20.15

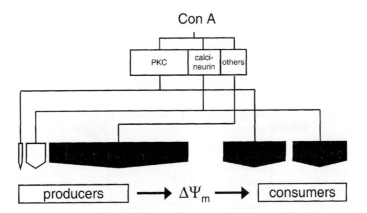

Fig. 20.1. Modular control analysis of cell signalling by the mitogen Con A, adapted from Krauss and Brand (2000). The signal is partitioned into three blocks: PKC, calcineurin and others. The size of the block indicates how much of the signal is transmitted by that block. The signalling blocks then act on metabolic blocks to increase the metabolic rate. Positive (stimulatory) effects are shown in black and negative (inhibitory) in white.

$$I\varepsilon^i_{\text{Con } A} = \frac{\Delta J_i}{J_i} - \sum_{\text{all } x}\left(\varepsilon^i_x \cdot \frac{\Delta x}{x}\right) \qquad (20.14)$$

$$IR^a_{\text{Con } A} = \sum_{\text{all } i}(C^a_i \cdot I\varepsilon^i_{\text{Con } A}) \qquad (20.15)$$

where $I\varepsilon$ is the integrated elasticity of reaction block i ($\Delta\Psi_m$ producer or consumer block) to the step change in the effector Con A. J_i is the flux (reaction rate) through reaction block i, x is the intermediate ($\Delta\Psi_m$). IR is the integrated response of system variable a (respiration rate) to the step change in the effector Con A. C is the control coefficient of reaction block i over variable a (respiration rate). Equation 20.14 calculates the integrated elasticity of each block to Con A as solely the direct effect of the effector on the block, subtracting any indirect effects such as those that are mediated through a change in membrane potential. A Monte Carlo analysis was used to estimate the statistical significance of the coefficients.

The analysis revealed that the PKC signalling block had a stimulatory effect on the $\Delta\Psi_m$ consumers and a small inhibitory effect on the $\Delta\Psi_m$ producers (Fig. 20.1). The calcineurin signalling block acted similarly. Since membrane potential was not changed significantly during stimulation, the unidentified signalling block must have a considerable stimulatory effect on the $\Delta\Psi_m$ producers.

6. Regulation analysis of cell signalling in pathways that initiate metabolic depression

6.1. Overview

Various signals, such as pH, opioids, and oxygen availability, as well as various components of the signalling pathway, such as kinases and phosphatases, have been found to play a role in metabolic depression. Exactly how such changes interlink with the observed changes in metabolic rate, and the relative importance of different signalling pathways, remains obscure. Bishop and colleagues obtained a quantitative overview of the various cellular processes that play a role in metabolic depression in snails (Bishop and Brand, 2000; Bishop et al., 2002; Guppy et al., 2000). A quantitative overview of the signalling pathways that result in the observed changes in cellular processes is required to determine what signals cause the activity of substrate oxidation to decrease during aestivation. It is likely that the depression of metabolism observed is due to an even and widespread downregulation. It is important to maintain the metabolites (such as ATP) at constant, or near constant levels, so that the cells can continue to function and are able to return to pre-depression levels of metabolic activity once the environmental conditions allow.

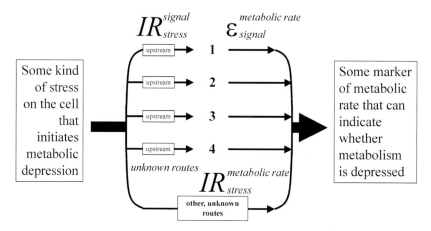

Fig. 20.2. Structure of a system that could be investigated using modular regulation analysis to determine the relative importance of signalling routes by which a stress-induced depression of metabolism could be measured. The four numbers each represent a signalling protein, the activity of which is regulated by phosphorylation and dephosphorylation. Integrated response coefficients represent the change in phosphorylation of each of these proteins in response to the stress. The effect of a change in phosphorylation of each signalling protein on the metabolic rate of the cells is represented by elasticity coefficients. The corresponding integrated and elasticity coefficients for each signalling protein are multiplied to give partial response coefficients. These represent the amount of change in metabolic rate that is transmitted by that signalling protein. The amount of signal carried by unknown signalling routes in response to the stress is represented by an integrated response coefficient (lower block) and is calculated by subtracting the sum of the partial response coefficients from the total observed change in metabolic rate when metabolic depression is initiated.

Regulation analysis could look at the contributions of individual signalling proteins in initiating metabolic depression. The analysis would determine the relative importance of different signalling molecules in metabolic depression. The structure of the system is shown in Fig. 20.2 and the analysis is summarised in Fig. 20.3. A stress on the cells changes the activity of various signalling proteins, such as kinases and phosphatases (described using integrated response coefficients). These signalling molecules then act downstream to change the metabolic rate (quantified with elasticity coefficients). The integrated response coefficients (how much the activity of the signalling proteins change) are multiplied by the elasticity coefficients (how each signalling protein affects the metabolic rate) to give a set of partial response coefficients (Eq. 20.9). These describe how much of the change in metabolic rate goes through each signalling protein. The analysis also accounts for any change in the metabolic rate that does not go through one of the specified signalling molecules. Thus, there is a block for "unknown" signalling mechanisms, described with an integrated response coefficient.

This can be calculated by subtracting the sum of the partial response coefficients from the total observed change in metabolic rate (Eq. 20.19). Figure 20.2 illustrates a system with only four specified signalling proteins, but clearly the more there are the better, as this makes the "unknown" block smaller and is, therefore, more informative.

6.2. Practicalities

The signalling proteins could be measured either as reaction rates or as concentrations. We suggest that the phosphorylated/dephosphorylated ratio of some signalling molecules could be measured using antibodies. Because the phosphorylation and dephosphorylation of proteins can act as an on/off switch, the ratio of phosphorylation states indicates the relative activity of the protein. Measuring the precise kinase or phosphatase activities of a series of signalling proteins would be difficult, so using antibodies is a much better alternative.

For a regulation analysis of metabolic depression, the output from the system should be some marker of metabolic rate. This could be the rate of

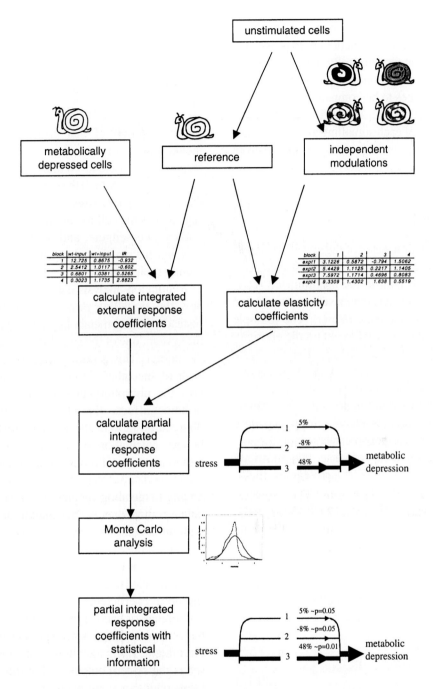

Fig. 20.3. Overview of regulation analysis of cell signalling pathways that initiate metabolic depression. The numbers shown are for illustrative purposes only.

respiration or of an enzyme, the rate of accumulation of a waste product, the rate of consumption of a substrate, or the expression level of one or more relevant genes.

A specific example of a model that could be used for this type of analysis is the isolated frog hepatocyte. These cells can be induced to enter a metabolically depressed state, and the extent of the depression can be monitored, for example, by measuring expression of relevant genes, heat output or respiration rate. Turtle hepatocytes could also be used, in this case using anoxia as the cue for metabolic depression. As an alternative to the *in vitro* approach, cells or organelles could be isolated from active versus metabolically depressed whole animals, allowing metabolic depression in different organs or tissues to be compared.

The integrated response coefficients (how the phosphorylation of each signalling protein changes in response to the stress) are easy to obtain from experimental data and Eq. 20.16.

$$IR_{\text{stress}}^{\text{signal}} = \frac{\Delta \text{signal}_{s-r}}{\text{signal}_r} \qquad (20.16)$$

The phosphorylation states of the specified signalling proteins and the chosen marker of metabolic rate are measured both in the unstimulated reference (*r*) cells and in cells that have been stimulated to suppress their metabolic rate (*s*). The multiple modulation method is used to obtain the elasticity coefficients (Eqs. 20.13 and 20.17). There is a wide range of signalling agonists and antagonists and other inhibitors that could be used. The exact choice of inhibitors is not important; what matters is that each modulation produces a different pattern of change in protein phosphorylation or dephosphorylation. Integrated response and elasticity coefficients for each signalling protein are multiplied together, giving partial response coefficients (Eq. 20.18) that describe how much of the metabolic response is transmitted by each signalling protein.

$$\frac{\Delta \text{metabolic rate}}{\text{metabolic rate}} = \sum_{\text{all signals}} \left(\frac{\Delta \text{signal}}{\text{signal}} \cdot \varepsilon_{\text{signal}}^{\text{metabolic rate}} \right) \qquad (20.17)$$

$${}^{\text{signal}}IR_{\text{stress}}^{\text{metabolic rate}} = IR_{\text{stress}}^{\text{signal}} \cdot \varepsilon_{\text{signal}}^{\text{metabolic rate}} \qquad (20.18)$$

Equation 20.19 calculates how much response passes through the block that contains the unknown signalling mechanisms (Fig. 20.2).

$${}^{\text{unknown routes}}IR_{\text{stress}}^{\text{metabolic rate}} = IR_{\text{stress}}^{\text{metabolic rate}} - \sum_{\text{all signals}} (IR_{\text{stress}}^{\text{signal}} \cdot \varepsilon_{\text{signal}}^{\text{metabolic rate}}) \qquad (20.19)$$

6.3. How to interpret results from this type of experiment

Each signalling block contains all the pathways upstream of the measured signalling molecule, so all signalling routes by which the amount of phosphorylation of that protein is changed are included within the block as part of its integrated response coefficient. Within the block are the mechanisms by which the cell senses or anticipates the environmental stress and transduces the message across the cell membrane, as well as any secondary messengers and any other signalling proteins that may be upstream of the measured signalling molecule.

Regulation analysis can be used to investigate how the signalling proteins regulate each other. Instead of using metabolic rate as the output from the system, one of the signalling proteins is used as the output. This protein then has an elasticity to each of the other signalling proteins, describing how the phosphorylation of the other signalling proteins affects the phosphorylation: dephosphorylation ratio of the chosen output protein. As before, calculation of the elasticity coefficients requires a series of modulations of the system, and the same data set that was used to investigate signalling to metabolic depression can be used to calculate control over each signalling protein in turn, without needing any more experiments. The integrated responses and elasticities for each signalling protein (other than the one that is the new output) are multiplied together to give partial integrated response coefficients that describe the relative importance of each signalling protein in effecting the change in the phosphorylation state of the output protein. Each signalling protein in turn can be set as the system output, and in this way the network of interactions between signalling proteins can be traced.

This example only refers to the measurement of phosphorylation states of signalling proteins to indicate relative activity, but not all can be measured using this method. Other types of signalling proteins are also involved in the known pathways of cell signalling. These include adenylyl cyclase, cGMP phosphodiesterase and phospholipases. If measures of the active and inactive forms of these other signalling molecules are available, then they could be removed from the "unknown" portion of the system and specified in their own right.

The analysis described here could be extended to investigate the enzymatic targets of the signalling proteins. Instead of using some marker of metabolic rate as the output, the effect of changes in the phosphorylation of each signalling protein on metabolic enzymes of interest could be investigated. This would require additional information about how much control each enzyme has over the metabolic rate (its control coefficient), but is a potential method to quantify the precise mechanisms of signalling that initiate metabolic depression.

7. Summary

Modular control analysis can be applied to many different systems to quantify control and regulation. It can be used to describe linear enzyme reaction systems, subsets of metabolism, whole cell metabolism, and cell signalling. The method described here for determination of the relative importance of different signalling routes involved in initiating metabolic depression in response to stress is generally applicable to many types of cellular and molecular stress and can also be applied to many other systems of interest.

References

Ainscow, E.K. and Brand, M.D. (1998). Errors associated with metabolic control analysis. Application of Monte Carlo simulation of experimental data. J. Theor. Biol. 194, 223–233.

Ainscow, E.K. and Brand, M.D. (1999a). Internal regulation of ATP turnover, glycolysis and oxidative phosphorylation in rat hepatocytes. Eur. J. Biochem. 266, 737–749.

Ainscow, E.K. and Brand, M.D. (1999b). Quantifying elasticity analysis: how external effectors cause changes to metabolic systems. Biosystems 49, 151–159.

Ainscow, E.K. and Brand, M.D. (1999c). The responses of rat hepatocytes to glucagon and adrenaline. Application of quantified elasticity analysis. Eur. J. Biochem. 265, 1043–1055.

Ainscow, E.K. and Brand, M.D. (1999d). Top-down control analysis of ATP turnover, glycolysis and oxidative phosphorylation in rat hepatocytes. Eur. J. Biochem. 263, 671–685.

Bishop, T. and Brand, M.D. (2000). Processes contributing to metabolic depression in hepatopancreas cells from the snail *Helix aspersa*. J. Exp. Biol. 203, 3603–3612.

Bishop, T., St-Pierre, J. and Brand, M.D. (2002). Primary causes of decreased mitochondrial oxygen consumption during metabolic depression in snail cells. Am. J. Physiol. 282, R372–R382.

Brand, M.D. (1996). Top down metabolic control analysis. J. Theor. Biol. 182, 351–360.

Brand, M.D. (1997). Regulation analysis of energy metabolism. J. Exp. Biol. 200, 193–202.

Brand, M.D. and Ainscow, E.K. (2000). Regulation of energy metabolism in hepatocytes. In: Technological and Medical Implications of Metabolic Control Analysis. (Cornish-Bowden, A. and Cardenas, M.L., Eds.), pp. 131–138. Kluwer Academic Publishers, Amsterdam.

Brand, M.D. and Curtis, R.K. (2002). Simplifying metabolic complexity. Biochem. Soc. Trans. 30 (2), 25–30.

Brand, M.D., Hafner, R.P. and Brown, G.C. (1988). Control of respiration in non-phosphorylating mitochondria is shared between the proton leak and the respiratory chain. Biochem. J. 255, 535–539.

Buck, L.T. and Hochachka, P.W. (1993). Anoxic suppression of Na(+)-K(+)-ATPase and constant membrane potential in hepatocytes: support for channel arrest. Am. J. Physiol. 265, R1020–R1025.

Cornish-Bowden, A. and Cardenas, M.L. (1999). Technological and medical implications of metabolic control analysis. In NATO Science series 3. High Technology, vol. 74. Dordrecht: Kluwer Academic Press.

Curtis, R.K. and Brand, M.D. (2001). Control analysis of gene expression. Biochemical Society 675th meeting, abstract A32.

Fell, D. (1997). Understanding the Control of Metabolism. Portland Press, London.

Fell, D. (1998). Increasing the flux in metabolic pathways: a metabolic control analysis perspective. Biotechnol. Bioeng. 58, 121–124.

Fell, D. and Sauro, H. (1985). Metabolic control and its analysis. Additional relationships between elasticities and control coefficients. Eur. J. Biochem. 148, 555–561.

Fell, D. and Thomas, S. (1995). Physiological control of metabolic flux: the requirement for multisite modulation. Biochem. J. 311, 35-39.

Giersch, C. (1995). Determining elasticities from multiple measurements of flux rates and metabolite concentrations. Application of the multiple modulation method to a reconstituted pathway. Eur. J. Biochem. 227, 194–201.

Groen, A.K., van Roermund, C.W., Vervoorn, R.C. and Tager, J.M. (1986). Control of gluconeogenesis in rat liver cells. Flux control coefficients of the enzymes in the gluconeogenic pathway in the absence and presence of glucagon. Biochem. J. 237, 379–389.

Guppy, M., Reeves, D.C., Bishop, T., Withers, P., Buckingham, J.A. and Brand, M.D. (2000). Intrinsic metabolic depression in cells isolated from the hepato-pancreas of estivating snails. FASEB J. 14, 999–1004.

Hafner, R.P., Brown, G.C. and Brand, M.D. (1990). Analysis of the control of respiration rate, phosphorylation rate, proton leak rate and protonmotive force in isolated mitochondria using the 'top-down' approach of metabolic control theory. Eur. J. Biochem. 188, 313–319.

Heinisch, J. (1986). Isolation and characterization of the two structural genes coding for phosphofructokinase in yeast. Mol. Gen. Genet. 202, 75–82.

Heinrich, R. and Rapoport, S.M. (1974). A linear steady state treatment of enzymatic chains. General properties, control and effector strength. Eur. J. Biochem. 42, 89–95.

Hofmeyr, J.-H.S., Cornish-Bowden, A. and Rohwer, J.M. (1993). Taking enzyme kinetics out of control; putting control into regulation. Eur. J. Biochem. 212, 833–837.

Hofmeyr, J.-H.S., Kacser, H. and van der Merwe, K.J. (1986). Metabolic control analysis of moiety-conserved cycles. Eur. J. Biochem. 155, 631–641.

Kacser, H. (1983). The control of enzyme systems *in vivo*: Elasticity analysis of the steady state. Biochem. Soc. Trans. 11, 35–40.

Kacser, H. and Acerenza, L. (1993). A universal method for achieving increases in metabolite production. Eur. J. Biochem. 216, 361–367.

Kacser, H. and Burns, J.A. (1973). The control of flux. Symp. Soc. Exp. Biol. 27, 65–104.

Kacser, H. and Burns, J.A. (1979). Molecular democracy: who shares the controls? Biochem. Soc. Trans. 7, 1149–1160.

Kacser, H., Burns, J.A. and Fell, D. (1995). The control of flux: 21 years on. Biochem. Soc. Trans. 23, 341–366.

Kesseler, A. and Brand, M.D. (1994a). Effects of cadmium on the control and internal regulation of oxidative phosphorylation in potato tuber mitochondria. Eur. J. Biochem. 225, 907–922.

Kesseler, A. and Brand, M.D. (1994b). Localisation of the sites of action of cadmium on oxidative phosphorylation in potato tuber mitochondria using top-down elasticity analysis. Eur. J. Biochem. 225, 897–906.

Kesseler, A. and Brand, M.D. (1994c). Quantitative deter-

mination of the regulation of oxidative phosphorylation by cadmium in potato tuber mitochondria. Eur. J. Biochem. 225, 923–935.

Kholodenko, B.N., Cascante, M., Hoek, J.B., Westerhoff, H.V. and Schwabber, J. (1998). Metabolic design: how to engineer living cells to desired metabolite concentrations and fluxes. Biotechnol. Bioeng. 59, 239–247.

Krauss, S. and Brand, M.D. (2000). Quantitation of signal transduction. FASEB J. 14, 2581–2588.

Krauss, S., Buttgereit, F. and Brand, M.D. (1999). Effects of the mitogen concanavalin A on pathways of thymocyte energy metabolism. Biochim. Biophys. Acta 1412, 129–138.

Land, S.C., Buck, L.T. and Hochachka, P.W. (1993). Response of protein synthesis to anoxia and recovery in anoxia-tolerant hepatocytes. Am. J. Physiol. 265, R41–R48.

Murphy, M.P. (2001). How understanding the control of energy metabolism can help investigation of mitochondrial dysfunction, regulation and pharmacology. Biochim. Biophys. Acta 1504, 1–11.

Oliver, S.G., Winson, M.K., Kell, D.B. and Baganz, F. (1998). Systematic functional analysis of the yeast genome. Trends Biotechnol. 16, 373–378.

Raamsdonk, L.M., Teusink, B., Broadhurst, D., Zhang, N., Hayes, A., Walsh, M.C., Berden, J. A., Brindle, K.M., Kell, D.B., Rowland, J.J., Westerhoff, H.V., van Dam, K. and Oliver, S.G. (2001). A functional genomics strategy that uses metabolome data to reveal the phenotype of silent mutations. Nature Biotechnol. 19, 45–50.

Reder, C. (1988). Metabolic control theory: a structural approach. J. Theor. Biol. 135, 175–201.

Rolfe, D.F.S. and Brand, M.D. (1996). Proton leak and control of oxidative phosphorylation in perfused, resting rat skeletal muscle. Biochim. Biophys. Acta 1276, 45–50.

Sauro, H., Small, J.R. and Fell, D. (1987). Metabolic control and its analysis. Extensions to the theory and matrix method. Eur. J. Biochem. 165, 215–221.

ter Kuile, B.H. and Westerhoff, H.V. (2001). Transcriptome meets metabolome: hierarchical and metabolic regulation of the glycolytic pathway. FEBS Lett. 500, 169–171.

Teusink, B., Baganz, F., Westerhoff, H.V. and Oliver, S.G. (1998). Metabolic control analysis as a tool in the elucidation of the function of novel genes. Meth. Microbiol. 26, 297–336.

Sensing, Signaling and Cell Adaptation. Edited by K.B. Storey and J.M. Storey
© 2002 Elsevier Science B.V. All rights reserved.

CHAPTER 21

Evolution of Physiological Adaptation

Douglas L. Crawford
Division of Molecular Biology & Biochemistry, University of Missouri, Kansas City, Kansas, USA

Abbreviations for the 11 glycolytic enzymes:
hexokinase: HK; phosphoglucoisomerase (gluco-phosphate isomerase) PGI; phosphofructokinase: PFK; aldolase: ALD; triosephosphate isomerase: TPI; glyceraldehyde-3-phosphate dehydrogenase: GAPDH; phosphoglycerate kinase: PGK; phosphoglycerate mutase: PGM; enolase: ENO; pyruvate kinase: PK; lactate dehydrogenase: LDH

1. Introduction

Organisms faced with environment challenges (variable salinity, lower oxygen concentration, low food availability) often will alter the activities of metabolic enzymes (Hazel and Prosser, 1974; Hochachka and Somero, 1984; Prosser, 1986; Pierce and Crawford, 1997b). These changes can be reversed in an organism's lifetime and thus are physiological acclimations. Alternatively, these changes may be due to genetic differences among organisms and are evolutionary adaptations if they evolved by natural selection. Notice, the ability to achieve physiological acclimation is genetic and most likely an adaptation. The molecular mechanisms responsible for these changes are vast and variable and include protein specific degradation, reversible phosphorylation, or changes in gene expression. Some of these mechanisms are more cost effective than others. For example, phosphorylation uses one or a few ATPs to change the activity of an enzyme, much less energy than is required for the ATP-dependent degradation or synthesis of a protein. Subsequent reactivation by dephosphorylation saves an enormous amount of

energy compared to *de novo* protein synthesis. Despite this, the physiological regulation of many proteins is dependent on degradation pathways (Yamao, 1999; Hardewig et al., 2000; Kornitzer and Ciechanover, 2000), and *de novo* synthesis of phosphorylatable enzymes occurs often (Granner and Pilkis, 1990). The fact that the optimal (most cost effective) physiological processes are not always the solutions used by organisms reflects evolutionary processes. Similar to the *Panda's Thumb* (Gould, 1992), evolutionary processes do not engineer the best solution and instead adapt using the available variation. Thus, if a selective advantage for altering enzyme activity is present on a physiological time scale and the only genetic variation for this phenotype is variation in enzyme concentration, then changes in concentration will evolve rather than more economical mechanisms. Presented here is a personal perspective of the evolutionary mechanisms responsible for physiological adaptations. The major point is that considerable evolutionary variation occurs in physiological processes and in the molecular mechanisms used to respond to physiological insults. This is somewhat surprising: instead of one or a few physiological solutions, many different solutions can be demonstrated. Importantly, not all solutions are the most cost-effective.

2. Evolutionary physiology: appropriate comparisons

Evolutionary studies of physiological processes require the appropriate comparison and statistical approaches. If we are to understand why a specific physiological process is used to respond to changes in the environment, we need to know how it evolves. If a physiological process is evolutionarily pliable (versus conserved), the appropriate comparison will be among closely related species (e.g. within a family or genera) (Crawford et al., 1999). If the physiological process is conserved among a diversity of organisms, then comparisons should be among different orders, classes, or phyla.

Surprisingly little is known about the variability in the molecular mechanisms responsible for physiological responses to environmental changes. For example, in response to a change in temperature, do many organisms utilize similar molecular mechanisms for physiological compensation? Some conserved molecular mechanisms regulate cellular physiology. Thus, virtually all organisms initiate the expression of Hsp70 when faced with short-term heat stress. Yet Hsp70 induction does not occur in all fishes, and among closely related mussels and other organisms, the induction of stress response varies (Hofmann and Somero, 1996; Feder and Hofmann, 1999; Hofmann et al. 2000; Place and Hofmann, 2001). The point of this chapter is that understanding the considerable variation in the molecular mechanisms responsible for physiological responses requires comparisons among these closely related taxa.

The appropriate statistical analysis is a second and important factor in examining molecular mechanisms regulating physiological responses. If we are interested in the evolution of physiological responses, we need to acknowledge that some changes are not adaptive (a derived trait evolved by natural selection). The variation in these traits can have little or no effect on fitness of an organism. Notice, this does not preclude a measurable phenotypic variation. Instead, no one phenotype has a sufficient selective coefficient to overcome neutral drift (Nei, 1987). [Selection coefficients have to be greater than $1/n_e$, where n_e is the effective population size, to overcome the effect of genetic drift. Thus, organisms that typically have small populations are less likely to have adaptive changes.] For example, the enzyme triosephosphate isomerase (TPI) catalyzes a near equilibrium reaction. A near equilibrium reaction is a reaction where the substrate and product have nearly the same free energy, and thus the reaction goes backward nearly as easily as it goes forward. This enzyme has been characterized as being nearly perfect because its catalytic rate (specifically k_{cat}) is at a maximum and is limited only by the rate of substrate diffusion (Stroppolo et al., 2001). Thus, because it can quickly catalyze a near-equilibrium reaction, a substantial change in the activity of this enzyme may have little affect on net forward reaction; it may have little effect on the phenotype. One can measure a change in activity, but this change may not be adaptive, i.e., may not have evolved by natural selection. Instead, many changes may have evolved by neutral evolutionary processes. Acknowledging that many differences may not be adaptive provides an appropriate null hypothesis: quantitative differences between taxa are neutral. Rejecting this null hypothesis requires a demonstration that the change in a trait is significantly different than what would be expected by random drift. Evidence that cellular and molecular differences are due to natural selection in response to different environments would include differences that are much greater than random differences or that occur multiple times in phylogenetically independent taxa (Pierce and Crawford, 1997a; Crawford et al., 1999). Notice this requires that we have a specific expectation concerning how much variation exists within and between organisms and knowledge about their phylogenetic distribution.

Knowledge about phylogenetic distributions is important because comparisons among species need to take their evolutionary relationships into consideration. Two common oversights were pointed out by Felsenstein (1985) and amplified by Garland and colleagues (Garland and Adolph, 1994; Garland and Carter, 1994; Garland et al., 1992; Garland and Janis, 1993): (1) between two taxa, one should expect differences that are unrelated to a specific adaptive hypothesis, and (2) comparisons among many taxa that ignore phylogenetic relationships are statistically biased toward finding significant differences. In the context of recent theoretical frameworks and statistical methods, comparative biologists can test more specifically whether patterns of change among taxa are most likely due to non-random evolutionary changes, e.g., see (Graves and Somero, 1982; Garland and Janis, 1993; Pierce and Crawford, 1997a). These approaches are similar to regression methods used to account for variation in body mass.

3.   Variation in physiological and biochemical traits

Significant differences in physiology and biochemistry occur within a species and among closely related taxa. For example, acclimation responses vary among different populations of the lizard *Sceloporus occidentalis* (Tsuji, 1988a,b). When acclimated to cold temperatures, colder northern populations of these lizards tended to reduce metabolism (inverse acclimation) whereas warmer southern populations show a compensatory increase in metabolism. Within grasshopper species, heritable differences in the cuticular lipid composition affect water conservation (Gibbs and Crowe, 1991; Gibbs et al., 1991; Gibbs and Mousseau, 1994). In the teleost fish *Fundulus heteroclitus*, Place and Powers (1979, 1984) have demonstrated significant differences between populations for the Michaelis–Menten constant (K_m), inhibition constant and catalytic rate constants of the two alleles of heart-type lactate dehydrogenase (LDH-B). Importantly, these allelic variations correlate with variation in swimming speed, developmental rate, metabolism and survival (DiMichele and Powers, 1982a,b, 1991; Paynter et al., 1991). Similarly, Watt et al. (1986) demonstrated that kinetic differences among phosphoglucose isomerase (PGI) alleles in Colias butterflies affect flight capability and ultimately reproductive fitness. Allelic variation in PGI also affects variation in glucose metabolism in sea anemones (Zamer and Hoffmann, 1989) and carbon fixation in primroses (Kruckeberg et al., 1989). These studies demonstrate that there is significant variation in physiological and biochemical traits within or among closely related species. However, without knowledge of the ancestral state, we are left

wondering if these changes are a recent adaptation (occurring within a species in response to new environmental challenges) or represent a more ancestral polymorphism.

Data on LDH-A (muscle-specific) kinetic variation in vertebrates suggest that biochemical changes are common (Graves and Somero, 1982; Graves et al., 1983) and occur in a diversity of vertebrates (Hochachka and Somero, 1984). This work has demonstrated that vertebrates living in different thermal habitats have similar LDH-A K_m values for pyruvate when measured at biologically relevant temperatures and pHs. Of these examples, the most definitive is the pattern of kinetic variation among closely related barracudas (Graves and Somero, 1982). When measured under standard conditions, barracuda LDH-As have different K_m values for pyruvate that compensate for the differences in environmental temperature. Thus when measured at the biologically appropriate temperature and pH, barracuda LDHA K_ms are conserved (Fig. 21.1). Similar results were found in studies on a wider diversity of vertebrates (Hochachka and Somero, 1984). The variation in K_m pyruvate values for LDH-A among barracudas is best described as due to evolution by natural selection because the phylogenetic distribution demonstrates the independent and convergent changes in LDH-A K_ms. Importantly, variations in barracuda K_ms are due to only a few amino acid replacements and, in at least one case, a single amino acid change (Holland et al., 1997). Interestingly, these amino acid substitutions are not near the active site or other regions directly involved in catalysis (Holland et al., 1997). These data suggest that there may be many more sites (in comparison to the limited number of amino acids that are directly involved in catalysis) that can effect a kinetic change. Thus, because the replacement of a few amino acids and many different amino acid sites can effect a change, adaptive kinetic change could occur readily. Support for this hypothesis is provided by the data above and other data by Somero and colleagues (Hochachka and Somero, 1984; Somero, 1996). These results indicate substantial variation in K_ms even among closely related taxa and that differences of a few degrees in environmental temperatures are

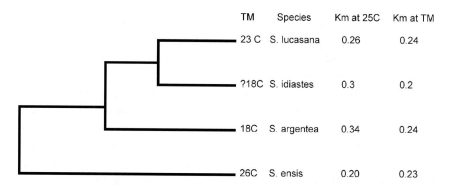

Fig. 21.1. Barracuda LDH-A kinetics, phylogeny and mean annual temperature. Horizontal distance is proportional to Nei's genetic distance. Michaelis–Menten constants (K_m) for pyruvate are in millimolar at two different temperatures: (1) 25°C and (2) mean annual temperature, TM. Notice, that K_m values are more readily explained by TM and not by genetic distance. Data from Graves and Somero (1982).

responsible for significant differences in enzyme kinetic constants.

The variation in these enzymes and species illustrates two important points: (1) comparisons among closely related species, or within a species, can more definitively demonstrate that functional differences evolved by natural selection, and (2) because natural selection has resulted in K_m variation, the variation in K_ms must be biologically important even though it is relatively small. Convincing evidence that a trait has evolved by natural selection is a persuasive argument that this trait is biologically important (affecting health, longevity, or probability of reproduction).

3.1. Enzyme expression

King and Wilson (1975) theorized that because the morphological and anatomical differences between chimpanzees and humans could not be associated with protein polymorphisms or antigenic differences "*that evolutionary changes in anatomy and way of life are more often based on changes in the mechanisms controlling the expression of genes than on the sequence changes in proteins*". Variation in gene expression resulting in changes in enzyme concentration can have a larger effect on enzyme activity and may be evolutionarily more important than variation in protein sequence that affects enzyme kinetic parameters (Wilson, 1976; Laurie-Ahlberg, 1985; Crawford and Powers, 1989, 1992; Heinstra, 1993). For

example, in *F. heteroclitus*, the variations in *Ldh-B* expression caused by changes in transcription rates have a much larger effect on reaction rates than do the variations in LDH-B kinetic constants (Crawford and Powers, 1989, 1992). Adaptive changes in transcription rates may be evolutionarily more important because mutations in promoter regions (gene regulatory sequences) are not constrained by codon usage or the effects of amino acid replacements on the tertiary or quaternary structures of a protein. That is, amino acid substitutions that could have a positive effect on enzyme activity (e.g., stability or kinetic constants) also may have a negative effect on other attributes of an enzyme's function (Somero, 1978, 1995). In contrast, promoter sequence variation only affects the linear array of nucleotides that may affect protein:DNA binding. Thus, there may be many more acceptable nucleotide substitutions that do not have negative epistatic interactions, and individually or in combination, these substitutions could affect transcription. This hypothesis is supported by analysis of adaptive change among *Escherichia coli* isolates subjected to glucose limited environments (Treves et al., 1998). All six independent replicate populations of *E. coli* that evolved acetate cross-feeding polymorphisms adapted by over expression of the enzyme acetyl-CoA synthetase. Variations in the 5′ regulatory region were responsible for all these changes. The independent evolution of adaptive enzyme expression suggests that changes in transcription are evolutionarily more probable than other mechanisms that affect enzyme activity.

The general or prevalent importance of altering enzyme activity by altering gene expression was most clearly shown by microarray studies on the experimental evolution of yeast to reduced carbohydrates (Ferea et al., 1999). Microarrays are thousands of 150–250 micron spots of DNA bound to microscope slides in a precise and known pattern (Ramsay, 1998; Schena et al., 1998). Each DNA spot quantitatively hybridizes to a specific mRNA

so that expression of thousands of individual genes can be measured simultaneously.

In the experimental analysis of yeast evolutionary adaptation, three strains were subjected to low glucose for hundreds of generations (Ferea et al., 1999). The remarkable finding was the consistent and widespread changes in mRNA expression. Genes with altered expression in the three evolved strains included genes involved in glycolysis, the tricarboxylic acid cycle, oxidative phosphorylation, and metabolite transport (Fig. 21.2).

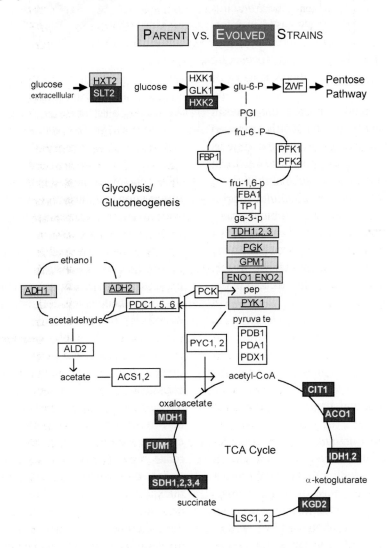

Fig. 21.2. Adaptive evolutionary change in yeast gene expression. Comparison of the quantitative changes in mRNA expression between parental strain and strains selected in a low glucose environment. All enzymes are boxed and have upper case letters. Substrates have lower case letters. Dark boxes with light italic letters indicate a substantial increase in mRNA. Gray boxes with underscored letters indicate a substantial decrease in mRNA. All changes are found in the three replicate evolved populations of yeast subjected to low glucose relative to the parental strain. Adapted from Fera et al. (2000).

These results indicate that increased fitness is acquired by altering the expression of mRNA for genes involved in central metabolism. These yeast microarrays measured the expression of mRNA for all 6,400 genes. Thus, these authors can rightfully claim that metabolic evolutionary adaptation involves change in mRNA expression for many genes.

3.2. *Fundulus*

My laboratory uses teleost fish in the *Fundulus* clade. The phylogeny of this clade has been determined using molecular, allozyme, and morphological data, so the relationships among the species are known with a high degree of confidence (Fig. 21.3) (Wiley, 1986; Cashner et al., 1992; Bernardi and Powers, 1995). Importantly, a number of species in this clade are distributed along the North American Atlantic Coast which has a steep thermal cline as well as a pronounced seasonal variation. The most parsimonious explanation for the distribution of Atlantic coast species is that four *Fundulus* species independently colonized the steep thermal gradients along the North American Atlantic seacoast because each one has a sister species located in the Gulf of Mexico or the southeastern United States, and more basal taxa in this clade are located in the Gulf of Mexico (Fig. 21.3). The clinal variation in temperature has produced genetic divergence in enzyme expression and metabolism within and among species of *Fundulus* (Crawford and Powers, 1990, 1992; Pierce and Crawford 1996, 1997a,b; Crawford et al., 1999a,b; Podrabsky et al., 2000). Details concerning these adaptations are discussed below.

Among species of *Fundulus*, fish from colder populations show an increase in ten of the eleven glycolytic enzymes (including LDH) (Crawford et al., 1999a). Surprisingly, most of these changes (seven of the enzymes) are significantly correlated with evolutionary distance (Pierce and Crawford, 1997a). This is the neutral expectation: taxa separated by greater amounts of time have more opportunity to accumulate random neutral mutations, some of which will affect enzyme expression. These changes in expression may be adaptive, but

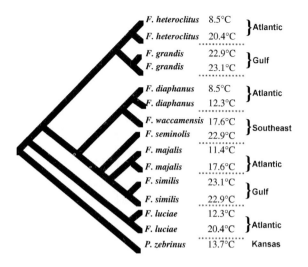

Fig. 21.3. *Fundulus* phylogeny. Phylogeny, distribution and mean annual temperatures of populations within and among species of *Fundulus*. Phylogenetic relationships are based on morphological characters (Wiley, 1986), allozymes (Cashner et al., 1992) and mitochondrial DNA (Bernardi and Powers, 1995).

there is no evidence that they are. Furthermore, the observation that the amount of change is correlated with evolutionary distance supports the null hypothesis: differences among taxa are neutral. Thus, without other compelling evidence, we should accept that these substantial changes in enzyme expression are not adaptive and instead are neutral.

Three changes are not consistent with the null hypothesis of neutral evolution. Three enzymes, glyceraldehyde-3-phosphate dehydrogenase (GAPDH), LDH, and pyruvate kinase (PK), have greater activities in colder populations that are not explained by evolutionary distance (Fig. 21.4). These data are indicative of evolution by natural selection for two compelling reasons. First, enzyme activity and mean annual temperature correlate significantly and independently of evolutionary distance (Fig. 21.4). Among 15 taxa, this is unexpected and unlikely to occur randomly. Second, the four species that independently invaded the colder northern water of the Atlantic all have greater enzyme activity in the northern populations than the corresponding southern populations indicating that increased enzyme evolved independently four times. These data indicate an adaptive evolution of glycolytic enzyme expression, and

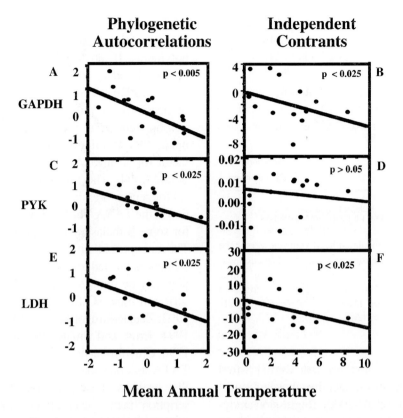

Mean Annual Temperature

Fig. 21.4. Phylogenetically-independent relationships between the concentrations of three enzymes and mean annual temperature. Autocorrelation residuals represent taxon-specific variation not explained by phylogenetic distance (Cheverud and Dow, 1985; Cheverud et al., 1985; Gittleman and Kot, 1990). Independent contrasts represent rates of evolutionary change (Garland and Janis, 1993). Least squares regression slopes are shown in the figure. One-tailed probabilities were calculated with standard probability tables. (A & B) GAPDH correlation with temperature using autocorrelation residuals ($r = -0.624$, $p < 0.005$) or independent contrasts ($r = -0.576$, $p < 0.025$). (C & D) PK correlation with temperature using autocorrelation residuals ($r = -0.565$, $p < 0.025$) or independent contrasts ($r = 0.408$, $p > 0.05$). (E & F) LDH correlation with temperature using autocorrelation residuals ($r = -0.601$, $p < 0.025$) or independent contrasts ($r = -0.541$, $p < 0.025$).

thus, changes in these three enzyme are biologically important (Pierce and Crawford, 1997a). The significant correlations between temperature and enzyme concentrations suggest that (i) temperature is an important selective factor influencing enzyme concentration, (ii) these enzymes affect metabolic flux, and (iii) the interspecific variation in the other enzymes are not direct adaptations to temperature. Notice that two of these enzymes are typically classified as equilibrium enzymes (enzymes that catalyze reactions where the substrate and product have similar free energies). Our data indicate that these equilibrium enzymes are adaptively important and affect metabolic flux.

The evolutionary analysis indicating that the variation in G3PDH, PK and LDH are adaptations

indicates that they are biologically important. To provide further support for this conclusion, we measured oxygen consumption in isolated heart ventricles (Podrabsky et al., 2000) to assess how the change in expression affects biological function. Fish from the northern and southern populations of *F. heteroclitus* were acclimated to a common temperature for 4–6 weeks, sacrificed, and their heart ventricles removed. Blood was expelled and oxygen consumption measured during a 3-min period in a glucose-only Ringer solution. After determination of oxygen consumption, heart ventricles were rapidly frozen, homogenized and the maximum activity of all ten glycolytic enzymes and LDH was measured as an index of enzyme concentration. The metabolic rates of the

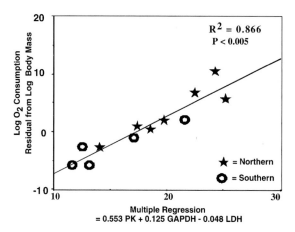

Fig. 21.5. Metabolism of isolated heart ventricle versus the regression equation for the amount of enzyme. Equation was determined using a step-wise multiple regression. Three enzymes and only these three are significant covariates in the multivariate equation. Data from (Podrabsky et al., 2000).

heart ventricles from northern fish were 1.8-fold larger and significantly greater than those of southern fish (Podrabsky et al., 2000). Regression analyses indicate that three enzymes, and only these three enzymes, are significant covariates with glucose usage and explain 87% of the variation in glucose metabolism (Fig. 21.5). These three enzymes are the same enzymes that are adaptively important: G3PDH, PK and LDH. Thus, the evolutionary and functional analyses indicate that alteration in enzyme expression is an important mechanism for physiological adaptation.

What is the molecular mechanism responsible for the adaptive changes in glycolytic enzyme expression? One of these enzymes is *Ldh-B*. In *F. heteroclitus* the differential expression of this enzyme is due to changes in the transcription rate (Crawford and Powers, 1992). This change in transcription rate reflects changes in the DNA regulatory sequences that affect RNA transcription.

Transcription is regulated by proteins interacting with specific DNA sequences (Guarente and Bermingham-McDonogh, 1992; Goodrich et al., 1996). One region of DNA that effects a subtle change in enzyme expression is the proximal promoter that occurs within a few hundred base pairs (bp) of the start of transcription (Ernst and Smale,

1995b). DNA sequences in the proximal promoter bind transcription factors, accurately position RNA polymerase II and direct the rate of polymerase initiation (Roeder, 1991; Smale, 1994; Ernst and Smale, 1995b; Yean and Gralla, 1997). One of the sequences in the proximal promoter is TATAA at −30 bp (relative to the start of transcription) that binds the transcription factor TFIID, an essential factor for all mRNA transcription (Roeder, 1991; Smale, 1994). The TATAA sequence can be critical for gene expression. For example, a point mutation in the TATAA box is thought to be responsible for some β-thalasemia (Antonarakis et al., 1984). The unexpected diversity in proximal promoter sequences has led to the suggestion that this variation may be important for subtle changes in transcription rate (O'Shea-Greenfield and Smale, 1992; Zenzie-Gregory et al., 1992; Javahery et al., 1994; Ernst and Smale, 1995b; Goodrich et al., 1996; Crawford et al., 1999a; Segal et al., 1999). In TATA-less promoters, interactions among TFIID, SP1 proteins and other proximal promoter transcription factors can effect subtle variation in expression (Wiley et al., 1992; Azizkhan et al., 1993; al-Asadi et al., 1995; Yean and Gralla, 1997). This is in contrast to upstream enhancers/ promoter elements responsible for developmental or tissue specific patterns of expression which typically effect a large difference (>10 ×) in transcription or "on/off" scenarios. Upstream enhancers and tissue specific regulatory elements are important in determining when or where a gene is expressed (Patel, 1994; Ernst and Smale, 1995a; Novina and Roy, 1996). Sequence variation in these temporal and spatial control elements either result in disruption of function (Vincent and Wilson, 1989; Gumucio et al., 1993, 1994, 1996; McKenzie et al., 1994; Patel, 1994; Ross et al., 1994; Lowe and Wray, 1997; Singh et al., 1998), or, because of redundancy in these elements, have no appreciable effect (Fang and Brennan, 1992; McKenzie et al., 1994; Ludwig and Kreitman, 1995). Thus, the proximal promoter, which is not redundant and where sequence variation can effect small changes in transcriptional processes, may more likely evolve changes producing adaptive variation in enzyme expression.

In *F. heteroclitus*, the *Ldh-B* proximal promoter is TATA-less, has Inr sequences associated with multiple start sites, a cluster of Sp1-like sequences (Segal et al., 1996, 1999) and considerable DNA sequence variation [average number of nucleotide substitutions per site between populations (Dxy) 0.027 or an average of 7.9 nucleotide differences between populations]. The differences in the proximal promoter (Fig. 21.6) appear to be the result of evolution by natural selection (Crawford et al.,

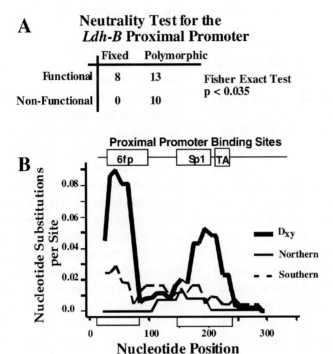

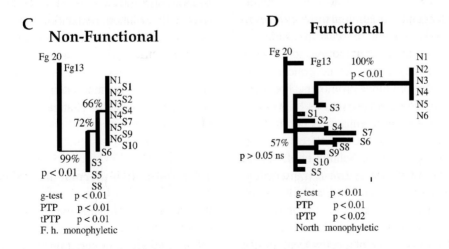

Fig. 21.6. Evolution of *Ldh-B* proximal promoter. (A) Neutrality test (McDonald and Kreitman, 1991) based on the proportion of polymorphism within a population versus between populations for functional (DNA:protein binding) or non-functional sequences. (B) Graphic representation of the per site nucleotide polymorphism for (1) between populations, (2) within the northern, or (3) within the southern populations. (C) & (D) Evolutionary relationship within and between species. Fg: *F. grandis*. N and S are northern and southern populations. Bootstrap values and maximum likelihood probabilities are given next to branches with bootstrap values >50%.

1999b) because the variation between populations in the functional or protein:DNA binding sites is so much larger than the variation within populations or between *F. heteroclitus* southern warm water populations and *F. grandis* (a sister taxa of *F. heteroclitus*). That is, for functional DNA sequences (sequences that bind transcription factors *in vitro* and *in vivo* and when deleted or mutated alter transcription) one expects either much less variation compared to non-functional (neutral) sequences or proportionally the same variation across taxa compared to that found within a population (McDonald and Kreitman, 1991). Yet, the DNA binding regions in the *Ldh-B* promoter have much more variation between populations than within populations (Fig. 21.6) (Crawford et al., 1999b). This can be seen in Fig. 21.6B where there is a peak of polymorphism in the functional sequence between populations but not within a population. Importantly, these changes in the functional DNA sequence are a derived condition in the northern population (Fig. 21.6C,D). The functional sequences are only unique in the northern population and the southern population and F. grandis are evolutionarily indistinguishable (Fig. 21.6D) whereas the non-functional sequences group all *F. heteroclitus* as evolutionarily distinct from *F. grandis* (Fig. 21.6C).

Although sequence variation in the *Ldh-B* proximal promoter is thought to be responsible for much of the variation in *Ldh-B* expression (Crawford et al., 1999a,b), some debate concerns the importance of the proximal promoters. Proximal promoters have been characterized as "slaves" of upstream or tissue specific enhancers because enhancer elements have such strong effects on transcription. More specifically, Schulte et al. (1997) suggest that the difference in *Ldh-B* transcription between populations of *F. heteroclitus* is due to the effect of thermal acclimation on an upstream, hormonally regulated enhancer. These data have been supported by, for lack of a better term, in situ transfections (Schulte et al., 2000). Together these data sets suggest that multiple evolutionary changes may affect *Ldh-B* transcription rates.

Changes in *Ldh-B* expression among the other species of *Fundulus* from colder water also involve changes in mRNA (Kolell and Crawford, unpublished). Specifically, among five *Fundulus* species, populations in the colder northern waters of the Atlantic have significantly greater amounts of *Ldh-B* mRNA compared to southern populations or their sister taxa. Among *Fundulus*, changes in mRNA for *Ldh-B* seem to be common. However, is this common? Are changes in mRNA expression a frequent mechanism for adaptation? I have argued that changes in mRNA expression are more likely to evolve than the biochemical reaction mechanisms of enzymes (e.g., K_m, k_{cat}). We recently demonstrated a meaningful amount of variation in gene expression within and among populations: 18% of approximately 1000 genes examined in *Fundulus* have significant differences in expression among individuals within a population (Oleksiak et al., 2002). Importantly, the difference among individuals within a population is not due to experimental error: with eight replicates per individual the standard deviation for the vast majority of genes is less than 10% of the mean (i.e., CV for 95% of the genes is less than 10%). Most of this variation is best explained by neutral evolution because it is correlated with evolutionary distance. However, distinct from the neutral expectation, there are 26 genes where the northern population had a significantly different level of expression relative to the southern population and *F. grandis*. To begin to understand the evolutionary importance of these measures of gene expression will require appropriate studies among closely related species and supporting evidence demonstrating the functional consequence of changes in mRNA expression.

4. Evolution of thermal acclimation

Ectothermic organisms experience variation in body temperature daily, seasonally, or over their habitat range. These variations in temperature can affect metabolism with potentially detrimental consequences. Physiological acclimation is one way to ameliorate the effect of temperature changes on metabolism (Prosser, 1986). Acclimation responses often are seen as a change in enzyme concentration or activity. The effects of acclimation on enzymes have a rich and complex history (see reviews by Hazel and Prosser, 1974; Shaklee et al., 1977; Hochachka and Somero, 1984; Johnston and Dunn, 1987).

The cellular and molecular bases for acclimation have been well studied. For example, respiration rates in a variety of fish tissues increase in cold-acclimated individuals compared with tissues acutely exposed to lower temperatures; this has been demonstrated in striped bass liver slices (Stone and Sidell, 1981), goldfish liver homogenates (Kanungo and Prosser, 1959), brook trout liver homogenates (Hochachka and Hayes, 1962), and rainbow trout hepatocytes (Hazel and Prosser, 1974). Note that because these are isolated tissues, differences in respiration occurred in the absence of circulating hormones or influence from the central nervous system. Thus, acclimation involves changes not readily reversed by tissue preparation and is most likely due to changes in long-lived proteins. Among the different metabolic pathways, glycolysis is generally found to be more important at lower temperatures. This is based on the observation that glycolytic rate is relatively independent of the acclimation temperature whereas lipid catabolism is temperature dependent and thus contributes less to overall metabolism in cold-acclimated individuals (Hochachka and Hayes, 1962; Moerland and Sidell, 1981; Stone and Sidell, 1981). The relative temperature independence of glycolysis indicates that there are physiological mechanisms, such as increased enzyme activity, maintaining a constant metabolic flux.

In the glycolytic pathway one of the more potent regulatory enzymes is phosphofructokinase (PFK). PFK has allosteric kinetics and is modulated by a large number of activators and inhibitors (Hochachka and Somero, 1984). The evolution of the PFK kinetic complex complements the experimental data (Carpenter and Hand, 1986; Galazzo and Bailey, 1990; Somero and Hand, 1990; Van Der Veen et al., 1995; Wegener, 1996) indicating the regulatory importance of the non-equilibrium enzyme PFK (non-equilibrium enzymes are enzymes that catalyze a reaction where the product has a much lower free energy than the substrate,

and thus they are not readily reversible). However, the presence of non-equilibrium regulatory enzymes does not exclude the possibility that equilibrium enzymes influence flux (Torres et al., 1988; Cornish-Bowden and Cardenas, 1990; Melendez-Hevia et al., 1990). This may be especially true for long-term physiological changes like thermal acclimation where there is enough time to re-establish different enzyme concentrations. Yet, for most acclimation responses it is uncertain which enzymes are involved. In some cases an acclimation response has been demonstrated for a select few enzymes within a pathway (Yaumuchi et al., 1975; Moerland and Sidell, 1981; Kleckner and Sidell, 1985). Moerland and Sidell (1981) did not find an acclimation response by PFK or PK in *Fundulus heteroclitus* livers, despite changes in glycolytic rate. It is possible that changes in the activities of these enzymes were mediated by allosteric modulators *in vivo* or that acclimation responses occurred in other enzymes that were not examined.

To determine which enzymes are involved in acclimation, the maximum activities of all ten glycolytic enzymes and LDH were measured in heart ventricles among *Fundulus* species acclimated to two temperatures (Pierce and Crawford, 1997b). Maximum activity is used as a index of enzyme concentration (Pierce and Crawford, 1994, 1996). Seven closely related species of *Fundulus* were examined to determine evolutionary history of acclimation responses. Glycolysis was chosen to examine the effect of acclimation on enzyme maximal activity because it is a well characterized pathway containing classically defined rate-limiting enzymes (typically HK, PFK and PK) and near-equilibrium enzymes (PGI, ALD, TPI GAPDH, PGK, PGM, ENO and LDH) (Rolleston, 1972; Stryer, 1981).

Acclimation responses are highly variable within the *Fundulus* clade and are not phylogenetically conserved (Fig. 21.7). Unique acclimation responses are seen in two Atlantic species, *F. heteroclitus* and *F. majalis*, and involve four different enzymes. These responses are absent in their sister species, with the possible exception of a PK response (marginally insignificant) that might be

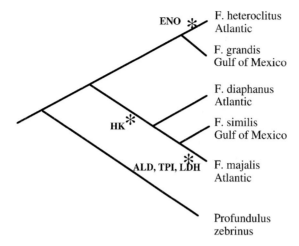

Fig. 21.7. Evolution of physiological acclimation. Acclimation response was characterized by measuring the maximal activity of all the glycolytic enzymes and LDH. Maximum activity is used solely as a measure of enzyme concentration. Asterisks mark acclimation responses with the abbreviation of the enzyme involved. For example, three species have an acclimation response involving the enzyme HK.

shared between *F. majalis* and *F. similis*. Only one response, in HK, is conserved within a subgenus (Fig. 21.7). Thus, physiological acclimation responses that alter enzyme concentration are evolutionarily labile (i.e., evolve independently in different lineages and involve different glycolytic enzymes in these lineages).

Equilibrium enzymes that had a significant acclimation response increased at lower temperatures in the Atlantic species. Since these responses arose after their divergence from other *Fundulus* species, they are relatively recent and may be in response to modern conditions such as present day temperatures. Thus, for the equilibrium enzymes, the acclimation response is in a direction which would ameliorate the effect of temperature variation.

The HK response is more complex. An acclimation response only occurs within a single subgenus, yet HK concentrations decrease at the lower temperature in one Gulf coast species, *F. similis*, and in a single Atlantic coast species, *F. diaphanus*, and increase in the other Atlantic species of this subgenus, *F. majalis*. Thus, an acclimation response may have arisen in an ancestor, and since the divergence of the species, the response has been modified

within the subgenus. The observation that HK maximal activity changes with acclimation conditions in one subgenus, while the direction of the response is variable among the species in the subgenus, may be indicative of the sensitivity of this enzyme's expression to physiological conditions. Because the direction of the HK response is variable, it is unclear whether this physiological sensitivity is due to selection or random, genetic drift. One solution to this issue is to determine cardiac glycolytic flux among the species of this subgenus and correlate the change in HK with change in glycolytic flux.

Ignoring the usual criteria for significant differences, these data show other trends. Let's define a meaningful difference between acclimation temperatures as a difference between means plus or minus one standard error. In two species, most of the glycolytic enzymes that were different when using these liberal criteria have a larger value at 10°C than at 20°C. Six of six and eight of ten enzymes that were different are greater at 10°C than 20°C in *F. heteroclitus* and *F. majalis*, respectively. In two other species, most of the glycolytic enzymes that were different when using these liberal criteria have the opposite trend. Eight of eight and nine of ten enzymes that were different are lower at 10°C than at 20°C for *F. similis* and *F. diaphanus*, respectively. In the fifth species, *F. grandis*, there are approximately an equal number of enzymes that increase or decrease (three of eight are larger at 10°C that at 20°C). One can statistically test if these patterns are unlikely. That is, similar to tossing a coin where any one result (heads) is not significant, one can ask whether these patterns of change are significant. These patterns (most of the enzymes are greater or lesser at 10°C) are significant in three of the five species (*F. heteroclitus*, *F. diaphanus*, and *F. similis*) using a sign-rank test; $P < 0.05$ (Siegel, 1956). These trends may suggest that (1) within these species the expression of each glycolytic enzyme is not completely independent, (2) in two species most of the loci have a compensatory increase in enzyme concentration, and (3) in two other species most the glycolytic loci decrease at low acclimation temperatures. Assuming these are meaningful trends, they suggest that acclimation response involves all enzymes in a pathway, not just a few rate-limiting enzymes. A positive or negative trend is not restricted to a specific clade, suggesting that acclimation is evolutionarily labile. However, it should be stressed that most of these differences and patterns among acclimation conditions are for individual enzymes that are not statistically significant. Thus, it is difficult to define the importance of these patterns.

Which enzymes respond to acclimation would depend on whether only a few rate-limiting enzymes or equilibrium enzymes also influence physiological processes. ENO concentration increased at lower temperatures in *F. heteroclitus*. The concentrations of ALD, TPI, and LDH increased at lower temperatures in *F. majalis*, another coastal Atlantic species. These four enzymes are typically classified as equilibrium enzyme. The one non-equilibrium enzyme, HK, had a variable response to acclimation in three species, *F. majalis*, *F. similis* and *F. diaphanus*, all belonging to the same subgenus. There is no obvious pattern of acclimation among these data and thus no one enzyme or set of enzymes that always evolves an acclimation response.

The observation that equilibrium enzymes change is counter to predictions of classical control theories (Crabtree et al., 1972; Crabtree and Newsholme, 1972, 1987; Newsholme and Crabtree, 1973) but is in agreement with metabolic control theories. Metabolic control theories predict that many enzymes can modulate flux (Cornish-Bowden and Cardenas, 1990). Experimental data on a variety of organisms have supported this idea. Variation in PGI activity affects fitness in Colias butterflies and carbohydrate metabolism in sea anemones (Watt et al., 1986; Zamer and Hoffmann, 1989). In primroses, the control coefficient (a quantitative measure of how metabolic flux changes in response to variation in enzyme activity) of PGI was significant and varied with environmental conditions (Kruckeberg et al., 1989). The activity of another near-equilibrium enzyme, ENO, modulates metabolism in the bacteria *Zymomonas mobilis* and in rat hearts (Kashiwaya et al., 1994; Verrijzer et al., 1995). In rat hearts, the flux control coefficient for ENO

ranged from 0.27 to 0.55 depending on physiological conditions. In *F. heteroclitus*, ENO acclimated in a compensatory manner while in *F. majalis*, other enzymes have compensatory responses. These data suggest that control of metabolic flux may not reside in any one particular enzyme and instead may be variably distributed among enzymes depending on physiological conditions and evolutionary opportunities.

Are these changes in enzyme concentration meaningful? That is, could these patterns arise due to neutral or random changes that increase an enzyme's concentration at colder temperatures (e.g., a random amino acid substitution resulting in a greater relative protein stability at low temperatures)? Yes, given the variation among taxa it is possible that these measures of acclimation (i.e., change in enzyme activity) could represent random variation. Yet this explanation does not seem satisfactory. First, if neutral processes are responsible, an equal number of enzymes should increase or decrease at low temperatures: this does not occur (Pierce and Crawford, 1997b). Second, if random changes in enzyme expression are responsible for the acclimation responses, there are two expectations depending on whether neutral changes in enzyme expression occur readily or rarely. If neutral changes occur readily, then most taxa should have acclimation responses and no phylogenetic signal would be observed. If neutral changes occur rarely, then acclimation should be phylogenetically restricted and occur in a few enzymes. The data do not support a neutral evolutionary hypothesis: for the equilibrium enzymes compensatory acclimation occurs often (in four of the eight enzymes) but only in two phylogenetically unrelated species. Thus, the pattern of enzyme expression is unusually variable and, although it cannot be ruled out, seems to be due to evolutionary processes other than neutral evolution.

4.1. Molecular mechanism in Fundulus acclimation

The molecular mechanisms responsible for *Ldh-B* expression during acclimation have been examined. The original observation was that within a single population of *F. heteroclitus* (New Jersey, NJ) the increase in LDH-B enzyme at low temperature was due to an increase in mRNA expression (Crawford and Powers, 1989, 1990). This study was expanded to two other populations, one in the north (Maine, ME) and the other in the south (Georgia, GA). In all three populations more LDH-B enzyme occurs at the 10°C acclimation temperature than at the 20°C acclimation temperature (Segal and Crawford, 1994). However, the mechanisms responsible for these changes at different temperatures in different populations are not the same. At 20°C in all populations the amount of LDH-B is regulated by altering the transcription rate of mRNA expression (Crawford and Powers, 1992). Additionally, in both NJ and GA at 10°C, greater LDH-B enzyme was significantly correlated with *Ldh-B* mRNA concentration. These data indicate that at both 10 and 20°C similar molecular mechanisms affect LDH-B enzyme expression. However, at 10°C in the northern population (ME) there is no relationship between *Ldh-B* mRNA and the levels of the enzyme it encodes (Segal and Crawford, 1994). It seems unlikely that at one temperature (20°C) in all populations LDH-B would be regulated by mRNA expression, and in two of three populations that all *Ldh-B* genotypes would similarly be regulated but at one of the temperatures in one population (north, ME), a different mechanism would be involved. We did not believe this data originally, and thus this experiment was repeated three times with the same result.

5. Conclusions

Within *Fundulus* an adaptive difference in glycolytic enzyme expression affects metabolism. The molecular mechanisms responsible for one of these enzymes, LDH-B, are due to recent changes of a few nucleotides in the *Ldh-B* gene within species, suggesting that changes in gene expression that are biologically important can evolve readily. Acclimation responses among *Fundulus* are evolutionarily labile—no change in enzyme expression is conserved throughout the genus, and responses unique to one species are observed in four enzymes. These responses are primarily in equilibrium enzymes and are achieved by altering enzyme concentration in a potentially compensatory direction. Surprisingly, the molecular mechanism for these acclimation responses is quite variable: it is not the same in all populations at all temperatures. With the advent of microarray technology and the ability to develop microarrays for many different species (Crawford, 2001), we are in a powerful position to examine the quantitative differences in mRNA expression for most genes. These studies will provide hypotheses into the cellular mechanisms responsible for adaptation and acclimation. It will be the responsibility of physiologists to provide the functional importance to these data and thus the necessary data to inform Functional Genomics.

Acknowledgement

The ideas, theories and data presented in this article would not have been possible without my collaboration with Valerie A. Pierce, Jeff A. Segal, Margie F. Oleksiak and Kevin J. Kolell. I have also benefited from the exchange of ideas with R. Frank Rosenzweig, Frank G. Berger and Martin Kreitman. NSF provide much of the moneys needed for any research.

References

Al-Asadi, R., Yi, E.C. and Merchant, J.L. (1995). Sp1 affinity for GC-rich elements correlates with ornithine decarboxylase promoter activity. Biochem. Biophys. Res. Commun. 214, 324–330.

Antonarakis, S.E., Irkin, S.H., Cheng, T.C., Scott, A.F., Sexton, J.P., Trusko, S.P., Charache, S. and Kazazian, H.H. (1984), Beta-thalassemia in American Blacks: novel mutations in the "TATA" box and an acceptor splice site. Proc. Natl. Acad. Sci. USA 81, 1154–1158.

Azizkhan, J.C., Jensen, D.E., Pierce, A.J. and Wade, M. (1993). Transcription from TATA-less promoters: dihydrofolate reductase as a model. Crit. Rev. Eukaryotic Gene Express. 3, 229–254.

Bernardi, G. and Powers, D.A. (1995). Phylogenetic relationships among nine species from the genus *Fundulus* (Cyprinodontiformes, Fundulidae) inferred from sequences of the cytochrome B. Copeia, 1995, 469–473.

Carpenter, J.F. and Hand, S.C. (1986). Comparison of pH-dependent allostery and dissociation for phosphofructokinases from Artemia embryos and rabbit muscle nature of the enzymes acylated with diethylpyrocarbonate. Arch. Biochem. Biophys. 248, 1–9.

Cashner, R.C., Rogers, J.S. and Grady, J.M. (1992) Phylogenetic studies of the genus *Fundulus*. In: Systematic, Historical Ecology and North American Freshwater Fishes (Mayden R.L., Ed.), pp 421–437. Stanford University Press, Stanford, CA.

Cheverud, J.M. and Dow, M.M. (1985). An autocorrelation analysis of genetic variation due to lineal fission in social groups of rhesus macaques *Macaca mulatta*. Am. J. Phys. Anthro. 67, 113–122.

Cheverud, J.M., Dow, M.M. and Leutenegger, W. (1985). The quantitative assessment of phylogenetic constraints in comparative analysis sexual dimorphism in body weight among primates. Evolution 39, 1335–1351.

Cornish-Bowden, A. and Cardenas, M.L. (1990). Control of Metabolic Processes. Plenum Press, New York

Crabtree, B., Higgins, S.J. and Newsholme, E.A. (1972). The activities of pyruvate carboxylase, phosphoenolpyruvate carboxylase and fructose diphosphatase in muscles from vertebrates and invertebrates. Biochem. J. 130, 391–396.

Crabtree, B. and Newsholme, E.A. (1972). The activities of phosphorylase, hexokinase, phosphofructokinase, lactate dehydrogenase and the glycerol 3-phosphate dehydrogenases in muscles from vertebrates and invertebrates. Biochem. J. 126, 49–58.

Crabtree, B. and Newsholme, E.A. (1987). The derivation and interpretation of control coefficients. Biochem. J. 247, 113–120.

Crawford, D.L. (2001). Functional genomics does not have to be limited to a few select organisms. GenomeBiology http://www.genomebiology.com/2001/2/1/interactions/1001/

Crawford, D.L. and Powers, D.A. (1989). Molecular basis of evolutionary adaptation at the lactate dehydrogenase-B locus in the fish *Fundulus heteroclitus*. Proc. Natl. Acad. Sci. USA 86, 9365–9369.

Crawford, D.L. and Powers, D.A. (1990). Molecular adaptation to the thermal environment: genetic and physiological mechanisms. In: Molecular Evolution (Clegg M.T. and O'Brien S.J., Eds.), pp 213–222. Wiley-Liss, New York.

Crawford, D.L. and Powers, D.A. (1992). Evolutionary adaptation to different thermal environments via transcriptional regulation. Mol. Biol. Evol. 9, 806–813.

Crawford, D.L., Pierce, V.A. and Segal, J.A. (1999a). Evolutionary physiology of closely related taxa: analyses of enzyme expression. Amer. Zool. 32, 389–400.

Crawford, D.L., Segal, J.A. and Barnett, J.L. (1999b). Evolutionary analysis of TATA-less proximal promoter function. Mol. Biol. Evol. 16, 194–207.

DiMichele, L. and Powers, D.A. (1982a). LDH-B genotype-specific hatching times of *Fundulus heteroclitus* embryos. Nature 296, 563–564.

DiMichele, L. and Powers, D.A. (1982b) Physiological basis for swimming endurance differences between LDH-B genotypes of *Fundulus heteroclitus*. Science 216, 1014–1016.

DiMichele, L. and Powers, D.A. (1991). Allozyme variation developmental rate and differential mortality in the teleost *Fundulus heteroclitus*. Physiol. Zool. 64, 1426–1443.

Ernst, P. and Smale, S.T. (1995a). Combinatorial regulation of transcription II: The immunoglobulin mu heavy chain gene. [Review]. Immunity 2, 427–438.

Ernst, P. and Smale, S.T. (1995b). Combinatorial regulation of transcription. I: General aspects of transcriptional control. Immunity 2, 311–319.

Fang, X.M. and Brennan, M.D. (1992). Multiple cis-acting sequences contribute to evolved regulatory variation for Drosophila Adh genes. Genetics 131, 333–343.

Feder, M.E. and Hofmann, G.E. (1999). Heat-shock proteins, molecular chaperones, and the stress response: Evolutionary and ecological physiology [Review]. Annu. Rev. Physiol. 61, 243–282.

Ferea, T.L., Botstein D., Brown, P.O. and Rosenzweig, R.F. (1999). Systematic changes in gene expression patterns following adaptive evolution in yeast. Proc. Natl. Acad. Sci. USA 96, 9721–9726.

Galazzo, J.L. and Bailey, J.E. (1990). Fermentation pathway kinetics and metabolic flux control in suspended and immobilized *Saccharomyces cerevisiae*. Enz. Microb. Technol. 12, 162–172.

Garland, T. and Janis, C.M. (1993). Does metatarsal-femur ratio predict maximal running speed in cursorial mammals? J. Zool. 229, 133–151.

Gibbs, A. and Crowe, J.H. (1991). Intra-individual variation in cuticular lipids studied using Fourier transform infrared spectroscopy. J. Insect Physiol. 37, 743–748.

Gibbs, A. and Mousseau, T.A. (1994). Thermal acclimation and genetic variation in cuticular lipids of the lesser migratory grasshopper (*Melanoplus sanguinipes*): effects of lipid composition on biophysical properties. Physiol. Zool. 67, 1523–1543.

Gibbs, A., Mousseau, T.A. and Crowe, J.H. (1991). Genetic and acclimatory variation in biophysical properties of insect cuticle lipids. Proc. Natl. Acad. Sci. USA 88, 7257–7260.

Gittleman, J.L. and Kot, M. (1990). Adaptation statistics and a null model for estimating phylogenetic effects. Systematic Zool. 39, 227–241.

Goodrich, J.A., Cutler, G. and Tjian R. (1996). Contacts in context: promoter specificity and macromolecular interactions in transcription. Cell 84, 825–830.

Gould, S.J. (1992). The Panda's Thumb. W.W. Norton & Company, New York/London, pp. 1–19.

Granner, D. and Pilkis, S. (1990), The genes of hepatic glucose metabolism. J. Biol. Chem. 265, 10173–10176.

Graves, J.E., Rosenblatt, R.H. and Somero, G.N. (1983). Kinetic and electrophoretic differentiation of lactate dehydrogenase of teleost species-pairs from the Atlantic and Pacific coast of Panama. Evolution 37, 30–37.

Graves, J.E. and Somero, G.N. (1982) Electrophoretic and functional enzyme evolution in four species of eastern Pacific barracudas from different thermal environments. Evolution 36, 97–106.

Guarente, L. and Bermingham-McDonogh, O. (1992). Conservation and evolution of transcriptional mechanisms in eukaryotes. Trends Genet. 8, 27–32.

Gumucio, D.L., Shelton, D.A., Bailey, W.J., Slightom, J.L. and Goodman, M. (1993) Phylogenetic footprinting reveals unexpected complexity in trans factor binding upstream from the epsilon-globin gene. Proc. Natl. Acad. Sci. USA 90, 6018–6022.

Gumucio, D.L., Shelton, D.A., Blanchard-McQuate, K., Gray, T., Tarle, S., Heilstedt-Williamson, H., Slightom, J.L., Collins, F. and Goodman, M. (1994). Differential phylogenetic footprinting as a means to identify base changes responsible for recruitment of the anthropoid gamma gene to a fetal expression pattern. J. Biol. Chem. 269, 15371–15380.

Gumucio, D.L., Shelton, D.A., Zhu, W., Millinoff, D., Gray, T., Bock, J.H., Slightom, J.L. and Goodman, M. (1996). Evolutionary strategies for the elucidation of cis and trans factors that regulate the developmental switching programs of the beta-like globin genes. Mol. Phylogen. Evol. 5, 18–32.

Hardewig, I., van Dijk, P.L., Leary, S.C. and Moyes, C.D. (2000). Temporal changes in enzyme activity and mRNA levels during thermal challenge in white sucker. J. Fish Biol. 56,, 196–207.

Hazel, J.R. and Prosser, C.L. (1974), Molecular mechanisms of temperature compensation in poikilotherms. Physiol. Rev. 54, 620–677.

Heinstra, P.W.H. (1993), Evolutionary genetics of Drosophila alcohol dehydrogenase gene-enzyme system. Genetica 92, 1–22.

Hochachka, P.W. and Hayes, F.R. (1962). The effect of temperature acclimation on pathways of glucose metabolism in the trout. Can. J. Zool. 40, 261–270.

Hochachka, P.W. and Somero, G.N. (1984). Biochemical Adaptation. Princeton University Press, Princeton, NJ.

Hofmann, G.E., Buckley, B.A., Airaksinen, S., Keen, J.E. and Somero, G.N. (2000). Heat-shock protein expression is absent in the Antarctic fish *Trematomus bernacchii* (Family Nototheniidae). J. Exp. Biol., 203, 2331–2339.

Hofmann, G.E. and Somero, G.N. (1996). Interspecific variation in thermal denaturation of proteins in the congeneric mussels *Mytilus trossulus* and *M. galloprovincialis*: Evidence from the heat-shock response and

protein ubiquitination. Marine Biol. 126, 65–75.

Holland, L.Z., McFall-Ngai, M. and Somero, G.N. (1997). Evolution of lactate dehydrogenase-A homologs of barracuda fishes (genus Sphyraena) from different thermal environments: Differences in kinetic properties and thermal stability are due to amino acid substitutions outside the active site. Biochemistry 36, 3207–3215.

Javahery, R., Khachi, A., Lo, K., Zenzie-Gregory, B. and Smale, S.T. (1994). DNA sequence requirements for transcriptional initiator activity in mammalian cells. Mol. Cell. Biol. 14, 116–127.

Johnston, I.A. and Dunn, J. (1987). Temperature acclimation and metabolism in ectotherms with particular reference to teleost fish. In: Temperature and Animal Cells (Bowler, K. and Fuller, B.J., Eds.), pp. 67–93, Company of Biologists, Cambridge.

Kanungo, M.S. and Prosser, C.L. (1959). Physiological and biochemical adaptation of goldfish to cold and warm temperatures. II Oxygen consumption of liver homogenates. J. Cell. Comp. Physiol. 54, 265–274.

Kashiwaya, Y., Sato, K., Tsuchiya, N., Thomas, S., Fell, D.A., Veech, R.L. and Passonneau, J.V. (1994). Control of glucose utilization in working perfused rat heart. J. Biol. Chem. 269, 25502–25514.

King, M.C. and Wilson, A.C. (1975). Evolution at two levels: molecular similarities and biological differences between humans and chimpanzees. Science 188, 107–116.

Kleckner, N.W. and Sidell, B.D. (1985). comparison of maximal activities of enzymes from tissues of thermally acclimated and naturally acclimatized chain pickerel *Esox niger*. Physiol. Zool. 58, 18–28.

Kornitzer, D. and Ciechanover, A. (2000). Modes of regulation of ubiquitin-mediated protein degradation. J. Cell. Physiol. 182, 1–11.

Kruckeberg, A.L., Neuhaus, H.E., Feil, R., Gottlieb, L.D. and Stitt, M. (1989). Decreased-activity mutants of phosphoglucose isomerase In the cytosol and chloroplast of *Clarkia xantiana*, impact on mass-action ratios and fluxes to sucrose and starch and estimation of flux control coefficients and elasticity coefficients. Biochem. J. 261, 457–468.

Laurie-Ahlberg, C.C. (1985). Genetic variation affecting the expression of enzyme-coding genes in Drosophila: an evolutionary perspective. Isozymes: Current Topics in Biological Medical Research 12, 33–88.

Lowe, C.J. and Wray, G.A. (1997). Radical alterations in the roles of homeobox genes during echinoderm evolution. Nature 389, 718–721.

Ludwig, M.Z. and Kreitman, M. (1995). Evolutionary dynamics of the enhancer region of even-skipped in Drosophila. Mol. Biol. Evol. 12, 1002–1011.

McDonald, J.H. and Kreitman, M. (1991), Adaptive protein evolution at the Adh locus in Drosophila. Nature 351, 652–654.

McKenzie, R.W., Hu, J. and Brennan, M.D. (1994). Redundant cis-acting elements control expression of the *Drosophila affinidisjuncta* Adh gene in the larval fat body. Nucleic Acids Res. 22, 1257–1264.

Melendez-Hevia, E., Torres, N.V. and Sicilia, J. (1990). A generalization of metabolic control analysis to conditions of no proportionality between activity and concentration of enzymes. J. Theor. Biol. 142, 443–451.

Moerland, T.S. and Sidell, B.D. (1981). Characterization of metabolic carbon flow in hepatocytes isolated from thermally acclimated killifish *Fundulus heteroclitus*. Physiol. Zool. 54, 379–389.

Nei, M. (1987). Molecular Evolutionary Genetics. Columbia University Press, New York, NY.

Newsholme, E.A. and Crabtree, B. (1973). Metabolic aspects of enzyme activity regulation. Symp. Soc. Exp. Biol. 27, 429–460.

Novina, C.D. and Roy, A.L. (1996). Core promoters and transcriptional control. Trends Genet. 12, 351–355.

O'Shea-Greenfield, A. and Smale, S.T. (1992). Roles of TATA and initiator elements in determining the start site location and direction of RNA polymerase II transcription. J. Biol. Chem. 267, 6450.

Oleksiak, M.F., Churchill, G. and Crawford, D.L. (2002). Polymorphism in gene expression. Submitted.

Patel, N.H. (1994). Developmental evolution: insights from studies of insect segmentation. Science 266, 581–590.

Paynter, K.T., Dimichele, L., Hand, S.C., Powers, D.A. (1991) Metabolic implications of Ldh-B genotype during early development in *Fundulus heteroclitus*. J. Exp. Zool. 257, 24–33.

Pierce, V.A. and Crawford, D.L. (1994). Rapid enzyme assays investigating the variation in the glycolytic pathway in field-caught populations of *Fundulus heteroclitus*. Biochem. Genet. 32, 315–330.

Pierce, V.A. and Crawford, D.L. (1996). Variation in the glycolytic pathway—the role of evolutionary and physiological processes. Physiol. Zool. 69, 489–508.

Pierce, V.A. and Crawford, D.L. (1997a). Phylogenetic analysis of glycolytic enzyme expression. Science 275, 256–259.

Pierce, V.A. and Crawford, D.L. (1997b). Phylogenetic analysis of thermal acclimation of the glycolytic enzymes in the genus *Fundulus*. Physiol. Zool. 70, 597–609.

Place, A.R. and Powers, D.A. (1979). Genetic variation and relative catalytic efficiencies: lactate dehydrogenase B allozymes of *Fundulus heteroclitus*. Proc. Natl. Acad. Sci. USA 76, 2354–2358.

Place, A.R. and Powers, D.A. (1984). Kinetic characterization of the lactate dehydrogenase (LDH-B4) allozymes of *Fundulus heteroclitus*. J. Biol. Chem. 259, 1309–1318.

Place, S.P. and Hofmann, G.E. (2001). Temperature interactions of the molecular chaperone Hsc70 from the eury-

thermal marine goby *Gillichthys mirabilis*. J. Exp. Biol., 204, 2675–2682.

Podrabsky, J.E., Javillonar, C., Hand, S.C. and Crawford, D.L. (2000). Intraspecific variation in aerobic metabolism and glycolytic enzyme expression in heart ventricles from *Fundulus heteroclitus*. Am. J. Physiol. 279, R2344–2348.

Prosser, C.L. (1986). Adaptational Biology: From Molecules to Organisms. Wiley, New York.

Ramsay, G. (1998). DNA chips: state-of-the art. Nature Biotechnology 16, 40–44.

Roeder, R.G. (1991). The complexities of eukaryotic transcription initiation: regulation of preinitiation complex assembly. Trends Biochem. Sci. 16, 402–408.

Rolleston, F.S. (1972). A theoretical background to the use of measured concentrations of intermediates in study of the control on intermediary metabolism. Curr. Top. Cell. Regul. 5, 47–75.

Ross, J.L., Fong, P.P. and Cavener, D.R. (1994). Correlated evolution of the cis-acting regulatory elements and developmental expression of the Drosophila Gld gene in seven species from the subgroup melanogaster. Devel. Genet. 15, 38–50.

Schena, M., Heller, R.A., Theriault, T.P., Konrad, K., Lachenmeier, E., and Davis, R.W. (1998). Microarrays: biotechnology's discovery platform for functional genomics. Trends Biotechnol. 1, 301–306.

Schulte, P.M., Glemet, H.C., Fiebig, A.A. and Powers, D.A. (2000). Adaptive variation in lactate dehydrogenase-B gene expression: Role of a stress-responsive regulatory element. Proc. Natl. Acad. Sci. USA 97, 6597–6602.

Schulte, P.M., Gomezchiarri, M. and Powers, D.A. (1997). Structural and functional differences in the promoter and 5′ flanking region of Ldh-B within and between populations of the teleost *Fundulus heteroclitus*. Genetics 145, 759–769.

Segal, J.A., Barnett, J.L. and Crawford, D.L. (1999). Functional analyses of natural variation in Sp1 binding sites of a TATA-less promoter. J. Mol. Evol. 49, 736–749.

Segal, J.A. and Crawford, D.L. (1994). LDH-B enzyme expression: the mechanisms of altered gene expression in acclimation and evolutionary adaptation. Am. J. Physiol. 267, R1150–R1153.

Segal, J.A., Schulte, P.M., Powers, D.A. and Crawford, D.L. (1996). Descriptive and functional characterization of variation in the *Fundulus heteroclitus* Ldh-B proximal promoter. J. Exp. Zool. 275, 355–364.

Shaklee, J.B., Christiansen, J.A., Sidell. B.D., Prosser, C.L. and Whitt, G.S. (1977). Molecular aspects of temperature acclimation in fish: contributions of changes in enzyme activities and isozyme patterns to metabolic reorganization in the green sunfish. J. Exp. Zool., 201, 1–20.

Siegel, S. (1956). Nonparametric Statistics for Behavioral Sciences. McGraw-Hill, New York.

Singh, N., Barbour. K.W. and Berger, F.G. (1998). Evolution of transcriptional regulatory elements within the promoter of a mammalian gene. Mol. Biol. Evol. 15, 312–325.

Smale, S.T. (1994). Core promoter architecture for eukaryotic protein-coding genes. In: Transcription: Mechanisms and Regulation (Conaway, R.C. and Conaway, J.W., Eds.), pp. 63–81. Raven Press, New York.

Somero, G.N. (1978). Temperature adaptation of enzymes: biological optimization through structure-function compromises. Ann. Rev. Ecol. System. 9, 1–29.

Somero, G.N. (1995). Proteins and temperature. Annu. Rev. Physiol. 57, 43–68.

Somero, G.N. (1996). Temperature and proteins: Little things can mean a lot. News Physiol. Sci. 11, 72–77.

Somero, G.N. and Hand, S.C. (1990). Protein assembly and metabolic regulation: physiological and evolutionary perspectives. Physiol. Zool. 63, 443–471.

Stone, B.B. and Sidell, B.D. (1981). Metabolic responses of striped bass (*Morone saxatilis*) to temperature acclimation: 1. Alterations in carbon sources for hepatic energy metabolism. J. Exp. Zool. 218, 370–380.

Stroppolo, M.E., Falconi, M., Caccuri, A.M. and Desideri, A. (2001). Superefficient enzymes. Cell. Mol. Life Sci. 58, 1451–1460.

Stryer, L. (1981). Biochemistry. W.H. Freedman, San Francisco.

Torres, N.V., Mateo, F., Sicilia, J. and Melendez-Hevia, E. (1988). Distribution of the flux control in convergent metabolic pathways: theory and application to experimental and simulated systems. Intl. J. Biochem., 20, 161–165.

Treves, D.S., Manning, S. and Adams, J. (1998). Repeated evolution of an acetat-crossfeeding polymorphism in long-term populations of *Escherichia coli*. Mol. Biol. Evol. 15, 789–797.

Tsuji, J.S. (1988a). Seasonal profiles of standard metabolic rate of lizards *Sceloporus occidentalis* in relation to latitude. Physiol. Zool. 61, 230–240.

Tsuji, J.S. (1988b). Thermal acclimation of metabolism in *Sceloporus* lizards from different latitudes. Physiol. Zool. 61, 241-253.

Van Der Veen, P., Ruijter, G.J. and Visser, J. (1995) An extreme creA mutation in *Aspergillus nidulans* has severe effects on D-glucose utilization. Microbiology 141, 2301–2306.

Verrijzer. C.P., Chen, J.L., Yokomori, K. and Tjian, R. (1995). Binding of TAFs to core elements directs promoter selectivity by RNA polymerase II. Cell 81, 1115–1125.

Vincent. K.A. and Wilson, A.C. (1989). Evolution and transcription of old world monkey globin genes. J. Mol. Biol., 207, 465–480.

Watt,W.B., Carter, P.A. and Donohue, K. (1986). Females' choice of good genotypes as mates is promoted by an insect mating system. Science 233, 1187–1190.

Wegener, G. (1996). Flying insects: model systems in exercise physiology. Experientia 52, 404–412.

Wiley, E.O. (1986). A study of the evolutionary relationships of *Fundulus* topminnows (Teleostei: Fundulidae). Am. Zool. 26, 121–130.

Wiley, S.R., Kraus, R.J. and Mertz, J.E. (1992). Functional binding of the "TATA" box binding component of transcription factor TFIID to the –30 region of TATA-less promoters. Proc. Natl. Acad. Sci. USA 89, 5814–5818.

Wilson, A.C. (1976). Molecular Evolution. Sinauer, MA, Sunderland.

Yamao, F. (1999). Ubiquitin system: selectivity and timing of protein destruction. J. Biochem. 125, 223–229.

Yaumuchi, T., Stegeman, J.J. and Goldberg, E. (1975). The effects of starvation and temperature acclimation on pentose phosphate pathway dehydrogenase in brook trout liver. Arch. Biochem. Biophys. 167, 13–20.

Yean, D. and Gralla, J. (1997). Transcription reinitiation rate: a special role for the TATA box. Mol. Cell. Biol. 17, 3809–3816.

Zamer, W.E. and Hoffmann, R.J. (1989). Allozymes of glucose-6-phosphate isomerase differentially modulate pentose-shunt metabolism in the sea anemone *Metridium senile*. Proc. Natl. Acad. Sci. USA 86, 2737–2741.

Zenzie-Gregory, B., O'Shea-Greenfield, A. and Smale, S.T. (1992). Similar mechanisms for transcription initiation mediated through a TATA box or an initiator element. J. Biol. Chem. 267, 2823–2830.

Sensing, Signaling and Cell Adaptation. Edited by K.B. Storey and J.M. Storey
© *2002 Elsevier Science B.V. All rights reserved.*

CHAPTER 22

Dynamic Use of cDNA Arrays: Heterologous Probing for Gene Discovery and Exploration of Organismal Adaptation to Environmental Stress

Sean F. Eddy and Kenneth B. Storey*
Institute of Biochemistry and Department of Chemistry, Carleton University, Ottawa, Ontario, Canada

1. Introduction

DNA microarrays are the hottest new technology available for gene discovery and are rapidly being put to use in many novel applications. The underlying methodology behind DNA microarrays has been prevalent in molecular biology for the past quarter century but only recently has the technology been developed to allow the production and quantification of arrays containing thousands of gene elements. For example, arrays containing 31,500 elements (cDNAs and expressed sequence tags or ESTs) have been used to screen kidney cancer cell lines in an attempt to discover all the genes associated with the development of cancer (Boer et al., 2001). This new use of DNA hybridization has also been used to study the effects of aging (Helmberg, 2001) and in the food agriculture and biotechnology sector to uncover the effects on genes in digestive systems that are exposed to genetically modified foods (van Hal et al., 2000). Other studies are analyzing signaling pathways in plants (Hugouvieux et al., 2001) and new uses in neuropharmacology are underway (Marcotte et al., 2001). DNA array screening has recently been coupled with subtractive PCR hybridization to investigate differentially expressed transcripts (Wu, 2001). Each of these applications

underscores the great diversity and seemingly limitless potential of research applications for cDNA microarrays.

Studies to date have used homologous probing—that is, the cDNA immobilized on the arrays are from the same species as the samples that are being screened. This guarantees cross-reaction with all gene elements on the array so that a complete profile of the responses of all genes can be made. To date, arrays have been produced for several of the major mammalian model species (human, rat, mouse) as well as a few other common model species such as *Arabidopsis thaliana, Caenorhabditis elegans, Drosophila melanogaster, Escherichia coli* and *Saccharomyces cerevisiae,* but what of other organisms for which there are no species-specific cDNA arrays? The time, expertise and expense that would be required to produce homologous arrays for every species of interest would be daunting and, if required, would place the power of cDNA array analysis outside the grasp of most scientists. Heterologous probing may be a viable alternative. Arrays produced with genes from one species could potentially provide useful information when used to screen samples from closely related, or even distantly related, species. Heterologous probing of cDNA arrays of closely related species could certainly provide an important first step in gene discovery.

Recently, our lab became interested in the use of heterologous probing as a means of identifying genes that are up-regulated during mammalian

---
*Corresponding author

hibernation in two species, thirteen-lined hibernating ground squirrels, *Spermophilus tridecemlineatus* (Hittel and Storey, 2001) and in little brown bats, *Myotis lucifugus* (Eddy and Storey, 2001). Using rat or human cDNA arrays, we found that heterologous probing was successful as a means of gene discovery in both species. Other studies in our laboratory have also used heterologous probing to assess stress-induced changes in gene expression over even greater phylogenic distances; for example, we are applying the technique to analyses of freeze-induced gene expression in freeze-tolerant frogs and insects, and anoxia-induced genes in marine snails (Larade and Storey, 2002). These studies provide valuable clues to the types of genes that are expressed in a given species under a particular stress.

Conventional methods of gene discovery include differential cDNA library screening and differential-display polymerase chain reaction (dd-PCR) as well as newer methods such as quantitative reverse transcription PCR (qRT-PCR) or serial analysis of gene expression (SAGE). Heterologous probing of cDNA arrays cannot replace these conventional methods but provides a valuable supplemental technique. In particular, the technique is useful as an initial screening tool to allow assessment of the responses of broad families of genes (e.g., signal transduction systems, glycolytic enzymes, molecular transporters, transcription factors) to an imposed stress. The success of heterologous probing depends, of course, on a high degree of sequence identity between the cDNA probes on the arrays and the same genes in other species. In the case of hibernating mammals, a high percent identity at the cDNA level is very likely between human and squirrel/bat genes and sufficient to allow hybridization of squirrel or bat cDNAs with most of the cDNAs found on the arrays. Indeed, in other studies, we have confirmed high sequence identity for various individual ground squirrel genes with their human or rat homologues, for example, 97.5% for ventricular myosin light chain 1 and >93% for cytochrome c oxidase 1 in *Spermophilus lateralis* (Fahlman et al., 2000) and >90% for fatty acid binding protein in *S. tridecemlineatus* (Hittel and Storey, 2001).

For non-mammalian species, we are attempting to show that there are enough genes with sufficiently high sequence identity to their mammalian counterparts to make heterologous screening a practical supplementary technique for gene discovery, particularly for the purposes of assessing changes in gene expression in response to changes in physiological state. The remainder of this review provides a summary of the development and use of DNA array technology, a discussion of the practical considerations involved in heterologous probing, and a summary of our initial application of human and rat cDNA arrays to analyze changes in gene expression in mammalian hibernators.

2. Differential gene expression: the early years

Historically, the most common method of analyzing differential gene expression has been cDNA library screening followed by Northern blot analysis. In recent years a number of techniques based on the polymerase chain reaction (PCR) have gained popularity, each with their own advantages and disadvantages. These have included semi-quantitative relative RT-PCR, dd-PCR, qRT-PCR and SAGE. A number of these techniques combine PCR steps with library construction (Timblin et al., 1990) in an attempt to reduce the number of false positives obtained and decrease the length of time required for screening. Varying degrees of success are achieved with these techniques when using nontraditional animal model systems in which cDNA sequences are unknown.

cDNA library screening with Northern blotting is a standard in molecular biology for analyzing differential gene expression. These procedures are reliable and frequently allow researchers to obtain full-length cDNA clones of genes from the species in question. These procedures are of particular importance when dealing with unusual animal species and/or with experimental systems or stresses that may stimulate expression of novel genes. Indeed, cDNA library screening remains the only way to detect the expression of novel or previously unidentified genes. For example, cDNA library screening has uncovered multiple novel genes (and gene products) that aid freezing survival in freeze-tolerant frogs and snails (Cai and Storey, 1997; Larade et al., 2001; McNally et al., 2002). Since cDNA library screening is a well-known standard technology in molecular biology, the chances for its success when applied to any new system/stress are high but a main drawback to the technology is the large amount of time and resources that it requires.

The newest addition to the range of techniques available for studying differential gene expression is cDNA arrays. The main advantages to using cDNA arrays as screening tools are speed, reliability/reproducibility and the ability to simultaneously screen hundreds or thousands of genes producing comprehensive analyses of major subsections of cellular metabolism.

3. DNA arrays: a brief history

Since the advent of the Southern blot in the 1970s, researchers have found a number of ways to use nucleic acids on solid supports, including nylon membranes or glass. DNA macroarrays on nylon membranes are a recent development and DNA microarrays on glass slides are the newest addition to this growing field. DNA array technology is still much in its infancy but already numerous reviews and original research papers have been produced on the subject (Cunningham, 2000; Jain, 2000; Burgess, 2001; Blohm and Guiseppi-Elie, 2001; Lobenhofer et al., 2001). There are two major procedures for producing commercially available DNA arrays. The first is the method of spotting PCR amplified cDNAs onto solid supports. The first example of this was with nylon membranes to produce macroarrays and subsequently Patrick Brown and colleagues at Stanford University spotted cDNAs onto treated glass microscope slides to develop the first cDNA microarrays (Schena et al., 1995), now the most commonly used type of array. The second method of array production, developed by Stephen Fodor and colleagues (Pease et al., 1994), involves synthesizing short, specifically chosen oligonucleotides (<25 in length) *in situ* using light-directed combinatorial chemistry. Affymetrix developed this procedure and microarrays of this type are commonly referred to as high-density oligonucleotide arrays.

Nylon macroarrays arrays provide a good starting point for many smaller research labs because few changes need to be made to the current molecular biology techniques that are present in most labs in order to use these arrays. These arrays can contain thousands of genes and use common materials and radioactively labeled probes made by the same probe-synthesizing procedures as for cDNA library screening. Nylon arrays are also reusable, making them attractive to smaller labs that need to get maximum return for their investment. Only recently, have some reusable microarrays become available (Dolan et al., 2001). The only inherent problem associated with nylon macroarrays is that a comparison of control versus experimental sample situations cannot be performed on a single array. Paired arrays must be used, one hybridized with control probe and one hybridized with stressed probe. Gene expression on each array must then be normalized to the expression of one or more control or housekeeping genes on the same array before a comparison of normalized gene expression levels can be made between the two arrays. This is a tedious process that has several potential sources of error, not the least of which is the housekeeping genes chosen as controls. To act as a control, the chosen housekeeping genes must remain constitutively and constantly expressed under a variety of cellular conditions but the possibility exists that the expression of these genes might actually change under certain circumstances in different tissues/species/stresses. This potential problem is of particular importance if heterologous probing is to be undertaken.

Because there are intrinsic size limitations in the number of cDNAs that can be spotted and probed on nylon arrays, labs that wish to initiate a broader study of gene expression will find cDNA microarrays more suitable for their purposes. Microarrays now contain thousands of cDNAs and, because of recent sequencing efforts with a variety of genomes, many of these represent identified genes. Others are ESTs that currently have unknown cellular functions. cDNA microarrays produced by spotting cDNAs on treated glass microscope slides can contain up to 31,500 non-redundant cDNAs (Boer et al., 2001). Slight changes to standard molecular biology techniques are required for the detection of differential gene expression using cDNA microarrays. ^{32}P-labeled probes are not used. Instead, cDNA from control and experimental samples are labeled with Cy3 and Cy5 dyes. Common Cy-labeled dyes used for microarray experiments are conjugated to dCTP and thus are incorporated into the first strand cDNA. Both cDNA samples (control versus experimental) are hybridized to the same array. The binding of Cy3 labeled cDNA and Cy5 labeled cDNA is detected by measuring the fluorescence of each at different wavelengths. Cy3 labeled cDNA has a maximum absorption at 550 nm and emits at 570 nm whereas Cy5 labeled cDNA absorbs maximally at 649 nm and emits at 670 nm. The fluorescence intensity at each emission wavelength is proportional to the amount of Cy3 or Cy5 bound. Continual reduction in costs of producing and purchasing arrays coupled with the small amount of sample material required increases the appeal of these arrays to many labs. However, some problems may occur with glass slide microarrays due to batch-to-batch variation when the coatings on the slides are not uniform.

The Affymetrix method was developed to attain highly specific sequences arrayed at a high density on a solid support. This technique uses photolithography and chemical protecting groups to synthesize oligonucleotides directly onto a solid support, a procedure that results in a high degree of quality control and specificity associated with Affymetrix GeneChips. In order to synthesize oligonucleotides on the GeneChips, the actual cDNA or partial cDNA sequence of the genes must be known. This information is not known for the vast majority of organisms whose genomes have not been sequenced, as is the case with our hibernator species. The high cost of producing and purchasing GeneChips has, to date, detracted from their wide acceptance in the academic field.

As with any new technique, the cost of running array-screening experiments has greatly decreased over the past couple of years and currently methods are being developed that will make microarray screening even more cost effective by producing re-usable microarrays (Dolan et al., 2001). Experimental variability has been a concern because, to date, microarrays can only be used for a single hybridization experiment (control versus stressed targets on a single microarray), and hence, there is no way to determine experimental variability. Currently, variability is reduced by running duplicate experiments with reciprocal labeling of control and stressed cDNA probes (e.g., control samples are Cy3-labeled in one run and Cy5-labeled in other and vice versa for experimental samples) as well as by running repeated trials (Wildsmith et al., 2001). The initial high costs of purchasing fluorescent dyes, microarrays and microarray readers has also meant that this technology has taken a few years to become feasible as a viable alternative to classic techniques. Over time, however, microarrays will likely come to dominate the field, and just as the technical simplicity of macroarrays out-competes cDNA library screening for many applications, so to will dye-based microarray screening soon take over from ^{32}P-based macroarray technology. Apart from the shift away from radioactive technology, another major benefit of using microarrays over macroarrays is the reduction in amount of cell or tissue sample material that is required, which can be limiting in many experimental situations.

4. Evaluation of mammalian hibernation via cDNA array screening

Our lab has recently begun to use cDNA array technology for heterologous screening of cDNA macroarrays and microarrays as a means of assessing differential gene regulation in a number of animal models, particularly mammalian hibernators. Hibernating mammals provide natural models for a number of stress states relevant to medical science including ischemia, hypoxia and hypothermia. Skeletal muscle is of particular interest because of the phenomenon of disuse atrophy, which occurs rapidly when muscles of most mammals go unused for an extended period, but does not seem to greatly affect hibernators even though they remain torpid for many months.

Initial studies began with a traditional approach to studying differential gene expression. A cDNA library was constructed from skeletal muscle of hibernating bats, *M. lucifugus*, and after primary screening with probes made from muscle extracts of control versus hibernating animals, approximately 60 candidate clones were found that were potentially up-regulated in torpid animals. However, further characterization of these clones, by reverse Northern blotting failed to confirm that any of the genes were up-regulated, which was quite unexpected. In other trials, dd-PCR was used to make a comparable assessment of gene expression in skeletal muscle of control versus hibernating golden-mantled ground squirrels, *S. lateralis*. Data from these experiments revealed three potentially up-regulated genes, but subsequent sequence analysis showed that the three genes were unidentifiable and Northern blotting failed to confirm up-regulation. After these trials, we turned our attention to cDNA arrays. Our first experiments used ATLAS™ nylon macroarrays (both 588 and 1196 gene arrays) because of the ease of probe preparation and hybridization that used protocols and radioactive materials already available in the lab. More recently, we have also screened for hibernation-specific gene expression using human 19K cDNA microarrays.

There are a number of commercial software programs available for the manipulation and quantification of the data output from cDNA macroarrays. Because of the high cost associated with such programs, we explored the possibility of using shareware or software already present in our lab for data manipulation and quantification. In our first trials with macroarrays, we identified control/hibernation pairs that showed elevated signal during hibernation by visual inspection of autoradiograms. We then employed Imagequant (Molecular Dynamics) software to quantify spot intensities on control and hibernator arrays, first normalizing the intensity of each spot to selected housekeeping genes and other surrounding local cDNAs that showed no change in spot intensity. Relative spot intensities were then compared

between the two metabolic states. This proved quite useful when dealing with a handful of changes on macroarrays that contained 588 spotted cDNAs but the program showed limitations when dealing with arrays containing a larger number of cDNAs (1196 gene ATLAS™ rat arrays or human 19K cDNA microarrays). This was because each cDNA had to be normalized to local or surrounding cDNAs by manually selecting spots of interest using Imagequant and normalizing them relative to a number of housekeeping genes as well as normalizing cDNA signals relative to local or surrounding cDNAs. This is done to account for preferential binding on different subsections of the arrays, a daunting task with numerous putative changes in gene expression seen on the larger arrays. For further macroarray analysis, we wanted a program that had the ability to quantify the complete set of macroarray data simultaneously. Further cDNA ATLAS™ macroarray analysis was performed using the EST array analysis program (http://www.uni-mainz.de/FB/Medizin/Tumorgenetik/array/download.htm). This program overlays images obtained from control and experimental samples and then produces a composite image of the two arrays. The intensity of each is analyzed simultaneously providing an analysis of spot intensities over the entire array. Genes of interest must still be normalized relative to a number of control/housekeeping genes. An advantage of the EST program is its ability to identify subtle changes that may go unnoticed with visual inspection and analysis using Imagequant.

In our first trials with ATLAS™ cDNA macroarrays to analyze hibernation-specific changes in gene expression with *M. lucifugus* and *S. tridecemlineatus* skeletal muscle, we found that cross-reaction of hibernator cDNAs with the nylon arrays was good (Fig. 22.1). Indeed, the percentage of cDNAs that cross-reacted was 93% for *S. tridecemlineatus* and 73% for *M. lucifugus*. It is possible that *S. tridecemlineatus* probes had a higher percentage of cross-reacting cDNAs with ATLAS™ rat arrays than *M. lucifugus* probes because ground squirrels and rats are both members of the Order Rodentia, whereas bats are members of Order Chiroptera, a distinctly different

A.

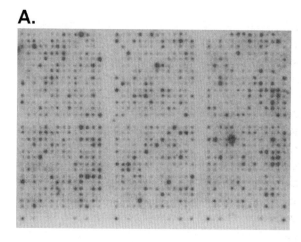

B.

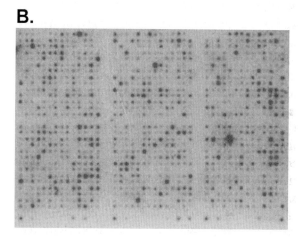

Fig. 22.1. Typical cross-reactivity displayed using ATLAS™ Rat 1.2 cDNA arrays hybridized with ^{32}P-probe from (A) euthermic skeletal muscle of *M. lucifugus*, and (B) euthermic skeletal muscle of *S. tridecemlineatus*.

hibernation-specific changes in gene expression in a number of other tissues from both *M. lucifugus* and *S. tridecemlineatus* including heart, brown adipose tissue, kidney and liver. Results with these tissues showed comparable rates of cross-reaction to those mentioned above. Screening resulted in the identification of selected genes that are up-regulated during hibernation. Prominent among

A.

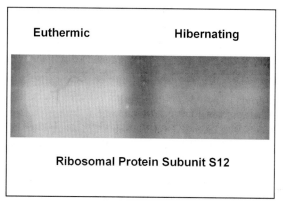

B.

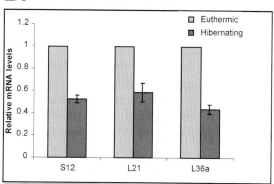

Fig. 22.2. Relative RT-PCR analysis of the expression of three ribosomal protein subunits in skeletal muscle of euthermic (control) versus hibernating *S. tridecemlineatus*. Genes were chosen for evaluation based on results from ATLAS™ cDNA nylon array screening showing putative down-regulation of these genes during hibernation. (A) Representative 1% TAE gel displays relative cDNA levels after RT-PCR amplification of ribosomal protein subunit S12 from normalized first strand cDNA. Equal amounts of euthermic and hibernating mRNA were used for first strand synthesis and S12 expression levels were normalized against alpha tubulin and GAPDH expression. (B) Bar graph showing mean result from $n = 3$ trials comparing relative mRNA levels in hibernator samples compared with controls that were normalized to 1.

group of mammals. Of the cross-reacting genes, one group was of particular interest. A group of cDNAs that encode for components of the small and large ribosomal subunits were consistently down-regulated during hibernation in *S. tridecemlineatus;* these included L19, L21, L36a, S17, S12 and S29. Relative RT-PCR was used to confirm the three that showed the greatest hibernation-specific down-regulation on the arrays. Figure 22.2 shows results for S12, L36a and L21 and demonstrates that mRNA transcript levels for these three genes in hibernating muscle were reduced, on average, to 45–55% of the corresponding levels in euthermic control animals (S. Eddy, unpublished data). Arrays have also been used to screen for these was the very strong up-regulation of heart and adipose tissue isoforms of fatty acid binding protein (H-FABP and A-FABP, respectively) during hibernation (Hittel and Storey, 2001). Hibernation-induced up-regulation of H-FABP and A-FABP was also found in *M. lucifugus* (S. Eddy, unpublished data).

When glass slide microarrays containing 19K spotted cDNAs became available commercially, we re-examined gene expression in hibernator skeletal muscle. The human 19K cDNA slides from the Ontario Cancer Institute (http://www.uhnres.utoronto.ca/services/microarray/index.html) were chosen for several reasons. The first was their relatively low cost. Secondly, there are a high number of unique transcripts found on this array. Thirdly, with 19,200 unique transcripts present it is likely that we are screening a substantial percentage of the genes found in the human genome, which is thought to contain approximately 30,000–40,000 genes (Venter et al., 2001). With upwards of 80% sequence identity between homologous genes in placental mammals, heterologous probing with the 19K microarray could potentially assess the status during hibernation of approximately 50% of the genes in the squirrel or bat genomes. The human 19K cDNA slides are produced by arraying PCR amplified clones with a spot to spot spacing of 170 μm. Because there are a high number of ESTs found in animal databases, many of the cDNAs found on the arrays are ESTs. Identifying when these ESTs are expressed or overexpressed will undoubtedly lead to the identification of the function of the genes that these represent. Furthermore, hibernation-induced changes in the expression of selected ESTs could indicate the occurrence of novel genes that are hibernation-specific, genes that could then be pulled out of a cDNA library and characterized with the potential for identification of unique proteins that support hibernation.

In optimizing the use of the 19K microarrays for heterologous probing, several factors were considered: (1) the use of total RNA versus poly(A)$^+$RNA as the source material, (2) washing temperature, (3) hybridization time, and (4) type of polymerase employed. The latter two factors were without

effect; there was little difference in fluorescent signal intensity when hybridization times of 16 hours versus 24 hours were considered and no noticeable difference between using Superscript (BRL) or Superscript II (BRL). However, in a comparison between the use of 10–40 μg of total RNA versus 1 μg of poly (A)$^+$RNA for first strand cDNA synthesis, we found that total RNA worked best for hybridization when using at least 15 μg and best when using 30–40 μg. Other groups have reported similar results (Wildsmith et al., 2001). The additional purification step involved in obtaining poly(A)$^+$RNA did reduce some of the background on the arrays but we observed a significant drop in overall fluorescent signal when purified poly(A)$^+$RNA was used for probe preparation. Hybridizations were typically carried out for 20 hours in a slide chamber. For washing steps, we initially washed slides at 50°C for 30 min as indicated in the protocol supplied with the slides but this resulted in very low rates of cross-reaction, just 15–25% (Eddy and Storey, 2001). However, by reducing the washing temperature to room temperature (~21°C), we were able to increase cross-reaction up to 85%. Figure 22.3 shows a representative example (brain from *M. lucifugus*) of an array hybridized under lower washing temperature conditions. Since a key aim with heterologous probing is to maximize the number of cross-reacting genes, this simple adjustment of washing temperature was a highly successful modification of the procedure.

When using cDNA microarrays we shifted from the EST analysis program to the Scanalyze program developed in Michael Eisen's lab at the Lawrence Berkeley National Lab (http://rana.lbl.gov/EisenSoftware.htm) (Eisen et al., 1998). We chose this program because of its relatively easy user interface. A number of other array analysis programs are also being developed that may provide improvements, including Matarray (Wang et al., 2001) and AMADA (Xia and Xie, 2001).

Initial experiments on the 19K cDNA arrays compared gene expression in skeletal muscle of euthermic versus hibernating *M. lucifugus*. The first use of microarrays hybridized with cDNA prepared from muscle of hibernating versus euthermic

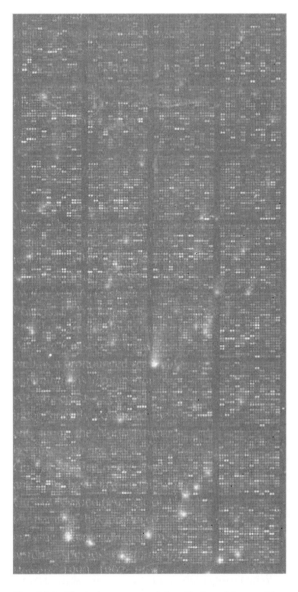

Fig. 22.3. Typical cross-reactivity displayed by probing human 19K cDNA arrays with first strand cDNA synthesized from 15 μg of total RNA from brain of euthermic (control) versus hibernating *M. lucifugus*. Euthermic cDNA was labeled with Cy3 and hibernating cDNA was labeled with Cy5. View this image in color at http://www.carleton.ca/~kbstorey/batarray.jpg

elevation in transcript levels that is considered to be the cut-off by most researchers (Hedge et al., 2000). Changes of lesser magnitude are typically within the range of variation and noise of DNA microarrays. In subsequent trials we also examined hibernation-dependent changes in gene expression in *M. lucifugus* brain (Fig. 22.3) and heart. Initial analysis indicated approximately 30 genes (out of 19,200) that were putatively up-regulated in each organ, most of them being ESTs. The images for these 30 were then observed on an individual basis to determine whether or not the up-regulation appeared valid. This reduced the number from 30 to five for brain and three for heart (Table 22.1). Various candidates were eliminated because of variable background levels around some of the spotted cDNAs and oddities in the actual spots of

Table 22.1. Putatively up-regulated genes identified from our initial experiments probing human 19K cDNA microarrays with *M. lucifugus* cDNA prepared from three organs. cDNAs were considered to be up-regulated if the increase in expression in hibernator versus control samples was two-fold or greater, as determined by Scanalyze analysis.

Gene	Accession number	Tissue
Amyloid beta precursor protein	W05611	Muscle
Glucosyl transferase (putative)	N20259	Muscle
Insulin-like growth factor binding protein 7 (partial match)	H49910	Muscle
EST	AA210692	Muscle
EST	W37120	Muscle
Carboxypeptidase E	AA001138	Brain
DEAD/H (Asp-Glu-Ala-Asp/His) box polypeptide 5	AA001454	Brain
Synaptosomal-associated protein, 25kD	AA017038	Brain
Putative translation initiation factor	R85732	Brain
EST	N40782	Brain
Interferon-related developmental protein 2	R48311	Heart
EST	R48060	Heart
EST	H20335	Heart

bats resulted in higher than anticipated background, which limited the ability to analyze some sections of the microarrays. Of the sections that were analyzed, we found only five cDNAs that could be considered to be up-regulated during hibernation, showing the minimum two-fold interest. The putatively up-regulated cDNAs will require further confirmation with northern blot analysis and/or quantitative RT-PCR to determine the legitimacy of these initial results. It is interesting to consider, however, that the difference between the euthermic (high metabolic rate, 37°C body temperature) and hibernating (metabolic rate reduced by ~100 fold, 5°C body temperature) states may come down to changes in expression patterns of just a very small percentage of the genes in the genome.

A major problem with microarrays that is still not fully resolved is what contributes to background noise levels. A number of researchers have looked at this problem (Diehl et al., 2001; Mills and Gordon, 2001; and references therein) but improvements are still needed and these will likely come as more and more researchers use microarray technology.

5. Reproducibility and reliability

Any new technology must first prove its worth in order to become a viable alternative to previously accepted protocols and techniques. At present, researchers remain skeptical as to what constitutes reliable data in microarray experiments despite the mounting evidence demonstrating the validity of array data (Yue et al., 2001). As with any experiment, the number of repetitions will increase confidence in results and some researchers have found that numerous replicates should be performed to attain reliable data (Wittes and Friedman, 1999; Wildsmith et al., 2001).

Recent studies have shown that microarrays give comparable results to those found using Northern blots (Bartosiewicz et al., 2000). Researchers have also found that reciprocal labeling of samples can produce variation even though this method was initially proposed as an important way to improve data. However, discrepancies did not occur with replicates that maintained the use of Cy3 labeled cDNA for control and of Cy5 for experimental labeled samples. In some cases, Northern blots indicate more up-regulated genes than do the microarrays. For example, a comprehensive study by Taniguchi et al. (2001) quantitatively comparing DNA microarrays and Northern blots found that for four of 46 genes compared, DNA arrays did not detect expression changes that were evident on Northern blots. This suggests that the sensitivity of microarrays is inferior to that of Northern blotting. However, for the majority of the genes that were induced or suppressed, DNA microarray results appeared to be quantitatively similar to those found on Northern blots with the exception of one gene where DNA microarrays detected a four-fold induction but Northern blotting revealed a 60-fold increase (Taniguchi et al., 2001).

For our purposes, when we are undertaking analysis of a new system/tissue, we combine the RNA from three independent RNA isolations from different tissue samples into each final control or experimental sample. This significantly decreases the variability between arrays and these pooled samples provide a more global average of the differences in gene expression between control and experimental samples than do comparable arrays run with individual samples, even when the results of three such individual runs are combined.

6. Outlook

First trials at using cDNA arrays for heterologous screening to assess gene expression in hibernating mammals proved effective. We achieved good fluorescent signal intensity on the spotted cDNAs and, after optimizing washing temperature, excellent cross-reactivity of up to 85%. This bodes well for the future use of heterologous screening as a mechanism of gene discovery. The advent of DNA arrays has allowed researchers to bypass the long and arduous tasks involved in differential screening of cDNA libraries. Although library screening is a trusted method of gene discovery, it consumes time and resources that can be saved by moving to a DNA array screening protocol. Our lab has encountered few problems with using commercially available DNA arrays to investigate gene expression in other mammalian species. For example, up-regulation of H-FABP and A-FABP during hibernation in *S. tridecemlineatus* was identified via the use of Clontech ATLAS™ arrays (Hittel and Storey, 2001), demonstrating the value of cDNA arrays for gene discovery via heterologous probing.

When it comes to the question of the reliability of cDNA arrays, our experience is that DNA arrays are an excellent means to initially target genes that respond to various environmental stresses and physiological states. However, further characterization and quantification must still be done with northern blots, relative RT-PCR and qRT-PCR. DNA microarray technology is still much in its infancy, but it is quickly changing the methodology of many labs. With functional expression of entire genomes at the fingertips of many researchers, complex patterns of gene expression are now being studied. A number of researchers are also seeking to develop public domains that house raw data from array analysis (Geschwind, 2001; Becker, 2001; Miles, 2001) much like GenBank stores DNA sequences. This data would be available to anyone wishing to use and analyze it in any number of ways such as to track the expression of a particular gene under a variety of experimental stresses. Development of these public databases of expression data will undoubtedly assist in categorizing expression patterns associated with numerous cellular responses. With public databases, new areas of bioinformatics will emerge for those interested in deciphering complex gene expression patterns in response to an abundance of stimuli. The use of cDNA arrays can only benefit the entire scientific community as we push forward to elucidate cellular and molecular responses to stress.

References

Bartosiewicz, M., Trounstine, M., Barker, D., Johnston, R. and Buckpitt, A. (2000). Development of a toxicological gene array and quantitative assessment of this technology. Arch. Biochem. Biophys. 376, 66–73.

Becker, K.G. (2001). The sharing of cDNA microarray data. Nature Rev. Neurosci. 2, 438–440.

Blohm, D.H. and Guiseppi-Elie, A. (2001). New developments in microarray technology. Curr. Opin. Biotechnol. 12, 41–47.

Boer, J.M., Huber, W.K., Sultmann, H., Wilmer, F., von Heydebreck, A., Haas, S., Korn, B., Gunawan, B., Vente, A., Fuzesi, L., Vingron, M. and Poustka, A. (2001). Identification and classification of differentially expressed genes in renal cell carcinoma by expression profiling on a global human 31,500-element cDNA array. Genome Res. 11, 1861–1870.

Burgess, J.K. (2001). Gene expression studies using microarrays. Clin. Exp. Pharmacol. Physiol. 28, 321–328.

Cai, Q. and Storey, K.B. (1997). Upregulation of a novel gene by freezing exposure in the freeze-tolerant wood frog (*Rana sylvatica*). Gene 198, 305–312.

Cunningham, M.J. (2000). Genomics and proteomics: the new millennium of drug discovery and development. J. Pharmacol. Toxicol. Meth. 44, 291–300.

Diehl, F., Grahlmann, S., Beier, M. and Hoheisel, J.D. (2001). Manufacturing DNA microarrays of high spot homogeneity and reduced background signal. Nucleic Acids Res. 29, e38.

Dolan, P.L., Wu, Y., Ista, L.K., Metzenberg, R.L., Nelson, M.A. and Lopez, G.P. (2001). Robust and efficient synthetic method for forming DNA microarrays. Nucleic Acids Res. 29, e107.

Eddy, S.F. and Storey, K.B. (2001). Gene expression in hibernation: testing skeletal muscle of little brown bats, *Myotis lucifugus*, using commercially available cDNA microarrays. Proc. Virtual Conference in Genomics and Bioinformatics. October 15–16.

Eisen, M.B., Spellman, P.T., Brown, P.O. and Botstein, D. (1998). Cluster analysis and display of genome-wide expression patterns. Proc. Natl. Acad. Sci. USA, 95, 14683–14688.

Fahlman, A., Storey, J.M. and Storey, K.B. (2000). Gene up-regulation in heart during mammalian hibernation. Cryobiology 40, 332–342.

Geschwind, D.H. (2001) Sharing gene expression data: an array of options. Nature Rev. Neurosci. 2, 435–438.

Hegde, P., Qi, R., Abernathy, K., Gay, C., Dharap, R., Gaspard, R., Hughes, J.E., Snesrud, E., Lee, N. and Quakenbush, J. (2000). A concise guide to cDNA microarray analysis. Biotechniques 29, 548–562.

Helmberg, A. (2001). DNA-microarrays: novel techniques to study aging and guide gerontologic medicine. Exp. Gerontol. 36, 1189–1198.

Hittel, D.S. and Storey, K.B. (2001). Differential expression of adipose- and heart-type fatty acid binding proteins in hibernating ground squirrels. Biochim. Biophys. Acta 1522, 238–243.

Hugouvieux, V., Kwak, J.M. and Schroeder, J.I. (2001). An mRNA cap binding protein, ABH1, modifies early abcisic acid signal transduction in *Arabidopsis*. Cell 106, 477–487.

Jain, K.K. (2000). Applications of biochip and microarray systems in pharmacogenomics. Pharmacogenomics 1, 289–307.

Larade, K., Nimigan, A. and Storey, K.B. (2001). Transcription pattern of ribosomal protein L26 during anoxia exposure in *Littorina littorea*. J. Exp. Zool. 290, 759–768.

Larade, K. and Storey, K.B. (2002). A profile of the metabolic responses to anoxia in marine invertebrates. In: Cell and Molecular Responses to Stress. (Storey, K.B. and Storey, J.M., Eds.). Vol. 3, Chap. 3. Elsevier, Amsterdam.

Lobenhofer, E.K., Bushel, P.R., Afshari, C.A. and Hamadeh, H.K. (2001). Progress in the application of DNA microarrays. Environ. Health Perspect. 109, 881–892.

Marcotte, E.R., Srivastava, L.K. and Quirion, R. (2001). DNA microarrays in neuropsychopharmacology. Trends Pharm. Sci. 22, 426–436.

McNally, J.D., Wu, S., Sturgeon, C.M. and Storey, K.B. (2002). Identification and characterization of novel freezing inducible gene, *Li16*, in the wood frog, *Rana sylvatica*. FASEB J., 16, 902–904.

Miles, M.F. (2001). Microarrays: lost in a storm of data? Nature Rev. Neurosci. 2, 441–443.

Mills, J.C. and Gordon, J.I. (2001). A new approach for filtering noise from high-density oligonucleotide microarray datasets. Nucleic Acids Res. 29, e72.

Pease, A.C., Solas D., Sullivan, E.J., Cronin, M.T., Holmes, C.P. and Fodor, S.P. (1994). Light-generated oligonucleotide arrays for rapid DNA sequence analysis. Proc. Natl. Acad. Sci. USA 91, 5022–5026.

Schena, M., Shalon, D., Davis R.W. and Brown P.O. (1995). Quantitative monitoring of gene expression patterns with a complementary DNA microarray. Science 270, 467–470.

Taniguchi, M., Miura, K., Iwao, H. and Yamanaka, S. (2001). Quantitative assessment of DNA microarrays—comparison with Northern blot analyses. Genomics 71, 34–39.

Timblin, C., Bergsagel, P.L. and Kuehl, W.M. (1990). Identification of consensus genes expressed in plasmacytomas but not B lymphomas. Curr. Top. Microbiol. Immun. 166, 141–147.

van Hal, N.L.H., Vorst, O., van Houwelingen, A.M.M.L., Kok, E.J., Peijnenburg, A., Ahoroni, A., van Tunen, A.J. and Keijer, J. (2000). The application of DNA microarrays in gene expression analysis. J. Biotech. 78, 271–280.

Venter, J.C., Adams, M.D., Myers, E.W., Li, P.W., Mural, R.J., Sutton, G.G., et al. (2001). The sequence of the human genome. Science 291, 1304–1351.

Wang, X., Ghosh, S. and Guo, S-W. (2001). Quantitative quality control in microarray image processing and data acquisition. Nucleic Acids Res. 29, e75.

Wildsmith, S.E., Archer, G.E., Winkler, J.A., Lane, P.W. and Bugelski, P.J. (2001). Maximization of signal derived from cDNA microarrays. Biotechniques 30, 202-206.

Wittes, J. and Friedman, H.P. (1999). Searching for evidence of altered gene expression: a comment on statistical analysis of microarray data. J. Natl. Cancer Inst. 91, 400–401.

Wu, T.D. (2001). Analysing gene expression data from DNA microarrays to identify candidate genes. J. Pathol. 195, 53–65.

Xia, X. and Xie, Z. (2001). AMADA: analysis of microarray data. Bioinformatics 17, 569–570.

Yue, H., Eastman, P.S., Wang, B.B., Minor, J., Doctolero, M.H., Nuttal, R.L., Stack, R., Becker, J.W., Montgomery, J.R., Vainer, M. and Johnston, R. (2001). An evaluation of the performance of cDNA microarrays for detecting changes in global mRNA expression. Nucleic Acids Res. 29, e41.

Index

Abiotic stress 155, 159–161, 163
Abscisic acid
– and Ca^{2+} signaling 97
– and freeze tolerance 128
– in plant cold acclimation 123, 128, 159
– signal transduction 129
– stomatal closure 98
Acclimation 244
– membrane adaptation 124
– to cold in plants 123
Actin 52
Acyl–lipid desaturases 142
Adaptation 297
Adenosine 47, 110
– ischemia preconditioning 2, 52
– metabolism 47
Adenosine deaminase 47
Adenosine kinase 47
Adenosine receptor
– agonists 49
– antagonists 49
– down-regulation 51
– expression 50
– knock-out mice 50
Adenosine transporter inhibitors 49
Adenylyl cyclase 51
– ADP-ribosyl cyclase 95
Adipogenesis 270
Adrenaline 289
Adult hippocampal ischemia 49
Aerotaxis 243
Aestivation in snails 288
Allosteric enhancers 49
Alveolar endothelial cells 13-22
Alzheimer's disease 70
Amino acids 189, 199
– arginine synthetase control 194
– free pools 190
– adaptation to protein deficiency 191
– regulation of gene expression 192, 193, 195
– control of protein metabolism 196, 197, 199
– inhibition of autophagy 207
– in insulin signaling 212
– metabolism 189
– mTOR activation 211
– regulation of 43S pre-initiation complex formation 198
– regulation of p70s6k activity 200

– regulation of the eIF4F translation initiation complex 198
– signaling in pancreatic beta cells 214
– starvation 192
– stimulation of glycogen synthesis 207
– stimulation of initiation and elongation factors 209
– stimulation of S6 phosphorylation 207
AMP-dependent protein kinase 213
Androstane receptor 224
Angiogenesis 47
Anoxia 293
Anoxia tolerance
– carbohydrate metabolism 28–31
– cDNA array 37–38
– cGMP 30–31, 39–40
– gene expression 36–39
– MAPKs 41
– molluscan 27–43
– mRNA pools 35–36
– oxygen sensing 39–40
– protein synthesis 31–35
– transcription factors 40–41
– turtles 7,8
Antioxidant response element 221, 224
Antisense oligonucleotides 51
Aplysia 95, 99
Apoptosis 1, 5, 19, 55, 61, 235
– of brown adipocytes 277
ArcB sensor kinase 245
Aryl hydrocarbon receptor 223
Asparagine synthetase 194
Asthma 65
ATF-4 195
ATP-gated ion channels 108
ATP sensor 213
ATP turnover 289
Autonomic nerves 49
Autophagy 197, 207

Bacterial infection and immune responses 99
Bacterial transport 244
Betaines 124
Blood flow 51
Bradykinin 110
Brain ischemia 1
– preconditioning 3
Brown adipose tissue 258, 269
– thermogenic and adipogenic differentiation 270, 271

Budding yeast 75

cADPR *see* cyclic ADP-ribose
Calcineurin 290
Calcium
   – intracellular stores 94
   – involvement in temperature sensing 129
   – mobilization 93
   – signaling 91, 93, 95
Cancer 68, 70
Cardiac nociception 115
Cardiovascular disease 69, 70
Caseinolytic proteases 146
CD11/CD18 54
CD38 95, 100
cDNA array
   – analysis software 319, 321
   – anoxia tolerance 36–38
   – glass microarrays 317–322
   – heterologous probing 36–38, 315, 316, 321
   – history 317, 318
   – hibernation, 318–323
   – mollusc 36–38
   – nylon macroarrays 317–320
   – optimization 321
   – reproducibility 323, 324
cDNA library screening 316
Cell adaptation, consequences of heat stress 173
Cell culture models, development of 5
Cell signaling 290
   – control analysis of 284
Cellular energy status 245
Cellular recognition and migration 235
Cellular stress 61
Chaperonin 179
Chemical-induced stress 221
Chemokines 64
Chemopreventive agents 221
Chemotaxis 243
Chloroplastic electron transport 244
CHOP 193
Closed PSII reaction centers 247
Cold acclimation 123, 128, 130, 250
   – adjustment of photosynthesis 250
   – biochemical changes 123
   – gene expression 125
   – membranes 124
   – soluble sugars 124
Cold adaptation 302
Cold-inducible genes 125, 140, 141, 143–145
Cold-inducible RNA helicases 146
Cold responsive element 126
Concanavalin A 290
Concentration control coefficients 285
Conserved sequence motifs 156
Control analysis 283
COR genes 126, 167

Coronary vasodilation 51
c-Rel 61
CREB 271
Cryoprotective properties of dehydrins 167
Cyanobacteria 139, 175, 245
Cyclic ADP-ribose (cADPR) 91
   – and abscisic acid 97
   – and plant response to environmental stress 97
   – and sponge response to heat stress 98
   – Ca^{2+} mobilization 93
   – in bacterial infection and immune responses 99
   – structure of, 91
   – synthesis 95
CYP enzymes 222
Cystic fibrosis 65
Cyt b_6f complex 247
Cytochrome P450 222
Cytokines 64
Cytotoxicity 237

Dehydration in plants 122, 155
Dehydrins 155
   – cell and tissue distribution 164
   – conformation 159
   – cryoprotective properties 167
   – drought-induces 163
   – gene expression 159
   – in seeds 164
   – low temperature 160
   – properties 158
   – protective effects 165
   – salinity 163
   – tissue distribution 165
Desaturases 142
Desiccation tolerance 164
Disease 61
Drought 252
   – adjustment of photosynthesis 250
   – dehydrins 163
   – responsive element 126
   – signal transduction 129
Drug metabolizing enzymes 221
   – cytochrome P450 222
   – aryl hydrocarbon receptors 223
   – peroxisome proliferator-activated receptors 223
   – steroid xenobiotic receptors 224

Ecto-5′ nucleotidases 49
Elastase 54
Elasticity coefficients 285
Electromagnetic radiation
   – depth of transmission in tissue 234
   – and human health 235
Electron transport 244
ENaC 14
Endogenous adenosine 47
Endometriosis 67

Endothelium 49
Energy sensing 243
Energy taxis 243
Energy transfer 246
Environmental stress 75
 – plant response to 97
Enzymatic synthesis and degradation 95
Enzyme expression 300
E-selectin 55
Evolution of physiological acclimation 308
Evolutionary physiology 297
Extracellular regulated kinases 272

Fatty acid desaturases 140, 143, 144
Ferredoxin 247
Ferritin 38
 – levels following IR radiation 238
Fibroblasts 50, 237
Fission yeast 75, 80
Flux control coefficients 284
FNR 247
Free amino acids pools 190
Free radical reactions and light production 235
Freezing injury in plants 122
Freeze tolerance 121, 127
 – COR genes 127, 167
 – esk1 gene 128
 – mutational analysis 127
 – role of abscisic acid 128
 – signal transduction 129
Fundulus 299, 302
 – acclimation 309
 – phylogeny 302

Gastro-intestinal disease 66
Gene expression 61, 300
 – analytical methods 36–38
 – anoxia-induced 36–38
 – antioxidant response element 225
 – ATF4 and amino acid signaling 195
 – *Arabidopsis*
 – cDNA arrays 36–38, 317–323
 – cGMP-mediated 39, 40
 – CHOP
 – cold acclimation in plants 125
 – cold and desiccation induced in plants 98
 – cold-induced in cyanobacteria 140
 – COR genes 127, 167
 – dehydrins 159
 – desaturases 142
 – ferritin 238
 – hibernation-induced 316, 319–323
 – HIF 20
 – lactate dehydrogenase 300
 – lung hypoxia 13-22
 – methods 316
 – regulation by amino acids 189

 – responses to UVA and IR radiation 237
 – ribosomal protein 320
 – RNA binding proteins 145
 – stress induced 155
 – uncoupling proteins 270
 – VEGF 18
Genomics 284
Glucagon 289
Glucose transporters 17, 275
Glycogen phosphorylase 31
Glycolytic enzymes 18, 29, 285, 299, 302, 307
Greenhouse effect 233
GroEL 180

H+-dependent ATP synthase 247
Heat shock 80, 81, 83
 – membrane fluidity during 178
 – proteins (HSPs) 173, 238, 173
 – HSP27 52
 – HSP70 5, 183
Heat stress 98
 – chaperonins 179
 – GroEL 180
 – HSPs 173
 – in sponges 98
 – perception and transduction of 174
 – role of membranes in 173
 – small HSPs 180
Helix aspersa 288
Heme oxygenase 238
Hepatocyte metabolism 289
Hepatopancreas cells 288
Hibernation 6
 – bats 319–323
 – cDNA arrays screening 318–323
 – cDNA library screening 316
 – fatty acid binding protein 321
 – gene expression 319–323
 – ground squirrels 7, 319–323
 – NMDA receptor 8
 – ribosomal protein 320
 – suppressed traumatic tissue response 7
HIF-1 69
Histidine kinase 77, 131, 150
Histidine phosphotransfer protein 77
Hog1 75
Hog1 MAPK cascade 76, 77, 79
Homeostasis 243
HSP27 52
HSP70 5, 183, 298
Hyperosmolarity 79
Hypoxia 13, 68
 – adhesion molecules 29
 – in hibernation 6,7
 – effect on adhesion molecules expression 20
 – effect on glucose transport 17
 – effect on glycolytic enzymes 18

– effect on sodium transport proteins 14, 15
– effect on proteins involved in glucose metabolism 16, 17
– effect on vascular growth factor expression 18, 19
– glucose metabolism in lung 16
– HIF 20, 69
– lung gene expression 13-22
– NMDA receptor 8
– sodium transport proteins 14
– turtles 7-9
– VEGF 18, 69
Hypoxia-inducible factor, 36, 39, 69
– in regulation of gene expression 20, 21

IκB 62
IκB kinase 64
Implantation 67
Infarct 50
Inflammation 61, 112
– chronic airway 65
Inflammatory diseases 70
– of the reproductive tract 67
Inflammatory stress response 64
Infrared radiation 234
– sensing 235
– sources of high dose 236
– sources of low dose 236
– spectrum 233
– survival of skin fibroblasts 237
Inosine 47
Insulin 208, 212, 272
Insulin growth factor 272
Interleukin 55, 56
Ion channels
– ATP-gated 108
– heat-gated 106
– K-ATP channels 63
– proton-gated 108
– voltage activated sodium 109
Iron deficiency 252
Irritable bowel syndrome 114
Ischemia 1, 47
– adenosine role 52
– brain preconditioning 3
– cell culture models 5
– hibernation 7
– myocardial preconditioning 1
– reperfusion 47, 54

JNK 81

K-ATP channels 63
– ischemia preconditioning 2, 53
– hibernation 7

Lactate dehydrogenase 299
Late embryogenesis abundant (LEA) proteins 155

Lck 52
Leukocytopenia 7
Lhca 246
Lhcb 246
Light absorption 246
Littorina littorea 27, 31, 33–35
Low temperature 160
– dehydrin expression 160
– stress in cyanobacteria 140
– sensors 150
– *see also* Cold
Lung 13-22
– adhesion molecules 20
– HIF 20
– glucose transporters 17
– glycolytic enzymes 18
– sodium transport proteins 14
– VEGF 18

Macrophage inflammatory protein-2 (MIP-2) 56
Macrophages 55
MAPK 75, 227
– activation by phenolics and isothiocyanates 228
– anoxia tolerance 41
– brown adipose proliferation 272
– docking site 84
– drug metabolism 221, 227
– SAPKs 75
– Spc1 in fusion yeast 80, 81, 83
Mast cells 49
Membrane association of heat-shock proteins 182
Membrane domains 174
Membrane fluidity 147
– changes 131
– definition 173
– desaturases 142
– during heat shock 178
Membrane lipid composition and physical state of 174
Membrane lipids 140
Membrane order, definition 173
Membrane permeability 174
Membrane-located sensor 184
Membrane-perturbating signal 174
Membrane potential 244
Membranes as cellular thermometers 174, 175
Menstruation 67
Metabolic control analysis 283
Metabolic rate depression 288, 291
– anoxia tolerance 28–31
– molluscs 28–31, 288, 291
– pH effects 28–31
– reversible phosphorylation 29–31
Mitochondria
– anion carriers 257, 259
– K-ATP-channels 53
– membrane potential 288, 290
– proton gradient 265

– respiration 290
– respiration in aestivating snails 288
– transport 244
– uncoupling proteins 257
Mitogen 290
Mn-superoxide dismutase 54
Modular approach 287
Modular control analysis 287, 290
Modular regulation analysis 284
Mollusc
– aestivating snails 288, 291
– anoxia tolerance 27–43
– carbohydrate metabolism 28–31
– gene expression 36–39
– metabolic control analysis 288, 291
– protein synthesis 31–35
Monte Carlo analysis 290
mTOR (mammalian target of rapamycin) 196, 208, 211
– activation by amino acids 211
– as an ATP sensor 213
– insulin effects 212
– regulation of protein breakdown 197
– kinase activity by amino acids 200
Multistep phosphorelay 82
Myocardial preconditioning 1
Myotis lucifugus 316, 319–322
Mytilus edulis 27

NADPH 245
Naturally occurring trait variation 167
Nerve growth factor 112
Neurology 69
Neutrophil adhesion 54
Nicotinic acid adenine dinucleotide phosphate (NAADP) 91
– Ca^{2+} mobilization 93
– synthesis 95
– structure 93
Nitric oxide 53
NMDA receptor 4
Nociception 106
– cardiac 115
– ion channels involved in 106-109
– visceral 113
Nociceptor sensitisation 109-112
– hyperalgesia 114
Non-bilayer phase 182
Non-lamellar phases
– definition 173
Non-photochemical quenching 249
Noradrenaline 272
Nuclear export signal (NES) 85
Nuclear Factor κB (NF-κB) 50, 56, 61, 67, 68, 69
– activation 61
– asthma 65
– gastro-intestinal disease 68
– inflammatory response 64
– reproductive function 67

– rheumatoid arthritis 65
NF-κB1 (p105/p50) 61
NF-κB2 (p100/p52) 61
Nuclear translocation 77, 78
Nucleocytoplasmic trafficking 84
Nutrient limitations 252
Nutritional limitation 80

Open PSII reaction centers 247
Orphan nuclear receptors 223
Osmotic stress 77, 83
– Hog1 76
Over-excitation 249
transgenic mice 262
Ovulation 67
Oxidative stress 68, 80, 83, 240
– signaling in fission yeast 82
Oxygen sensing 21, 39, 40
– role of cGMP 40

p38 MAPK 52, 75, 81, 275
p70S6 kinase 200, 208
Pancreas 214
Pain detection 105, 113, 115
Partial response coefficients 286
PAS domains 245
PAS/PAC 83
Pathological disorders 174
Peroxisome proliferator-activated receptors 223, 270
Pertussis 52
PGE2 110
Phosphatase 292
Phosphatidylinositol 3-kinase 208, 276
Phosphofructokinase 29, 31, 285
Phospholipase A_2 182
Phospholipase C 52
Phospholipase D 52
Photoacclimation 249
Photoautotrophs 243
Photodamage 249
Photoinhibition 249
Photo-oxidation 246
Photoprotection 248, 249, 251
Photorespiration 252
Photostasis 243, 248
Photosynthesis 243
Photosystem I and II 245
Phototaxis 243
Plants
– cold acclimation 123
– dehydrins 155
– environmental stress 97
– freeze tolerance 121
– gene expression 125
– mutational analysis 127
Plastoquinol 247
Plastoquinone 247

PP2C 81
Preconditioning 1, 49, 52
   – adenosine role 52
   – brain ischemia 3
   – focal
   – global
   – myocardial ischemia 1
   – rabbit isolated cardiomyocytes 53
   – remote 54
Primary sensor 173
Primary visceral neurons 115
Proliferation of brown adipocytes 272
Proline and cold stress in plants 125
   – esk1 mutants 130
Protective protein induction 237
Protein degradation 196
Protein induction 237, 239
Protein kinase A 110, 272
Protein kinase B 210
Protein kinase C 3, 52, 110, 290
Protein kinase G 30, 31, 39, 40
Protein phosphatase 212
Protein phosphorylation 29
   – eIF2 33
   – glycogen phosphorylase 31
   – oxygen sensing 40
   – phosphofructokinase 31
   – protein kinase G 30, 31, 39, 40
   – pyruvate kinase 39–31
Protein synthesis 31–35
   – eIF2 33
   – mammalian hibernation 7
   – polysomes 34
   – ribosomal protein 34, 35
Protein tyrosine phosphatases 78
Protein undernutrition 191
Proton-gated ion channels 108
Proton gradient 244, 265
Proton leak 289
Proton motive force 244
Purine salvage 47
Pyruvate kinase 39–31

Quinone pool 244

Reaction center pigments, P680 and P700 246
Reactive oxygen species 53, 54, 68, 265
   – light radiation 235
   – role in bio-regulation 235
Receiver domain 77
Redox sensing 243, 247, 265
Reductive pentose phosphate cycle 247
Regulation analysis 284, 292
RelA (p65) 61
RelB 61
Reperfusion 1, 54
Reprogramming of carbon metabolism 252

Respiration 244
Response coefficients 286
Response regulator 77
Rheumatoid arthritis 65
Ribonucleoprotein 35, 36
Ribosomal protein 34, 35, 146
Ribosomes
   – eIF2 33, 198, 209
   – initiation and elongation factors 198, 209
   – internal ribosome entry point 192
   – L26 protein 38, 39
   – polysome dissociation 34
   – S6 protein regulation 197, 207
RNA-binding proteins 145
RNA helicase 146

Saccharomyces cerevisiae 75, 261
Salinity 163
Sceloporus occidentalis 299
Schizosaccharomyces pombe 75
Serotonin 110
Signal transduction 221
Snails, aestivating 288
Sodium transport proteins 14
Soluble sugars
   – accumulation during cold acclimation 124
Spc1 81, 83, 84, 85
Spc1 MAPK cascade in fission yeast 80, 81, 83
Spermophilus tridecemlineatus 316, 319, 320
Sponge response to heat stress 98
State transitions 250
Steroid xenobiotic receptor/pregnane X receptor 224
Stress-activated protein kinases (SAPKs) 75
Stress-induced genes 155
Stress response 61, 129
Supraventricular tachycardia 56
Synechococcus 139
Synechocystis 131, 139, 175

Teleost fish 302
Temperature sensors 131, 173
   – histidine kinases 131
   – in cyanobacteria 131
   – in plants 133
   – membrane fluidity 131, 147
Thermal acclimation 306, 307, 309
Thermogenesis 258, 264, 269
Thylakoid membrane 245, 246
Thymocyte metabolism 290
Top-down approach 287
Transcription factors
   – ATF-2 193
   – CBF in plants 126, 167
   – C/EBP 193, 271
   – CHOP 193
   – CREB 271
   – HIF 20, 69

– Nrf2 221
– PPARs 271
– TNF-α 61, 276
– VEGF 18, 69
Transcriptional activation of stress regulated genes 173
Tumor necrosis factor-alpha (TNF-α) 56, 61, 276

Uncoupling proteins 257
– knockout 263, 269
– overexpression 262
– stUCP 257
– UCP1 257, 269
– UCP2 257, 261
– UCP3 257, 261
Uterine cycle 67

UV irradiation 80
Vascular endothelial growth factor (VEGF) 18, 69
Vasodilation 47, 51
VCAM-1 55

Viral infection 69
Visceral afferent neurons 113
Visceral hyperalgesia 114
Voltage-gated sodium channels 109

Water deficit 163

Xanthophyll cycle 249
Xenobiotic metabolizing enzymes 222

Yeast 75, 261
– evolutionary adaptation
– gene expression and adaptive evolution 301
– Hog1 76
– MAPKs 75
– osmotic stress 76
– overexpression of desaturase 175
– Spc1 cascade 80
– stress pathways 175
– uncoupling protein 261

CELL AND MOLECULAR RESPONSES TO STRESS
Edited by Kenneth B. Storey *and* Janet M. Storey

Volume 1. Environmental Stressors and Gene Responses

Chapter 1. Cell Homeostasis and Stress at Year 2000—Two Solitudes and Two Research Approaches – Peter W. Hochachka. 1

Chapter 2. Quantitative Design of Muscle Energy Metabolism for Steady-State Work – Raul K. Suarez 17

Chapter 3. Adaptation and Divergence in Stressful Environments – Michael Travisano. 29

Chapter 4. Stress and the Geographic Distribution of Marine and Terrestrial Animals – Steven L. Chown and Andrew Clarke. 41

Chapter 5. The Evolution of Thermal Sensitivity in Changing Environments – George W. Gilchrist 55

Chapter 6. Adaptations of the Cell Membrane for Life in Extreme Environments – Jack L.C.M. van de Vossenberg, Arnold J.M. Driessen and Wil N. Konings 71

Chapter 7. Cell and Molecular Responses to Hypoxic Stress – Enbo Ma and Gabriel G. Haddad 89

Chapter 8. Molecular and Cellular Stress Pathways in Ischemic Heart Disease: Targets for Regulated Gene Therapy – Keith A. Webster. 99

Chapter 9. Cellular and Molecular Basis of Stress Heart – Dipak K. Das and Nilanjana Maulik 113

Chapter 10. Transcriptional Response to Hyperosmotic Stress – Robin L. Stears and Steven R. Gullans 129

Chapter 11. The Activation of Trans-Acting Factors in Response to Hypo- and Hyper-Osmotic Stress in Mammalian Cells – Kuang Yu Chen, Jiebo Lu and Alice Y.-C. Liu 141

Chapter 12. Osmotic Regulation of DNA Activity and the Cell Cycle – Dietmar Kültz. 157

Chapter 13. Life Without Water: Responses of Prokaryotes to Desiccation – Daniela Billi and Malcolm Potts. 181

Chapter 14. Stress Response in Marine Sponges: Genes and Molecules Involved and Their Use as Biomarkers – Werner E.G. Müller, Claudia Koziol, Matthias Wiens and Heinz C. Schröder 193

Chapter 15. The Effects of Bioenergetic Stress and Redox Balance on the Expression of Genes Critical to Mitochondrial Function – S.C. Leary and C.D. Moyes 209

Chapter 16. The Heat Shock Response to Tropical and Desert Fish (genus Poeciliopsis) – Carol E. Norris and Lawrence E. Hightower 231

Chapter 17. The Molecular Biological Approach to Understanding Freezing-Tolerance in the Model Plant, Arabidopsis thaliana – Gareth J. Warren, Glenn J. Thorlby and Marc R. Knight. 245

Chapter 18. Molecular Regulation of Insect Diapause – David L. Denlinger 259

Chapter 19. How do Deep-Sea Microorganisms Respond to Changes in Environmental Pressure? – Chiaki Kato, Kaoru Nakasone, Mohammed Hassan Qureshi and Koki Horikoshi 277

Chapter 20. Signaling in Copper Ion Homeostasis – Zhiwu Zhu, Roslyn McKendry and Christopher L. Chavez. . . . 293

Volume 2. Protein Adaptations and Signal Transduction

Chapter 1. Signal Transduction and Gene Expression in the Regulation of Natural Freezing Survival – Kenneth B. Storey and Janet M. Storey. 1

Chapter 2. Drosophila as a Model Organism for the Transgenic Expression of Antifreeze Proteins – Bernard P. Duncker, Derrick E. Rancourt, Michael G. Tyshenko, Peter L. Davies and Virginia K. Walker. 21

Chapter 3. Cold-adapted Enzymes: An Unachieved Symphony – Salvino D'Amico, Paule Claverie, Tony Collins, Georges Feller, Daphné Georlette, Emmanuelle Gratia, Anne Hoyoux, Marie-Alice Meuwis, Laurent Zecchinon and Charles Gerday 31

Chapter 4. The Role of Cold-shock Proteins in Low-temperature Adaptation – Jeroen A. Wouters, Frank M. Rombouts, Oscar P. Kuipers, Willem M. de Vos, and T. Abee 43

Chapter 5. Hibernation: Protein Adaptations – Alexander M. Rubtsov 57

Chapter 6. Aquaporins and water stress – Alfred N. Van Hoek, Yan Huang and Pingke Fang 73

Chapter 7. Gene Expression Associated with Muscle Adaptation in Response to Physical Signals – Geoff Goldspink and Shi Yu Yang 87

Chapter 8. Early Responses to Mechanical Stress: From Signals at the Cell Surface to Altered Gene Expression – Matthias Chiquet and Martin Flück 97

Chapter 9. Fasting and Refeeding: Models of Changes in Metabolic Efficiency – Stephen P.J. Brooks 111

336

Chapter 10. *Nutritional Regulation of Hepatic Gene Expression* – Howard C. Towle 129
Chapter 11. *The AMP-activated/SNF1 Protein Kinases: Key Players in the Response of Eukaryotic Cells to*
Metabolic Stress – D. Grahame Hardie 145
Chapter 12. *Cellular Regulation of Protein Kinase C* – Alexandra C. Newton and Alex Toker. 163
Chapter 13. *Mitogen-activated protein kinases and stress* – Klaus P. Hoeflich and James R. Woodgett 175
Chapter 14. *How to Activate Intrinsic Stress Resistance Mechanisms to Obtain Therapeutic Benefit* –
Prasanta K. Ray, Tanya Das and Gaurisankar Sa 195
Chapter 15. *Regulation of Ion Channel Function and Expression by Hypoxia* – Chris Peers 203
Chapter 16. Ca^{2+} *Dynamics Under Oxidant Stress in the Cardiovascular System* – Tapati Chakraborti, Sudip Das,
Malay Mandal, Amritlal Mandal and Sajal Chakraborti. 213
Chapter 17. *Role of NF-E2 Related Factors in Oxidative Stress* – David Bloom, Saravanakumar Dhakshinamoorthy,
Wei Wang, Claudia M. Celli and Anil K. Jaiswal 229
Chapter 18. *Signal Transduction Cascades Responsive to Oxidative Stress in the Vasculature* – Zheng-Gen Jin
and Bradford C. Berk. 239
Chapter 19. *Oxidative Stress Signaling* – Hasem Habelhah and Ze'ev Ronai 253
Chapter 20. *Antioxidant Defenses and Animal Adaptation to Oxygen Availability During Environmental Stress* –
Marcelo Hermes-Lima, Janet M. Storey and Kenneth B. Storey 263